Mine Water Hydrogeology
and Geochemistry

Geological Society Special Publications
Society Book Editors

A. J. FLEET (CHIEF EDITOR)
P. DOYLE
F. J. GREGORY
J. S. GRIFFITHS
A. J. HARTLEY
R. E. HOLDSWORTH
A. C. MORTON
N. S. ROBINS
M. S. STOKER
J. P. TURNER

Special Publication reviewing procedures

The Society makes every effort to ensure that the scientific and production quality of its books matches that of its journals. Since 1997, all book proposals have been refereed by specialist reviewers as well as by the Society's Books Editorial Committee. If the referees identify weaknesses in the proposal, these must be addressed before the proposal is accepted.

Once the book is accepted, the Society has a team of Book Editors (listed above) who ensure that the volume editors follow strict guidelines on refereeing and quality control. We insist that individual papers can only be accepted after satisfactory review by two independent referees. The questions on the review forms are similar to those for *Journal of the Geological Society*. The referees' forms and comments must be available to the Society's Book Editors on request.

Although many of the books result from meetings, the editors are expected to commission papers that were not presented at the meeting to ensure that the book provides a balanced coverage of the subject. Being accepted for presentation at the meeting does not guarantee inclusion in the book.

Geological Society Special Publications are included in the ISI Science Citation Index, but they do not have an impact factor, the latter being applicable only to journals.

More information about submitting a proposal and producing a Special Publication can be found on the Society's web site: www.geolsoc.org.uk.

GEOLOGICAL SOCIETY SPECIAL PUBLICATION No. 198

Mine Water Hydrogeology and Geochemistry

EDITED BY

P.L. YOUNGER
University of Newcastle upon Tyne, UK

and

N.S. ROBINS
British Geological Survey, UK

2002
Published by
The Geological Society
London

THE GEOLOGICAL SOCIETY

The Geological Society of London (GSL) was founded in 1807. It is the oldest national geological society in the world and the largest in Europe. It was incorporated under Royal Charter in 1825 and is Registered Charity 210161.

The Society is the UK national learned and professional society for geology with a worldwide Fellowship (FGS) of 9000. The Society has the power to confer Chartered status on suitably qualified Fellows, and about 2000 of the Fellowship carry the title (CGeol). Chartered Geologists may also obtain the equivalent European title, European Geologist (EurGeol). One fifth of the Society's fellowship resides outside the UK. To find out more about the Society, log on to www.geolsoc.org.uk.

The Geological Society Publishing House (Bath, UK) produces the Society's international journals and books, and acts as European distributor for selected publications of the American Association of Petroleum Geologists (AAPG), the American Geological Institute (AGI), the Indonesian Petroleum Association (IPA), the Geological Society of America (GSA), the Society for Sedimentary Geology (SEPM) and the Geologists' Association (GA). Joint marketing agreements ensure that GSL Fellows may purchase these societies' publications at a discount. The Society's online bookshop (accessible from www.geolsoc.org.uk) offers secure book purchasing with your credit or debit card.

To find out about joining the Society and benefiting from substantial discounts on publications of GSL and other societies worldwide, consult www.geolsoc.org.uk, or contact the Fellowship Department at: The Geological Society, Burlington House, Piccadilly, London W1J 0BG: Tel. +44 (0)20 7434 9944; Fax +44 (0)20 7439 8975; Email: enquiries@geolsoc.org.uk.

For information about the Society's meetings, consult *Events* on www.geolsoc.org.uk. To find out more about the Society's Corporate Affiliates Scheme, write to *enquiries@geolsoc.org.uk*.

Published by The Geological Society from:
The Geological Society Publishing House
Unit 7, Brassmill Enterprise Centre
Brassmill Lane
Bath BA1 3JN, UK
(*Orders*: Tel. +44 (0)1225 445046
Fax +44 (0)1225 442836)
Online bookshop: http://bookshop.geolsoc.org.uk

The publishers make no representation, express or implied, with regard to the accuracy of the information contained in this book and cannot accept any legal responsibility for any errors or omissions that may be made.

©The Geological Society of London 2002. All rights reserved. No reproduction, copy or transmission of this publication may be made without written permission. No paragraph of this publication may be reproduced, copied or transmitted save with the provisions of the Copyright Licensing Agency, 90 Tottenham Court Road, London W1P 9HE. Users registered with the Copyright Clearance Center, 27 Congress Street, Salem, MA 01970, USA: the item-fee code for this publication is 0305-8719/00/$15.00.

British Library Cataloguing in Publication Data
A catalogue record for this book is available from the British Library.

ISBN 1-86239-113-0
Typeset by Alden Bookset, UK
Printed by The Cromwell Press, Wiltshire, UK.

Distributors

USA
AAPG Bookstore
PO Box 979
Tulsa
OK 74101-0979
USA
Orders: Tel. + 1 918 584-2555
Fax +1 918 560-2652
E-mail *bookstore@aapg.org*

India
Affiliated East-West Press PVT Ltd
G-1/16 Ansari Road, Daryaganj,
New Delhi 110 002
India
Orders: Tel. +91 11 327-9113
Fax +91 11 326-0538
E-mail *affiliat@nda.vsnl.net.in*

Japan
Kanda Book Trading Co.
Cityhouse Tama 204
Tsurumaki 1-3-10
Tama-shi
Tokyo 206-0034
Japan
Orders: Tel. +81 (0)423 57-7650
Fax +81 (0)423 57-7651

Contents

YOUNGER, P.L. & ROBINS, N.S. Challenges in the characterization and prediction of the hydrogeology and geochemistry of mined ground — 1

BOOTH, C.J. The effects of longwall coal mining on overlying aquifers — 17

WOLKERSDORFER, C. Mine water tracing — 47

WHITWORTH, K.R. The monitoring and modelling of mine water recovery in UK coalfields — 61

DUMPLETON, S. Effects of longwall mining in the Selby Coalfield on the piezometry and aquifer properties of the overlying Sherwood Sandstone — 75

ADAMS, R. & YOUNGER, P.L. A physically based model of rebound in South Crofty tin mine, Cornwall — 89

ROBINS, N.S., DUMPLETON, S. & WALKER, J. Coalfield closure and environmental consequence – the case in south Nottinghamshire — 99

MCKELVEY, P., BEALE, G., TAYLOR, A., MANSELL, S., MIRA, B., VALDIVIA, C. & HITCHCOCK, W. Depressurization of the north wall at the Escondida Copper Mine, Chile — 107

KUMA, J.S., YOUNGER, P.L. & BOWELL, R.J. Hydrogeological framework for assessing the possible environmental impacts of large-scale gold mines — 121

BANWART, S.A., EVANS, K.A. & CROXFORD, S., Predicting mineral weathering rates at field scale for mine water risk assessment — 137

BOWELL, R.J. The hydrogeochemical dynamics of mine pit takes — 159

SAALTINK, M.W., DOMENECH, C., AYORA, C. & CARRERA, J. Modelling the oxidation of sulphides in an unsaturated soil — 187

GANDY, C.J. & EVANS, K.A. Laboratory and numerical modelling studies of iron release from a spoil heap in County Durham — 205

BANKS, D., HOLDEN, W., AGUILAR, E., MENDEZ, C., KOLLER, D., ANDIA, Z., RODRIGUEZ, J., SÆTHER, O.M., TORRICO, A., VENEROS, R. & FLORES, J. Contaminant source characterization of the San José Mine, Oruro, Bolivia — 215

NUTTALL, C.A. & YOUNGER, P.L. Secondary minerals in the abandoned mines of Nenthead, Cumbria as sinks for pollutant metals — 241

YOUNGER, P.L., The importance of pyritic roof strata in aquatic pollutant release from abandoned mines in a major, oolitic, berthierine–chamosite–siderite iron ore field, Cleveland, UK — 251

LEBLANC, M., CASIOT, C., ELBAZ-POULICHET, F. & PERSONNÉ, C. Arsenic removal by oxidizing bacteria in a heavily arsenic-contaminated acid mine drainage system (Carnoulès, France) — 267

BROWN, M.M.E., JONES, A.L., LEIGHFIELD, K.G. & COX, S.J. Fingerprinting mine water in the eastern sector of the South Wales Coalfield — 275

BANKS, D., PARNACHEV, V.P., FRENGSTAD, B., HOLDEN, W., VEDERNIKOV, A.A. & KANNACHUK, O.V. Alkaline mine drainage from metal sulphide and coal mines: examples from Svalbard and Siberia — 287

HATTINGH, R.P., PULLES, W., KRANTZ, R., PRETORIUS, C. & SWART, S. Assessment, prediction and management of long-term post-closure water quality: a case study–Hlobane Colliery, South Africa — 297

NUTTALL, C.A., ADAMS, R. & YOUNGER, P.L. Integrated hydraulic–hydrogeochemical assessment of flooded deep mine voids by test pumping at the Deerplay (Lancashire) and Frances (Fife) Collieries — 315

LOREDO, J., ORDÓÑEZ, A. & PENDÁS, F. Hydrogeological and geochemical interactions of adjoining mercury and coal mine spoil heaps in the Morgao catchment (Mieres, NW Spain)   327

NEUMANN, I. & SAMI, K. Structural influence on plume migration from a tailings dam in the West Rand, Republic of South Africa   337

JOHNSON, K.L. & YOUNGER, P.L., Hydrogeological and geochemical consequences of the abandonment of Frazer's Grove carbonate-hosted Pb/Zn fluorspar mine, north Pennines, UK   347

HOLTON, D., KELLY, M. & BAKER, A. Paradise lost? Assessment of liabilities at a Uranium mine in the Slovak Republic: Novoveska Huta   365

REES, S.B., BOWELL, R.J. & WISEMAN, I. Influence of mine hydrogeology on mine water discharge chemistry   379

It is recommended that reference to all or part of this book should be made in one of the following ways:

YOUNGER, P.L. & ROBINS, N.S. (eds) 2002. *Mine Water Hydrogeology and Geochemistry*. Geological Society, London, Special Publications, **198**.

ADAMS, R. & YOUNGER, P.L. 2002. A physically based model of rebound in South Crofty tin mine, Cornwall *In*: YOUNGER, P.L. & ROBINS, N.S. (eds). *Mine Water Hydrogeology and Geochemistry*. Geological Society, London, Special Publications, **198**, 89–97.

Referees

The Editors are grateful to the following people for their assistance with the reviewing of papers submitted to this Special Publication

Dr R Adams	University of Newcastle, UK
Prof A C Aplin	University of Newcastle, UK
Dr C Ayora	CSIC, Barcelona, Spain
Mr D Banks	Holymoor Consulting, Chesterfield, UK
Dr S A Banwart	University of Sheffield, UK
Dr L C Batty	University of Newcastle, UK
Dr J P Bloomfield	British Geological Survey, Wallingford, UK
Prof C Booth	Northern Illinois University, DeKalb, Illinois, USA
Dr R J Bowell	SRK Consulting, Cardiff, UK
Mr M M E Brown	Camborne School of Mines, University of Exeter, Redruth, UK
Dr S P Burke	University of Sheffield, UK
Dr S Dumpleton	British Geological Survey, Keyworth, UK
Dr K Evans	University of Sheffield, UK
Prof G Fleming	University of Strathclyde, UK
Dr R Fuge	University of Wales, Aberystwyth, UK
Dr P J Hayes	Water Management Consultants Ltd, Shrewsbury, UK
Dr J A Heathcote	Entec UK Limited, Shrewsbury, UK
Dr A W Herbert	Environmental Simulations International, Shrewsbury, UK
Mr D B Hughes	University of Newcastle, UK
Dr K L Johnson	University of Newcastle, UK
Dr D G Kinniburgh	British Geological Survey, Wallingford, UK
Dr B A Klinck	British Geological Survey, Keyworth, UK
Prof L Lövgren	University of Umeå, Sweden
Prof D A C Manning	University of Newcastle, UK
Prof J Mather	Royal Holloway, University of London, Egham, UK
Mr P McKelvey	Water Management Consultants Ltd, Shrewsbury, UK
Dr A E Milodowski	British Geological Survey, Keyworth, UK
Mr B D R Misstear	Trinity College, Dublin, Ireland
Ms K L Morton	K L M Consulting Services, Johannesburg, South Africa
Dr C A Nuttall	University of Newcastle, UK
Dr D W Peach	British Geological Survey, Wallingford, UK
Prof M Sauter	Universität Jena, Germany
Dr J M Sherwood	Environment Agency, Exeter, UK
Dr P L Smedley	British Geological Survey, Wallingford, UK
Mr R G Smith	Environment Agency, Nottingham, UK
Mrs J A Walker	Scott Wilson Kirkpatrick, Chesterfield, UK
Mr D C Watkins	Camborne School of Mines, University of Exeter, Redruth, UK
Dr N R Walton	University of Portsmouth, UK
Dr R S Ward	Environment Agency, Solihull, UK
Dr C Wolkersdorfer	Technische Universität Bergakademie Freiberg, Germany

Challenges in the characterization and prediction of the hydrogeology and geochemistry of mined ground

P. L. YOUNGER[1] & N. S. ROBINS[2]

[1] *Hydrogeochemical Engineering Research and Outreach, Department of Civil Engineering, University of Newcastle, Newcastle upon Tyne NE1 7RU, UK*
[2] *British Geological Survey, Maclean Building, Wallingford Oxfordshire OX10 8BB, UK*

Abstract: Although mining is no longer a key industry in the UK, the international mining industry continues to expand. One of the principal legacies of past mining in Britain is water pollution emanating from abandoned mine voids and waste rock depositories. This has necessitated many expensive technical evaluations and remedial programmes in recent years, from which important lessons may be drawn for the still-growing mining industry overseas. Perhaps the single most important lesson is that there can never be too much information on mine hydrogeology and geochemistry available at the post-closure phase. As this phase is also the longest in the overall life cycle of any mine, it should be given appropriate consideration from the outset. The post-closure studies described in this paper and in this volume (as well as elsewhere) highlight the dearth of hydrological data that are usually available when compared with the wealth of geometric information available from mine abandonment plans. It is advocated that the collection of appropriate environmental data is built into the initial mine development plan and that monitoring commences from the green field site onwards. The uncertainties related to predictive modelling of mine water arisings are considerable, whilst those of predicting mine water quality are even greater. Numerous pointers towards robust mine water management strategies are identified, and a call for 'defensive mine planning' is made, in which relatively modest investments in hydrogeochemical control measures during the exploration and exploitation phases of the mine life cycle will yield dividends in the post-closure phase. With such measures in place, and enhanced monitoring data to hand, the conjunctive application of physical and geochemical evaluations will eventually provide much-needed predictive tools to inform site management decisions in the future.

Exploitation of the Earth's mineral wealth by mining has been an unrelenting pursuit of mankind since the earliest days. Underground flint mines of Neolithic age are still accessible in southeast England (Holgate 1991), and by the Bronze Age underground metal mines were well established in many parts of western and southern Europe (Shepherd 1993). Mining activity accelerated with the dawn of the Industrial Revolution when coal came to occupy a central place in the resource inventory, and mining has continued apace ever since. Although mining has long been in decline in the much-worked areas of Europe, at the global scale the mining industry continues to expand, at present mainly in the form of open-pit mining. The recent closure of some very large surface mines in Europe (Fig. 1) and North America (Bowell 2002) provides valuable lessons for the mining industry worldwide. With many of the larger open pits currently nearing their feasible limits (which often reflect groundwater-related slope stability constraints; see McKelvey *et al.* 2002), a future renaissance in deep mining is to be expected within the next few decades. Hence, the lessons to be learned from recently closed deep mines in Europe will retain applicability elsewhere for decades to come, and will indeed prove invaluable in the development of other uses of underground voids, such as disposal of high- and medium-level radioactive wastes in deep repositories. It is the purpose of this paper to outline some of the key findings of recent research into the hydrogeological and geochemical characterization of active and abandoned mine systems, and to identify challenges for future research.

The British experience: some lessons learned the hard way

Even the newest mining economies will eventually come to share the superannuated status

(a)

(b)

Fig. 1. (a) The main working pit (western basin) of the Westfield Opencast Coal Site, Fife, Scotland, as it appeared in the height of its productivity in 1974 (after Robins 1990). The far side of the pit is the highwall, 245 m in height, the outermost benches of which lie along the Ochil Fault, through which considerable quantities of groundwater entered the mine, as explained in the text. (Photograph: British Geological Survey, catalogue number D 1748, used with permission.) (b) The Westfield site viewed from approximately the same position on 12 December 2001, showing the pit lakes that have developed in the mine voids since coaling ceased. (Photograph: P Younger.)

which now characterizes much of the European mining sector. When large systems of interconnected mines finally come to be abandoned, the experiences of the environmental consequences of various closure strategies gained in the former mining districts of Europe should prove invaluable to developing intelligent closure strategies. Nowhere are the long-term environmental consequences of mining and mine closure more evident than in Britain, from which some of the best-documented cases of mine water pollution are recorded (e.g. Robins 1990; NRA 1994; Banks *et al.* 1997; Younger 1994; 1998; 2001; 2002; Brown *et al.* 2002).

Although the British Empire dominated much of the world until the early twentieth century, ensuring that Britain could therefore obtain mined commodities from a wide range of sources, there remains a substantial legacy of mining within the UK itself, for two reasons:

- mining was carried out in the '*Cassiterides*' (widely considered to refer to the tin mining district of Cornwall) from the earliest eras of recorded history. Copper mining was underway in the Bronze Age, most notably in North Wales, and both Roman and Mediaeval workings for argentiferous galena are scattered through much of the uplands of Wales, northern England and the Isle of Man (with a few mines also in Scotland). Hence, there is a substantial environmental legacy from very ancient mines;
- while it is, indeed, true that the British Empire was able to exploit mineral resources the world over, the heavier bulk minerals were expensive to transport and were therefore largely sourced from home. It is only since the mid twentieth century that the bulk of UK demand for ferrous and base metal ores has been satisfied by imports (e.g. Goldring & Juckes 2001), and the demise of the domestic coal mining industry is ongoing even as we write, battered by a combination of cheap imports of coal and changes in governmental policy in favour of other sources of energy (Powell 1993).

When mining was the major source of employment in much of northern and western Britain, environmental damage was considered a small price to pay (Fig. 1a). The benefits of full employment and municipal prosperity more than compensated for any misgivings over polluted air and water. The magnificent Georgian architecture of the city centres of Manchester, Glasgow and Newcastle upon Tyne are testimony to the industrial wealth created by the Victorian coal mining industry. However, the unsightly oil shale bings of West Lothian, the uneven surface of the M1 motorway between Nottingham and Sheffield (due to coal mine subsidence) and the numerous flooded areas of former farmland in Cheshire (due to solution mining of halite) are other less desirable consequences of past mining activity. Less apparent to the casual observer is the damage that past mining has done and is now doing to the aqueous environment. Large volumes of Carboniferous grits and sandstones are contaminated with sulphate-rich groundwaters, which cause serious ecological degradation where they discharge to surface watercourses. Current estimates are that some 600 km of British streams and rivers are damaged by polluted discharges from abandoned mines (about 400 km from coal mines (Table 1), with the remainder from metal mines; Younger 2000*a*). The costs associated with responsible remediation of such pollution are often so substantial that they beg the question as to whether they might have been minimized had our forefathers pursued extraction and mine waste management a little differently (e.g. Younger & Harbourne 1995). While hindsight is of little value for the owners of existing problematic abandoned mine sites, it has some value for owners of other mines which are currently in the planning or extractive phases, and is therefore indulged in to some degree in the examples that follow.

One of the most problematic aspects of effective mine site management is the mismatch between a keen twenty-first century environmental consciousness and the (often undocumented)

Table 1. *Ferruginous discharges from coal mines in UK (after Younger 2000a)*

Coalfields	Number of identified discharges	Mean Fe concentrations in discharges (mg/l)	Total length of rivers and streams contaminated by ochre (km)
South Wales, North Wales	90	17	109
Warwickshire, Nottinghamshire, Derbyshire, Staffordshire	5	–	2.5
Yorkshire, Durham and Northumberland	120	17	61
Lancashire, Cumbria	25	20	35
Scotland–Midland Valley	167	27	180

realities of site evolution over periods of decades or even centuries. It is, perhaps, obvious that this problem is at its most marked in areas where mining has been underway for thousands of years. However, even where a mineral deposit has only been mined within the last few decades, the legacy of mining decisions made by previous owners many years ago can act as serious restraints on post-closure management and re-use of a site. Such problems are further exacerbated where frequent changes in site management have resulted in the loss of all non-statutory records of mining and mine waste disposal practices. The Westfield Opencast Coal Site (OCCS) in Fife, Scotland, provides a revealing example of these sorts of problems. The Westfield OCCS worked a sequence of Carboniferous strata which was unusually rich in coal: the site had an overburden to coal ratio of only 2.2, which compares with values in the range 15–30 at most other Scottish opencast sites. Westfield commenced production in the 1950s, and in its main productive period between 1961 and 1986 it yielded more than 25 million tonnes of coal (Grimshaw 1992). Figure 1a shows the main working pit as it appeared in 1974, in the period of peak output. The highwall of the pit reached a total height of 245 m, making Westfield the deepest opencast coal mine ever worked in Europe (Grimshaw 1992). One of the few drawbacks of this phenomenally rich site was its poor mine water quality, with total iron concentrations reaching $1250\,\text{mg}\,\text{l}^{-1}$ (Robins 1990). It seems likely that the principal source of such high iron concentrations is active pyrite oxidation in and above the zone of water table fluctuation in the large masses of waste rock and washery fines that were tipped into the pit some decades ago. However, the absence of detailed records of mine waste disposal practices in the 1960s and early 1970s prevents the *a priori* identification of acid-generation 'hotspots' within the bodies of mine waste. Another problem arising from previous mining decisions is the unnecessarily high water make of the open pit, which was occasioned by incautious holing through the Ochil Fault in the 1970s, inducing inflow of good quality groundwater from Devonian strata which lie to the north of the site (Robins 1990). Now that coaling has finally ceased at Westfield, the former open pits have partly flooded to form pit lakes (Fig. 1b). The current site owners are implementing a number of measures to prevent the site posing any problems to the adjoining river system. By judicious positioning of a major rock bund within the void, the relatively weakly mineralized water crossing the Ochil Fault has been retained to form a pit lake of good quality.

Meanwhile, state-of-the-art facilities have been installed to control water levels and discharged water quality in the other pit lake, which remains markedly acidic at the time of writing. The current after-use plans for the site are visionary and wholly in line with the modern philosophy of sustainable development, that is that the site is being developed in a manner which meets the needs of the present which both improves environmental quality and does not limit the scope for future generations to make their own decisions about the best use of the site. However, the options open to today's engineers and planners are unquestionably limited by the actions of previous generations of miners whose work pre-dated the emergence of the concept of sustainable development.

The most infamous case of acid mine water pollution to have arisen in Britain within the last few decades is without doubt that associated with the 1991 closure of Wheal Jane, an underground tin–zinc mine located in the Carnon Valley, Cornwall. ('Wheal' is a Celtic (Cornish) word meaning 'mine'.) The sudden release of an estimated 50 million litres (Ml) of highly polluted water and sludge from an old adit resulted in highly visible staining of the Fal estuary and the Western Approaches of the English Channel (Bowen *et al.* 1998). A flow of contaminated water ranging between 8 and 15 Ml day^{-1} has continued ever since. This pollution prompted immediate public expenditure of some £8.3 million for amelioration works, and a further sum of similar magnitude for the identification of a long-term strategy to deal with the discharge (Knight Piésold & Partners Ltd 1995). Current total expenditure at this one site had reached £20 million at the time of writing, a few months before the 10th anniversary of the original outburst.

The Wheal Jane case illustrates some interesting points of wider interest.

- Although there has been a risk to ground and surface waters from mining for a very long time, the scale of modern mining methods and the gradual change in environmental consciousness have worked together to exacerbate risks to the point of unacceptability. Metal mining began in the area around Wheal Jane as long ago as 2000 BC, and peaked in the Victorian era, before ceasing altogether in the Carnon Valley in 1913. The modern workings began in 1969, only to close again in 1978 for hydrogeological reasons (i.e. the excessive costs of dewatering which fell upon Wheal Jane after the adjoining

Mount Wellington Mine closed in April 1978; Davis & Battersby 1985). For 12 years the mine lay abandoned, and will certainly have been discharging acid water to the Carnon River, although there are no records of any environmental outcry. It is only the post-closure pollution following the 1980–1991 period of working that has finally given rise to environmental agitation. Is this because the spectacular 1991 outbreak was truly unique, or because the environmental damage that was once acceptable when mining was the major source of employment in the area is no longer acceptable in a rapidly greening Carnon Valley?

- The environmental costs of mine water pollution are illustrated in bald cash terms by the Wheal Jane case, which probably represents an upper-bound estimate of the costs of remedial–preventative pump-and-treat interventions. Although its onset was heralded by an instantaneous and seemingly unexpected outburst at surface, the ongoing generation of acidity as the water table fluctuates in shallow workings has been the main focus of remedial action, and this is precisely the process which gives rise to most concern and most remedial costs at the majority of other mine water pollution sites in the UK.

These two observations have particular significance in the current Europe-wide debate on whether a more specific regulatory framework is needed to control pollution and related hazards associated with catastrophic spills from mine sites. The heightened consciousness that has given rise to this debate has been fostered by two spectacular tailings dam failures, at Aznalcóllar in Spain (Grimalt *et al.* 1999) and at Baia Mare in Romania (UNEP 2000). Following these two events, there have been widespread calls for tighter regulation of mine waste, resulting in consultation by the European Commission for a new EU Mine Waste Directive. Laudable as this intention may be, it is nevertheless the case that, in the longer term, more environmental damage is wrought by polluted drainage from abandoned workings than from waste rock/tailings deposits. For instance, a survey of all known mine water discharges in Scotland found that abandoned workings were responsible for around 98% of cases, with spoil heaps and tailings repositories contributing only about 2% of the total (Younger 2001). Hence, launching new regulatory legislation in response to a spectacular tailings dam failure runs the risk of overlooking the main problem in the long run. The reason for this is simple: even the largest spoil heap receives recharge from a very small area compared to the hundreds of square kilometres underlain by old workings.

The full range of problems associated with the flooding of abandoned mine workings in Britain has been catalogued by Younger & Adams (1999) as follows, arranged in approximate order of decreasing incidence/environmental severity.

a. *Surface water pollution*, after voids have eventually flooded up to ground level, leading to acidic and/or ferruginous discharges (Table 1) (which sometimes also contain elevated levels of other ecotoxic metals, most notably zinc) into previously 'clean' streams (e.g. NRA 1994). The consequences of such pollution can extend to abandonment of public water supply intakes, fish deaths and impoverishment of aquatic flora and fauna (Kelly 1999), poisoning of land animals that drink the water, and removal of a critical food source for birds and mammals which prey on freshwater fish.

b. *Localized flooding* of agricultural, industrial or residential areas, particularly where structures have been inadvertently constructed over former mine entries (e.g. Younger & LaPierre 2000; Younger 2002).

c. *Temporary loss of dilution* for other pollutants in surface waters as former pumped discharges cease to augment flows (e.g. Banks *et al.* 1996).

d. *Overloading of sewers by mine water inflows, exacerbated by clogging by ochre precipitates* (see Younger 2002).

e. *Pollution of overlying aquifers* by upward migration of mine water (e.g. Robins & Younger 1996; Younger & Adams 1999).

f. *Temporarily accelerated mine gas emissions*, driven ahead of the rising water table (e.g. Robinson 2000). While media attention is usually paid to the risk of flammable methane emissions, methane is such a light gas that it will have normally vented to surface anyway if an upward pathway exists. In this context, the more worrying prospect is of 'stythe' (carbon dioxide), which is more dense than air, and will thus lurk above the water table until hydraulically forced upwards.

g. *The risk of subsidence* as rising waters weaken previously dry, open, shallow old workings, primarily through slaking of seat-earths underlying pillars of coal left to provide roof support (e.g. Smith & Colls 1996).

Looking world-wide, it is important to add the specific problem of acidic pit lakes to the above list of issues. Although there are few examples of polluted pit lakes formed in former open-pit mine voids in the UK (the Westfield site mentioned above being a case in point), they are widespread in other countries, where they may be a principal mine water management issue (Bowell 2002; Holton et al. 2002). These pit lakes can form sinks for regional groundwater flow (by evaporative water loss) or else they may be 'flow-through lakes', which give rise to plumes of polluted water in the surrounding aquifers down the hydraulic gradient (Neumann & Sami 2002; see Younger et al. 2002 for further discussion of this matter). Polluted pit lakes pose environmental hazards of their own, such as exposing migratory waterfowl to toxins if they rest on the lake (Bowell 2002) and/or releasing harmful gases (such as radon), which may accumulate in the closed depression of the pit and pose a health risk to rock climbers and others who may informally use the pit lake and surroundings for recreational purposes (e.g. Holton et al.2002).

Towards robust management strategies

On first inspection, this list of possible impacts of water level rise after mine closure (a process termed 'mine water rebound') is bewilderingly broad. In reality, the full range of impacts is unlikely to be experienced in any one case, and thus the key to effective management is to employ a characterization and prediction strategy based on sound hydrogeological principles (e.g. Kuma et al. 2002), preferably executed within a risk assessment framework, such as RBCA (risk-based corrective action) (e.g. Banwart & Malmström 2001; Banwart et al. 2002; Holton et al. 2002). On this basis, a social–environmental cost–benefit analysis should be developed, which provides the principal tool for deciding on the desirability of alternative remedial actions. The costs which need to be evaluated include (Younger & Harbourne 1995):

- environmental damage, such as loss of aquatic habitat and damaging the livelihoods of terrestrial animals that feed on aquatic plants and animals;
- damage to property and commerce, such as rendering further abstraction of water for public or industrial use impossible, damage to commercial fisheries, and the disincentive to urban redevelopment posed by visible stream pollution, etc.;
- damage to society, such as loss of visual amenity, loss of sport fisheries, etc.;
- the expenditure associated with various remedial strategies.

It is frequently found that the cost of a preventative strategy in relation to post-closure water pollution from mines is small compared with that of the potential damage and cost of amelioration (Younger & Harbourne 1995; Robins & Younger 1996). Nevertheless, the necessary investment has not always been forthcoming, and it has only been since the advent of the UK Coal Authority in 1994 that a rolling programme to address the worst of the abandoned coal mine discharges in the UK has been established. No comparable programme yet exists for metalliferous mines.

The foundation of a successful cost–benefit evaluation of remedial options for a polluting mine water discharge is a source characterization initiative, which is largely hydrogeological and geochemical in technical content (e.g. Knight Piésold & Partners Ltd., 1995). However, where funds for remedial actions have been scarce, finance to support independent scientific research into mine water hydrogeology and geochemistry was virtually non-existent before the mid-1990s. Then in 1996 the Environment Agency and Northumbrian Water Ltd launched a joint programme of research into 'improved modelling of abandoned coalfields' (see Younger & Adams 1999 for details). Subsequently, the Natural Environment Research Council and the Coal Authority have funded significant research into various aspects of the topic (see Banwart et al.2002; Gandy & Evans 2002; Nuttall et al 2002; Robins et al.2002; Whitworth 2002). The bulk of research to date has concerned the rate of mine water rise and prediction of outflow locations and flow rates, with geochemical work being as yet far less well developed. A future objective is the conjunction of investigations into both the physical and geochemical evaluation at the same time, as advocated by Johnson & Younger (2002).

Predictive physical modelling tools developed by UK-based researchers since the mid-1990s have included:

- semi-distributed, lumped parameter models applicable at fairly large scales. The best example of this genre is the Fortran code 'GRAM' (Groundwater Rebound in Abandoned Mineworkings;

Sherwood 1997). The code and its application has been widely discussed (e.g. Younger et al. 1995; Sherwood & Younger 1997; Younger & Adams 1999; Burke & Younger 2000; Adams & Younger 2001). One of the shortcomings of GRAM is its inability to cope with declines in head-dependent inflow as rebound proceeds; modifications of the concept to incorporate this phenomenon have been developed by Banks (2001) and Whitworth (2002).

- full physically based models, in which a pipe network model (representing major mine roadways, etc.) is routed through a variably saturated porous media (the VSS-NET code; Younger & Adams 1999; Adams & Younger 2001, 2002);
- (three-dimensional) 3-D visualization software (VULCAN) that allows identification of mine water flow paths and outflow zones, which has been successfully linked to various hydrological simulation codes (Dumpleton et al. 2001; Robins et al. 2002) including GRAM.

Predictive geochemical models have a much weaker pedigree to date. Simple assay techniques popular in surface mining applications, such as 'acid–base accounting' (ABA) in which the sulphide content of a waste rock is compared with the content of buffering minerals (principally calcite), provide some information on the pollutant generation potential of strata under highly specific conditions. While these conditions may be satisfied to some degree in opencast coal mining situations (Kleinmann 2000), ABA is generally far more difficult to apply to other settings with more complicated hydrogeological characteristics, such as many base metal mines and virtually all deep mines (for coal or metals). Add to this the usual absence of historic ABA results for the strata formerly disturbed by deep mine workings that have long been abandoned but continue to pose a pollution risk, and the need for an alternative to simple ABA is imperative (e.g. Banwart & Malmström 2001). While advances in this regard have been made for the case of pit lake wall rocks (Bowell 2002) the geometric disposition of reactive minerals in disturbed strata surrounding deep mine voids is far harder to characterize/conceptualize. Standard geochemical modelling codes, such as those belonging to the PHREEQE family, have found a certain amount of tentative application in such cases (e.g. Younger et al. 1995), but severe problems of parameterization have inhibited their widespread uptake. Simpler mass-balance models have been coupled to GRAM by Sherwood (1997), and applied in 'stand-alone' mode by Banwart & Malmström (2001). A post-processing water quality prediction protocol was developed for application to rebound predictions yielded by VSS-NET (Younger & Adams 1999; Younger 2000b). Although this has proved to be fairly robust in field applications, it cannot be regarded as being anything more than a 'first approximation' (Younger 2000b). More recently, initial steps have been taken to couple numerical simulation codes to a novel object-oriented code simulating pyrite oxidation and the attenuation of acidity by a variety of reactions identified from field and laboratory studies (e.g. Gandy & Evans 2002). Full-scale application of kinetic geochemical modelling to the derivation of management plans for real problem sites remains a state-of-the-art venture, which has so far been attempted in very few recorded instances: a good early example is provided by Hattingh et al. (2002).

It is clear that significant areas of uncertainty remain with regard to hydrogeological and geochemical processes in and around mined systems, and these must be addressed if the challenge of developing robust mine water management strategies is to be realized. In the following section we attempt to identify some of the key hydrogeological and geochemical uncertainties in this context. Although all the papers in this volume contribute to the struggles to diminish these uncertainties and rise to these challenges, much remains to be done.

Hydrogeological uncertainties and challenges

Uncertainties

Much of the discussion that follows derives from our experience in addressing the principal hydrogeological question that arises from the cessation of dewatering of mines: i.e. where and when will mine waters decant to the surface environment or adjoining aquifers?

Obviously, the discharge points from any groundwater system tend to be topographic lows (for surface decant) or stratigraphic/structural lows (for decant to adjoining aquifers). There are two basic approaches to identifying such features:

- a priori identification by applied geological mapping (e.g. Younger et al. 1995; Younger & Adams 1999);
- automated identification of decant points using distributed, physically based groundwater flow models coupled with digital

terrain models (e.g. Adams & Younger 2001) and/or digital geological structure models (e.g. Dumpleton et al. 2001).

The question of when such decants will become operative can only be addressed by some form of predictive hydrogeological modelling. Although suitable simulation codes now exist (e.g. Sherwood & Younger 1997; Younger & Adams 1999; Banks 2001; Whitworth 2002) the parameterization of models for real systems remains fraught by issues of system complexity and the general inadequacy of basic information.

There are three principal areas of hydrogeological uncertainty surrounding any investigation of mined ground to determine optimum post-closure management. These may be grouped under three headings:

- shortcomings in data needed to characterize the hydrogeology of mine systems;
- scarcity of information on hydraulic properties and processes;
- problems of scale.

These problems have all been identified by previous workers (e.g. Younger et al. 1995; Sherwood & Younger 1997; Adams & Younger 2001; Whitworth 2002), although rarely discussed in any detail. We will now consider each of these sources of uncertainty in turn, with the aim of identifying priorities for future data collection.

Shortcomings in data needed to characterize the hydrogeology of mine systems

With regard to the availability of data, Sherwood & Younger (1997) have noted that predictive hydrogeological modelling of mined systems typically suffers from both a lack of certain types of information (principally hydrological data) and a super abundance of other information (mine plans). It is extremely rare to find extensive groundwater monitoring networks outside the operating areas of deep mines. Surface mines, being more conspicuous, have often had to fight harder for mining permits and will thus usually be better served with monitoring wells. However, there are a number of common problems with these monitoring wells:

- They are rarely situated in the most optimal locations. This is because mine owners rarely have property rights in the most suitable locations outside of their curtilage, and may not receive the most cordial of welcomes from aggrieved neighbours to whom they make requests for access for purposes of borehole drilling. (Opponents of a surface mine development may well be suspicious that the borehole is a clandestine exploration borehole rather than a benign monitoring well.) Such limitations on the ability of minor operators to actually monitor the impacts of their dewatering operations are a key issue in relation to demands from regulators for extensive monitoring networks; in some cases, effective monitoring may require that regulators exercise their powers of compulsory access (although we are not aware of any cases where this has occurred in the UK).
- To save money, monitoring wells around quarries are often drilled by the company's blast-hole drillers during periods when their ordinary workload is temporarily light. Notwithstanding their drilling prowess, blast-hole drillers are rarely au fait with the intricacies of hydrogeological monitoring well completion, so that the resultant monitoring wells too often yield data of little or no use.

In the face of absent or poor piezometric data, the hydrogeologist is often faced with having to make unsubstantiated (and potentially drastic) assumptions about piezometric head distributions in and around mine workings. Given the importance of head-dependent inflows in governing the rate of flooding of workings after abandonment (for further discussion of this point, see Younger & Adams 1999; Banks 2001; Johnson & Younger, 2002; Whitworth 2002), this is a serious source of uncertainty.

At the other extreme, a retrieval of all available mine plans for even a relatively small area of multiple-horizon mine workings will generally yield a bewildering volume of paper. The challenges associated with transforming these plans into digitally available, hydrogeologically relevant data suitable for modelling purposes are substantial (e.g. Dumpleton et al. 2001; Robins et al. 2002). The modeller therefore often struggles to simplify the vast amounts of information to the point of identifying a few critical features (e.g. points of potential hydraulic connection between otherwise separate volumes of mine workings; Sherwood 1997). There is always a risk of over simplification on the one hand, or the retention of unwieldy degrees of detail on the other, and only an experienced practitioner is likely to strike the right balance with any alacrity.

In addition to these 'hard' sources of information, we have found that it is often necessary and desirable to fill substantial gaps in data availability for areas with a long mining history by resorting to anecdotal information. Miners usually remember vividly which parts of the workings were wettest, for instance, as these will have been the subject of repeated 'wet working claims' (i.e. requests for wage supplements, payable where the working was rendered particularly uncomfortable by the presence of dripping or lying water). Officials who had responsibility for mine safety assessments will also have been keenly aware of the positions of aquifers or bodies of old flooded workings in relation to active workings, and may well have been involved in drilling boreholes to assess the water pressures in such workings. However, anecdotal information needs to be treated with caution, especially where the information is second hand ('My Grandfather told me ...'). A classic case of misleading anecdotal information is presented by Adams & Younger (2002), who report a case in which some supposed ancient workings were widely believed by miners to exist, but which ultimately proved never to have been excavated when the rising mine water reached them and responded as it would to unworked crystalline bedrock of very low storativity (Adams & Younger 2002).

Given the variable sources of data used in the hydrogeological assessment of mined systems, it is always valuable to establish the sensitivity of the predictions made to the quality of the data used. Where numerical models have been constructed such sensitivity analyses can be undertaken using conventional methods (e.g. Sherwood & Younger 1997; Burke & Younger 2000; Dumpleton et al. 2001). Where computational resources permit, it is probably best to face the uncertainties squarely and undertake the modelling in probabilistic mode, using Monte Carlo methods or some other stochastic simulation protocol (e.g. Younger et al. 1995; Sherwood 1997; Sherwood & Younger 1997). From our experience with such modelling efforts, we would rank the critical data needs for mine water modelling in the following order of decreasing sensitivity:

- the geological framework, and thus the hydraulic relationship of mined ground to adjacent country rock including aquifer units;
- mine geometry (defined by abandonment plans), and estimates of void space volume and interconnectivity between void spaces, areas of goaf, etc., which are derived from these;
- sources and piezometric characteristics of lateral head-dependent inflows into mine voids;
- operational mine dewatering records;
- post-closure water level monitoring data;
- local rainfall recharge and the ways in which this is transmitted to voids at depth;
- historical evidence of pre-mining piezometry (from shallow wells and springs);
- anecdotal evidence of mine water occurrence during and after working.

Scarcity of information on hydraulic properties and processes

With the best data-set in the world uncertainties would still remain, not least because of the difficulties inherent in assessing and conceptualizing flow processes in the complex of voids and country rock which together constitute a mined hydrogeological system. These difficulties are manifest in problems such as:

- water budgets that cannot be brought into balance over annual time-steps (e.g. Robins et al. 2002);
- substantial over-or underprediction of the rates of mine water rise in an abandoned system (e.g. Adams & Younger 2002);
- difficulties in predicting the required pumping rate for a new (or renewed) dewatering system (e.g. Davis & Battersby 1985; Nuttall et al. 2002).

The detailed reasons for these difficulties are multiple and vary from case to case, but they all ultimately derive from uncertainties in hydraulic properties of the mine voids and their surrounding strata. There is a fundamental paradox inherent in the determination of the hydraulic properties of mined ground, which may be summarized as follows:

- we know that mining changes the saturated hydraulic conductivity and storativity of the enclosing strata;
- drainage of the affected strata into the dewatered mine voids leaves the strata in an unsaturated condition;
- it is not possible to measure the saturated hydraulic conductivites of unsaturated voids;
- while these can be measured once the strata are saturated once more, by the time this happens the original motivation for predictive modelling will have been overtaken by events.

In rare cases, hydraulic tests undertaken in boreholes drilled into still-saturated strata overlying active mine workings have revealed the temporal variations in transmissivity (and, to a lesser degree, storativity), which are induced by deformation of the rock mass due to void closure/goafing (e.g. Booth 2002; Dumpleton 2002). The hydraulic properties of abandoned flooded voids can also be assessed using various test-pumping approaches. The predominance of turbulent flow in flooded voids can be inferred by numerical simulation of the time–drawdown data obtained from such tests (e.g. Adams & Younger 2001, 2002; Nuttall et al. 2002), and can also be revealed by means of tracer tests between shafts (e.g. Wolkersdorfer 2002) and by changes in natural water quality (Nuttall et al. 2002).

Problems of scale

Even where direct measurements exist from which to characterize mine void flow regimes and the hydraulic properties of the adjoining strata, they are highly unlikely to be available at sufficient spatial density to support meaningful interpolation of values for unmeasured points in the system. While this problem is by no means unique to mined groundwater systems, it is especially vexing in this context both because measurement points tend to be somewhat more sparse than in public supply aquifers and because extreme variations in hydraulic conductivity and storativity typically occur over very short distances in mined ground. Very real problems, therefore, exist in attempting to conceptualize and model the hydrogeological behaviour of mined ground over the full range of scales relevant to many management decisions (i.e. from a single dewatering well to an entire inter connected coalfield; e.g. Robins et al. 2002).

To overcome these problems it is possible to apply the various modelling tools summarized earlier in this paper, each of which has been deliberately developed for application at a given scale (e.g. Sherwood 1997; Younger & Adams 1999; Adams & Younger 2001; Banks 2001). At the very finest scales, it may be necessary to develop physically based models in which individual mine roadways/shafts are resolved and represented by one or more elements of a pipe network (e.g. Adams & Younger 2001). At intermediate scales, 'ponds' of highly interconnected workings can be identified and modelled using semi-distributed water balance models (e.g. Sherwood & Younger 1997; Banks 2001; Robins et al. 2002; Whitworth 2002). At the coarsest of scales, very simple models (which may be no more than manual calculations relating mined void volumes to average annual recharge rates) may provide as good information as any more sophisticated approach. Whichever approach is adopted, the customary lack of data will inevitably necessitate frequent updating of rebound predictions in the light of the latest data. A good example of repeated updating of rebound timing predictions is provided for the case of Whittle Colliery, Northumberland, by Parker (2000).

Geochemical uncertainties and challenges

For all of the uncertainties that beset assessments of mine water quantities, the position in relation to prediction of mine water quality is many times worse. The difficulties in applying standard geochemical modelling packages to real mine water cases has already been mentioned. These difficulties have prompted two extreme responses: the first is to eschew superfluous detail and model the release of pollutants from weathered mine waste using simple mass-balance calculations (e.g. Banwart & Malmström 2001; Banwart et al. 2002; Loredo et al. 2002). At the other extreme, attempts have been made to numerically model the generation and multispecies transport of acid mine waters and spoil leachates through both the source bodies of mine waste rock (Gandy & Evans 2002; Hattingh et al. 2002) and in the aquifers infiltrated by such leachates (Saaltink et al. 2002). Such efforts have revealed the persistence of the following problems, which serve as barriers to further advances in predictive geochemical modelling:

- Lack of information on the nature and kinetics of several key sink–source reactions for the major mine water pollutant species (e.g. Bowell 2002; Saaltink et al. 2002). For instance, recent results suggest that the precipitation-dissolution kinetics of potassium jarosite (which help to buffer pH around 1.8 in very acidic mine waters) are hysteretic, with dissolution occurring much more slowly than precipitation (e.g. Saaltink et al. 2002).
- The inherent difficulties in assigning appropriate 'active surface geometries' to those fractions of the rock mass (such as pyrite) that act as pollutant sources (e.g. Younger et al. 1995). Standard 'shrinking core' models have been found to produce adequate results for highly comminuted waste rocks (e.g. Gandy & Evans 2002) and simple reacting fraction representations for pit lake wall rocks have also proved to have some predictive

power (Bowell 2002). However, the same cannot truly be stated for flooded or partially flooded deep mines (e.g. Hattingh et al. 2002; Nuttall et al. 2002), and significant progress in this regard is unlikely to be made before a wider range of comprehensive hydrogeochemical data-sets have been acquired for representative mine systems (along the lines of those presented by Nuttall et al. 2002; and Hattingh et al. 2002) and analysed using inverse geochemical modelling techniques.

- Numerical limitations, which arise when the transport of a large number of species is simulated using a multispecies geochemical modelling code. Run times increase markedly in proportion to both the dimensionality of the transport domain (i.e. if a 1-D flow field is replaced with a 3-D flow field) and to the number of pollutant species considered. Even with the very latest versions of popular reactive transport codes, single runs can easily take 12–18 h where six or seven species are simulated for more than 20 time-steps in a 2-D flow field. With such slow run times, the analysis of uncertainty by means of Monte Carlo analysis is generally out of the question. While further advances in computer technology can be confidently expected to ease this problem over time, it may well be several years before stochastic, multi species 3-D transport modelling can be applied routinely for geochemical systems characterized by rapid reactions, such as those which dominate the evolution of mine water chemistry (e.g. Banwart et al. 2002).

Given that predictive geochemical modelling is still subject to such considerable practical limitations, it will be many more years before pollution exposure risk assessments will be able to advance beyond the current state-of-the-art as presented by Holton et al. (2002).

Towards defensive mine planning

Given the myriad problems associated with polluted waters draining from abandoned mine voids, the question must be asked: can new mine workings be designed so that such problems will be averted in future when these workings are finally abandoned? We would argue that the various experiences gained recently in relation to abandoned workings provide the basis for a new paradigm in mine design, which we term 'defensive mine planning'. As we conceive it, defensive mine planning starts from the acknowledgement that closure is the longest part of the mine life cycle, and is, therefore, worthy of respect from the very start of a mining project. An appropriate level of respect for the ultimate post-closure phase would be manifest in different ways at different stages in the mine life cycle. Table 2 summarizes some preliminary suggestions in this regard, together with their rationale.

Of course 'defensive mine planning' has long been practised for health and safety purposes: for instance, entire bodies of regulations have long governed mining below water bodies (e.g. Orchard 1975) and the safe design and monitoring of mine waste tips. We propose that compliance with these regulations be augmented by the implementation of relatively simple additional measures, along the lines proposed in Table 2, in order to minimize post-closure environmental management problems. We do not envisage that the measures we propose will increase the cost burdens of mining companies; rather, they will substitute a little extra expenditure during the exploratory and working phases of the mine life cycle for substantially greater expenditures that would otherwise be incurred after closure (when cash flows are generally at their most negative).

While the simple suggestions made in Table 2 are highly preliminary in nature, and may be viewed as idealistic when one considers the considerable inherited legacies that beset the opening of new mines in areas which have previously been subject to intense historic/prehistoric mining (e.g. Davis & Buttersby 1985), we think that the general approach of basing environmentally-defensive mine planning on hydrogeological and geochemical principles has considerable potential for the future. It is beyond the scope of this volume to fully realize this potential, but we hope the readers will find many signposts in the pages that follow.

This volume arises from a meeting of the Hydrogeological Group of the Geological Society of London, which was co-sponsored by the Applied Mineralogy Group of the Mineralogical Society, that was held at the headquarters of the British Geological Survey, Keyworth, on 15 February 2001, and includes a number of additional papers on related themes. Our own insights represent the fruit of much research funded by NERC, EPSRC, the Coal Authority, the Environment Agency, UK Coal, Scottish Coal and many other mining companies.

Table 2. *Examples of 'defensive mine planning' measures (based on hydrogeological and geochemical principles) applicable at different stages in the life cycle of a mine*

Stage in mine life cycle	Proposed measures	Relevance to long-term environmental performance of mined system
Exploration	(i) Assay the overburden for long-term pollutant release potential; (ii) ensure adequate after-use of exploration boreholes, either by: (a) efficient back-filling and sealing; or (b) by equipping them for hydrogeological monitoring purposes (see Kuma et al. 2002)	(i) Allows minimization of long-term water quality liabilities by careful handling; (ii) (a) minimization of long-term water make, and, therefore, of liabilities associated with site drainage; (b) acquisition of pre- and syn-mining hydrogeological data to allow full assessment and planning of mitigation measures for any water management problems arising from dewatering and/or mine abandonment
Detailed design	Plan pillar locations and geometry of major mine access features to facilitate easy blocking of potential post-abandonment hydraulic pathways	Minimization of deep circulation of waters after mine closure, which should help to limit rock–water interaction residence times and therefore keep salinities as low as possible
Site preparation	(i) Construct mine access features consistent with detailed design; (ii) locate mineral processing and tailings/waste rock storage facilities in those portions of the site least likely to give rise to environmental pollution in emergency situations	(i) Minimizes post-closure costs to achieve management objectives; (ii) achieves compliance with standard water quality protection policies of regulators; will also make eventual decommissioning less expensive to achieve
Main phase of extraction	(i) Careful design of panels/pillars/benches to minimize the inducement of excess water inflow from surrounding strata; (ii) local 'over-dosing' with calcite stone dust and/or topical grouting of high-S–high-K zones to minimize later pollutant mobilization	(i) Minimizes water make and all associated costs; (ii) minimizes mobilization of acidic ions in mine water after mine is flooded
Mine waste management	Make provisions for selective handling/careful disposal of the most pollution-generating waste rock (using methods such as co-mingling with reductants/alkalis, O_2 exclusion, by means of water covers/dry covers, etc.)	Pre-empts possible future water quality liabilities, which would likely be very long term in nature.

Mine abandonment	(i) Engineer any long-term preferred drainage routes for 'permanence';	(i) Ensure long-term drainage routes are predictable and reliable;
	(ii) seal major mine access features at or just below anticipated climax water table position;	(ii) minimize deep circulation (and therefore salinization) of mine waters;
	(iii) consider the installation of (?replaceable) *in situ* reactive media in main shafts/declines to provide treatment of polluted drainage prior to surface discharge;	(iii) maximize the potential for emergence of good quality water at the ground surface;
	(iv) make sure facilities are in place for monitoring of rebound and climax water table positions	(iv) secures long-term monitoring to allow early identification of any problems/demonstration of system stability to third parties
Restoration	(i) Ensure that hydrological issues are given suitable prominence in restoration plans for underground voids and mine waste depositories;	(i) Minimization of long-term pollutant release through restricting access to acid-generating materials by O_2 and/or H_2O;
	(ii) involve all relevant stakeholders in financial and institutional arrangements for post-closure site maintenance and monitoring activities;	(ii) establishment of a secure socio-economic foundation for long-term site management/after-use;
	(iii) implement any short-term intensive water treatment measures during the 'first flush' (see Younger 2000*b, c*)	(iii) avoids any legal problems during the period of most elevated pollutant concentrations
After-care	(i) Implement post-closure site maintenance and monitoring activities;	(i) Achieve stable post-closure water management system;
	(ii) implement long-term (?passive) water treatment measures as appropriate (see Younger 2000*c*)	(ii) ensure long-term attainment of water quality objectives in receiving watercourse

References

ADAMS, R. & YOUNGER, P.L. 2001. A strategy for modelling ground water rebound in abandoned deep mine systems. *Ground Water*, **39**, 249–261.

ADAMS, R. & YOUNGER, P.L. 2002. A physically based model of rebound in South Crofty tin mine, Cornwall. In: YOUNGER, P.L. & ROBINS, N.S. (eds) *Mine Water Hydrogeology and Geochemistry*. Geological Society, London, Special Publications, **198**, 89–97.

BANKS, D. 2001. A variable-volume, head-dependent mine water filling model. *Ground Water*, **39**, 362–365.

BANKS, D., YOUNGER, P.L. & DUMPLETON, S. 1996. The historical use of mine-drainage and pyrite-oxidation waters in central and eastern England, United Kingdom. *Hydrogeology Journal*, **4**, 55–68.

BANKS, D., YOUNGER, P.L., ARNESEN, R.-T., IVERSEN, E.R. & BANKS, S.D. 1997. Mine-water chemistry: the good, the bad and the ugly. *Environmental Geology*, **32**, 157–174.

BANWART, S.A. & MALMSTRÖM, M.E. 2001. Hydrochemical modelling for preliminary assessment of minewater pollution. *Journal of Geochemical Exploration*, **74**, 73–97.

BANWART, S.A., EVANS, K.E. & CROXFORD, S. 2002. Predicting mineral weathering rates at field scale for mine water risk assessment. In: YOUNGER, P.L. & ROBINS, N.S. (eds) *Mine Water Hydrogeology and Geochemistry*. Geological Society, London, Special Publications, **198**, 137–157.

BOOTH, C.J. 2002. The effects of longwall coal mining on overlying aquifers. In: YOUNGER, P.L. & ROBINS, N.S. (eds) *Mine Water Hydrogeology and Geochemistry*. Geological Society, London, Special Publications, **198**, 17–45.

BOWELL, R.J. 2002. The hydrogeochemical dynamics of mine pit lakes. In: YOUNGER, P.L. & ROBINS, N.S. (eds) *Mine Water Hydrogeology and Geochemistry*. Geological Society, London, Special Publications, **198**, 159–185.

BOWEN, G.G., DUSSEK, C. & HAMILTON, R.M. 1998. Pollution resulting from abandonment and subsequent flooding of Wheal Jane Mine in Cornwall, UK. In: MATHER, J., BANKS, D., DUMPLETON, S., FERMOR, M. (eds) *Groundwater contaminants and their migration*. Geological Society, London, Special Publications, **128**, 93–99.

BROWN, M.M.E., JONES, A.L., LEIGHFIELD, K.G. & COX, S.J. 2002. Fingerprinting mine water in the eastern sector of the South Wales Coalfield. In: YOUNGER, P.L. & ROBINS, N.S. (eds) *Mine Water Hydrogeology and Geochemistry*. Geological Society, London, Special Publications, **198**, 275–286.

BURKE, S.P. & YOUNGER, P.L. 2000. Groundwater rebound in the South Yorkshire coalfield: a first approximation using the GRAM model. *Quarterly Journal of Engineering Geology and Hydrogeology*, **33**, 149–160.

DAVIS, C.J. & BATTERSBY, G.W. 1985. Reopening of Wheal Jane mine. *Transactions of the Institution of Mining and Metallurgy (Section A: Mining Industry)*, **94**, A135–A147.

DUMPLETON, S. 2002. Effects of longwall mining in the Selby Coalfield on the piezometry and aquifer properties of the overlying Sherwood Sandstone. In: YOUNGER, P.L. & ROBINS, N.S. (eds) *Mine Water Hydrogeology and Geochemistry*. Geological Society, London, Special Publications, **198**, 75–88.

DUMPLETON, S., ROBINS, N.S., WALKER, J.A. & MERRIN, P.D. 2001. Mine water rebound in South Nottinghamshire: risk evaluation using 3-D visualization and predictive modelling. *Quarterly Journal of Engineering Geology and Hydrogeology*, **34**, 307–319.

GANDY, C.J. & EVANS, K.A. 2002. Laboratory and numerical modelling studies of iron release from a spoil heap in County Durham. In: YOUNGER, P.L. & ROBINS, N.S. (eds) *Mine Water Hydrogeology and Geochemistry*. Geological Society, London, Special Publications, **198**, 205–214.

GOLDRING, D.C. & JUCKES, L.M. 2001. Iron ore supplies to the United Kingdom iron and steel industry. *Transactions of the Institution of Mining and Metallurgy (Section A: Mining Technology)*, **110**, A75–A85.

GRIMALT, J.O., FERRER, M. & MACPHERSON, E. 1999. The mine tailing accident in Aznalcóllar. *Science of the Total Environment*, **242**, 3–11.

GRIMSHAW, P. N. 1992, *Sunshine Miners. Opencast Coalmining in Britain*, 1942–1992. British Coal Opencast, Mansfield, Nottinghamshie.

HATTINGH, R.P., PULLES, W., KRANTZ, R., PRETORIUS, C. & SWART, S. 2002. Assessment, prediction and management of long-term post-closure water quality: a case study – Hlobane Colliery, South Africa. In: YOUNGER, P.L. & ROBINS, N.S. (eds) *Mine Water Hydrogeology and Geochemistry*. Geological Society, London, Special Publications, **198**, 297–314.

HOLGATE, R. 1991. *Prehistoric Flint Mines*. Shire, Princes Risborough.

HOLTON, D., KELLY, M. & BAKER, A. 2002. Paradise lost? Assessment of liabilities at a uranium mine in the Slovak Republic: Novoveska Huta. In: YOUNGER, P.L. & ROBINS, N.S. (eds) *Mine Water Hydrogeology and Geochemistry*. Geological Society, London, Special Publications, **198**, 365–377.

JOHNSON, K.L. & YOUNGER, P.L. 2002. Hydrogeological and geochemical consequences of the abandonment of Frazer's Grove carbonate-hosted Pb–Zn fluorspar mine, North Pennines, UK. In: YOUNGER, P.L. & ROBINS, N.S. (eds) *Mine Water Hydrogeology and Geochemistry*. Geological Society, London, Special Publications, **198**, 347–363.

KELLY, M.G. 1999. Effects of heavy metals on the aquatic biota. In: PLUMLEE, G.S., LOGSDON, M.J. (eds) *The Environmental Geochemistry of Mineral Deposits. Part A: Processes, Techniques and Health Issues. Reviews in Economic Geology, Volume 6A*. Society of Economic Geologists, Littleton, CO, 363–371.

KLEINMANN, R.L.P. (ed.) 2000. *Prediction of Water Quality at Surface Coal Mines.* Manual prepared by members of the Prediction Workgroup of the Acid Drainage Technology Initiative (ADTI). National Mine Land Reclamation Centre, West Virginia University, Morgantown, West Virginia, USA.

KNIGHT PIÉSOLD & PARTNERS LTD. 1995. *Wheal Jane Minewater Study. Environmental Appraisal and Treatment Strategy.* Report to the National Rivers Authority, South Western Region, Exeter.

KUMA, J., YOUNGER, P.L. & BOWELL, R.J. 2002. Hydrogeological framewrok for assessing the possible environmental impacts of large-scale gold mines. In: YOUNGER, P.L. & ROBINS, N.S. (eds) *Mine Water Hydrogeology and Geochemistry.* Geological Society, London, Special Publications, **198**, 121–136.

LOREDO, J., ORDÓÑEZ, A. & PENDÁS, F. 2002. Hydrogeological and geochemical interactions of adjoining mercury and coal mine spoil heaps in the Morgao catchment (Mieres, northwestern Spain). In: YOUNGER, P.L. & ROBINS, N.S. (eds) *Mine Water Hydrogeology and Geochemistry.* Geological Society, London, Special Publications, **198**, 327–336.

MCKELVEY, P., BEALE, G., TAYLOR, A., MANSELL, S., MIRA, B., VALDIVIA, C. & HITCHCOCK, W. 2002. Depressurization of the north wall at the Escondida Copper Mine, Chile. In: YOUNGER, P.L. & ROBINS, N.S. (eds) *Mine Water Hydrogeology and Geochemistry.* Geological Society, London, Special Publications, **198**, 107–119.

NRA, 1994. *Abandoned Mines and the Water Environment,* Report of the National Rivers Authority. Water Quality Series No. 14, HMSO, London.

NEUMANN, I. & SAMI, K. 2002. Structural influence on plume migration from a tailings dam in the West Rand, Republic of South Africa. In: YOUNGER, P.L. & ROBINS, N.S. (eds) *Mine Water Hydrogeology and Geochemistry.* Geological Society, London, Special Publications, **198**, 337–346.

NUTTALL, C. A., ADAMS, R. & YOUNGER, P. L. 2002. Integrated hydraulic–hydrogeochemical assessment of flooded deep mine voids by test pumping at the Deerplay (Lancashire) and Frances (Fife) collieries. In: YOUNGER, P.L. & ROBINS, N.S. (eds) *Mine Water Hydrogeology and Geochemistry.* Geological Society, London, Special Publications, **198**, 315–326.

ORCHARD, R.J. 1975. Working under bodies of water. *The Mining Engineer,* **170**, 261–270.

PARKER, K. 2000. Mine water – the role of the Coal Authority. *Transactions of the Institution of Mining and Metallurgy (Section A: Mining Technology),* **109**, A219–A223.

POWELL, D. 1993. *The Power Game: The Struggle for Coal.* Duckworth, London.

ROBINS, N.S. 1990. *Hydrogeology of Scotland.* HMSO, London.

ROBINS N.S. & YOUNGER P.L. 1996. Coal abandonment – mine water in surface and near surface environment: some historical evidence from the United Kingdom, In: *Proceedings of the Conference in Minerals Metals and the Environment II, Prague, 3–6 September 1996.* Institution of Mining and Metallurgy, 253–262.

ROBINS, N.S., DUMPLETON, S. & WALKER, J. 2002. Coalfield closure and environmental consequence – the case in South Nottinghamshire. In: YOUNGER, P.L. & ROBINS, N.S. (eds) *Mine Water Hydrogeology and Geochemistry.* Geological Society, London, Special Publications, **198**, 99–105.

ROBINSON, R. 2000. Mine gas hazards in the surface environment. *Transactions of the Institution of Mining and Metallurgy (Section A: Mining Technology),* **109**, A228–A236.

SAALTINK, M.W., DOMÈNECH, C., AYORA, C. & CARRERA, J. 2002. Modelling the oxidation of sulphides in an unsaturated soil. In: YOUNGER, P.L. & ROBINS, N.S. (eds) *Mine Water Hydrogeology and Geochemistry.* Geological Society, London, Special Publications, **198**, 187–204.

SHEPHERD, R. 1993. *Ancient Mining,* Institution of Mining and Metallurgy/Elsevier Applied Science, London.

SHERWOOD, J.M. 1997. *Modelling Minewater Flow and Quality Changes After Coalfield Closure.* Unpublished PhD thesis, University of Newcastle.

SHERWOOD, J.M. & YOUNGER, P.L. 1997. Modelling groundwater rebound after coalfield closure. CHILTON, J. *et al.* (eds) *Groundwater in the Urban Environment: Problems, processes and Management.* Balkema, Rotterdam, 165–170.

SMITH, J.A. & COLLS, J.J. 1996. Groundwater rebound in the Leicestershire Coalfield. *Journal of the Chartered Institution of Water and Environmental Management,* **10**, 280–289.

UNEP, *Cyanide spill at Baia Mare, Romania. Spill of Liquid and Suspended Waste at the Aural S.A. Retreatment Plant in Baia Mare. Assessment Mission Report,* United Nations Environment Programme, Office for the Co-ordination of Humanitarian Affairs, UNEP, Geneva.

WHITWORTH, K.A. 2002. The monitoring and modelling of mine water recovery in UK coalfields. In: YOUNGER, P.L. & ROBINS, N.S. (eds) *Mine Water Hydrogeology and Geochemistry.* Geological Society, London, Special Publications, **198**, 61–73.

WOLKERSDORFER, C. 2002. Mine water tracing. In: YOUNGER, P.L. & ROBINS, N.S. (eds) *Mine Water Hydrogeology and Geochemistry.* Geological Society, London, Special Publications, **198**, 47–60.

YOUNGER, P.L. 1994. Minewater pollution: the revenge of Old King Coal. *Geoscientist,* **4** (5), 6–8.

YOUNGER, P.L. 1998. Coalfield abandonment: geochemical processes and hydrochemical products. NICHOLSON, K. (ed) *Energy and the Environment. Geochemistry of Fossil, Nuclear and Renewable Resources.* Society for Environmental Geochemistry and Health, McGregor Science, Aberdeenshire, 1–29.

YOUNGER, P.L. 2000a. Iron. DARCY, B.J.D., ELLIS, J.B., FERRIER, R.C., JENKINS, A., DILS, R. (eds) *Diffuse Pollution Impacts,* Terence Dalton Publishers, Lavenham, for Chartered Institution of Water and Environmental Management, 95–104.

YOUNGER, P.L. 2000b. Predicting temporal changes in total iron concentrations in groundwater flowing from abandoned deep mines: a first approximation. *Journal of Contaminant Hydrogeology*, **44**, 47–69.

YOUNGER, P.L. 2000c. Holistic remedial strategies for short- and long-term water pollution from abandoned mines. *Transactions of the Institution of Mining and Metallurgy (Section A: Mining Technology)*, **109**, A210–A218.

YOUNGER, P.L. 2001. Mine water pollution in Scotland: nature, extent and preventative strategies. *Science of the Total Environment*, **265**, 309–326.

YOUNGER, P.L. 2002. The importance of pyritic roof strata in aquatic pollutant release from abandoned mines in a major, oolitic, berthierine–chamosite–siderite iron ore field, Cleveland, UK. In: YOUNGER, P.L. & ROBINS, N.S. (eds) *Mine Water Hydrogeology and Geochemistry*. Geological Society, London, Special Publications, **198**, 251–267.

YOUNGER, P.L., BANWART, S.A. & HEDIN, R.S. 2002. *Mine water: Hydrology, Pollution, Remediation*. Kluwer, Oordrecht.

YOUNGER, P.L. & HARBOURNE, K.J. 1995. To pump or not to pump: cost–benefit analysis of future environmental management options for the abandoned Durham Coalfield. *Journal of the Chartered Institution of Water and Environmental Management*, **9**, 405–415.

YOUNGER, P.L. & ADAMS, R. *Predicting mine water rebound*, Environment Agency R&D Technical Report W179, Enviromental Agency, Bristol.

YOUNGER, P.L. & LAPIERRE, A.B. 2000. Uisge Mèinne': mine water hydrogeology in the Celtic lands, from *Kernow* (Cornwall) to *Ceapp Breattain* (Cape Breton, Nova Scotia). ROBINS, N.S. & MISSTEAR, B.D.R. (eds) *Groundwater in the Celtic regions: Studies in Hard-Rock and Quaternary Hydrogeology*. Geological Society, London, Special Publications, **182**, 35–52.

YOUNGER, P.L., BARBOUR, M.H. & SHERWOOD, J.M. 1995. Predicting the Consequences of Ceasing Pumping from the Frances and Michael Collieries, Fife. BLACK, A.R. & JOHNSON, R.C. (eds) *Proceedings of the Fifth National Hydrogeology Symposium, British Hydrological Society, Edinburgh, 4–7 September 1995*, 2.25–2.33.

The effects of longwall coal mining on overlying aquifers

COLIN J. BOOTH

Department of Geology and Environmental Geosciences, Northern Illinois University, DeKalb, IL 60115, USA (e-mail: colin@geol.niu.edu)

Abstract: The hydrogeological effects of longwall mines are vertically zoned. The heavily fractured strata immediately above the mine dewater, but they are typically overlain by a zone of low permeability that prevents shallower aquifers from draining to the mine. However, shallow bedrock aquifers experience head changes caused by fracturing during subsidence. New fracture void space takes up water, causing large head drops especially in confined aquifers. Increased fracture permeability affects heads because upper aquifers in high relief areas lose water through fractured aquitards to lower aquifers, and because the higher permeabilities lower hydraulic gradients and up-gradient heads, and increase down-gradient discharge. In addition, a secondary drawdown spreads out laterally through transmissive aquifers from the potentiometric low in the subsiding zone. After undermining, water levels may recover due to closure of fractures and to recharge flowing back into the affected area. Studies at two active longwall mines in Pennsylvanian coal measures in Illinois support the conceptual model, with variations. Unconsolidated, unconfined aquifers were not significantly affected by mining. At one site, heads in a moderately transmissive sandstone declined due to mining but recovered fully afterwards. Increased permeability led to enhanced well yields, but water quality deteriorated, probably because of oxidation and mobilization of *in situ* sulphides during the unconfined and recovery phases. At the other site, heads in a poorly transmissive sandstone fell rapidly during subsidence and did not recover; hydrogeological responses varied at the site scale due to variations in bedrock–drift continuity. Predictions and monitoring schemes can be guided by the general conceptual model, but must consider local hydrogeological variations. Effects in shallow aquifers not in direct contact with the mine can be simulated using readily available flow models.

This paper discusses the generalized conceptual model of the hydrogeological effects of longwall coal mining on overlying aquifers and the shallow groundwater system that has evolved from numerous case studies since the 1970s. As an overview paper, it includes numerous references and summarizes previously published studies by the author. The main geographical focus is on the Carboniferous (Pennsylvanian) coalfields of the eastern and mid-western United States. The general principles apply to other areas, but specific impacts may differ substantially because of geological differences.

The paper is divided into two major sections. First, the mechanisms and impacts of the hydrogeological response to longwall mining are discussed, including information from previous studies in various coalfields. Second, a long-term investigation at two sites in Illinois, USA, is described, which illustrates the application of the general conceptual model but also shows the different responses that result from minor variations in the geological setting within the same coalfield.

Mechanisms of the hydrogeological effects of longwall mining

Longwall mining is an economic, efficient form of underground coal mining that is characterized by almost total extraction of large areas of the coal seam and by the resultant rapid, predictable, extensive subsidence of the overlying strata and ground surface. It contrasts with supported methods such as room-and-pillar (bord-and-pillar) mining, in which pillars of coal or re-stacked rock are left to support the mine roof, but which may produce irregular, localized subsidence long after mining has finished.

The hydrological impact of longwall mining operates through two quite distinct mechanisms: drainage and subsidence. All underground mines are low-pressure groundwater sinks during active mining. If the overlying strata are sufficiently permeable, groundwater will drain into the mine, creating inflow and drainage problems, and sometimes depletion of aquifer resources. Aquifers in close contact with the mine are generally dewatered. Shallow aquifers separated from the mine by confining layers may be unaffected by drainage to it.

After mining has ceased a mine may flood, producing a high permeability pathway that distorts regional groundwater flow, particularly if the mine is connected with others to form an extensive flow network. When entire mining areas are closed down, the rebound of regional water levels after long-term depression can create substantial mine water discharge problems, as observed over the last decade in coalfields in England (e.g. Younger 1998; Burke & Younger 2000). On the other hand, deep mines in low-permeability environments may remain dry for years after mining operations have ceased, as reported for some mines in Illinois (Cartwright & Hunt 1983).

Longwall mining also affects the groundwater system through subsidence and strata movements. These movements are expressed in the deformation of existing fractures and the creation of new ones, changing the hydraulic properties of the strata and, consequently, the hydraulic gradients, heads and groundwater flow patterns, independently of drainage to the mine itself.

Description of longwall subsidence and its hydrological effects

A primary engineering problem of any underground extraction is roof control, i.e. preventing damaging collapse of the roof strata that the extracted material formerly supported. In room-and-pillar mines, this is achieved by limiting the span of the openings such that the roof is supported between pillars of coal or re-stacked rock. In contrast, longwall mining completely extracts large areas of coal by using a deliberate, controlled collapse of the roof over the mined-out area, while keeping only a limited working section of the mine supported. Modern longwall mining, which is well described in the mining engineering literature (e.g. Bieniawski 1987), is more efficient, safe and economical than room-and-pillar mining. It became increasingly used in the USA through the 1970s and 1980s, first in the Appalachian coalfield, later in Illinois and more recently in Utah (e.g. Kadnuck 1994). Longwall mining has been widely used in the UK and adopted in various forms in many other areas, for example South Africa (Hodgson 1985), New South Wales (Holla 1991) and Nova Scotia (Reddish et al. 1994), and as short-wall mining under weak sandstone in Western Australia (Nikraz et al. 1994).

Longwall mining completely removes large rectangular areas (panels) of coal, typically several kilometres long by 150–300 m wide. The roof over the narrow strip of working area across the width of the panel is temporarily supported using moveable hydraulic jacks, which bear the load of the immediate roof strata. The coal is cut from the working face by a shearer and removed to pillar-supported side tunnels on a conveyor belt. The supports are advanced as the coal face is removed, and the unsupported roof behind them collapses into the mined-out void.

The strata above the immediate roof settle onto the collapsed material, and so on up through the overburden to the ground surface, where a subsidence trough develops that approximately outlines the mined-out panel. The ground at the advancing front of the trough undergoes a sequence of horizontal tension and compression described as a 'subsidence travelling wave' (Schmechel et al. 1979). Within the overburden strata, this sequence occurs as volumetric dilation followed by compression.

The strata movement is manifested as fracturing, bedding separation and changes in existing joint apertures, causing substantial changes in fracture porosity and permeability. These in turn lead to changes in potentiometric heads, groundwater flow patterns and well yields in shallow bedrock aquifers. Previous studies (e.g. Hill & Price 1983; Walker 1988; Matetic & Trevits 1991, 1992; Booth & Spande 1992) have demonstrated a typical response of rapid (but often temporary) decline in potentiometric heads, caused by the sudden increase in fracture porosity, and alteration of hydraulic gradients, caused by increased permeabilities. These effects are due primarily to *in situ* hydraulic property changes within the shallow bedrock aquifers, not to drainage into the mine. Despite subsidence, a zone of low-permeability strata generally maintains overall confining characteristics and prevents hydraulic connection between the shallow aquifers and the mine (Singh & Kendorski 1981; Coe & Stowe 1984; Booth 1986; Rauch 1989). Thus, there will be a hydrological response in a shallow aquifer even if the mine itself is dry.

Deformation zones above the mine. The dominant form of deformation over the longwall panel is vertical settlement. However, there is considerable variation in the nature of the stress and deformation at different levels in the overburden due to the characteristics of the mine geometry, the equipment and the lithology. For example, the hydraulic supports will bear the weight of roof strata only up to a certain height. Above that level, the overburden load is distributed in a pressure arch onto the side, front and rear abutments, including the caved area (US: 'gob'; UK: 'goaf'). Higher strata sag but are essentially self-supporting. Thus, several deformational zones have been identified over longwall panels (e.g. Peng 1986; Rauch 1987):

I – a caved zone, typically 2–8 times the height of the workings, in which roof material collapses directly into the mined-out longwall area;

II – a heavily fractured zone, typically 30–40 times the height of the workings, in which the strata break by vertical fractures and horizontal bedding-plane separations;

III – above that, a continuous deformation zone which subsides coherently with little extensive fracturing;

IV – a zone of well-defined and open fracturing at the ground surface and in the shallow strata, which can move more freely than the deeper strata.

The thicknesses of these zones vary with the lithology and from area to area, but provide a convenient framework for understanding the hydrological response. The most important hydrological division of the overburden is into three zones:

1. – a deep, intensely fractured, highly permeable zone, which corresponds to zones I and II above and dewaters into the mine;

2. – an intermediate zone, typically developed in a shale-dominated interval, which subsides coherently with little fracturing, maintains overall confining (low permeability) characteristics and corresponds to zone III above;

3. – a near-surface fractured zone, corresponding to zone IV above, in which aquifers are affected by *in situ* fracturing but not generally by drainage into the mine, from which they are separated by the intermediate confining layer.

The presence of the intermediate confining zone is a critical aspect of longwall hydrology. It was first noted (Fig. 1) by Singh & Kendorski (1981) and has subsequently been affirmed by many researchers (e.g. Coe & Stowe 1984; Booth 1986; Tieman & Rauch 1987; Rauch 1989; Hutcheson *et al.* 2000). The successful operation of longwall mines under lakes and the sea convincingly demonstrates that a confining zone normally exists and that highly permeable fractures directly connecting the mine to the surface generally do not occur.

Subsidence engineering guidelines dictate minimum separations between longwall mining and overlying productive aquifers or surface water bodies (Holla 1991; Singh & Kendorski 1981). Similarly, a minimum thickness of the relatively unfractured, confining layer is necessary to maintain hydraulic separation between the mine and shallow aquifers, although the critical value depends on lithology, structure and topography. For example, in their integrated model, which simulated elastic strains and resulting permeability changes above a longwall mine in a theoretical setting of the Appalachian Plateau type, Elsworth & Liu (1995) found that the critical separation between the mine and well bottoms, separating the wells with permanent water-level losses from those that were unaffected, recharged or only temporarily depressed, was of the order of 90 m under valleys and 150 m under hilltops and plateaux.

However, in the appropriate geological setting, the effective confining separation can be much less. For example, Van Roosendaal *et al.* (1995) observed that an overburden dominated by shales and glacial clay prevented the loss of water from shallow sand aquifers to a longwall mine only 60 m deep in Illinois, USA.

Time-related deformation at the edges of the panel. Deformational styles are also distinctly

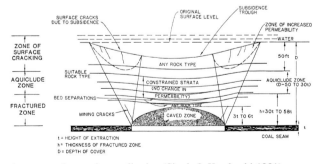

Fig 1. Strata deformation zones above a longwall mine (Singh & Kendorski 1981).

zoned laterally around the subsidence trough. For example, geotechnical studies by the Illinois State Geological Survey (ISGS) at the Jefferson County site in Illinois (see below) indicated maximum surface tensile and compressive strains at 22 and 60 m, respectively, inside the edge of a 183-m wide panel (Mehnert et al. 1994). Time-domain reflectometry (TDR) cable studies distinguished shear and extensional movements. The strata in the interior were dominated by vertical extensional separations, while the edges of the panel were largely characterized by subsurface shear fracturing and open extensional fractures at the ground surface. Elsworth & Liu's (1995) model simulated a zone of vertical separation in the lower strata directly over the panel, and zones of shear failure and increased vertical permeability (as a function of strain) in the abutment region and the near-surface zone at the sides of the panel.

Any point in the shallow strata above the panel will first experience tensional stresses and exhibit shear deformation as the face approaches. At the ground surface, tension cracks open up. Then, as the face undermines and passes the location, the ground subsides rapidly, and bedding separations and increased permeability occur within the strata. Subsequently, the location undergoes compressional stresses, and fractures in the interior of the subsidence trough partially close up, with a decrease of permeability down from the tensional peak.

Within the subsidence trough, stress differences should cause systematic spatial zonation of the hydraulic effects. Whereas the inner trough undergoes both tension (dilation) and then compression, the outer margins (rear, sides and final front) undergo only the tensional phase of the subsidence wave. They should thus have a greater residual increase in permeability than the inner zone, exhibit anomalous potentiometric levels and increased vertical leakage, and may act as a corridor of preferential groundwater flow. A post-mining groundwater flow model would, therefore, include an approximately rectangular inner area of slightly increased permeability surrounded by a border of more increased permeability (Booth et al. 1998).

Previous field studies of permeability changes

The permeability of a single fracture is approximately proportional to the aperture (width of the opening) squared, and of a set of fractures to the aperture cubed (as described in Domenico & Schwartz 1998). Thus, increases in fracture aperture or bedding-plane separation cause significant increases in fracture permeability. These in turn can create changes in the local hydraulic gradients, inter-aquifer leakage and the groundwater flow field, some of which may be permanent.

Changes in hydraulic properties vary substantially according to the strata position relative to the mine both vertically and laterally. Complex changes occur in the caved or heavily fractured zone immediately above the extraction and in the pillars and ribs surrounding the mine, significantly affecting water inflow to the mine. Aston & Singh (1983) and Aston et al. (1983) reviewed several studies conducted at English mines, which showed permeabilities increasing to transient peaks as the mine face approached, then, after the panel had passed, declining to a slightly increased residual level. Neate (1980), whose studies were included in the review, measured permeability changes at two sites at intervals above the seam as the mine face approached. At the Lynemouth site (mine depth 207 m, test intervals 15–55 m above the mine), permeability fluctuations began when the face reached within 70 m and continued mainly as discrete, temporary peaks. The hydraulic conductivities were initially about $1 \times 10^{-7}\,\mathrm{m\,s^{-1}}$, increased by about an order of magnitude, then declined to about $2 \times 10^{-7}\,\mathrm{m\,s^{-1}}$ after mining stopped 7 m short of the borehole. At Wentworth (mine depth 54 m), the hydraulic conductivities at various intervals between 20 and 47 m above the seam increased in steps from about $2 \times 10^{-9}\,\mathrm{m\,s^{-1}}$ to peak at about $9 \times 10^{-8}\,\mathrm{m\,s^{-1}}$, then declined to between 2.5×10^{-8} and $5.2 \times 10^{-8}\,\mathrm{m\,s^{-1}}$.

Holla & Buizen (1991) conducted borehole packer permeability tests above a 424-m deep mine in New South Wales, Australia. They observed increases in the number of fractures and in the permeability values, but found no correlation between the two. Regions of greater fracture dilation were separated by regions of lesser dilation or even compression, and the authors concluded that continuous vertical hydraulic connections from the ground surface to the mine were unlikely to develop.

Typical increases in permeability in shallow bedrock strata are one to two orders of magnitude, for example as measured by Whittaker et al. (1979) in shallow (up to 25 m deep) poorly permeable shale above a 54-m deep longwall mine in England. However, the behaviour can be inconsistent, as shown by several studies in the Appalachian coalfield. Matetic & Trevits (1991) conducted tests at two adjacent longwall panels 60–90 m deep in southeastern Ohio. Changes in the specific capacities of wells varied from a decrease of 0.8 times to an increase of 8.0 times due to the first panel, and an additional increase of up to 3 times from the second panel. The greatest increase, from 1.8 to 845 m^2 day^{-1}, was in a valley bottom well 46 m above the mine. Matetic & Trevits (1992) also studied two adjacent panels, 225–258 m deep, in central Pennsylvania. Tests in several boreholes nested to depths of 46 and 91 m showed that the valley bottom wells with the highest pre-mining specific capacities had decreases (by 0.4–0.8 times) due to mining, whereas wells with lower initial values had increases of between 2 and 4 times. Johnson (1992) found quite different results before and after mining at four sites in Appalachia, varying from no change at one site to increases in transmissivity of 5 to 13 times at others, the increases being greater for valley wells than for hilltop wells. Other studies that have demonstrated increased permeability or transmissivity over longwall panels in Appalachia include Johnson & Owili-Eger (1987) and Schulz (1988).

Thus, although changes in permeability might be expected to relate systematically to the mining subsidence regime both spatially and temporally, factors such as topographic relief and lithological variation complicate the relationship. In the Illinois studies reported later in this paper, topography is not a control, as the relief at all sites is low, but geological differences are important.

Dynamic changes in hydraulic head

The typical early potentiometric response of bedrock aquifers to undermining and subsidence is a drop in water level. There are several different possible causes for this head drop:

- direct drainage to the mine;
- increased fracture porosity;
- transmitted drawdown around the potentiometric low;
- leakage from upper to lower aquifers through fractured aquitards;
- changes in hydraulic gradients due to increased permeabilities.

Head drops due to direct drainage to the mine. Although shallower aquifers tend to be hydraulically isolated from the mine drain by confining layers, there are certainly situations in which a thinner and/or more permeable overburden permits direct drainage and potential head loss.

For example, study of the Lancashire No. 20 mine in Cambria County, Pennsylvania (Booth 1984, 1986), in the northern Appalachian coalfield, showed that averaged mine drainage quantities, which were in the range 13–20

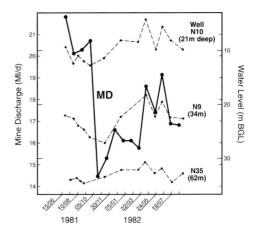

Fig. 2. Correlation of mine drainage (MD) with shallow groundwater levels, Lancashire No. 20 Mine, Pennsylvania (BGL = below ground level). From Booth (1986). Reproduced with the permission of the *Ground Water* journal.

million litres per day (Ml day^{-1}), corresponded both to greater rates of longwall mining and with seasonal water-level hydrographs of shallow wells (Fig. 2). The latter relationship clearly indicates that inflow was responsive to shallow groundwater recharge events. The connections were probably under the major valleys where the mine was only about 90 m deep. At least one localized section of the mine underlying a valley had persistent inflow problems (Wahler & Associates 1979), which were probably due to the combination of the thinner overburden and a permeable fracture zone. However, under the adjacent plateaux and ridges the mine was up to 240 m deep, and over most of the area the shallow sandstone aquifers were probably hydraulically separated from the mine workings. Their water levels responded to longwall undermining, with sudden drops in water level, and to adjacent longwall mining, with gradual, slight declines, but these can be attributed to causes other than direct drainage, as discussed below.

Head drops due to increases in fracture porosity. Various authors (e.g. Hill & Price 1983) have attributed the primary drop in water levels to the sudden creation of new fracture void space. However, the relationship of this head drop to the storage coefficient has not, to the author's knowledge, been explicitly explained.

The total increase in void space in a subsided area equals the volume of the extraction minus the volume through which the ground has subsided. Averaged over the whole overburden, this increase is unimpressive. For example, if we consider a hypothetical typical Illinois mine, an extraction thickness of about 3 m at a depth of about 200 m produces a ground subsidence of about 2 m. Thus, the total overburden volume increases only an additional 0.5%, and an initial overall porosity of (say) 10% (i.e. 20 m^3 initial void space in a 200 m high column of unit area) would increase only to 10.5%. Furthermore, much of the increased void space occurs in the lower caved and fractured zone, not in the shallow aquifers.

However, the potentiometric response to the remaining void-space increase is disproportionately large because of the low porosity of hydraulically effective fractures and the low storage coefficient of confined aquifers. The void-space increase is not diffuse, but is concentrated into a relatively small number of fracture dilations and bedding separations. Sudden drainage of water into this new void space causes the hydraulic head to drop. The new void space represents a large percentage of the fracture porosity. This is only a small proportion of total (fracture plus intergranular matrix) porosity, but the fractures respond quickly compared to the less permeable intergranular voids. Thus, the head drops occur quickly as void space increases. The magnitude of the head drop in each affected unit is expressed through the unit's storage coefficient (S) – the specific yield for an unconfined aquifer, the storativity for a confined aquifer – defined as the volume of water released (or taken up) (ΔV_w) per unit area (A) of aquifer per unit drop (or gain) in hydraulic head (Δh). Rearranging,

$$\Delta h = \Delta V_w / (A \times S)$$

where Δh is the change in head; and ΔV_w is the change in volume of water, which for a column of aquifer of unit area ($A = 1$) equals the change in volume of (fracture) void space.

Because of the enormous differences in magnitude between specific yields (which reflect actual drainage of pores) and storativities (which depend on the elastic compressibilities of water and the aquifer) the responses of unconfined and confined aquifers are very different. Neglecting the minor changes on the magnitudes of the overall storage coefficient caused by the increased fracture porosity, consider a unit area column of a sandstone aquifer in which the increase in bedding separation is a mere 0.01 m. For an unconfined aquifer with specific yield 0.1, the drop in head is only 0.1 m; but for a confined aquifer with storativity 10^{-5} the theoretical drop in head is 1000 m! In reality, the head simply drops very rapidly down to the top of the aquifer, which changes from a confined condition to a threshold unconfined condition. A truly unconfined state presumably depends on whether air can enter the pore spaces, which would be possible at least around a piezometer or well.

Thus, the expected potentiometric effect of the increase in fracture porosity in a confined aquifer is a rapid head drop to unconfined or threshold-unconfined conditions. Aquifers that are already unconfined are much less sensitive to this increase. Unconsolidated aquifers (e.g. sand and gravel) are also much less sensitive than those that are consolidated (e.g. sandstones) because they are characterized by intergranular porosity not by fractures. Our studies in Illinois showed that water levels in unconsolidated, unconfined aquifers (sand units within the glacial drift) did not respond significantly to longwall undermining, whereas bedrock units at the same site did. The only changes in water table in the drift aquifers were responses to subsidence-related changes in ground elevation (Booth et al. 1998).

Head drops due to the transmitted drawdown effect. The advancing potentiometric low in the subsiding area is surrounded by an elongate 'cone of depression' expanding as a 'drawdown' effect through the aquifer. This secondary effect spreads ahead of undermining and is therefore the first response seen in any individual well. The head drop is transmitted further and occurs gradually in more transmissive units, whereas in poorly transmissive units it occurs suddenly and closer to the site and time of undermining. The differences were clearly observed in our Illinois studies (see below), in which advance declines in head occurred early and gradually in the moderately transmissive aquifer at the Jefferson site, but sudden head drops just before mining occurred in the poorly transmissive aquifer at the Saline site.

Coalfield aquifers typically have relatively low transmissivities. Most studies in the Appalachian coalfield show that the head drops related to longwall mining are localized to within a few hundred metres of the mine panel. For example, Moebs & Barton (1985) reported that water levels in 46 m deep wells in a shallow aquifer overlying a 230 m deep mine declined >30 m over the panels, slightly on the edge, and not at all beyond a distance between 177 and 387 m. Tieman & Rauch (1987) plotted the extent of dewatering of wells adjacent to longwall panels at a site in southwestern Pennsylvania, where the overburden thickness was 200–335 m, the local relief 90–137 m, and the separations between the mine and well bottoms were at least 150 m.

The upper dewatering zone extended out at an 'angle of dewatering influence' of about 42°, to a distance of about 300 m. In another study in southwestern Pennsylvania, Walker (1988) found that the first water-level fluctuations occurred when the advancing face was 120–180 m distant (approximately equal to the thickness of the overburden), and concluded that wells are generally unaffected unless they are within the angle of draw.

In the author's study of the Lancashire No. 20 mine (Booth 1986), water levels in shallow wells were unaffected by supported undermining but declined in response to longwall mining. The most severe response (well dewatering) occurred directly over the panels, but the secondary (drawdown) effect was observed at a distance of 320 m, but not at 442 m.

Head drops due to draining of perched or upper-level aquifers. Local groundwater flow systems in coalfields with high topographic relief reflect a balance of conflicting controls (Booth 1988). The topography drives vertical flow, but the stratification favours lateral flow and hill-side discharges (Stoner 1983; Kipp & Dinger 1991). Subsidence-related increases in vertical permeability shift the balance toward the topographically driven system. Fracturing of aquitards can cause water to drain from perched or upper-level aquifers down to lower-level aquifers or to local discharge areas. Several studies in the Appalachian coalfield have attributed head drops to this mechanism (e.g. Leavitt & Gibbens 1992; Werner & Hempel 1992). A similar phenomenon was suggested by Hutcheson *et al.* (2000), based on their study of a longwall mine 91–244 m deep in an area of high relief (183 m) in Kentucky.

The loss of water from upper to lower aquifers may in fact offset the effect of mining on the latter. The study by Hutcheson *et al.* (2000) showed that, due to mining 400 m away, piezometric levels rose in some deeper piezometers below regional drainage level, which the authors attributed to increased recharge in the subsidence-fractured areas. Schmidt (1992) studied inflow and outflow zones in wells in a valley above a room-and-pillar mine in Pennsylvania, and found that the wells maintained their water levels because the water flowing into them from the shallow system exceeded the water draining from them down to the mine. Thus, paradoxically, water levels may remain steady in lower aquifers that are draining directly to the mine, but decline in upper aquifers that are not. It is this sort of complex interaction between position relative to the mine and position relative to topography, stratification and the local groundwater flow system that makes the generalized prediction of mining impacts so difficult to apply to an individual well or site.

Head drops due to decreased hydraulic gradients caused by increased permeabilities. Changes in permeability *per se* do not alter heads. However, the permeability (expressed as the hydraulic conductivity, K) is linked to hydraulic gradient (I) and specific discharge (q) through Darcy's Law:

$$q = -KI.$$

Thus, when the permeability is increased by subsidence-related fracturing, the hydraulic gradient in the affected area must decline and/or the specific discharge (throughput) must increase. Heads will, therefore, decline up-gradient, but down-gradient from the affected area, heads may rise (as observed in valley wells by Johnson 1992) and groundwater discharges may increase. Generally, the overall effect will be seen within a local watershed as an increase in stream and spring flow in the discharge areas (as reported by Tieman & Rauch 1987), and a lowering of heads and loss of stream flow in the upland recharge areas.

Recovery of water levels after mining

At least two principal, separate mechanisms produce water-level recovery after subsidence: compression and recharge.

Within the inner subsidence trough, an early partial recovery occurs because the extensional (dilational) phase of subsidence is followed by a compressional phase in which tension fractures close back up. Settlement of the beds may also cause some closure of the bedding separations. These closures partially reverse the head drops caused by the dilational increase in fracture porosity – in effect, squeezing the water back out of the fractures. They do not affect head drops produced by the other factors.

In addition, the potentiometric depression due to the changes in porosity is a transient (unsteady state) feature. The water levels in the affected area would be expected to recover as water flowed back to the potentiometric low along the temporary hydraulic gradients. Studies show that potentiometric levels and stream flows typically recover at least partially within a few months after mining (Tieman & Rauch 1987; Walker 1988; Matetic & Trevits 1990). However, this aspect of recovery is dependent on connection to sources of recharge and the ability of the aquifer to transmit water back into the affected area. At our study sites in Illinois (Booth 1999), full recovery after the cessation of mining occurred at a site (Jefferson County) with a moderately transmissive aquifer, but no recovery occurred at another site (Saline County) where the sandstone aquifer had very low transmissivity and restricted lateral pathways to sources of recharge.

In addition, delayed drainage of water from the intergranular pore spaces can contribute to long-term recovery, as noted by D. Banks (pers. comm./review 2001).

Unlike the porosity changes, changes in permeability permanently affect the groundwater flow system. Although the compressional closure of fractures may partially reverse the dilational increases in permeability, there is generally some residual net increase, especially along the tensional margins of the subsidence trough (Booth et al. 1998). As a result of the various factors of leakage through fractured aquitards, decreased gradients and altered groundwater flow paths, most sites will exhibit permanent head losses somewhere. This makes long-term recovery appear inconsistent and hard to predict. In a study by Cifelli & Rauch (1986), only one of 19 water supplies directly over total extraction mining had any significant recovery, whereas Johnson & Owili-Eger (1987) reported several case studies in which recovery typically occurred within a few months. Leavitt & Gibbens (1992) looked at 174 domestic water wells near longwall mines exploiting the Pittsburgh seam of the Appalachian Plateau coalfield, and found that 64% returned to service without the need for remedial action; valley wells were less affected by mining than upland wells. Similarly, Johnson (1992) determined that recovery was more likely in valleys than on hilltops. Werner & Hempel (1992) considered that recovery was likely in wells below the regional water table but not in perched aquifers, below which the aquitard had been fractured.

Sequence of Potentiometric Changes. Summarizing the above, the following composite model (expressed in Fig. 3) is adapted from Booth et al. (2000). The figure shows, from the bottom up, the longwall mine, the caved and heavily fractured zones, the major confining layer and a fractured shallow aquifer overlain by a shallow confining layer. It corresponds to typical sites in Illinois, but the potentiometric response sequence would apply to most areas, with appropriate allowance for local controls such as topography.

The mine is advancing towards the left in the figure. A point in the shallow subsurface experiences, more or less sequentially: extensional horizontal stresses (expressed as aligned tension cracks on the ground surface), shear deformation and opening of joints, rapid vertical subsidence and separation of bedding planes, horizontal compression which partially re-closes joints, and then continuing settlement of the strata which may partially re-close bedding-plane openings.

The changes in hydraulic heads reflect this sequence. The numbered sequence of observations can be considered either as separate features at different places at an instant of time, or as successive observations at the same point at successive times:

1. the pre-mining condition (which may already be affected by earlier mining);
2. first response to the transmitted secondary drawdown at the limit of influence;
3. a gradual decline in head in advance of mining due to the transmitted drawdown effect;
4. a rapid head drop during undermining, active subsidence, and the sudden increase in fracture porosity. The change from stage 3 to stage 4 approximately coincides with the onset of tensional fracturing, and is abrupt in space and time because it represents a sudden spatial and temporal discontinuity in the aquifer properties;
5. the potentiometric minimum, coincident with the maximum dilational (maximum fracture opening) phase;
6. a rapid partial rise in head due to fracture re-closure;
7. a gradual recovery of water levels as the site is recharged by water flowing back into the aquifer.

Hydrological studies of longwall mining in Illinois, USA

Northern Illinois University (NIU) and the Illinois State Geological Survey (ISGS) have conducted joint studies in hydrology and engineering geology over two active longwall mines in Illinois, in the Eastern Interior Coalfield, USA. These mines are referred to as the Jefferson County mine, located in south-central Illinois about 23 km southwest of the town of Mount Vernon, and the Saline County mine, located in southeastern Illinois about 8 km north of the town of Harrisburg. The field studies lasted 7 years.

The only previous hydrological study of longwall mining in Illinois was by Pauvlik & Esling (1987), who observed slight temporary variations in the permeability of an unconfined

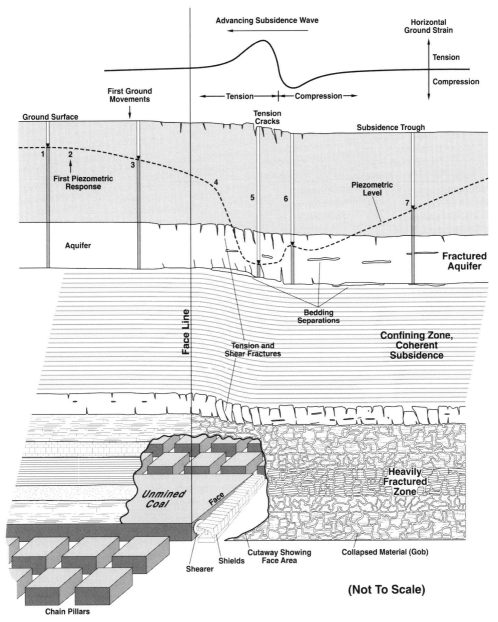

Fig. 3. Conceptual model of potentiometric response to longwall mining. From Booth *et al.* (2000). Reproduced with permission of the Association of Engineering Geologists (*Environmental and Engineering Geoscience*).

glacial–alluvial silt aquitard during subsidence. No permanent change was observed, which is reasonable considering that the aquitard was unconsolidated and, therefore, relatively insensitive to changes in fracture properties.

The Illinois Basin coalfield contains several thick coal seams of Pennsylvanian (Upper Carboniferous) age, within a typical coal-measure sequence of numerous alternating beds of shales, siltstones, sandstones, clays and limestones. Compared to the Appalachian coalfield, the stratigraphic sequence in Illinois is dominated more by low-permeability shales and less by permeable sandstones. In addition, the Illinois coalfield is largely overlain by Pleistocene glacial deposits, mostly tills, and the topographic relief is low. Groundwater flow systems are therefore sluggish, and brackish–saline water is encountered at depths as shallow as 50–200 m (Davis 1973; Cartwright & Hunt 1983). The permeable sandstones and limestones of the Lower Pennsylvanian and the Mississippian, which underlie the coal measures, form aquifers along the margins of the basin (Poole *et al.* 1989), but they are too deep and mineralized to be usable over most of the coalfield. The most productive aquifers are thick Pleistocene outwash sand-and-gravel units in the major valleys and buried valleys (Zuehls *et al.* 1981*a*, *b*). However, these are uncommon, and fresh groundwater resources are generally limited to minor sand-and-gravel units within the drift, till tapped by large-diameter wells, and shallow, poorly transmissive Pennsylvanian sandstones.

Because the aquifers are generally poor in yield and/or quality, public water supplies have largely shifted to surface-water reservoirs. However, shallow groundwater is still an important resource for rural residential and farm livestock supplies, and would surely become more so if global climate changes were to create frequent or extended drought conditions. The area has a humid continental climate with average annual rainfall around 100 cm, but serious droughts do occur (for example in 1988). Potential impacts of coal mining on groundwater supplies are therefore significant in a regional sense as well as for individual users.

The two sites studied have broadly similar characteristics. Both mines worked the Herrin (No. 6) Coal of the Pennsylvanian Carbondale Formation. At both sites, the overlying strata consist mainly of shales and siltstones, with minor thin limestones, sandstones, coals and clays. The bedrock is overlain by Pleistocene glacial deposits, mostly Illinoian till with discontinuous minor sand and gravel units, Wisconsinan glacilacustrine deposits at the Saline site and a cover of loess. The landscape is flat to gently rolling, with maximum local relief about 15 m, and the land use is mainly crop farming, with some pasture at the Jefferson County site. Most of the homes and farms located around the sites have private wells into the drift or bedrock, either still in use or abandoned, but still available for monitoring during the study.

Studies at each site were started before the mining of particular panels and continued for several years afterwards. The ISGS monitored vertical subsidence (using surveyed monuments), horizontal ground strains and subsurface strains (using extensometers and TDR cables in cored boreholes). The hydrological studies included potentiometric monitoring of piezometers and wells drilled on site and of existing residential and farm wells, and sampling for geochemical analysis. In addition, hydraulic properties were determined using packer, slug and pumping tests:

- Pre- and post-subsidence straddle-packer tests at all sites were conducted on the panel centre lines in NX (7.6 cm) diameter boreholes, core-drilled through the bedrock. US Department of the Interior (1981) procedures were followed. Boreholes were surged, flushed clean and geophysically logged, then water was injected through perforated pipe into vertical intervals isolated by inflatable packers. A sequence of step-up and step-down pressures was followed for each interval, to identify hysteretic or non-linear behaviour that could indicate permeability changes due to overpressure or clogging. The test data were analysed by the Hvorslev (1951) method.
- In the slug tests, a small volume of water was near-instantaneously added to each piezometer and the head response recorded using automatic pressure transducers. Data were analysed by the Bouwer & Rice (1976, revised Bouwer 1989) and Cooper *et al.* (1967) methods.
- Pumping tests were conducted in wells constructed over the centre lines of all three panels studied. Tests were conducted before mining at all the sites, but at the Saline County site the water levels did not recover sufficiently to permit pumping tests after mining. At the Jefferson County site, 16 tests of various durations and rates were conducted between 1988 and 1994 before and after mining. Analyses are discussed below.
- Approximately 500 samples of groundwater were taken from the two sites at various stages during the study. The geochemistry was discussed by Booth & Bertsch (1999). Sampling was by either pumping or bailer, but methods varied according to the nature of the well or piezometer sampled and the changing availability of equipment. Samples were tested in the field for pH, alkalinity, temperature and specific conductivity, and analysed in the laboratory for all major and several minor cations and anions.

Jefferson County site

At the Jefferson County site, four longwall panels, each about 183 m wide by over 1530 m long and separated by 61 m wide double-pillar

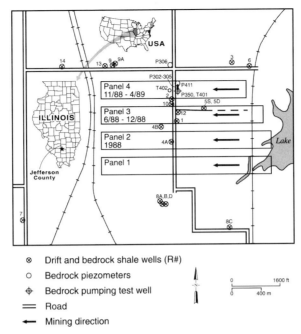

Fig. 4. Map of the Jefferson County site. From Booth & Bertsch (1999). Reproduced with the permission of Springer-Verlag (*Hydrogeological Journal* **7**, 561–575) (expect for the US public-domain components).

barriers, were mined between 1987 and 1989 (Fig. 4). Our study began during the mining of Panel 3 and focused on the final panel, No. 4. Previous mining by Panels 1 and 2 had already affected the potentiometric levels.

The mined coal seam at the study site was around 3 m thick at a depth of about 222 m. The overburden strata are relatively undeformed and flat lying, and are dominated by poorly permeable shales. However, about 174 m above the mine is the Mt Carmel Sandstone aquifer (Fig. 5), which is 23–25 m thick and is divided into a thin (3–5 m) upper bench and a thick (12–16 m) lower bench by a shale–siltstone unit. The top of the Mt Carmel is at a depth of about 24 m below ground, and is overlain by a shale confining unit 15–18 m thick, which is covered by 3–10 m of glacial till, sand and gravel, and loess.

Panel 4 began mining about 850 m from the study instrumentation, which it undermined in February 1989. Subsidence and strain behaviour were studied by ISGS (Mehnert *et al.* 1994; Trent *et al.* 1996). Ground subsidence at the panel centre line ultimately reached 2.1 m, of which about 2 m occurred within 2 months of mining. Subsidence was accompanied by considerable fracturing of the strata, particularly shear fractures in the marginal tension zones and

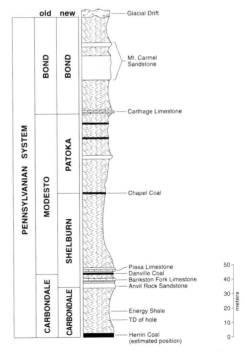

Fig. 5. Stratigraphic column at Panel 4, Jefferson County site. From Booth *et al.* (1997).

vertical bedding separation in the central trough area. The hydrological effects of the subsidence have been reported by Booth & Spande (1992) and Booth et al. (1997, 1998).

Hydraulic tests. Packer tests were conducted in centre-line boreholes T401 (pre-subsidence, 213 m deep) and T402 (drilled 6 months after subsidence 30 m west of T401, to a depth of 158 m, below which the strata were extremely fractured). The holes were tested in 6 m intervals in T401 and 3 m intervals in T402, completely through the Mt Carmel Sandstone and at selected intervals below that. Owing to subsidence, the geometric means of the sandstone permeabilities in the boreholes increased approximately one order of magnitude, from 4.7×10^{-7} to 9.1×10^{-6} m s^{-1}, (Booth & Spande 1992). The permeabilities of some individual horizons in the underlying shale and limestone increased by several orders of magnitude due to bedding-plane separations that were observed in the drilling behaviour and by geophysical logging.

Pre-mining boreholes drilled in 1988 included a 15 cm diameter test well (P350, open through the entire sandstone on the Panel 4 centre line), several piezometers (P302, P303, P304 screened into the lower sandstone, and P305 into a deep shale) in a transverse line across the panel and a piezometer (P306) off panel. The piezometers were used for slug tests and for observation during pump tests of P350. The piezometers over the panel became unusable because of subsidence damage, but well P350 survived, and two post-subsidence piezometers (P410 in the inner region, P411 in the tension zone) were drilled into the lower sandstone in 1992, and used in subsequent tests.

Sixteen pumping tests were conducted on well P350, at discharge rates varying from 0.30 to 0.94 l s^{-1} and durations ranging from less than 1 up to 24 h (Booth et al. 1998). In the 1988–1990 phase, the aquifer was unconfined, but from 1990 to 1995 the potentiometric level was above the top of the sandstone. The post-mining test results were analysed by several methods to examine possible behaviour as an unconfined, leaky confined, layered, bounded or double porosity aquifer. It was concluded that the aquifer as a whole behaved as a confined single-porosity system. However, pumping tests using a single packer demonstrated that the two sandstone benches were hydraulically separated and the lower bench behaved as a separate confined aquifer, and the upper bench was influenced by leakage from the overlying shale. The post-subsidence test responses were influenced by permeability discontinuity boundaries related to the subsidence trough.

Hydraulic conductivities determined from the slug and pumping tests are shown in Table 1 (Booth et al. 1998) The slug-test values for individual piezometers are the averages of several tests, and the model calibration values were obtained by calibration of pump-test simulations conducted using a MODFLOW (McDonald & Harbaugh 1988) numerical model for the bounded aquifer.

The natural lateral heterogeneity of the aquifer is indicated by a spatial variation of approximately two orders of magnitude for the pre-subsidence conductivity values determined using the same methods. For example, in slug tests ST1 (October 1988) the pre-subsidence conductivity varied from 1.4×10^{-8} m^{-1} s^{-1} in P306 (off-panel) to 3.0×10^{-6} m s^{-1} in P304 (inner panel region).

Values at the same piezometers also differed between the slug tests and pumping tests. Although different results from different methods can be caused by factors such as low-permeability skins in piezometers (Butler & Healey 1998), scale sampling effects within the heterogeneous aquifer were a more likely explanation in this case (Booth et al. 1998). Slug tests sample only the immediate vicinity of the borehole, whereas pumping tests sample much larger volumes of aquifer, and furthermore the differences in pumping rates and durations meant that different volumes of the aquifer were sampled by different pumping tests. Also, different vertical intervals were sampled; well P350 was open through the entire aquifer including the small confining unit, whereas the piezometers were screened in limited intervals in the lower bench.

Despite the heterogeneous and sampling variations, the effects of subsidence were clearly defined. Post-subsidence hydraulic conductivity values are greater than pre-subsidence values for almost all comparable tests. Conductivity values obtained from recovery analysis of the drawdowns in test well P350 varied between 2.0×10^{-8} and 2.0×10^{-7} m s^{-1} before subsidence to 3.5×10^{-7}–9×10^{-7} m s^{-1} from the earlier to the later post-subsidence tests. Considering tests of similar duration, the specific capacity of the well increased from 0.032 l s^{-1} m^{-1} in pre-subsidence test PT5 (239 min at 0.300 l s^{-1}) to 0.073 l s^{-1} m^{-1} in post-subsidence test PT8 (258 min at 0.328 l s^{-1}). Similar increases are apparent in the slug-test results for comparable areas, as well as in the packer tests noted earlier. In summary, the permeability of the aquifer increased by approximately one order of magnitude in the inner area of the subsidence

Table 1. *Hydraulic conductivity values (m s^{-1}) from slug (ST) and pumping (PT) tests, Panel 4, Jefferson County site*

Test	Date (month/year)	Centre line	Interior area		Tension zone	Outside
		P350	P304	P303	P302	P306
Pre-subsidence tests						
Slug	10/88	—	3.0×10^{-6}	1.8×10^{-7}	2.1×10^{-6}	1.4E-8
Slug	2/89	—	1.7×10^{-6}	1.8×10^{-7}	2.0×10^{-6}	—
PT4	10/88	1.7×10^{-7}	3.9×10^{-8}	1.3×10^{-6}	3.0×10^{-6}	—
PT5	1/89	3.3×10^{-8}	5.1×10^{-7}	2.9×10^{-7}	5.6×10^{-7}	—
Post-subsidence tests						
PT7	3/90	3.4×10^{-7}	—	—	—	—
PT8	3/90	3.3×10^{-7}	—	—	—	P306
			P410		P411	
PT11	7/92	5.4×10^{-7}	4.4×10^{-6}		1.0×10^{-4}	
Slug	1992	—	2.2×10^{-6}		1.4×10^{-5}	5.1×10^{-9}
Post-subsidenc tests		P350	P410	Transboundary from P410 and P350	P411	
PT13	8/93	9.6×10^{-7}	3.0×10^{-5}	1.7×10^{-4} and 3.8×10^{-5}	2.0×10^{-5}	
PT14	10/93	9.4×10^{-7}	3.0×10^{-5}	1.7×10^{-4} and 2.8×10^{-5}	2.1×10^{-5}	
PT15	3/94	Lower Sst only	2.5×10^{-5}	2.7×10^{-4}	3.0×10^{-5}	
PT16	3/94		3.0×10^{-5}	2.2×10^{-4}	2.0×10^{-5}	
Numerical model calibration		Inner area		Tension zone		Outside
East–west		8.6×10^{-7}		9.7×10^{-5}		7.6×10^{-8}
North–south		6.5×10^{-6}		72×10^{-4}		5.7×10^{-7}

trough and two orders in the tension zone at the margin. Storativity values similarly increased about one order of magnitude to around 10^{-3}.

Potentiometric changes. Water levels in the sandstone over Panel 3 had declined as the mine face approached, reached a minimum during the early tensional phase of subsidence and partially recovered after the face had passed (Mehnert *et al.* 1994). A similar response was observed in the well and piezometers over Panel 4 (Booth *et al.* 1997, 1998), as shown in Fig. 6. The water level was initially about 19 m below ground, but had probably already been affected by Panel 2 (400 m away). It declined to about 34 m with the nearest approach (152 m) of Panel 3, then slightly recovered and then declined gradually as Panel 4 face approached. When the site was undermined the sandstone water level fell rapidly to about 43 m below ground, approximately the top of the lower bench aquifer. Potentiometric levels recovered very quickly to about 35 m during the compressive phase, then gradually to about 12 m below ground by 4 years after the end of mining (Fig. 7).

The sandstone water level had thus demonstrated all the various responses of the conceptual model discussed earlier: transmitted drawdown due to adjacent and approaching mining; rapid decline in head due to increased fracture porosity during subsidence, with the aquifer becoming unconfined as a whole and threshold-unconfined for the lower bench; compressive partial recovery; and full long-term recharge recovery.

In contrast, there was little response in the drift water-table wells except for brief fluctuations during active ground movements and a slight potentiometric adjustment to the new topography created by the subsidence troughs. The differences between drift and bedrock responses are clearly shown by adjacent wells W5A (drift, 4 m deep) and W5B (shale, 19 m) located on the edge of Panel 3 (Fig. 8). The water level in W5B fell rapidly with Panel 3 subsidence in July 1988, whereas the water level in W5A was unaffected. The potentiometric response to fracturing is much greater in the confined bedrock than in the unconfined, unconsolidated drift.

Geochemical changes at the Jefferson site. Geochemical changes were described by Booth *et al.* (1998) and Booth & Bertsch (1999). The water from the drift wells was fresh (total dissolved solids (TDS) less than $600\,\mathrm{mg\,l^{-1}}$) and of mixed cation, sulphate-dominant type. Nitrate levels were generally high, most probably because of contamination from agricultural

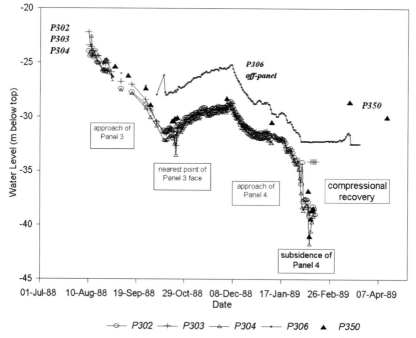

Fig. 6. Potentiometric levels in the Mt Carmel Sandstone during mining, Panel 4, Jefferson County site. From Booth *et al.* (1997).

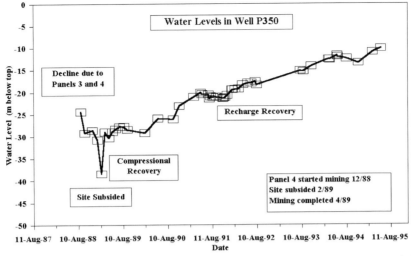

Fig. 7. Long-term water-level response in well P350, Panel 4, Jefferson County site. From Booth (1999).

fertilizers and cattle feedlots. The water from the upper shale was a brackish (TDS less than 2000 to more than 4000 mg l^{-1}), mixed cation, sulphate-dominant type. There was no apparent change in the drift water due to mining. The shale water in some wells exhibited a slight reduction in salinity, probably due to increased leakage from the drift.

Significant changes occurred in the groundwater chemistry of the Mt Carmel Sandstone aquifer. Major ion proportions are shown in a Piper trilinear diagram in Fig. 9. The native (pre-mining) water is represented by samples from R18, a 60 m deep sandstone well in an unmined area approximately 5 km east of the site, and pre-mining samples from P350. These were slightly brackish–fresh (900–1200 mg l^{-1}) and sodium bicarbonate dominant; the Na-HCO$_3$ facies is typical of shallow sandstones in much of the Illinois Basin (Graf et al. 1966; Booth & Saric 1987; Poole et al. 1989). Sulphate was almost zero in R18, and about 200 mg l^{-1} in P350.

During the post-mining recovery, the water in P350 became more brackish (TDS in the range 1990–2620 mg l^{-1}) with an increase in sulphate to 800–1272 mg l^{-1} (Fig. 10). Sodium increased from about 400 to about 600 mg l^{-1}, but bicarbonate levels remained approximately the same. The changes were greater in P350, open through the whole aquifer, than in piezometer P411, which was screened only in the lower bench of the sandstone. Two possible sources for the increase in sulphate were identified: leakage

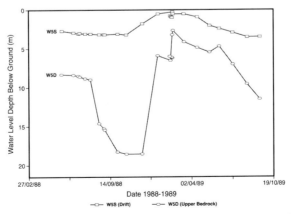

Fig. 8. Water levels in drift well W5S and bedrock well W5D, edge of Panel 3, Jefferson County site. From Booth & Spande (1992). Reproduced with the permission of the *Ground Water* journal.

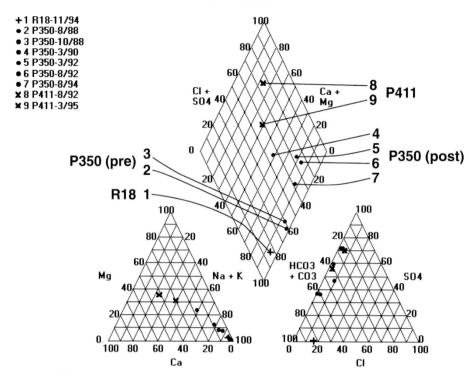

Fig. 9. Piper trilinear diagram of major ion proportions in the Mt Carmel Sandstone, Jefferson County site. From Booth & Bertsch (1999). Reproduced with the permission of Springer-Verlag (*Hydrogeological Journal* **7**, 561–575) (expect for the US public-domain components).

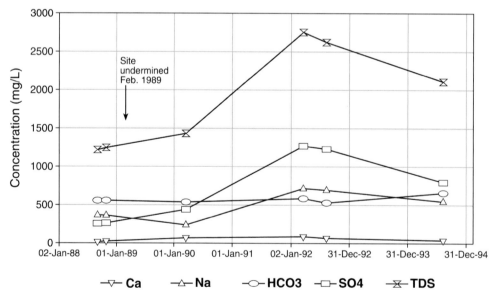

Fig. 10. Change in major ions in Mt. Carmel Sandstone well P350, Jefferson County site. From Booth & Bertsch (1999). Reproduced with the permission of Springer-Verlag (*Hydrogeological Journal* **7**, 561–575) (expect for the US public-domain components).

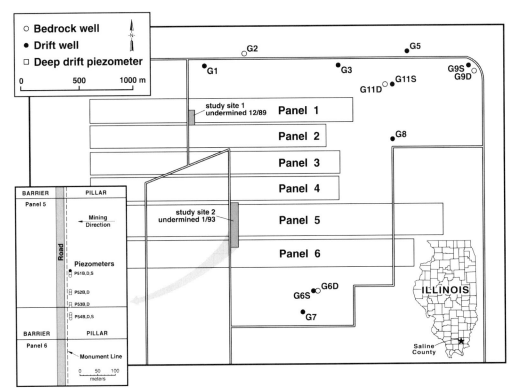

Fig. 11. Map of the Saline County site, with inset showing instrumentation over Panel 5. From Booth & Bertsch (1999). Reproduced with the permission of Springer-Verlag (*Hydrogeological Journal* **7**, 561–575) (expect for the US public-domain components)

from the overlying shale into the upper bench of the aquifer; and the mobilization, by water flowing back through the sandstone during recovery, of sulphate originating from sulphides oxidized during the period in which the sandstone was unconfined.

Saline County site

At the Saline County site, six adjacent westward-driven longwall panels of widths varying from 188 to 287 m, and lengths from 2286 to 3130 m, were mined successively north to south between 1989 and 1994 (Fig. 11). Our studies concentrated on Panels 1 (1989–1990) and 5 (1992–1993). The mined coal seam was about 2 m thick and overlain by Pennsylvanian bedrock consisting of shales and siltstones with thinner coals, clays, limestones and several thin to discontinuous sandstones. The principal sandstones are the Gimlet, typically about 3 m thick at a height of about 35 m above the mined seam, and the Trivoli, approximately 6 m thick and 60–90 m above the seam. The bedrock strata dip gently northwards at about 23 m km^{-1}, so that the mine is about 122 m deep at Panel 1 and 97 m deep at Panel 5 (Fig. 12). The bedrock is covered by unconsolidated deposits consisting mainly of Illinoian till containing minor sand and gravel units, overlain by Wisconsinan glacial lake deposits, capped by loess. The drift cover is 18–27 m thick at panel 1 and 12–18 m at Panel 5.

Farms and homes around the site generally have either shallow, large-diameter (1–2 m), water-table wells into the upper 6–8 m of drift, or drilled wells into sandstone. Yields of all aquifers are poor, and most homes are now connected to city water supplies.

Results of the hydrological studies at the Saline County site have been reported by Booth *et al.* (1994, 1997, 2000) and of the geochemical studies by Booth & Bertsch (1999).

Subsidence and strata deformation at Panel 1 (ISGS studies). Panel 1 was 204 m wide with an average mined-out height of 2.0 m and a roof depth of 122.5 m at the centre of the study site. The transverse line of monuments was undermined in December 1989. The eventual centre-line subsidence was 1.44 m, and the maximum

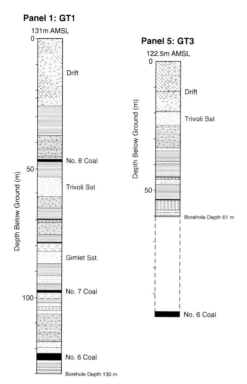

Fig. 12. Stratigraphic column above Panels 1 and 5, Saline County site. From Booth & Bertsch (1999). Reproduced with the permission of Springer-Verlag (*Hydrogeological Journal* **7**, 561–575) (expect for the US public-domain components).

extensional and compressional horizontal strains were located, respectively, about 16 and 41 m inside the panel edge. The ISGS geotechnical and hydrological investigations at Panel 1 were reported by Kelleher *et al.* (1991), Van Roosendaal *et al.* (1990, 1994) and Trent *et al.* (1996).

Two boreholes were drilled on the centreline before and after mining to depths of 130 and 84 m, respectively at angles of 10° to the vertical to maximize fracture interception. Bedrock cores showed an increase from 1.05 to 1.64 fractures per m from before to after subsidence, with a change from almost no high-angle fractures to many, especially in the Trivoli Sandstone. TDR monitoring in the pre-mining borehole showed that subsidence-induced differential shear and extensional displacement worked up through the strata from a first-break depth of about 53 m. Borehole geophysical logging indicated fracturing and bedding separation. Differential displacement, i.e. the difference between the amounts of vertical subsidence between the top and bottom of vertical intervals, was shown by multiple borehole extensometers to have occurred between different levels of the overburden strata, mainly during the 5 days around undermining. However, out of a total subsidence of 1.44 m, only about 0.15 m was differential displacement, of which 0.10 m was in the drift. Thus, the overburden subsided largely as a single coherent mass.

Hydraulic tests at Panel 1. Piezometers were installed into the lower drift sand aquifer (screened at depths between 19 and 22 m), the Trivoli Sandstone (various intervals between 42 and 60 m) and one piezometer into the Gimlet Sandstone (81–84 m). Slug tests and pumping-test observations in the sandstone piezometers before mining indicated hydraulic conductivities in the 10^7 m s^{-1} range. Subsidence damage restricted post-mining testing, but the limited results showed only minor increases within the same range of magnitude.

Permeability values were also obtained from pre- and post-subsidence packer injection tests conducted in the cored boreholes on the panel centre line. Generally, the strata were tight and would not accept water at valid injection pressures. Before mining, limited intakes were achieved in only a few intervals, including a 6 m interval in the Trivoli Sandstone where a hydraulic conductivity of 6×10^{-8} m s^{-1} was determined. Post-subsidence intakes were more sustained over more intervals, but the Trivoli conductivities remained in the 10^8 m s^{-1} range.

Potentiometric responses at Panel 1. Water levels in the deep (confined) drift piezometers over Panel 1 were about 7 m below ground before mining, fluctuated 2–4 m during subsidence, then stabilized at between 8 and 11 m below ground. Water levels in the sandstone piezometers were initially about 12 m below ground, had very little advance response to the approach of the mine and declined to depths of 49–55 m in the period from just before undermining to the time of maximum tension. In some piezometers, water levels then rose between 6 and 9 m, but no further recovery was observed in either the surviving bedrock piezometer on the barrier pillar or the replacement centre-line piezometer during 2 years of monitoring following mining. Figure 13 shows the response of one centre-line piezometer (BP3) and its post-subsidence replacement BPPS.

The water level in a 42 m deep bedrock residential well, located about 300 m north of the panel, declined rapidly from a pre-mining depth of about 11 m in 1989 to a stable post-mining depth of about 33.5 m, recovering only about 1.5 m in the winter of 1994–1995. However, shallow drift wells at the same distance had no discernible response to mining.

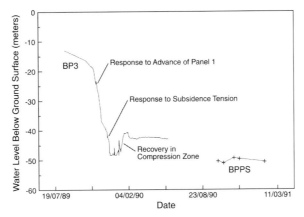

Fig. 13. Potentiometric response of Trivoli Sandstone to mining of Panel 1, Saline County site. From Booth et al. (1997).

The response to longwall mining at Saline Panel 1 clearly differed from that at the Jefferson site. The Trivoli Sandstone at Saline was only a quarter as thick and at least an order of magnitude less permeable than the Mt. Carmel Sandstone, and therefore the transmitted advance head drop was much less but the head drop during undermining was more abrupt. The sandstone was also isolated from potential recharge sources by thick confining units, continued mining up-dip, and its own low transmissivity. Long-term recovery was, therefore, negligible.

A different response was expected at the Panel 5 site, located further south (up-dip), where the Trivoli Sandstone is only about 20 m deep and occurs at the bedrock surface in contact with the glacial drift. In this case, the sandstone was expected to be more permeable initially, to exhibit greater increases in fracture permeability with subsidence, to be recharged more readily, and to have a more gradual head drop and more rapid recovery. The actual responses did not occur as predicted. They also varied substantially across the site at a local scale in both the sandstone and the drift units.

Subsidence and strata deformation at Panel 5 (ISGS studies). Panel 5 was mined from April 1992 to April 1993 a distance of 3131 m on a face width of 287 m at a depth of about 97 m. The Trivoli Sandstone was less than 7 m thick and was encountered approximately at the bedrock surface at about 20 m below ground. Final Panel 6, separated to the south by a 40-m chain–pillar barrier, was mined from May 1993 to July 1994 a distance of 2606 m on a face width of 280 m, and was nearest to the Panel 5 instrumentation in march 1994.

The ISGS geotechnical investigations were reported by P. J. DeMaris and N. Kawamura in Booth et al. (1997). A transverse line of subsidence monuments was monitored for 32 months starting in November 1992 and undermined on 2 January 1993. Ground subsidence primarily occurred during the period when the mine face was between −46 and +60 m of the line, and reached 1.37 m on the centre line by the end of monitoring in 1995. The horizontal ground strains were initially extensional and then compressional. The maximum transverse extensional and compressional strains occurred about 15 and 41 m in from the edge of the panel, the extensional strain being largely expressed as longitudinal tension cracks several centimetres wide and from 1 to over 100 m long.

Subsurface strain monitoring using the borehole TDR cable indicated several small extensional displacements in the shallow (<45 m deep) bedrock at contacts between different lithologies, beginning when the face was 47 m away. The final break was in the drift at a depth of 16.6 m several days after undermining.

Hydraulic tests at Panel 5. One sandstone test well and 12 piezometers into the drift and Trivoli Sandstone were constructed across the panel in 1992 and undermined at New Year 1993. Slug tests of the piezometers over the panel were conducted before, after and, in some cases, during subsidence. The hydraulic results are summarized in Table 2 (Booth et al. 2000). The two shallow drift piezometers, which were screened in thin sand and gravel within 6 m of the land surface, had hydraulic conductivities in the range 10^{-7}–10^{-6} m s^{-1}, which did not change significantly after mining. The four deep drift piezometers

Table 2. *Hydraulic conductivities ($m\,s^{-1}$) from Panel 5 slug tests, Saline County site*

Piezometer	Screened interval (mBGL)	Before (8/92–12/92)	During (12/92–1/93)	After subsidence (2/93–6/93)
		Hydraulic conductivity ($m\,s^{-1}$)		
Trivoli Sandstone				
P51B	20.9–23.9	$1.3 \times 10^{-8} – 3.6 \times 10^{-8}$	$1.8 \times 10^{-8} – 2.4 \times 10^{-8}$ 7.3×10^{-6} (1/2/93)	7.8×10^{-8}
P52B	20.4–23.5	3.5×10^{-8}	–	$5.4 \times 10^{-7} – 6.7 \times 10^{-7}$
P53B	17.1–20.1	4.1×10^{-9}	$5.2 \times 10^{-9} – 7.7 \times 10^{-9}$	2.3×10^{-9}
P54B	15.6–18.7	$1.7 \times 10^{-8} – 4.5 \times 10^{-8}$	1.0×10^{-8}	1.1×10^{-8}
Deep drift				
P51D	15.2–16.8	$1.2 \times 10^{-9} – 7.0 \times 10^{-9}$	–	$1.1 \times 10^{-9} – 1.4 \times 10^{-9}$
P52D	14.0–15.5	$1.0 \times 10^{-8} – 2.0 \times 10^{-7}$	–	$1.2 \times 10^{-8} – 6.7 \times 10^{-8}$
P53D	12.2–13.7	2.3×10^{-8}	5.6×10^{-9}	$8.9 \times 10^{-9} – 1.7 \times 10^{-8}$
P54D	10.9–12.5	$5.5 \times 10^{-7} – 2.1 \times 10^{-6}$	–	–
Shallow drift				
P51S	7.3–8.3	$1.0 \times 10^{-7} – 2.9 \times 10^{-7}$	–	3.1×10^{-7}
P54S	4.6–5.5	$3.2 \times 10^{-6} – 3.8 \times 10^{-6}$	–	–
P5CD	7.3–8.8	1.3×10^{-6}	–	–

were screened just above the bedrock surface. Their hydraulic conductivities did not change significantly with subsidence, but varied considerably with position across the site due to lithological differences. The lower drift over the panel varied from a clay till in the centre to a sandy till at the southern edge, with conductivities in the range 10^{-9}–10^{-8} m s^{-1}. Over the southern barrier (piezometer P54D) it was a sand-and-gravel unit with conductivities in the range 10^{-7}–10^{-6} m s^{-1}.

In the Trivoli Sandstone, pre-subsidence slug-test conductivities were mostly of the order of 10^{-8} m s^{-1}. During active subsidence, the conductivity increased two orders of magnitude at inner piezometer P51B, which was subsequently damaged by subsidence. However, in the post-mining replacement piezometer at this site, the conductivity dropped back to 7.8×10^{-8} m s^{-1}. Only piezometer P52B, located 46 m inside the panel, showed a permanent increase in permeability (one order of magnitude to 6.7×10^{-7} m s^{-1}). No significant changes were observed in the tension zone (P53B) or barrier (P54B) piezometers.

Permeabilities were also obtained from straddle-packer tests in NX (7.6-cm diameter) boreholes GT3 (6 months before mining) and GT4 (7 months after mining) drilled vertically just south of the Panel 5 centre line. In GT3, cored to 61 m, the subdrift bedrock surface was in shale at a depth of 19.2 m, overlying the top of the Trivoli Sandstone at 19.8 m. The Trivoli Sandstone had an apparent conductivity in the range 3×10^{-8}–4×10^{-8} m s^{-1}, but most of the sequence was too tight to accept water. In GT4, located only 5 m away from GT3, the top of bedrock was 1.8 m lower (at 21 m) and directly in the Trivoli Sandstone. GT4 was stopped at 40 m at a high-angle fracture within limestone. Open fracturing or bedding separation was indicated at depths of 32.9 and 33.5 m. The only measurable water intakes in GT4 were in a deeper fractured shale–limestone interval and in the Trivoli Sandstone, which had an apparent conductivity in the range 1×10^{-8}–4×10^{-8} m s^{-1} and erratic intake behaviour probably related to the filling of discrete fractures, another manifestation of the low porosity of the fractures and low permeability of the matrix.

Potentiometric response at Panel 5. The water table in shallow drift wells 200–300 m off the panels did not respond to mining. Water levels in the shallow drift piezometers at the panel fluctuated considerably during undermining, but rapidly stabilized and returned to a normal seasonal range of 0.9–4.0 m below ground.

The pre-mining potentiometric surface of the Trivoli Sandstone was flat, the depth to water level varying from 2 to 6 m according to the local ground elevation. The sandstone water levels started to decline in December 1992 when the mine face was -300 m away, and fell rapidly as the mine approached, a total of 17 m over the panel and 12 m over the barrier by the time of undermining (Fig. 14). Over the panel, the sandstone became unconfined and showed virtually no recovery except for slight (3 m) rises in the winter of 1994–1995. In the southern barrier piezometer (P54D) the water level recovered quickly by about 4 m in early 1993 and thereafter remained high, probably due to recharge from the overlying drift sand aquifer.

The response of the deep drift differed significantly between the clay till over the inner panel, the sandy till in the outer zone and sand on the barrier. The water levels in the outer and barrier areas (P53D and P54D) declined in response to mining concurrently with the bedrock potentiometric response, and reached minima (12 m deep in P53D, 9 m in P54D) during the tensional phase in early January. The barrier piezometer head eventually stabilized at 7.6 m depth. In contrast, water levels in the inner piezometers declined only slightly (less than 3 m) in December, then fluctuated erratically during the undermining period, including a rise of about 15 m above their previous minima and overflowing at the ground surface immediately before undermining. The anomalous piezometric rises, discussed by Booth *et al.* (1999), are a feature of low-permeability units that are unable to drain quickly, and reflect localized pore-water compression possibly due to shear deformation at the leading edge of subsidence (Van Roosendaal *et al.* 1995). After mining, these interior water levels declined gradually to a depth of about 4.5 m.

Figure 15 (from Booth *et al.* 2000) is a transverse section across Panel 5 and illustrates the relationship of the various potentiometric surfaces, geological units and subsidence. Pre- and post-mining conditions are represented by the potentiometric levels for November 1992 and November 1994. The pre-mining potentiometric surface of the Trivoli Sandstone (line b–b) is flat. The post-mining surface (line b'-b') was lowered everywhere, but the greatest head drop was over the centre line of the panel where the sandstone became unconfined. The potentiometric surface of the deep drift was 2–4 m below ground and sloped south before mining (line d–d); the post-mining surface (line d'-d') had declined most at the edge and barrier, where the head relationships indicate leakage from the sandy drift down to the sandstone, depressing the heads in the drift but cushioning the head drops in the sandstone. In contrast, the deep drift potentiometric level in the

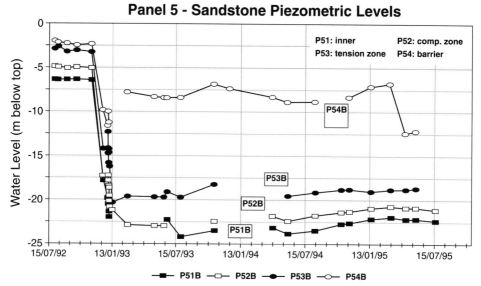

Fig. 14. Potentiometric response of Trivoli Sandstone to mining of Panel 5, Saline County site. From Booth *et al.* (2000). Reproduced with permission of the Association of Engineering Geologists (*Environmental and Engineering Geoscience*).

clay till over the central panel responded to mining with only slight declines. The greatest head drop in the deep drift was in the tension zone at P53D, probably due to increased vertical leakage through fractures. Three years after mining, the potentiometric levels had not recovered. The steep hydraulic gradient at the edge of the panel indicates that it is the low transmissivity of the sandstone, rather than any influence of later mining to the south, that is most responsible for the lack of recovery.

Geochemical results at the Saline County site. Water in the shallow drift wells off the panel was fresh (TDS < 1000 mg l^{-1}) with mixed cations and bicarbonate- or sulphate-dominated anions. In the shallow piezometers over Panel 5 it was slightly more mineralized (TDS 600–1600 mg l^{-1}), and the only apparent change due to mining was a slight increase in bicarbonate relative to sulphate. The deep drift wells over Panel 5 had slightly brackish water with TDS in the range 1200–1900 mg l^{-1}, mixed cations and sulphate- to bicarbonate-dominant anions; there was no obvious change due to mining. Nitrates were commonly present at low levels in the deep drift and moderate to high in the shallow drift.

Samples were obtained from the Trivoli piezometers over the panel before and after mining, and after mining from off-panel private wells in the Trivoli and Gimlet Sandstones (Booth & Bertsch 1999). Cation proportions from all sandstones plot on a Piper diagram (Fig. 16) in a straight line from 100% Na to a point of 40% Ca, 35% Mg and 65% Na, almost identical to the cation evolution trend observed in the sandstone at the Jefferson County site. The water in the Gimlet Sandstone ranges from a sodium chloride facies with TDS > 2800 mg l^{-1} in a 91 m deep well, to a mixed cation, sulphate facies with TDS 1300–1900 mg l^{-1} in a 43 m deep well, which is more likely in a zone of active groundwater circulation. The water in the Trivoli Sandstone was a marginally brackish (1100–1500 mg l^{-1}) sodium bicarbonate type, with relatively higher sulphate in the 20 m deep piezometers than in a 67 m deep residential well. The variation in sandstone water facies with different depths at this site corresponds well to the successive facies in sandstones progressively down-dip observed by Poole *et al.* (1989). After mining, the water in the Trivoli piezometers over the panel had increases in nitrates, and in calcium, magnesium and sulphate relative to sodium, bicarbonate and chloride, all suggesting increased downward leakage from the overlying drift. However, geochemical changes were less than at the Jefferson County site, where much more water flowed back in to the aquifer during the fuller recovery.

Conclusions from the Illinois studies

The Jefferson County site, where the mine was approximately 220 m deep, exhibited hydraulic

EFFECTS OF LONGWALL COAL MINING ON AQUIFERS 39

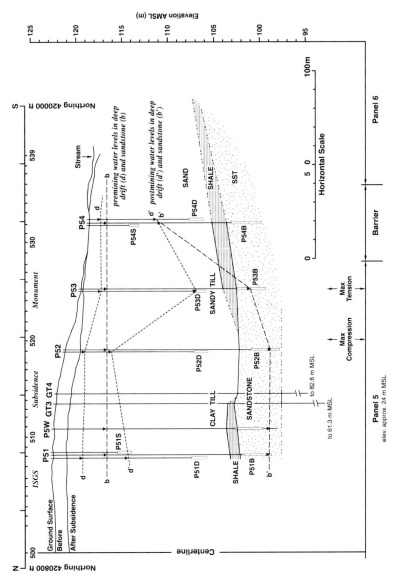

Fig. 15. Section across Panel 5, Saline County site, showing pre- and post-mining potentiometric levels. From Booth et al. (2000). Reproduced with permission of the Association of Engineering Geologists (*Environmental and Engineering Geoscience*).

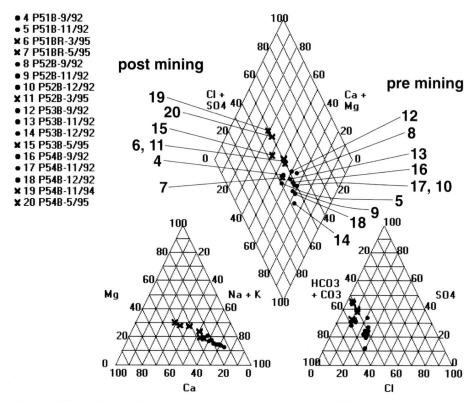

Fig. 16. Piper trilinear diagram of major ion proportions in sandstone water, Saline County site. From Booth & Bertsch (1999). Reproduced with the permission of Springer-Verlag (*Hydrogeological Journal* **7**, 561–575) (expect for the US public-domain components).

and potentiometric behaviour in accordance with the conceptual model. The permeability and storativity of the shallow, moderately transmissive Mt Carmel Sandstone increased substantially due to subsidence-related fracturing over the longwall panel. The increases were one order of magnitude in the interior of the subsidence trough over the panel, and two orders of magnitude along the tensional margins. During the active subsidence phase, potentiometric levels in the sandstone fell rapidly as new fracture space was opened up. The primary potentiometric low generated a secondary drawdown effect which spread outwards and was observed as a gradual decline in head as the mine face approached. Following undermining, the levels recovered slightly in the compressive phase, then fully over a period of 4 years as recharge flowed laterally back through the aquifer. The aquifer as a whole was initially confined, became unconfined during mining and returned to confined conditions during long-term recovery.

The combination of increased hydraulic properties and recovery of water levels represent a physical enhancement of the aquifer, with improved well yields. However, this was accompanied by a deterioration in water quality in the form of increased TDS and sulphates, which may have been partly due to leakage from the overlying fractured shale but was probably mostly due to oxidation of *in situ* sulphides when the aquifer became unconfined, followed by solution during the potentiometric recovery.

The Saline County site was similar geologically and topographically to the Jefferson County site, except that the sandstone units were thinner and less permeable, and thus about an order of magnitude less transmissive than the sandstone at Jefferson. The subsidence at Panel 1 produced only slight increases in permeability, despite the relatively shallow depth to the mine (122 m). The potentiometric response was modified by the lower transmissivity of the sandstone such that the secondary drawdown in advance of mining was slight, but the heads dropped rapidly to the top of aquifer with undermining. There was a slight compressive recovery but minimal long-term recharge recovery.

At Panel 5, the Trivoli Sandstone was at the bedrock surface in contact with the overlying drift. It was expected that the permeabilities would be higher and that potentiometric recovery after mining would be rapid because of easier recharge. However, the behaviour was similar to that at Panel 1. The initial permeability of the Trivoli Sandstone was low, and subsidence-related fracturing caused only localized minor increases. The potentiometric level in the sandstone declined rapidly with undermining, and no significant recovery over the panel was observed through to the end of monitoring more than 2 years later. The critical factor appeared to be the low transmissivity of the sandstone. However, the existence of a sand unit along the edges of the panel provided some recharge that locally cushioned the potentiometric impact in the sandstone.

At both Jefferson & Saline sites, the response of shallow, unconsolidated, unconfined aquifers to mining was negligible. Water tables fluctuated briefly as they accommodated to the changing topography of the ground surface, but there was no long-term head loss.

General conclusions

Longwall mining affects overlying aquifers by several different mechanisms. Potentiometric levels can sometimes be lowered as a response to direct drainage to the mine, but this is typically prevented because intermediate levels of the overburden retain confining properties. However, *in situ* fracturing caused by longwall mine subsidence creates multiple, interacting effects in the aquifer. The sudden increase in fracture porosity causes a transient head drop that is particularly pronounced in confined fractured aquifers because of the low magnitude of the storativity. A secondary drawdown effect is transmitted through the aquifer outwards from the potentiometric low in the subsidence zone. Increased fracture permeabilities cause decreased gradients, up-gradient head drops and down-gradient increased discharge. In high relief areas, this effect is amplified by the increased drainage of upper-level aquifers to lower aquifers and discharge areas through fractured aquitards, which enhances the topographically driven local flow system.

The most significant difference between the Illinois and Appalachian coalfields is the relief. In low-relief Illinois, the effects of increased permeability on the flow systems are more subdued: there is less opportunity for the increased drainage from upper aquifers to lower aquifers and discharge zones, and less significance to the effects of increased permeability on system gradients and throughput that are more pronounced where the natural hydraulic gradients are steeper. The potentiometric effects at the Illinois sites were focused on the areas over and immediately adjacent to the longwall panel.

Nevertheless, the transmitted potentiometric effect is still controlled by the magnitude of the transmissivity, and in both coalfields the lateral limit of head drop is within a few hundred metres. This limit should not be directly controlled by the depth of mining as it is a response to an *in situ* head drop in the shallow aquifer itself, not an effect transmitted up from the mine. The potentiometric drops are clearly greater over the mined panel, but in our studies these drops did not obviously correlate with the magnitude of the subsidence strains or increases in permeability. Essentially, the subsidence-related fracturing caused the confined bedrock aquifer to become unconfined, or nearly so. How suddenly or gradually this condition developed depended on the transmissivity of the aquifer.

Recovery should be the normal state of affairs after the transient potentiometric depression caused by mining, but it can be delayed or prevented by several factors:

- low transmissivity of the aquifer, which restricts lateral recharge through the aquifer;
- natural barriers to hydraulic continuity with sources of recharge, such as tight confining units and lateral boundaries in the aquifer;
- the location of continued mining operations, which can interrupt the physical or hydraulic continuity between the affected aquifer and recharge sources;
- continuing loss of water to the mine, or, in high-relief areas, through fractured aquitards to lower aquifers or down gradient to discharge zones.

Otherwise, the most important factor in the response to mining subsidence is the local hydrogeology. Variations in the hydrogeological properties, continuity and geometry of the hydrogeological units on scales as broad as the groundwater flow system and as local as the panel strongly affect the initial potentiometric response and critically control the recovery. Generalized monitoring schemes and predictions that do not adequately consider these local characteristics are liable to be erroneous.

Future research

Our understanding of the hydrological processes and effects of longwall mining has advanced greatly over the past two decades. However, translating the scientific understanding into direct engineering practice, so that specific predictions of the effects of planned mining at individual mine sites can be made, remains a problem. Geological variation between sites makes it difficult to apply generalized concepts or generic formulae to predict the exact occurrence, location, magnitude and timing of potentiometric changes and long-term recoveries on a site-specific basis.

It is a truism that our knowledge is never perfect and that more field data are always needed. However, three areas seem particularly in need of further work:

- more field data and expanded conceptual models to account for differences in the hydrogeological settings of the new coalfields in which longwall mining is being carried out;
- more research on the long-term recovery of groundwater levels and associated changes in groundwater chemistry on overlying aquifers after mining, which are less well documented than the short-term impacts;
- ways in which numerical modelling can be applied to individual sites to predict impacts and explain hydrological responses.

Truly integrated numerical models of the hydrological impact of longwall mining require the linking of the strata deformation to the mining (mining engineering and rock mechanics), the changes in hydraulic properties to the deformation (rock mechanics and hydraulics) and the changes in hydraulic properties to the changes in groundwater flow (groundwater hydrology). Attempts to develop such models about two decades ago (e.g. Booth 1984) were limited by the lack of empirical data from field studies, inadequate conceptual understanding of the hydrological processes, unavailability of ready-to-use, validated numerical models for the individual stages of the model and by the lack of adequate computer facilities (the need to reserve time and space on main frames made model development and calibration impractically time-consuming). However, there is now a great deal of empirical field data and we have a much better understanding of the hydrological mechanisms and effects. There are still no 'off-the-shelf' models available for the whole integrated process of the hydrological effects of longwall mining, but there are sophisticated, versatile, well-tested numerical models available for the individual components of groundwater flow, fracture characteristics and rock mechanics, and powerful personal computers capable of the modelling demands are widely available.

Increasingly sophisticated coupled numerical models that link the mining, deformation, hydraulic property changes and groundwater flow (including in some cases variable saturation and inflow to the mine) have been developed over the past ten years by research groups such as those at Penn State University (Bai & Elsworth 1991; Elsworth & Liu 1995; Liu & Elsworth 1997; Kim *et al.* 1997). These models are being applied to increasingly complex theoretical situations, although it will probably remain difficult to translate them from the academic research level to the off-the-shelf engineering and hydrological working world. The research group at Newcastle University has followed a somewhat different approach to modelling groundwater rebound in abandoned underground mine systems, by applying different forms of mass-balance models appropriate to pipe conduits, larger-scale connected mine ponds, or Darcian porous media to the different system components (Burke & Younger 2000; Adams & Younger 2001).

Consulting engineers and hydrogeologists with immediate application needs are likely at the moment to use the readily available commercial variants of well-validated groundwater flow models such as MODFLOW (McDonald & Harbaugh 1988). Using such Darcian, porous media models to model mine inflow and groundwater flow in the intensely fractured, variably saturated regions above the mine is problematical, and generally quite inappropriate for the non-Darcian flow through the mine openings themselves. However, such models should generally be acceptable for the shallow overlying aquifers considered separately from the deeper units and the mine inflow boundary. MODFLOW variants, readily available as visual interfaces such as Visual MODFLOW (Waterloo Hydrologic, Ontario, Canada), GMS (BOSS International Inc., Madison, Wisconsin, USA) and Graphic Groundwater (S. Esling, Southern Illinois University at Carbondale, Illinois, USA) are already widely used to simulate groundwater flow in complex, heterogeneous geological settings of almost all types. Their application is one means by which the scientific research into the hydrological effects of longwall mining can be applied to practical engineering problems.

The Lancashire No. 20 study was conducted under the supervision of R. R. Parizek and supported by the Pennsylvania State University Mineral Conservation Section and the Shell Companies Foundation. Contributors to the Illinois research were ISGS personnel R. A. Bauer, B. B. Mehnert, P. J. Demaris, D. Van Roosendaal and J. Kelleher, and NIU graduate assistants E. D. Spande, C. T. Pattee, J. D. Miller, A. M. Curtiss and L. P. Bertsch. The research was funded under the Illinois mine Subsidence Research Program of the Illinois Department of Energy & Natural Resources (1988–1992) and by the Office of Surface Mining Reclamation and Enforcement of the US Department of the Interior (1991–1995: Contract Agreement GR 196171). This support does not constitute an endorsement by either DENR or OSM of any views expressed in this article, which are solely those of the author.

References

ADAMS, R. & YOUNGER, P.L. 2001. A strategy for modelling ground water rebound in abandoned mine systems. *Ground Water*, **39**, 249–261.

ASTON, T.R.C. & SINGH, R.N. 1983. A reappraisal of investigations into strata permeability changes associated with longwall mining. *International Journal of mine Water*, **2**, (4), 1–14.

ASTON, T.R.C., SINGH, R.N. & WHITTAKER, B.N. 1983. The effect of test cavity geology on the *in situ* permeability of Coal measures strata associated with longwall mining. *International Journal of Mine Water*, **2**, (4), 19–34.

BAI, M. & ELSWORTH, D. 1991. Modeling of subsidence and groundwater flow due to underground mining. *Proceedings of the 34th Annual meeting of Association of Engineering Geologists, Chicago*, Association of Engineering Geologists, College Station, TX 981–990.

BIENIAWSKI, Z.T. 1987. *Strata Control in Mineral Engineering*; John Wiley, New York.

BOOTH, C.J. 1984. *A Numerical Model of Groundwater Flow Associated with an Underground Coal mine in the Appalachian Plateau, Pennsylvania* Ph.D. Dissertation, The Pennsylvania State University.

BOOTH, C.J. 1986. Strata movement concepts and the hydrogeological impact of underground coal mining. *Ground Water*, **24**, 507–515.

BOOTH, C.J. 1988. Interpretation of well and field data in a heterogeneous layered aquifer setting, Appalachian Plateau. *Ground Water*, **26**, 596–606.

BOOTH, C.J. 1999. Recovery of groundwater levels after longwall mining. In: FERNANDEZ-RUBIO, R. (ed.) *Mine, Water and Environment: Proceedings of the I.M.W.A. International Congress, Sevilla, Spain, September 1999*, IMWA, 35–40.

BOOTH, C.J. & SARIC, J.A. 1987. The effects of abandoned underground mines on ground water, Saline County, Illinois. In: *Proceedings of the National Symposium on Mining, Hydrology, Sedimentology and Reclamation, Springfield, Illinois, December, 1987*. University of Kentucky, Bulletin, **BU 145**, 243–248.

BOOTH, C.J. & SPANDE, E.D. 1992. Potentiometric and aquifer property changes above subsiding longwall mine panels, Illinois Basin coalfield. *Ground Water*, **30**, 362–368.

BOOTH, C.J. & BERTSCH, L.P. 1999. Groundwater geochemistry in shallow aquifers above longwall mines in Illinois, USA. *Hydrogeology Journal*, **7**, 561–575.

BOOTH, C.J., CURTISS, A.M. & MILLER, J.D. 1994. Groundwater response to longwall mining, Saline County, Illinois, USA. In: REDDISH, D.J. (ed.) *Proceedings of the Fifth International Mine Water Congress, Nottingham, September 1994*. **1**, IMWA, 71–81.

BOOTH, C.J., CARPENTER, P.J. & BAUER, R.A. 1997. *Aquifer Response to Longwall Mining, Illinois*; US Department of the Interior, Office of Surface Mining, Library Report, **637** Grant/Coop. Agreement GR196171.

BOOTH, C.J., SPANDE, E.D., PATTEE, C.T., MILLER, J.D. & BERSTCH, L.P. 1998. Positive and negative impacts of longwall mine subsidence on a sandstone aquifer. *Environmental Geology*, **34**, 223–233.

BOOTH, C.J., CURTISS, A.M., DEMARIS, P.J. & VAN ROOSENDAAL, D.J. 1999. Anomalous increases in piezometric levels in advance of longwall mining subsidence. *Environmental and Engineering Geology*, **4**, 407–418.

BOOTH, C.J., CURTISS, A.M., DEMARIS, P.J. & BAUER, R.A. 2000. Site-specific variation in the potentiometric response to subsidence above active longwall mining. *Environmental and Engineering Geology*, **6**, 383–394.

BOUWER, H. 1989. The Bouwer and Rice slug test – an update. *Ground Water*, **27**, 304–309.

BOUWER, H. & RICE, R.C. 1976. A slug test for determining hydraulic conductivity of unconfined aquifers with completely or partially penetrating wells. *Water Resources Research*, **12**, 423–428.

BURKE, S.P. & YOUNGER, P.L. 2000. Groundwater rebound in the South Yorkshire Coalfield: a first approximation using the GRAM model. *Quarterly Journal of Engineering Geology and Hydrogeology*, **33**, 149–160.

BUTLER, J.J. & HEALEY, J.M. 1998. Relationship between pumping test and slug test parameters: scale effect or artifact? *Ground Water*, **36**, 305–313.

CARTWRIGHT, K. & HUNT, C. S. 1983. *Hydrogeologic Aspects of Coal mining in Illinois: An Overview*. Illinois State Geological Survey, Environmental Geology Notes, **EGN 90**.

CIFELLI, R.C. & RAUCH, H.W. 1986. Dewatering effects from selected underground coal mines in north-central West Virginia. In: PENG, S.S. (ed.) *Proceedings of the Second Workshop on Surface Subsidence Due to Underground mining, Morgantown, West Virginia, August 1986*. West Virginia University, Morgantown, WV, 249–263.

COE, C.J. & STOWE, S.M. 1984. Evaluating the impact of longwall coal mining on the hydrologic balance. In: *Proceedings of the National Water Well Association Conference on The Impact of mining on Ground Water, Denver, Colorado*, National Water Well Association, Columbus, OH, 348–359.

COOPER, H.H., BREDEHOEFT, J.D. & PAPADOPULOS, I.S. 1967. Response of a finite diameter well to an instantaneous charge of water. *Water Resources Research*, **3**, 263–269.

DAVIS, R.W. 1973. Quality of near-surface waters in southern Illinois. *Ground Water*, **11**, 11–18.

DOMENICO, P.A. & SCHWARTZ, F.W. 1998. *Physical and Chemical Hydrogeology*; 2nd edn. John Wiley, New York.

ELSWORTH, D. & LIU, J. 1995. Topographic influence of longwall mining on ground water supplies. *Ground Water*, **33**, 786–793.

GRAF, D. L., MEENTS, F. W., FRIEDMAN, I. & SHIMP, N. F. 1966. *The Origin of Saline Formation Waters III: Calcium Chloride Waters*. Illinois State Geological Survey, Circular, **397**.

HILL, J.G. & PRICE, D.R. 1983. The impact of deep mining on an overlying aquifer in western Pennsylvania. *Ground Water Monitoring Review*, **3**, 138–143.

HODGSON, F.D.I. 1985. Hydrological disturbances associated with increased underground extraction of coal. *Transactions of the Geological Society of South Africa*, **88**, 541–544.

HOLLA, L. 1991. Some aspects of strata movement relating to mining under water bodies in New South Wales, Australia. In: NORTON, P.J. & VESELIC, M. (eds) *Proceedings of the Fourth I.M.W.A. Congress, Lubljana–Pörtschach, September 1991*. **1**, 233–244.

HOLLA, L. & BUIZEN, M. 1991. The ground movement, strata fracturing and changes in permeability due to deep longwall mining. *International Journal of Rock Mechanics, Mining Science, and Geomechanics Abstracts*, **28**, 207–217.

HUTCHESON, S.M., KIPP, J.A., DINGER, J.S., SENDLEIN, L.V.A., CAREY, D. I. & SECRIST, G. L. 2000. *Effects of Longwall Mining on Hydrology, Leslie County, Kentucky: Part 2: During Mining Conditions*. Kentucky Geological Survey, Report of Investigations **4**, Series XII.

HVORSLEV, M.J. 1951. *Time Lag and Soil Permeability in Groundwater Observations*. Bulletin **36**, Waterways Experiment Station, US Army Corps of Engineers, Vicksburg, MI.

JOHNSON, K.L. 1992. Influence of topography on the effects of longwall mining on shallow aquifers in the Appalachian coal field. In: PENG, S.S. (ed.) *Proceedings of the Third Workshop on Surface Subsidence Due to Underground Mining, Morgantown, West Virginia, June 1992*, West Virginia University, Morgantown, WV, 197–203.

JOHNSON, K.L. & OWILI-EGER, A.S.C. 1987. Hydrogeological environment of full extraction mining. In: *Proceedings of the Longwall USA Conference, Pittsburgh, Pennsylvania, June 1987*, 147–158.

KADNUCK, L.L.M. 1994. *Response of Springs to Longwall Coal Mining at the Deer Creek and Cottonwood Mines, Wasatch Plateau, Utah*. US Bureau of mines, Information Circular, **9405**.

KELLEHER, J.T., VAN ROOSENDAAL, D.J., MEHNERT, B.B., BRUTCHER, D.F. & BAUER, R.A. 1991. Overburden deformation and hydrologic changes due to longwall coal mine subsidence in the Illinois Basin. In: JOHNSON, A. I. (ed.) *Land Subsidence: Proceedings of the Fourth International Symposium on Land Subsidence, Houston, Texas, May 1991*. IAHS Publication **200**, 195–204.

KIM, J.-M., PARIZEK, R.R. & ELSWORTH, D. 1997. Evaluation of fully-coupled strata deformation and groundwater flow in response to longwall mining. *International Journal of Rock Mechanics and Mining Science*, **34**, 1187–1199.

KIPP, J.A. & DINGER, J.S. 1991. Stress relief fracture control of ground water movement in the Appalachian Plateau. Kentucky Geologic Survey, Series X1, Reprint 30.

LEAVITT, B.R. & GIBBENS, J.F. 1992. Effects of longwall coal mining on rural water supplies and stress relief fracture flow systems. In: PENG, S.S. (ed.) *Proceedings of the Third Workshop on Surface Subsidence Due to Underground Mining, Morgantown, West Virginia, June 1992*, West Virginia University, Morgantown, WV, 228–236.

LIU, J. & ELSWORTH, D. 1997. Three-dimensional effects of hydraulic conductivity enhancement and desaturation around mined panels. *International Journal of Rock mechanics and mining Science*, **34**, 1139–1152.

MATETIC, R.J. & TREVITS, M.A. 1990. Case study of longwall mining effects on water wells. *SME (Society of Mining, Metallurgy and Exploration) Annual Meeting, Salt Lake City, Utah, 26 February–1 March, 1990*. Preprint No. **90–141**.

MATETIC, R.J. & TREVITS, M.A. 1991. Does longwall mining have a detrimental effect on shallow ground water sources? In: *Proceedings of the Fifth National Outdoor Action Conference on Aquifer Restoration, Ground Water Monitoring, and Geophysical Methods*. AGWSE (Association of Ground Water Scientists and Engineers) Ground Water Management Book 5.

MATETIC, R.J. & TREVITS, M.A. 1992. Hydrologic variations due to longwall mining. In: PENG, S.S. (eds) *Proceedings of the Third Workshop on Surface Subsidence Due to Underground Mining, Morgantown, West Virginia, June 1992*, West Virginia University, Morgantown, WV, 204–213.

MCDONALD, M. & HARBAUGH, A. 1988. *A Modular Three-dimensional Finite-difference Ground-water Flow Model*. US Geological Survey, Techniques of Water Resources Investigations, Book 6, Chapter A1.

MEHNERT, B.B. VAN ROOSENDAAL, D.J., BAUER, R.A., DEMARIS, P.J. & KAWAMURA, N. 1994. *Final Report of Subsidence Investigations at the Rend Lake Site, Jefferson County, Illinois*. Illinois State Geological Survey, **IMSRP-X** (plus Appendices published as ISGS Open File Report, **1997-7**).

MOEBS, N.M. & BARTON, T.M. 1985. Short-term effects of longwall mining on shallow water sources. In: *Mine Subsidence Control*. US Bureau of Mines Information Circular, **IC 9042**, 13–24.

NEATE, C.J. 1980. *Effects of Mining Subsidence on the Permeability of Coal measure Rocks* Ph.D. Dissertation, University of Nottingham, UK.

NIKRAZ, H.R., EVANS, A.W. & PRESS, M.E. 1994. Effects of mine dewatering on poorly consolidated sandstones. In: REDDISH, D.J. (ed.) *Proceedings of the Fifth International Mine Water Congress, Nottingham, September 1994*, **1**, IMWA, 143–153.

PAUVLIK, C.M. & ESLING, S.P. 1987. The effects of longwall mining subsidence on the groundwater conditions of a shallow unconfined aquitard in southern Illinois. In: *Proceedings of the National Symposium on Mining, Hydrology, Sedimentology & Reclamation, Springfield, Illinois, December 1987*. University of Kentucky Bulletin, **BU 145**, 189–196.

PENG, S.S. 1986. *Coal Mine Ground Control*, 2nd ed. John Wiley, NewYork.

POOLE, V. L., CARTWRIGHT, K. & LEAP, D. 1989. Use of geophysical logs to estimate water quality of basal Pennsylvanian sandstones, southwest Illinois. *Ground Water*, **27**, 682–689.

RAUCH, H. W. 1987. Ground water impacts from surface and underground coal mining. In: *West Virginia Ground Water 1987: Status and Future Directions* West Virginia University, August 1987, No. **XXV**.

RAUCH, H. W. 1989. A summary of the ground water impacts from underground mine subsidence in the north central Appalachians. In: *Coal Mine Subsidence Special Institute, Eastern Mineral Law Foundation, Pittsburgh, Pennsylvania*, Chapter 2, 2.01–2.31.

REDDISH, D.J., YAU, X.L., BENBIA, A., CAIN, P. & FORRESTER, D.J. 1994. Modelling of caving over the Lingan and Phalen mines in the Sydney Coalfield, Cape Breton. In: REDDISH, D.J. (ed.) *Proceedings of the Fifth International Mine Water Congress, Nottingham, September 1994*, IMWA, **1**, 105–124.

SCHMECHEL, F.W., EICHELD, W.F. & SANTY, W.P. 1979. Automated data acquisition for subsidence characterization. In: *AIME Annual Meeting, New Orleans, Lousiana, February 1979*, Preprint Society for Mining, Metallurgy and Exploration, Littleton, CO **79–132**.

SCHMIDT, R.D. 1992. Factors affecting residential water well yields in the vicinity of room and pillar mines. In: PENG, S.S. (ed.) *Proceedings of the Third Workshop on Surface Subsidence Due to Underground Mining, Morgantown, West Virginia, June 1992*, West Virginia University, Morgantown, WV, 244–252.

SCHULZ, R.A. 1998. *Ground-water Hydrology of Marshall County, West Virginia, with Emphasis on the Effects of Longwall Coal Mining*; US Geological Survey, Water Resources Investigation Report, **88-4006**.

SINGH, M.M. & KENDORSKI, F.S. 1981. Strata disturbance prediction for mining beneath surface water and waste impoundments. In: *Proceedings of the First Conference on Ground Control in Mining, Morgantown, West Virginia, July 1981*, 76–88.

STONER, J.D. 1983. Probable hydrologic effects of subsurface mining. *Ground Water Monitoring Review*, **3**, 126–137.

TIEMAN, G.E. & RAUCH, H.W. 1987. Study of dewatering effects at an underground longwall mine site in the Pittsburgh seam of the northern Appalachian coalfield. *Eastern Coal Mine Geomechanics, Proceedings of the Bureau of Mines Technology Transfer Seminar, Pittsburgh, Pennsylvania, 19 November 1986*; US Bureau of mines Information Circular, **9137**, 72–89.

TRENT, B.A., BAUER, R.A., DEMARIS, P.J. & KAWAMURA, N. 1996. *Findings and Practical Applications from the Illinois Mine Subsidence Research Program, 1985–1993*. Illinois Mine Subsidence Research Program, Illinois State Geological Survey, **IMSRP-12**.

US Department of the Interior, 1981. *Ground Water Manual: A Water Resources Technical Publication*; (rev. edn.); Water and Power Resources Service, John Wiley, New York.

VAN ROOSENDAAL, D.J., BRUTCHER, D.F., MEHNERT, B.B., KELLEHER, J.T. & BAUER, R.A. 1990. Overburden deformation and hydrologic changes due to longwall mine subsidence in Illinois. In: CHUGH, Y.P. (ed.) *Proceedings of the Third Conference on Ground Control Problems in the Illinois Coal Basin*, Southern Illinois University, Carbondale, IL, 73–82.

VAN ROOSENDAAL, D.J., MEHNERT, B.B. KAWAMURA, N. & DEMARIS, P.J. 1994. *Final Report of Subsidence Investigations at the Galatia Site, Saline County, Illinois*. Illinois State Geological Survey, **IMSRP-XI**. (plus 12 Appendices published as ISGS Open File Report, **OFS 1997–9**).

VAN ROOSENDAAL, D.J., KENDORSKI, F.S. & PADGETT, J.T. 1995. Application of mechanical and groundwater-flow models to predict the hodyrogeological effects of longwall subsidence – a case study. In: PENG, S.S. (ed.) *Proceedings of the Fourteenth Conference on Ground Control in Mining, Morgantown, West Virginia, 1–3 August 1995*, West Virginia University, Morgantown, WV, 252–260.

W.A. WAHLER & ASSOCIATES. 1979. *Dewatering Active Underground Coal Mines: Technical Aspects and Cost Effectiveness*; Industrial Environmental Research Laboratory, US Environmental Protection Agency, Cincinnati, OH.

WALKER, J.S. 1988. *Case Study of the Effects of Longwall Mining Induced Subsidence on Shallow Ground Water Sources in the Northern Appalachian Coalfield*. US Bureau of Mines, Report of Investigations, **9198**.

WERNER, E. & HEMPEL, J.C. 1992. Effects of coal mine subsidence on shallow ridge-top aquifers in northern West Virginia. In: PENG, S.S. (ed.) *Proceedings of the Third Workshop on Surface Subsidence Due to Underground Mining, Morgantown, West Virginia, June 1992*, West Virginia University, Carbondale, IL, 237–243.

WHITTAKER, B.N., SINGH, R.N. & NEATE, C.J. 1979. Effect of longwall mining on ground permeability and subsurface drainage. In: ARGALL, G.O., BRAWNER, C.O. (eds) *Mine Drainage: Proceedings of the First Mine Drainage Symposium, Denver, Colorado*, Miller Freeman Publications, San Francisco, CA, 161–183.

YOUNGER, P.L. 1998. Hydrological consequences of the abandonment of regional mine dewatering schemes in the UK. In: *Hydrology in a Changing Environment, BHS (British Hydrological Society) Exeter Conference, 1998*. British Hydrological Society Occasional Paper **9**, 80–82.

ZUEHLS, E.E., RYAN, G.L., PEART, D.B. & FITZGERALD, K.K. 1981a. *Hydrology of Area 35, Eastern Region, Interior Coal Province, Illinois and Kentucky*. US Geological Survey, Water Resources Investigations Report, **81–403**.

ZUEHLS, E.E., RYAN, G.L., PEART, D.B. & FITZGERALD, K.K. 1981b. *Hydrology of Area 25, Eastern Region, Interior Coal Province, Illinois*. US Geological Survey Water Resources Investigations Open File Report, **81–636**.

Mine water tracing

CHRISTIAN WOLKERSDORFER

TU Bergakademie Freiberg, Lehrstuhl für Hydrogeologie, Gustav-Zeuner-Strasse 12, D-09599 Freiberg/Sachsen, Germany (e-mail: c.wolke@tu-freiberg.de)

Abstract: This paper describes how tracer tests can be used in flooded underground mines to evaluate the hydrodynamic conditions or reliability of dams. Mine water tracer tests are conducted in order to evaluate the flow paths of seepage water, connections from the surface to the mine, and to support remediation plans for abandoned and flooded underground mines. There are only a few descriptions of successful tracer tests in the literature, and experience with mine water tracing is limited. Potential tracers are restricted due to the complicated chemical composition or low pH mine waters. A new injection and sampling method ('LydiA'-technique) overcomes some of the problems in mine water tracing. A successful tracer test from the Harz Mountains in Germany with *Lycopodium clavatum*, microspheres and sodium chloride is described, and the results of 29 mine water tracer tests indicate mean flow velocities of between 0.3 and 1.7 m min^{-1}.

Hundreds of mines have been closed, due to economic or ecological reasons, in the recent past. In many cases, the groundwater table was allowed to rise again as the underground workings flooded. In some remote areas the pollution of surface or underground waters has been taken to be of no serious account, but in most of the developed countries (e.g. United States of America, Europe) polluted mine water must be treated. Mine water treatment, as has been shown by the examples of Picher, Oklahoma, USA (Sheibach *et al.* 1982; Parkhurst 1988), Wheal Jane, Cornwall, UK (Hamilton *et al.* 1997) or Königstein, Germany (Gatzweiler & Meyer 2000), can be extremely expensive, especially if acid mine water is involved. Therefore, the mine owners have to find new, innovative tools to control the hydrodynamic conditions within a flooded mine to minimize the pollution load (e.g. Scott & Hays 1975; Fernández-Rubio *et al.* 1987; Wolkersdorfer 1996; Lewis *et al.* 1997) and to introduce passive treatment techniques at the correct time after mine flooding (Younger 2000).

Techniques must be evaluated prior to applying natural attenuation or passive *in situ* remediation methods. Furthermore, after the chosen method has been applied, its reliability must be proven (Younger & Adams 1999), and tracer techniques are a useful tool to test the remediation method. Finally, tracer tests can also be used in addition to quantitative analyses of rock stability in flooded mines (Hunt & Reddish 1997).

Usually, tracer tests in the vicinity of mines are conducted to find connections between the ground surface and the mine, or vice versa. Typical examples are the tracer tests that accompany numerical models for performance assessment at potential radioactive waste disposal sites (e.g. Lee 1984). All tracer tests in mines can be grouped on the basis of their objectives:

- mine water inrushes/inundations (Skowronek & Zmij 1977; Goldbrunner *et al.* 1982; Wittrup *et al.* 1986; Wu *et al.* 1992; Lachmar 1994);
- optimize mining strategy (Adelman *et al.* 1960; Reznik 1990; Kirshner 1991; Kirshner & Williams 1993; Miller & Schmuck 1995);
- underground waste disposal (non-radioactive; Fried 1972; Himmelsbach & Wendland 1999);
- underground waste disposal (radioactive; Abelin & Birgersson 1985; Brewitz *et al.* 1985; Galloway & Erickson 1985; Cacas *et al.* 1990; Lewis 1990; Birgersson *et al.* 1992 (studies in the STRIPA mine, Sweden, since 1980–numerous papers have been published on the Stripa mine and the tracer tests conducted in the fractured crystalline rocks); Sawada *et al.* 2000);
- subsidence (Mather *et al.* 1969);
- remediation strategies (Aldous & Smart 1987; Doornbos 1989; Aljoe & Hawkins 1993, 1994; Davis 1994*a, b*; Wolkersdorfer

1996; Wolkersdorfer et al. 1997; Canty & Everett 1998).

Tracer tests have also been conducted to study 'heat mining' in geothermal projects (e.g. Gulati et al. 1978; Horne et al. 1987; Kwakwa 1989; Randall et al. 1990; Aquilina et al. 1998). Because their use is similar to studies in fractured rocks (see Himmelsbach et al. 1992; Käß 1998) and energy mining is usually carried out by the use of boreholes only, these studies will not be considered further.

Many tracer tests have not been published in the literature or are only available as academic theses (e.g. Anderson 1987; Bretherton 1989; Doornbos 1989; Diaz 1990; Wirsing 1995). This is often because they were either unsuccessful or the results confidential. Up to now, only a small number of tracer tests have been conducted in flooded underground mines to trace the hydrodynamic conditions within the flooded mine itself (Aljoe & Hawkins 1994; Wolkersdorfer et al. 1997; Wolkersdorfer & Hasche 2001). This was mainly due to the fact that no suitable method was available for injection of the tracer into the mine water at predetermined depths or without contaminating the mine water above the injection point. Most tests, therefore, injected the tracer at the surface to flow towards the mine through fractures (e.g. Lachmar 1994) or they used boreholes (e.g. Galloway & Erickson 1985; Cacas et al. 1990).

Aims of mine water tracer tests

Tracer tests are well established in groundwater studies where they are commonly used to investigate the hydraulic parameters or interconnections of groundwater flow (Käß 1998). Most of the techniques are well described and, depending on the aims of the tracer test and the hydrological situation, a range of tracers or methods can be chosen.

Published results of tracer tests in abandoned underground mines are not common, as already stated by Davis (1994a, b). Therefore, in mine water tracing, less experience exists and the expected results of an individual mine water tracer tests cannot always been found. The basic aims are as follows:

- testing the effectiveness of the bulkheads (dams);
- investigating hydrodynamic conditions;
- tracing connections between mine and surface;
- clarifying water inundations;
- investigating mass flow;

- estimating the decrease or increase of contaminants.

Historically, the first tracer tests conducted in mines were simply to reveal connections between ground or surface waters and the mine (e.g. Skowronek & Zmij 1977). One of the first tracer tests in a deep flooded underground mine to investigate the more complex hydrodynamic conditions was conducted in 1995 (Wolkersdorfer 1996).

In future studies, tracer tests should become a prerequisite for the evaluation of remediation strategies used for reclaiming abandoned mine sites. Kimball et al. (1999) have already provided an example of how to use tracer tests and synoptic sampling of trace metals in surface streams to evaluate the environmental impacts of mine effluents to watersheds.

Possible mine water tracers

Underground mines consist of a number of shafts, adits, raises and stopes, which are similar and comparable to the features found in karstic terrains. Therefore, a flooded mine can be looked on as a karst aquifer in a conceptual model, and numerical models of flow in karst aquifers (e.g. Liedl & Sauter 2000) might also describe the hydrodynamic situation in underground mines. Furthermore, the tracer techniques developed for karst aquifers (e.g. Habič & Gospodarič 1976; Käß 1998) should be appropriate also for flooded mines.

Several classes of tracers have been used in mine water tracing (Table 1). As, usually only successful tracer tests – if at all – are reported, little can be said about tracers that are unsuitable for mine water tracing. In the case of fluorescein, which easily adsorbs to organic materials, there are both successful tests as well as unsuccessful ones.

The chemical composition of mine waters very often tends to be extreme: total dissolved solids (TDS), pH (e.g. Iron Mountain: pH -3.6; Nordstrom et al. 2000) or metal concentrations are usually high (Banks et al. 1997) and consequently limit the number of tracers that could be used successfully. Even conservative tracers, such as fluorescein, might be unsuitable under certain mine water conditions (e.g. low pH, high suspension load, wooden supports, free chloride radicals). A detailed discussion about the effectiveness, strengths and weaknesses of tracers would go beyond the scope of this paper but can be found in Wolkersdorfer (1996) or Käß (1998).

Laboratory tests using the chosen tracer and typical mine water compositions are essential for

Table 1. *Artificial and natural tracers that have already been used in mine water tracing*

Artificial tracers	
Salts	Chloride (Mather *et al.* 1969; Aljoe & Hawkins 1994; Canty & Everett 1999; Wolkersdorfer & Hasche 2001)
	Bromide (Doornbos 1989; Kirshner 1991; Wu *et al.* 1992; Kirshner & Williams 1993; Aljoe & Hawkins 1993, 1994)
	Sulphur hexafluoride (Kirshner 1991; Kirshner & Williams 1993)
	Lithium (Wu *et al.* 1992)
	Iodide (Wu *et al.* 1992)
	Borate (Lewis 1990)
Dyes	Fluorescein, Uranine (Mather *et al.* 1969; Parsons & Hunter 1972; Goldbrunner *et al.* 1982)
	Rhodamine B (Skowronek & Zmij 1977; Aldous & Smart 1987; Davis 1994*a, b*)
	Rhodamine WT (Canty & Everett 1999)
Solid tracers	*Lycopodium* (Wolkersdorfer 1996; Wolkersdorfer *et al.* 1997)
	Microspheres (Wolkersdorfer & Hasche 2001)
Radioactive tracers	Lorenz 1973
Neutron activation analysis	Jester & Raupach 1987
Natural tracers	
Lead isotopes	Horn *et al.* 1995
Zinc variations	Bretherton 1989
Temperature	Anderson 1987; Wolkersdorfer 1996
Conductivity	Bretherton 1989; Reznik 1990; Wolkersdorfer 1996
Carbondioxide	Kirschner 1989
Tritium	Parsons & Hunter 1972; Goldbrunner *et al.* 1982; Diaz 1990; Pujol & Sanchez Cabeza 2000
Stable isotopes	Adelman *et al.* 1960; Klotz & Oliv 1982; Wittrup *et al.* 1986; Diaz 1990

obtaining positive results. All tracers have to be selected on their expected or known behaviour in mine water. An important consideration when conducting such tests is the duration of the laboratory test. Mine water commonly flows with mean velocities of $0.3-1.7 \text{ m min}^{-1}$ (95% confidence interval of 29 tracer tests investigated, excluding the maximum and minimum value; see Table 6), which can be used for a rough calculation of the expected residence time. The stability of the tracer in the mine water must be at least as long as the duration of the test. Preceding the Niederschlema–Alberoda tracer test, 45 days of laboratory tests, and preceding the Straßberg tracer test, 4 weeks of laboratory tests with regular tracer sampling and visual analyses were conducted to test the stability of the tracers within the mine water (Wolkersdorfer 1996; Wolkersdorfer & Hasche 2001).

The most suitable tracer is chosen on the results of the laboratory tests, the cost of the tracer material, and the cost of analysis. In addition, a detailed hydrogeological investigation is needed to clarify the most suitable injection and sampling points ('conceptual model of test site'). Many tracer tests, even if a suitable tracer has been chosen, have been unsuccessful due to a lack of knowledge about the hydrogeological situation and inappropriate selection of injection and sampling sites.

In the following sections a tracer test in a flooded underground mine will be used to explain the procedures necessary for a mine water tracer test. To understand the aims of the test, a short description of the mine and the hydrogeological situation at the time of the tracer test is given. Some conclusions are drawn, based on the test described, and other tracer tests conducted or described in the literature.

Case study: the Straßberg–Harz underground mine

Description of the mine

Located in the eastern Harz mountains, the Straßberg fluorspar mine is divided into three mining districts (from north to south: Brachmannsberg, Straßberg (Biwender), Glasebach) each connected by two underground adits but with different water chemistries (Table 2).

The surrounding area of the mine consists predominantly of Lower Devonian rocks. In the northern part Lower Carbonian rocks, that are partly influenced by metamorphism of the Ramberg pluton (intrusion age: $290 \pm 10 \text{ Ma}$),

Table 2. *Mean composition of the mine water in the Straßberg mine during the time of the tracer test (30 May – 27 July 2000) in mg L^{-1}. Li: <0.1 mg L^{-1}, NO$_3$: <0.5 mg L^{-1}*

Shaft	n	Na	K	Ca	Mg	Fe	Mn	Cl	SO$_4$	HCO$_3$	F
No 539	15	22	2	56	21	21	12	28	198	64	5
Fluor	11	15	2	140	29	22	6	17	387	77	8
Glasebach	9	14	5	178	32	10	13	17	385	184	7

crop out. Galena, sphalerite, pyrite, arsenopyrite, fluorspar and barite occur in veins. The country rock has a permeability of about $k_f = 10^{-6}$ m s^{-1}, and groundwater circulates through fissures, faults, karstic features and galleries (Wolkersdorfer & Hasche 2001). The natural hydrogeological situation has been substantially impacted by the many decades of mining. Pollutants are carried off the mine site by rainwater or drainage water and deposited into rivers, lakes and the groundwater. Mine flooding is believed to be the most economic method for redevelopment of underground mines.

Before its closure, the Straßberg fluorspar mine was the largest fluorspar mine in the German Democratic Republic. In 1991, economic and environmental reasons caused the closure of the Straßberg fluorspar mine, owned by the GVV (Gesellschaft zur Verwahrung und Verwertung von stillgelegten Bergwerksbetrieben mbH; Company for Remediation and Utilization of Abandoned Mines Ltd; Kuyumcu & Hartwig 1998). Situated in the Mid Harz Fault Zone of the eastern Harz Mountains (Fig. 1), approximately 30 km south of Quedlinburg and 6 km west of Harzgerode, the Straßberg mine was the most important producer of fluorite in the former GDR (Mohr 1978). Besides fluorite, the hydrothermal polymetallic mineralization of the vein structures comprises several ore minerals of Permian–Cretaceous age (e.g. pyrite, galena, sphalerite, chalcopyrite, arsenopyrite, wolframite, scheelite, siderite; Kuschka & Franzke 1974).

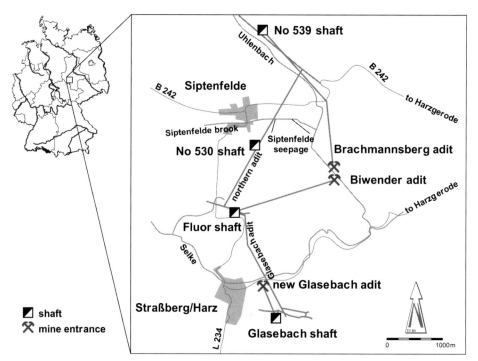

Fig. 1. Map of Germany with the location of the Straßberg mine in the eastern Harz mountains and its main galleries and shafts.

Mining was started more than 1000 years ago when silver, copper and lead were the targets of the miners. From the eighteenth century until 1990 mining focused on fluorspar, which was mainly found in the deeper parts of the mine (Bartels & Lorenz 1993). Sinking the Fluor dayshaft at the Straßberg pit in 1910 marked the start of the last production period, and between 1950 and 1970 the VEB Harzer Spatgrube joined the three most important deposits of the Straßberg mining district by driving two deep adits on the fifth and seventh levels (from north to south: Brachmannsberg pit, No. 539 shaft; Straßberg pit, Fluor shaft; and Glasebach pit, Glasebach shaft). Whilst the 3.5 km long Nordquerschlag (northern adit) connects the Brachmannsberg and Straßberg pit on the fifth level, the 1.5 km long Glasebachquerschlag (Glasebach adit) connects the Straßberg and Glasebach pits on the seventh level. Ultimately, when the ore reserves in the Brachmannsberg underground pit decreased in the 1980s, a dam was constructed in the northern adit, to separate the Brachmannsberg pit prior to flooding from the Straßberg pit.

On 31 May 1991 flooding of the Straßberg and Glasebach underground pits was started. Between July 1992 and August 1998, accompanying *in situ* temperature and conductivity measurements within the No. 539 shaft and the Fluor shaft (310 and 147 m deep, respectively) clearly showed that stratification within the water body was taking place (Kindermann 1998; Rüterkamp & Meßer 2000). Evidence for the stratification were differences in temperature, conductivity and metal-concentration between each of the water bodies (Table 3), with concentrations, particularly of iron, manganese and sulphate, increasing with depth.

Furthermore, the physico-chemical and chemical parameters of the water within the three parts of the mine respectively, showed significant differences, especially in TDS. After an intensive investigation the DMT (German Mining Technology) proposed to construct three new adits (the 'three-adit system') to drain the good quality water. They also predicted the long-term quality of the drainage water and, based on these results, a possible site of the final treatment plant was chosen (Rüterkamp & Meßer 2000). Almost immediately after the three adits were finished, stratification occurred as has been observed elsewhere (Wolkersdorfer 1996). Having analysed the new conditions, the consultant and the mine owners came to the conclusion that the water within the mine was flowing from the south to the north. Consequently, they thought that the mine water quality could be substantially improved by constructing a drainage pipeline from the Straßberg pit to the Glasebach pit.

At this stage, the Department of Hydrogeology at the Technical University Mining Academy Freiberg became involved in the project and proposed a tracer test in the mine. The Straßberg tracer test was designed to answer the following questions:

- How does the water flow between the three pits?
- Is the bulkhead between the Brachmannsberg and the Straßberg pit still effective?
- Does the sewage water from the Siptenfelde brook flow into the mine?
- What are the speeds of the mine water?
- Why did the stratification break down after the installation of the three-adit system?

After careful investigation of the hydrogeological and geochemical situation, as well as the accessibility of potential injection and sampling points, six injection and four sampling points were chosen. One of the injection points finally proved to be unsuitable, and the following injection points were actually used (from north to south): No. 539 shaft, Siptenfelde seepage, No. 530 shaft, Fluor shaft, Glasebach shaft

Table 3. *Selected constituents of the mine waters in the Flour and No 539 shaft in mg* l^{-1} *before and after the 3-adit system taken in use (after Rüterkamp & Meßer 2000)*

Depth* (mHN)	Fluor shaft						Depth* (mHN)	No. 539 shaft					
	25.8.1997			24.2.2000				26.8.1997			24.2.2000		
	Fe	Mn	SO$_4$	Fe	Mn	SO$_4$		Fe	Mn	SO$_4$	Fe	Mn	SO$_4$
c. 340	40	8	466	31	6	359	367	22	1	143	23	1	204
284	42	9	478	27	11	417	272	79	2	389	22	1	196
134, 204	52	19	600	50	18	525							

*mHN, meters above sea level (Kronstadt elevation).

(Fig. 1). It was important that tracers were injected as deep as possible within the No. 539, Fluor and Glasebach shafts because the three pits are connected to each other by adits at the shaft's deepest points.

As more than one injection point was needed for the test, only a multi-tracer test could be conducted. Owing to the volume of the mine, the use of salts would have required large amounts (e.g. 5–10 t of NaCl). Radioactive tracers would have been a good choice, but their detection is expensive and the German authorities do not allow these tracers to be used in large amounts. As has been shown elsewhere, fluorescent dyes seem to be unstable in mine water due to chemical and physical reactions and, as with salts, relatively large amounts have to be injected. Dyed club moss spores (*Lycopodium clavatum*, 30 µm diameter) and microspheres (15 µm diameter) were chosen because in both cases multiple colours can be used at the same time, and even small quantities contain billions of tracer particles that can be readily detected.

To guarantee reliable results, continuous sampling of the tracer is necessary. Unfortunately, microspheres and club moss spores cannot be sampled continuously, but only quasi-continuously using filters that have to be changed regularly (Käß 1998). Niehren & Kinzelbach (1998) presented an on-line microsphere counter (flow cytometer) for microspheres with a diameter of 1 µm and a flow rate of up to 1 ml min^{-1} to be used in groundwater studies. Because of the requirements of a tracer test in a flooded mine (rough underground conditions, high flow rates) and the conditions of the mine water itself (e.g. high suspension load), a flow cytometer was unfeasible. Therefore, the procedures and filter systems of the LydiA-technique (*Lycopodium Apparatus*; Wolkersdorfer *et al.* 1997) were selected to inject the tracers into the deep flooded mine at the predetermined depths without contamination and to sample the tracers. Owing to a pending patent no more details about the injection technique can be given here.

Tracer amount

It was calculated that 500–600 g of *Lycopodium* and 4×10^7 pieces of microspheres would be required at each injection point to get reasonable concentrations of tracers at the sampling points. In the case of the Siptenfelde Brook an amount of 20 000 l of saturated brine (*c.* 6.2 t of NaCl) was needed to raise the conductivity above background.

The calculations used to estimate the mass of tracer required are similar to those used in karst aquifer tracing (see Käß 1998). They can also be calculated on the assumption that the recovery rate of *Lycopodium* is about 2–7%, that of microspheres is 50–90% and that at least one tracer particle per litre of water must be present, assuming a total mixing of the mine water (Table 4).

During the tracer test, nearly 10 m^3 of water were pumped through the three filter systems that were installed and the filters changed on a 12 hour basis. Consequently, the tracers of 50–100 l of mine water were concentrated within the filter system and, thereafter, used for analyses as described below.

Tracer sampling and analyses

Owing to the characteristics of the tracers, two different sampling techniques were used. The sodium chloride was detected by continuous conductivity measurement at sampling points in the Uhlenbach brook (PIC GmbH, Munich/Germany), No. 539 shaft (LogIn GmbH, Gommern/Germany), Fluor shaft (LogIn GmbH, Gommern/Germany) and the Glasebach shaft (EcoTech GmbH, Bonn/Germany). Filter

Table 4. *Injection points, depth in shafts, and injection times of the seven tracers used*

Injection points (depth)	Tracers	Quantity[†]	Injection time
Siptenfelde seepage	NaCl brine	20 m^3 (6.2 t)	2 June: 9.08 – 10.15
No. 539 shaft (92 m)	Microspheres blue, 15 µm	4×10^7 pcs	5 June: 14.44*
No. 530 shaft (ca. 20 m)	Microspheres orange, 15 µm	4×10^7 pcs	5 June: 9.50 – 10.13
Fluor shaft (247 m)	Microspheres red, 15 µm	4×10^7 pcs	5 June: 12.18*
Fluor shaft (247 m)	Spores malachite green	264.9 g	5 June: 12.18*
Fluor shaft (247 m)	Spores saffron coloured	279.5 g	5 June: 12.18*
Glasebach shaft (4 m)	Microspheres green, 15 µm	4×10^7 pcs	5 June: 8.11*

Add 40 h to the times marked with an asterisk, as *LydiA* (*Lycopodium* Apparatus) opened approximately 40 h later.
[†] pcs, pieces.

systems, each with 100 and 15 μm filters (NY 100 HC, NY 15 HC; Hydro-Bios, Kiel/Germany) for collecting the solid tracers (microspheres, spores), were installed at the No. 539 shaft, Fluor shaft and Glasebach shaft. Mini piston pumps were used for sampling (Pleuger Worthington GmbH, Hamburg/Germany) and installed 5–10 m below the water surface. Every 12 hours the filter system was changed and the filters were stored in 500 ml brown glass bottles.

Most of the 147 filter samples contained noticeable amounts of Fe-oxides. Therefore, oxalic acid was added to remove both Fe-oxides and carbonates. Edetic acid, as recommended by Käß (1998), if used in chemically unbalanced quantities creates crystals that complicate the counting process (Wolkersdorfer et al. 1997). In the laboratory, after at least 1 day of reaction, the filters were carefully rinsed and the solids filtered through 8 μm cellulose nitrate filters (Sartorius, Göttingen, Germany) 47 mm in diameter, Nalgene plastic filters were used for membrane filtering using a hand vacuum pump. After each filtration the Nalgene filters, the filter unit and the working tables in the laboratory were cleaned to exclude any kind of contamination during sample preparation.

After drying and mounting the 147 cellulose nitrate filters on glass plates, the fluorescent microspheres and the spores were counted under a fluorescence microscope (Zeiss, Göttingen). Depending on the number of solid particles on the filters, an aliquot part of the whole filter was counted and the totals for the whole sample calculated.

Results

Sodium chloride. An increase in conductivity could only be detected at the Fluor shaft (Fig. 2). None of the other sampling points (Uhlenbach brook, No. 539 shaft and Glasebach shaft) showed a significant change in conductivity that could be attributed to the sodium chloride tracer. During the time of conductivity measurements, the Fluor shaft (30 May–31 July) discharged $3.2\,m^3\,min^{-1}$ of mine water resulting in a total tracer recovery of 39% (2.4 t of the 6.2 t injected). Considering the geological and tectonic conditions, this recovery rate is unexpectedly high, suggesting a good hydraulic connection between the Siptenfelde brook and the mine.

After some rainfall events (Fig. 2) the conductivity in the Fluor shaft increased significantly after about 1 day (e.g. 11 June, 7.49 pm; 11 June, 10.28 pm; 3 July, 9.34 pm). Each peak is seen to start quickly and tail out slowly (see inset in Fig. 2). Based on a distance of 2250 m between the Siptenfelde seepage and the Fluor shaft, a mean effective velocity of $1.5\,m\,min^{-1}$ can be calculated for the meteoric and mine water flowing between the brook and the shaft's

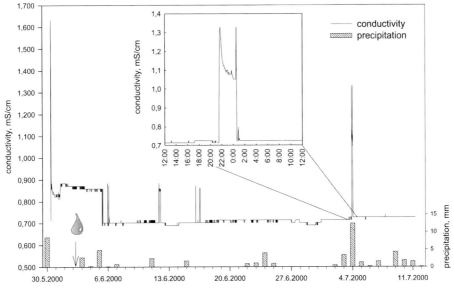

Fig. 2. Plot of precipitation (Siptenfelde station) and conductivity in the Fluor shaft. The arrow marks the time of the injection of the sodium chloride tracer into the Siptenfelde brook. Changes before 6 June are due to moving the conductivity probe upwards in the shaft by 5 m.

Table 5. *Mean effective velocity of mine water in the Straßberg mine. No tracer from No. 539 shaft could be detected anywhere*

From	To	Tracer	Velocity v_{eff} (m min^{-1})	Distance (m)
No. 530 shaft	Fluor shaft	Microspheres	0.1	1773
No. 530 shaft	Glasebach shaft	Microspheres	0.3	4798
Fluor shaft	Fluor shaft	Microspheres	0.2	238
Fluor shaft	Fluor shaft	Club moss spores	0.1	238
Fluor shaft	Glasebach shaft	Microspheres	0.3	3180
Fluor shaft	Glasebach shaft	Club moss spores	0.2–1.2	3180
Siptenfelde brook	Fluor shaft	NaCl–brine	1.5	2250

outflow (Table 5). None of these peaks were observed elsewhere, and reasons for these peaks other than the NaCl tracer can be excluded.

Club moss spores (Lycopodium clavatum). Club moss spores were only detected at the Fluor shaft and the Glasebach shaft (Fig. 3). A total of 323,220 spores in the Fluor shaft and 200,820 in the Glasebach shaft could be found after 8 June. Based on the ratio of the water pumped and the water flowing out of the three shafts, the recovery rate is as high as 6%.

Within the Flour shaft, the club moss spores peak reached 199,200 some 2.5 days after tracer injection, and in a relatively short time of 1.5 days decreased to nearly 4000 suggesting a low hydraulic dispersion. A second peak with 6500 spores occurred 6 days after tracer injection. From the injection point to the surface of the water, the spores have to flow 238 m, thus the mean effective velocity calculates to 0.1–0.2 m min^{-1} (Table 5).

It is possible that some of the filter nets used may have been contaminated from storage after an earlier tracer test. Therefore, the results cannot be interpreted easily and it is not clear whether the 1000–6000 spores are due to contamination or not. Nevertheless, 10.5 days after tracer injection is a clear peak with 25 000 spores and another one

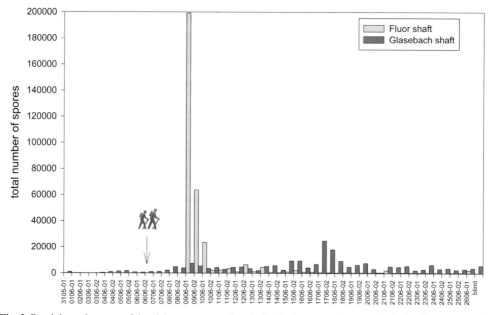

Fig. 3. Breakthrough curves of the club moss spores detected at the Fluor and Glasebach shafts. The arrow marks the time of tracer injection. The abscissa shows the dates of sampling. No spores were detected at the No. 539 shaft. Noticeable amounts of spores at the fluor shaft arrived 2.5 days, and at the Glasebach shaft 11 days, after tracer injection.

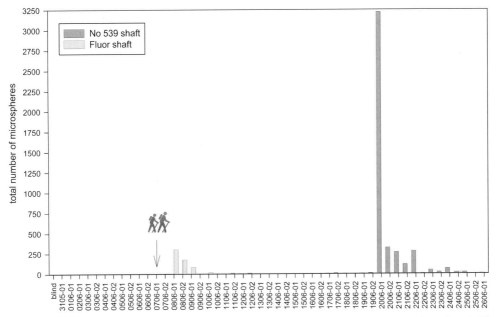

Fig. 4. Breakthrough curves of the microspheres detected at the Fluor shaft. (max.: 3219 microspheres). The abscissa shows dates of sampling. The arrow marks the time of tracer injection.

with 7400 spores 3 days after tracer injection. Once again, the maximum is reached very quickly, with a decline to background within 2 days. Between the injection point in the Fluor shaft and the detection point in the Glasebach shaft, the tracer had to flow 3180 m at the shortest pathway. Taking into consideration the two peaks and the shortest flow distance, the mean effective velocity is $0.2-1.2\,\mathrm{m\,min^{-1}}$ (Table 5).

Microspheres. From the microspheres injected in the No. 539, No. 530, Fluor and Glasebach shafts, only the microspheres from the No. 530 and Fluor shafts could be detected. It cannot be excluded that the LydiAs (*Lycopodium Apparatus*) lowered into the No. 539 and Glasebach shafts did not open properly.

In the Fluor shaft, microspheres from the Fluor shaft and the No. 530 shaft could be detected (Fig. 4). One day after the tracer injection 220 microspheres from the deep part of the Fluor shaft could be detected in the outflow from the shaft. A sharp peak was observed that tailed out within 1.5 days. The other peaks of microspheres from the Fluor shaft are negligible, but 13 days after tracer injection 3219 microspheres from the No. 530 shaft reached the sampling point at the Fluor shaft. A significant tracer signal could still be observed 2.5 and 4 days later. Based on the shortest distances of 238 and 1773 m from the injection to the sampling point, the mean effective velocities are $0.1-0.2\,\mathrm{m\,min^{-1}}$.

Only microspheres from the No. 530 shaft could be detected at the Glasebach shaft (Fig. 5). All the other microspheres, including those injected into the Glasebach shaft itself, were not sufficiently abundant to draw useful conclusions. Some 13 days after tracer injection, 9748 microspheres from the No. 530 shaft arrived at the sampling point at the Glasebach shaft. As already observed in the Fluor shaft, the peak tails out slowly and even 3 days later a significant amount of microspheres could still be detected. As the distance between the No. 530 and Glasebach shafts is 4798 m, a mean effective velocity of $0.3\,\mathrm{m\,min^{-1}}$ is indicated.

Table 5 summarizes the results of all successful tracer detections during the Straßberg tracer test. From the results obtained, the velocities are consistently around $\mathrm{dm\,min^{-1}}$, and it can be concluded that similar hydrodynamic conditions exist in the mine, independent of the location.

Straßberg mine: brief conclusions

The general flow direction throughout the tracer test was from north to south, as tracers injected were never found north of their injection point. All parts of the mine are hydraulically well connected, which explains the similar chemical

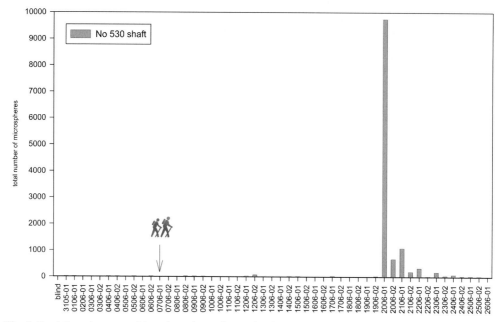

Fig. 5. Breakthrough curves of the microspheres detected at the Glasebach shaft (max.: 9748 microspheres). The abscissa shows dates of sampling. The arrow marks the time of tracer injection.

composition of the mine water in the Fluor and Glasebach shafts. Previous to the tracer test, it was believed that the general flow direction of the mine water was from south to north. Under the current flow regime, with the 3-adit system working, previously observed stratification cannot be re-established.

Furthermore, the sodium chloride tracer confirmed the assumption that there is a connection between the Siptenfelde seepage and the mine. The breakthrough curves clearly show that the hydraulic dispersion within the flow path through the partly unsaturated fissured aquifer and the drifts and shafts is small, and that the tracer is transported after rainfall events only. Because more than one third (39%) of the injected sodium chloride tracer was recovered within the six weeks of the tracer test, it must be assumed that there is a good connection between the Siptenfelde seepage and the northern adit. Finally, the tracer test's results confirm that the bulkhead at the northern adit is hydraulically inactive.

Outlook

Tracer tests in partly or fully flooded underground mines are rarely conducted and even less often published. Therefore, the experiences gained in the scientific world are limited until tracer tests in flooded mines become more widespread.

A critical prerequisite for a successful mine water tracer test is a conceptual model of the test site and extensive hydrogeological investigation of the mine and its surroundings, including the geological setting because tracers that are injected close to the surface have to flow through the unsaturated zone. Tracer velocities depend on the hydraulic head and the hydraulic properties of the rock. The mine geometry should be known reasonably accurately from mine plans, although former mine workers may have to be consulted to assist setting up the injection and sampling plan. In many cases, geochemical investigations of the mine water can help to understand the flow regime and trace elements might be useful to support the outcome of the tracer tests.

From the literature review, water velocities range between 0.001 and 11.1 m min^{-1} (Table 6), with a nearly log-normal distribution (24 tracer tests). Interestingly, 99% of all the mean measured velocities are between 0.2 and 1.5 m min^{-1}. Until now, the reasons for that have not been fully understood, but ongoing investigations will provide more details within the next few years.

Tracer tests in mines would benefit from the development of new innovative tracer techniques. Future work, therefore, has to focus on:

- tracer injection into predefined depths;
- measuring of the exact injection time;

Table 6. *Reported distances and mean velocities of worldwide tracer tests in underground mines. The table is given for comparison only, details concerning geological setting and hydraulic parameters are given in the literature cited. Mean of all 29 tracer tests: 0.3–1.7 m min^{-1} (95% confidence interval, excluding the maximum and minimum value)*

Distance (km)	Mean velocity (m min^{-1})	Author
0.2	0.001	Aljoe & Hawkins 1993*
0.044	>0.004	Aljoe & Hawkins 1993
0.077	0.01	Canty & Everett 1999
0.13	0.01	Aljoe & Hawkins 1994
0.780	0.01	Wolkersdorfer 1996
0.35	0.1	Mather et al. 1969
0.077	0.12	Canty & Everett 1999
0.077	0.14	Canty & Everett 1999
0.171	0.17	Canty & Everett 1999
0.283	0.1–0.2	This study
1.773	0.1–0.2	This study
1.7	0.1–0.3	Parsons & Hunter 1972
0.229	0.23	Canty & Everett 1999
3.180	0.2–1.2	This study
3.6	0.3	Aldous & Smart 1987
4.798	0.3	This study
0.15	0.4	Mather et al. 1969
0.172	0.4	Wolkersdorfer et al. 1997
0.216	0.5	Wolkersdorfer et al. 1997
0.220	0.5	Wolkersdorfer et al. 1997
0.2	0.6	Mather et al. 1969
0.5	1.3	Aldous & Smart 1987
2.250	1.5	This study
0.776	1.6	Wolkersdorfer et al. 1997
0.736	1.8	Wolkersdorfer et al. 1997
0.780	2.0	Wolkersdorfer et al. 1997
2.159	5.7	Wolkersdorfer et al. 1997
2.723	7.9	Wolkersdorfer et al. 1997
0.5	11.1	Aldous & Smart 1987

* Result probably wrong.

- tracer sampling and counting;
- new tracers, suitable for mine water.

I would like to express my gratitude to A. Hasche for the improvement of the LydiA-technique. H.-J. Kahmann, as well as K. Heinrich, of BST Mansfeld GmbH and the GVV for financial and personal support of the Straßberg tracer test. Also, I would like to thank the staff of the BST Mansfeld, especially the Köhn family, at the Straßberg mine, who spent a lot of time taking care of us. My colleagues at the TU Bergakademie Freiberg supported the work group whenever help was needed. This study had been financially supported by the BST Mansfeld, the DFG Graduate School 'Geowissenschaftliche und Geotechnische Umweltforschung' (speaker: Univ.-Prof. Dr B. Merkel) at the TU Bergakademie Freiberg and the EU Project PIRAMID (contract No. EVK1-CT-1999-00021). I would also like to thank P. Younger for his efforts to make me publish my results! This paper expresses the views held by the author and does not necessarily reflect those held by the BST Mansfeld, the GVV, or any authorities involved.

References

ABELIN, H. & BIRGERSSON, L. 1985. Migration experiments in the stripa mine, design and instrumentation. In: COME, B., JOHNSTON, P. & MÜLLER, A., (eds) *Joint CEC/NEA Workshop on Design and Instrumentation of in situ Experiments in Underground Laboratories for Radioactive Waste Disposal*. Balkema, Rotterdam, **CONF 8405165**, 180–190.

ADELMAN, F.L., BACIGALUPI, C.M. & MOMYER, F.E. 1960. *Final Report on the Pinot Experiment (Colorado)*; Lawrence Radiation Lab Report, Livermore **UCRL-6274**, California University.

ALDOUS, P.J. & SMART, P.L. 1987. Tracing groundwater movement in abandoned coal mined aquifers using fluorescent dyes. *Ground Water, Urbana*, **26**, 172–178.

ALJOE, W.W. & HAWKINS, J.W. 1993. *Neutralization of Acidic Discharges from Abandoned Underground Coal Mines by Alkaline Injection*; Report of Investigations, US Bureau of Mines, Washington, **9468**.

ALJOE, W.W. & HAWKINS, J.W. 1994. Application of aquifer testing in surface and underground coal mines. In: *Proceedings of the fifth International Mine Water Congress*, Nottingham, UK, IMWA **1**, 3–21.

ANDERSON, B.C. 1987. *Use of Temperature Logs to Determine Intra-well-bore flows, Yucca Mountain, Nevada*. Master's Thesis, Colorado School of Mines, Golden, 59.

AQUILINA, L., ROSE, P., VAUTE, L., BRACH, M., GENTIER, S., JEANNOT, R., JACQUOT, E., AUDIGANE, P., TRAN-VIET, T., JUNG, R., BAUMGÄRTNER, J., BARIA, R. & GÉRARD, A. 1998. A tracer test at the Soultzt-Sous-Forets hot dry rock geothermal site. In: *Proceedings of the 23rd Workshop on Geothermal Reservoir Engineering*; Stanford **SGP-TR-158**, 1–8.

BANKS, D., YOUNGER, P.L., ARNESEN, R.T., IVERSEN, E.R. & BANKS, S.B. 1997. Mine-water chemistry: the good, the bad and the ugly. *Environmental Geology*, **32**, 157–174.

BARTELS, C. & LORENZ, E. 1993. Die Grube Glasebach–ein Denkmal des Erz- und Fluoritbergbaus im Ostharz. *Der Anschnitt, Bochum*, **45**, 144–158.

BIRGERSSON, L., WIDEN, H., AAGREN, T. & NERETNIEKS, I. 1992. *Tracer Migration Experiments in the Stripa Mine 1980–1991*. Swedish Nuclear Fuel and Waste Management Co. Stockholm **STRIPA-TR-92-25**.

BRETHERTON, B. 1989. *Near-surface Acid Mine Water Pools and Their Implications for Mine Abandonment, Coeur d'Alene Mining District, Idaho* Master's Thesis, University of Idaho, Moscow.

BREWITZ, W., KULL, H. & TEBBE, W. 1985. In situ experiments for the determination of macro-permeability and RN-migration in faulted rock formations such as the oolithic iron ore in the Konrad Mine Design and instrumentation in underground laboratories for radioactive waste disposal. In: COME, B. JOHNSTON, P., & MULLER, A. (eds) *Design and instrumentation in underground laborataries for radioactive waste disposal*, Balkema, Rotterdam, 289–302.

CACAS, M.C., LEDOUX, E., DE MARSILY, G., TILLIE, B., BARBREAU, A., DURAND, E., FEUGA, B. & PEAUDECERF, P. 1990. Modeling fracture flow with a stochastic discrete fracture network: calibration and validation. 1. The flow model. *Water Resources Research*, **26**, 479–489.

CANTY, G.A. & EVERETT, J.W. 1998. Using tracers to understand the hydrology of an abandoned underground coal mine. In: *Proceedings of the Annual National Meeting*. American Society for Surface Mining and Reclamation, Princeton **15**, 62–72.

DAVIS, M.W. 1994a. Identification and Remediation of a Mine Flooding Problem in Rico, Dolores County, Colorado with a Discussion on the Use of Tracer Dyes, *Colorado Geological Survey Open-File Report, Denver* **91–1**, 15.

DAVIS, M.W. 1994b. The Use of Tracer Dyes for the Identification of a Mine Flooding Problem, Rico, Dolores County, Colorado. *Colorado Geological Survey Open File Report, Denver* **91–2**, 1–20.

DIAZ, W.D. 1990. *Hydrogeochemistry and Physical Hydrogeology of the Newfoundland Zinc Mine, Daniel's Harbour, Newfoundland*. Master's Thesis, St. John's, Newfoundland.

DOORNBOS, M.H. 1989. *Hydrogeologic Controls Upon the Water Quality of the Acid Mine Drainage in the Sand Coulee–Stochett Area, Montana*. Master's Thesis, Montana College of Mineral Science & Technology, Butte.

FERNÁNDEZ-RUBIO, R., FERNÁNDEZ-LORCA, S. & ESTEBAN ARLEGUI, J. 1987. Preventive techniques for controlling acid water in underground mines by flooding. *International Journal of Mine Water, Budapest*, **6** (3), 39–52.

FRIED, J.J. 1972. Miscible pollutions of groundwater – a study in methodology. In: BISWAS, A.K. (ed.) *Modelling of Water Resources Systems*. Harvest House, Montreal **2**, 520–529.

GALLOWAY, D.L. & ERICKSON, J.R. 1985. Tracer test for evaluating nonpumping intraborehole flow in fractured media. *Transactions of the American Nuclear Society*, **50**, 192–193.

GATZWEILER, R. & MEYER, J. 2000. Umweltverträgliche Stilllegung und Verwahrung von Uranerzbergwerken–Fallbeispiel WISMUT. In: *Proceedings Wismut 2000 Bergbäusanierung Internationale Konferenz, 11–14.7.2000, Schlema*; Wismat Guibit Chemmitz 1–16.

GOLDBRUNNER, J.E., RAMSPACHER, P., ZOJER, H., ZÖTL, J.G., MOSER, H., RAUERT, W. & STICHLER, W. 1982. Die Anwendung natürlicher und künstlicher Tracer in einem hochalpinen Magnesitbergbau. *Beiträge zur Geologie der Schweiz-Hydrologic, Bern*, **28**, 407–422.

GULATI, M.S., LIPMAN, S.C. & STROBEL, C.J. 1978. Tritium tracer survey at the geysers. *Geothermal Resource Council Annual Meeting: Geothermal Energy–A Novelty Becomes Resource, Hilo, Hawaii, 25–27 July 1978*. Geothermal Resource Council, Davis, CA, **2**, 237–239.

HABIČ, P. & GOSPODARIČ, R. 1976. *Underground Water Tracing–Investigations in Slovenia 1972–1975*. Institute Karst Research, Ljubljana.

HAMILTON, Q.U.I., LAMB, H.M., HALLET, C. & PROCTOR, J.A. 1997. Passive treatment systems for the remediation of acid mine drainage in Wheal Jane, Cornwall, UK. In: YOUNGER, P.L. (eds) *Mine Water Treatment Using Wetlands*. Chartered Institution of Water and Environmental Management, London, 33–56.

HIMMELSBACH, T. & WENDLAND, E. 1999. Schwermetalltransport in Sandsteinen unter Bedingungen einer hochsalinaren Porenwasserlösung–Laborversuch und Modellierung. *Grundwasser, Berlin*, **4**, 103–112.

HIMMELSBACH, T., HÖTZL, H., KÄSS, W., LEIBUNDGUT, C., MALOSZEWSKI, P., MEYER, T., MOSER, H., RAJNER, V., RANK, D., STICHLER, W., TRIMBORN, P. & VEULLIET, E. 1992. Transport phenomena in different aquifers (Investigations 1987–1992) – fractured rock test site Lindau/Southern Black Forest (Germany). *Steirische Beiträge zur Hydrogeologie, Graz*, **43**, 159–228.

HORN, P., HÖLZL, S. & NINDEL, K. 1995. Uranogenic and thorogenic lead isotopes (^{206}Pb, ^{207}Pb, ^{208}Pb) as tracers for mixing of waters in the Königstein uranium mine, Germany. In: *Proceedings of Uranium Mining and Hydrogeology, Freiberg, Germany*. GeoCongress, Freiberg **1**, 281–289.

HORNE, R.N., JOHNS, R.A., ADAMS, M.C., MOORE, J.N. & STIGER, S.G. 1987. The use of tracers to analyze the effects of reinjection into fractured geothermal reservoirs. In: *Proceedings of the fifth Geothermal Program Review, Washington*. Stanford University, CA 37–52.

HUNT, D.J. & REDDISH, D.J., 1997. The stability implications of groundwater recharge upon shallow, abandoned mineworkings in the UK. *Engineering Geology and the Environment*. In: MARINOS, P.G., KOUKIS, G.C., TSIAMBAOS, G.C. & STOURUARAS, G.C. (eds) Balkema, Rotterdam, **3**, 2425–2430.

IAEA, JESTER, W.A. & RAUPACH, D.C. 1987. Survey of applications of non-radioactive but neutron activatable groundwater tracers. In: *Isotope Techniques in Water Resources Development. International Symposium on the Use of Isotope Techniques in Water Resources Development, Wien*, IAEA 623–633.

KÄSS, W. 1998. *Tracing Technique in Geohydrology*. Balkema, Rotterdam.

KIMBALL, B.A., RUNKEL, R.L., BENCALA, K.E. & WALTON-DAY, K. 1999. Use of tracer Injections and Synoptic Sampling to Measure Metal Loading From Acid Mine Drainage, *Water Resources Investigations Report, Denver* **WRI 99-4018A (1)**, 31–36.

KINDERMANN, L. 1998. Kontrolle geochemischer Parameter beim Wiedereinbau von Reststoffen in ein stillgelegtes Bergwerk. *Wissenschaftliche Mitteilungen–Sonderheft, Freiberg*, **7**, 196–201.

KIRSCHNER, F.E. 1989. *Ground-surface Delineation of Fractures That Appear to be Connected to Underlying Mined-out Openings Using a Naturally Occurring Gaseous Tracer.* University of Idaho, Moscow.

KIRSHNER, F.E. 1991. *Hydrogeological Evaluation of an in-situ Leach Cell in Abandoned Mine and Mill Waste at the Kellogg, Idaho Superfund Site Using Tracers.* University of Idaho, Moscow.

KIRSHNER, F.E. & WILLIAMS, R.E. 1993. Hydrogeological evaluation of an in-situ leach cell in abandoned mine and mill waste at the Kellogg, Idaho Superfund site using tracers. In: *International Biohydrometallurgy Symposium, Jackson Hole*, The Minerals, Metals and Materials Society, Warrendale 77–87.

KLOTZ, D. & OLIV, F. 1982. Verhalten der Radionuklide ^{125}I, ^{85}Sr, ^{134}Cs und ^{144}Ce in drei typischen Sanden Norddeutschlands bei wasserungesättigtem Fließen. *GSF-Bericht, Neuherberg*, **R290**, 395–401.

KUSCHKA, E. & FRANZKE, H.J. 1974. Zur Kenntnis der Hydrothermalite des Harzes. *Zeitschrift für geologische Wissenschaften, Berlin*, **2**, 1417–1436.

KUYUMCU, M. & HARTWIG, H.-J. 1998. Aufgaben der GVV–Gesellschaft zur Verwahrung und Verwertung von stillgelegten Bergwerksbetrieben mbH. *Bergbau, Gelsenkirchen*, **49**, 18–22.

KWAKWA, K.A. 1989. Tracer results. In: PARKER, R.H. (eds) *Hot Dry rock Geothermal Energy; Phase 2B Final Report of the Camborne School of Mines Project.* Pergamon Press, Oxford, 1037–1099.

LACHMAR, T.E. 1994. Application of fracture-flow hydrogeology to acid-mine drainage at the Bunker Hill Mine, Kellogg, Idaho. *Journal of Hydrology*, **155**, 125–149.

LEE, W.L. 1984. Performance assessment of nuclear waste isolation systems Shaping our energy future. In: *Proceedings of the Eleventh Annual WATTec Conference and Exposition, Knoxville*, LTM Consultants 59.

LEWIS, R.A. 1990. *Borate Adsorption by Some Volcanic Tuffs; Evaluation of Borate as a Potential Tracer for Unsaturated Flow at Yucca Mountain Borate Adsorption by Some Volcanic Tuffs, Evaluation of Borate as a Potential Tracer for Unsaturated Flow at Yucca Mountain* PhD Thesis, Colorado School of Mines, Golden.

LEWIS, R.L., MCGOWAN, I.R., HERSHMAN, J.I. & TILK, J. 1997. Numerical simulation of bulkheads to reduce uncontrolled discharge from an underground copper mine. *Mining Engineering*, **49**, 68–72.

LIEDL, R. & SAUTER, M. 2000. Charakterisierung von Karstgrundwasserleitern durch Simulation der Aquifergenese und des Wärmetransports. *Grundwasser, Heidelberg*, **5**, 9–16.

LORENZ, P.B. 1973. *Radioactive Tracer Pulse Method of Evaluating Fracturing of Underground Oil Shale Formations; Bureau of Mines Report of Investigations: Washington*, **7791**, 33.

MATHER, J.D., GRAY, D.A. & JENKINS, D.G. 1969. The use of tracers to investigate the relationship between mining subsidence and groundwater occurrence at Aberfan, South Wales. *Journal of Hydrology*, **9**, 136–154.

MILLER, N.C. & SCHMUCK, C.H. *Use of a Tracer For in situ Stope Leaching Solution Containment Research; Bureau of Mines Report of Investigations, Spokane* **RI-9583**, 43.

MOHR, K. 1993. *Geologie und Minerallagerstätten des Harzes*, Schweizerbart, Stuttgart.

NIEHREN, S. & KINZELBACH, W. 1998. Artificial colloid tracer tests–development of a compact on-line microsphere counter and application to soil column experiments. *Journal of Contaminant Hydrology*, **35**, 249–259.

NORDSTROM, D.K., ALPERS, C.N., PTACEK, C.J. & BLOWES, D.W. 2000. Negative pH and extremly acidic Mine waters from Iron Mountain, California. *Environmental Sciences and Technology*, **34**, 254–258.

PARKHURST, D.L. 1988. *Mine-water Discharge, Metal Loading, and Chemical Reactions*, US Geological Survey, Reston. Open File Report, **OF 86-0481**, D5–D9.

PARSONS, A.S. & HUNTER, M.D. 1972. Investigation into the movement of groundwater from Bryn Pit above Ebbw Vale, Monmouthshire. *Water Pollution Control*, **71**, 568–572.

PUJOL, L. & SANCHEZ CABEZA, J.A. 2000. Use of tritium to predict soluble pollutants transport in Ebro River waters (Spain). *Environmental Pollution*, **108**, 257–269.

RANDALL, M.M., NICHOLLS, J., WILLIS-RICHARDS, J. & LANYON, G.W. 1990. Evaluation of jointing in the carnmenellis granite. In: BARIA, R. (ed) *Hot Dry Rock: Geothermal Energy.* Robertson Scientific Publications, Camborne, 108–123.

REZNIK, Y.M. 1990. Determination of groundwater paths using method of streaming potentials. In: GRAVES, D.H. (ed) *National Symposium on Mining: A New Beginning. Proceedings of the National Symposium on Mining.* University of Kensington, Lexington, 217–221.

RÜTERKAMP, P. & MEßER, J. 2000. *Untersuchungen zur hydraulischen und hydrochemischen Situation in den drei Teilrevieren der gefluteten Flussspatgrube Straßberg: 46.* Unpublished report, Deutsche Moutan-technologie (DMT).

SAWADA, A., UCHIDA, M., SHIMO, M., YAMAMOTO, H., TAKAHARA, H. & DOE, T.W. 2000. Non-sorbing tracer migration experiments in fractured rock at the Kamaishi Mine, Northeast Japan. *Engineering Geology*, **56**, 75–96.

SCOTT, R.L. & HAYS, R.M. 1975. Inactive and abandoned underground mines–water pollution prevention and control. In: MICHAEL & BAKER JR. Inc. (eds) US Environmental Protection Agency, Washington, DC **EPA-440/9-75-007**.

SHEIBACH, R.B., WILLIAMS, R.E. & GENES, B.R. 1982. Controlling acid mine drainage from the Picher Mining District, Oklahoma, United States. *International Journal Mine Water*, **1**, 31–44.

SKOWRONEK, E. & ZMIJ, M. 1977. Okreslenie pochodzenia wody wyplywajacej zza obmurza szybu znakowaniem rodamina B. [Determination of the origin of water flowing out from behind a shaft lining traced with Rhodamine B]. *Przegl Gorn, Gliwice, Poland*, **33**, 13–19.

WIRSING, G. 1995. Hydrogeologische Untersuchungen im Grubengebäude des Schauinslandbergwerks– Beurteilung des Grundwasservorkommens und Abgrenzung eines Wasserschutzgebietes für die Quelle 14 der Freiburger Energie- und Wasserversorgungs-AG: 45. Unpublished report Geologisches Laudesaut Buden-Württemberg.

WITTRUP, M.B., KYSER, T.K. & DANYLUK, T. 1986. *The Use of Stable Isotopes to Determine the Source of Brine in Saskatchewan Potash Mines*, Saskatchewan Geological Society, Regina, Special Publication, **8**, 159–165.

WOLKERSDORFER, Ch. 1996. Hydrogeochemische Verhältnisse im Flutungswasser eines Uranbergwerks–Die Lagerstätte Niederschlema/Alberoda. *Clausthaler Geowissenschaftliche Dissertationen, Clausthal*, **50**, 1–216.

WOLKERSDORFER, Ch. & HASCHE, A. 2001. Tracer test in the abandoned fluorspar mine Straßberg/ Harz Mountains, Germany. *Wissenschaftliche Mitteilungen, Freiberg*, **16**, 57–67.

WOLKERSDORFER, Ch., TREBUŠAK, I. & FELDTNER, N. 1997. Development of a tracer test in a flooded uranium mine using *Lycopodium clavatum*. *Tracer Hydrology, 97*. In: KRANJC, A. (ed.) Balkema, Rotterdam, **7**, 377–385.

WU, Q., JIN, Y., TIAN, B., LI, D. & XIA, Y. 1992. An application of artificial tracers in determining vertical groundwater recharge locations in a coal mine. In: *International Symposium on Water Tracing (SUWT), Balkema, Rotterdam*, **6**, 365–367.

YOUNGER, P.L. 2000. The adoption and adaptation of passive treatment technologies for mine waters in the United Kingdom. *Mine Water and the Environment*, **19** (2), 84–97.

YOUNGER, P.L. & ADAMS, R. 1999. *Predicting Mine Water Rebound*. Environment Agency, Bristol.

The monitoring and modelling of mine water recovery in UK coalfields

KEITH R. WHITWORTH

IMC Consulting Engineers, PO Box 18, Huthwaite, Sutton-in-Ashfield, Nottinghamshire, NG17 2NS, UK (e-mail: whitwork@imcgroup.co.uk)

Abstract: This paper draws together the information that has been obtained on mine water recovery since the large-scale closure of coal mines in the 1980s and 1990s. The data show that, following cessation of pumping, mine water recovery follows an exponential curve similar to the recovery of an aquifer following a pumping test. Several previously unpublished examples of mine water recovery data from around the UK are included in the paper and there is a detailed assessment of mine water recovery in the East Fife Coalfield in Scotland. The reasons for this type of mine water recovery are discussed and examples are given of the use of the data for both the interpretation and modelling of mine water recovery. In coal mining areas where no water-level recovery data are available, methods for the prediction of mine water inflow and recovery modelling are proposed and the problems associated with mine water recovery modelling are discussed.

The paper concludes that modelling of mine water recovery, based on mine water inflow and estimated void space, can be used to give reasonably accurate predictions of recovery times and flows, but that water level monitoring is essential for precise predictions.

The control of mine water during the period when coal mining was a nationalized industry was generally based on a safety first principle. This meant that when doubt existed about underground connections between modern mines and old abandoned areas of workings, mine water pumping always continued in the old areas. The result of this policy was a general lack of experience of mine water recovery and continuing doubt about underground connections.

The large-scale closure of mines in the 1980s and 1990s mean that in many cases whole coalfields were abandoned and that the pumping of mine water either completely stopped or was greatly reduced. Estimations of mine water recovery made by British Coal at the time of these closures were generally based on a water inflow related to the volume of water pumped from a mine and a residual void-space calculation. The void-space was calculated using roadway dimensions for supported excavations and a figure of 10% of the original extractions thickness for unsupported (total extraction) workings (National Coal Board 1972). Using this principle it was assumed that mine water recovery would proceed as a series of steps, with very little recovery when water was 'filling' a large void, followed by a period of more brisk recovery until the next large void was reached.

The monitoring of mine water recovery by IMC Consulting Engineers on behalf of the Coal Authority (the government agency set up to look after the non-privatized areas of coal mining) has shown that, at least at large scales, mine water recovery follows precise exponential curves that appear to be independent of the distribution of mining voids. These curves are very similar to the recovery curves observed following an aquifer pumping test.

Monitoring of mine water recovery

Since the coal mine closures of the 1980s and 1990s, monitoring of mine water recovery in several abandoned coalfields has shown that, in general, recovery follows an exponential curve with the rate of recovery reducing with time. Figure 1 shows the recovery curves for several mining units in the UK. The units vary from a single small mine (e.g. Whittle in Northumberland; Adams & Younger 2001) to large interconnected areas or whole coalfields (e.g. Sherwood 1997; Sherwood & Younger 1997; Burke & Younger 2000; Robins *et al.* 2002). The data suggest that water inflow rates are probably related to the difference in head between the water in the mine and the source of aquifers and the area that has been dewatered (cf. Banks 2001). It also implies a general interconnection of the mine workings within a block of ground that is governed by the hydrogeology of both the natural *in situ* strata and the changes to that

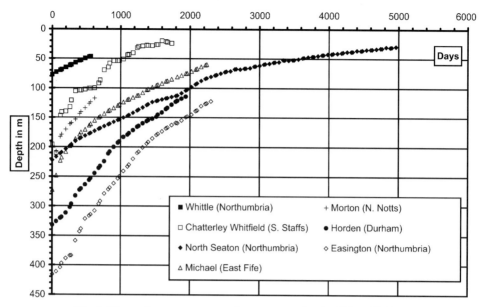

Fig. 1. Mine water recovery curves from a range of major UK coalfields.

strata caused by mining (Sherwood 1997). This interconnection of workings in coal mines could be anticipated from the way both natural and mining-induced permeability gave rise to the rules that governed the way coal can be mined in the UK. The regulations state that the total extraction of coal is not allowed within 60 m of an aquifer (including unconsolidated material likely to flow when wet) or within 45 m of other water-filled mine workings (National Coal Board 1956). These rules were brought in following experience of serious water inflows into mines when working too close to other bodies of water. The results of these regulations was that the active working blocks of a mine were kept 'dry' to avoid sterilization of reserves until all the coal had been extracted. Only where these blocks were isolated from the rest of the mine would old workings be allowed to flood, and then only to a level where the water recovery would not result in a risk of flooding to deeper, active workings. Detailed monitoring of mine water levels during the recovery period did not generally occur; only the start and end of the recovery period would be known, and the recovery 'curve' was assumed to be related to the filling of the void spaces. The occurrence of several major inflows of water into mine workings during the 1970s and 1980s resulted in a renewed interest in water movement around longwall faces. The resulting research tried to link the mechanical effects of total extraction coal mining with the subsidence data recorded at the surface and water inflow monitored underground.

Figure 2 shows how the natural stresses in the ground are altered around a typical longwall panel (National Coal Board 1978). The strains resulting from these stresses cause the opening of fractures and bedding planes, which can significantly increase the vertical and horizontal permeability.

East Fife coalfield

The East Fife Coalfield, situated to the north of the Firth of Forth, has been monitored by the Coal Authority since mine water pumping was abandoned in 1995 and provides a good example of the general principles of mine water recovery (e.g. Sherwood 1997; Sherwood & Younger 1997; Younger et al. 1995; Nuttall et al. 2002). Mining in the shallow, older areas of workings has been abandoned for a number of years. Initially, mine water had been pumped for safety reasons, but latterly these areas were allowed to recover and overflow to the deeper mines of Michael and Frances. Frances and Michael had closed in the 1970s but were pumped to control water levels up until 1995 (Younger et al. 1995). Figure 3 shows the general layout of the coalfield and the current monitoring sites. Since the cessation of pumping, mine water recovery in the workings has followed the typical exponential trend with, as yet, only minor deviations from the trend. These deviations occur on an annual cycle

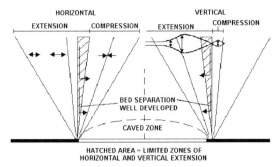

Fig. 2. Strain patterns around a typical longwall panel showing the zone of increased water flow and bed separation (after National Coal Board 1978).

and correlate with annual variations in rainfall. Figure 4 shows that as the mine water recovery reaches the level of each of the shallower blocks, water levels in these blocks take up the same recovery trend. The shallowest areas, monitored at Randolph, Muriespot and Dalginch, have yet to start recovery, although, as with the rest of the coalfield, a hydraulic gradient has developed (see Fig. 3). The hydraulic gradients currently developed in the East Fife Coalfield vary from about 1 in 200 (0.0005) to 1 in 600 (0.0016). Similar gradients of up to 1 in 1000 (0.001) are found in most monitored coalfields in the UK

The similar hydraulic gradients in different coalfields suggests that the various coal mines have similar hydraulic conductivities. The range of hydraulic gradients are probably related to the different types of mining and varying lithologies. The low hydraulic gradients will reflect open mine workings or fractured high-permeability strata, such as sandstone, adjacent to the mine workings. The higher hydraulic gradients will reflect workings that have closed or been backfilled and have only low-permeability strata, such as mudstones. adjacent to the workings.

The monitoring of water levels in East Fife is carried out using pressure transducers and data loggers recording a water level every 15 min. This allows very minor fluctuation in water level to be examined in detail. Figure 5 shows the tidal variations in water level monitoring at Frances. These oscillations take some 4 h to travel 2.5 km through the mine workings to Lochhead. The amplitude of the fluctuations of up to 0.4 m are caused by compression of the Coal Measures strata due to weight of water at high tide reducing the storativity of the aquifer by forcing water out of joints. The largest fluctuations of mine water due to tidal compression have been seen at Bates Colliery, Blyth where, during spring tides, fluctuations of up to 2.5 m have been recorded. The varying amplitudes of the fluctuations probably reflect the depth and extent of the under sea workings in an area.

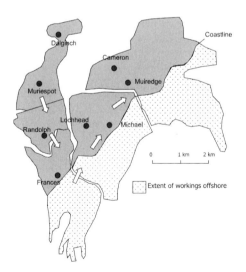

Fig. 3. General layout of the East Fife Coalfield (adapted after Sherwood 1997) showing major bodies of mine workings (shaded grey onshore, stippled where under sea) and positions of mine shafts and boreholes used for water level monitoring purposes, with arrows showing directions of water movement determined from hydraulic gradients and mine plan evidence.

Monitoring and mining connections

Monitoring of mine water recovery, as well as showing interconnection of workings, can be used to confirm the separation of mining units. A mining unit may comprise a single colliery or a group of interconnected collieries isolated from adjacent workings by Coal Measures, by an area where no mining has been carried out or by artificial barriers such as dams.

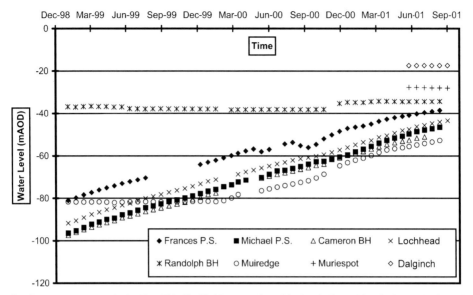

Fig. 4. Mine water recovery in the East Fife Coalfield, as monitored in the shafts and boreholes marked on Fig. 3.

Dams. Where only single roadways connected major mining blocks, dams were sometimes constructed in roadways to separate the mining units and prevent water migration. These dams were generally designed to withstand full hydrostatic head, but were rarely tested because pumping in the abandoned area usually continued and only a limited water head was allowed to build up against the dam. The effectiveness of the dams was therefore only proved with the total closure of mining in an area, sometimes many years after construction. Monitoring of water levels on the rise side of the dams has shown that the British Coal policy of continued pumping was entirely justified, as in nearly all instances where water levels on either side of a dam has been monitored, and a high hydrostatic head has built up on one side of the dam. Figure 6 shows an example from Durham where the mine water recovery in an area isolated by a dam was greater than the rest of the block. The resultant head caused the dam to fail. Since then the water level

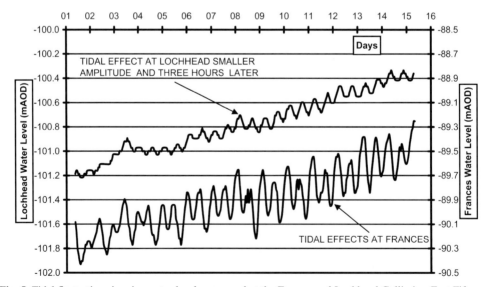

Fig. 5. Tidal fluctuations in mine water levels measured at the Frances and Lochhead Collieries, East Fife.

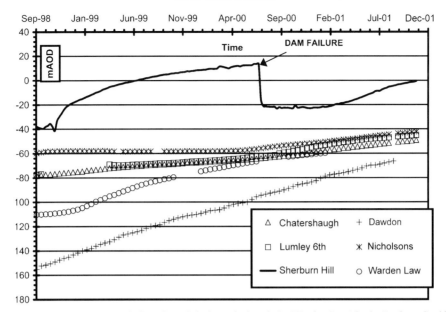

Fig. 6. Mine water recovery records for selected shafts and a borehole (Warden Law) in the Durham Coalfield to the east of the River Wear, showing the effect of failure of a dam on water levels in Sherburn Hill Colliery.

has fallen and harmonized with the recovery in the rest of the mining block.

Permeability of mining connections

The permeability of a mining connection varies greatly dependent on the type of opening, the lithology around the opening and the stresses acting on the opening. In general, mining connections have permeabilities up to several orders of magnitude greater than the permeabilities found in UK Coal Measures. Table 1 shows some typical Coal Measure permeability values obtained from drill stem testing (D.S.T.) in boreholes for the north east Leicestershire project by British Coal (1986). Based on personal observation of changes in flow it is believed that permeabilities in mine workings can revert to near natural permeability where vertical or horizontal stresses, combined with swelling mudstone lithologies, result in closure of a mine opening. In general, these 'closures' only occur in areas of total coal extraction and the edges of these areas usually retain a 'high' permeability. However, examples have been found where major underground roadways have become blocked with a large pressure difference across the blockage. The most striking example of this is the blockage of a 14 × 10 ft (4.26 × 3.05 m) roadway supported by steel arches at 1 m spacing, constructed between Barnsley Main and Monk Bretton Collieries at an elevation of −180 m OD (180 m below Ordnance datum). This roadway was driven specifically for the purpose of water drainage. Water levels were expected to recover at differing rates at Barnsley Main and Monk Bretton until the connection at −180 m OD was reached when the levels should have become harmonized. Figure 7 shows that as the water level at Barnsley Main neared −180 m OD the recovery decreased rapidly, as would be expected from an overflow at a major connection. However, within 6 months there was a sudden rapid recovery, then a return to the exponential curve precisely followed by the mine water recovery below the −180 m OD connection. The original monitoring borehole at Monk Bretton (Lundwood borehole) filled by British Coal in 1994 has been redrilled to confirm that mine water levels at Monk Bretton have not recovered. (This had always been suspected due to a flow of methane from the Monk Bretton mine.) The emission mine gases, either methane or oxygen-difficient air, stop when mine water levels rise above the insets or connections from the shafts to the workings. The new Lundwood borehole shows a static level of approximately −100 m OD, indicating an overflow to deeper workings. These water levels give a minimum of 80 m head difference across the roadway blockage between Barnsley Main and Monk Bretton. A recent borehole drilled by Alkane Energy at Monk Bretton has proved that the Barnsley Seam remains unflooded at a level

Table 1. Summary of DST results – North East Leicestershire Project

Strata	Total thickness tested (m)	Assessed permeability	Typical permeability ($m\ s^{-1}$)	(mD)	Maximum permeability (based on assumed uniform permeability) ($m\ s^{-1}$)	(mD)
Sherwood Sandstone Formation (Bunter)	22	High	10^{-5}	1000	2×10^{-5}	2200
Permian	22	Moderate	8×10^{-8}	8	9×10^{-8}	9
Middle Coal Measures above Deep Main (including upper sill)	232	Extremely low with occasional permeable joints	$<10^{-10}$	$<10^{-2}$	5×10^{-8}	5
Main Coal Seams (Deep Main, Parkgate and Blackshale) and adjacent strata	377	Extremely low with occasional low/permeability joints	$<10^{-10}$	$<10^{-2}$	2×10^{-9}	0.2
Lower Coal Measures (including lower sill below Blackshale)	243	Very low permeability	6×10^{-10}	0.06	10^{-9}	0.01
Namurian	50		4×10^{-10}	0.04		

below the connecting roadway level of −180 OD. Therefore, the Monk Bretton side of the roadway must be unsaturated and the head acting on the roadway is the difference between the level of the roadway (−180 m OD) and the current water level in Barnsley Main (+20 m OD), i.e. 200 m head of water.

Mine water recovery modelling

The modelling of mine water recovery is important for assessing the risks to aquifers from recovery of contaminated mine water, and for the prediction of the timing and the flow rate of potential surface discharges (e.g. Sherwood 1997; Younger & Adams 1999; Banks 2001; Adams & Younger 2001). It is also very useful in the assessment of coal mine methane (CMM) reserves.

Using mine water recovery curves

Modelling can simply take the form of forward projection of a monitored exponential mine water recovery curve (Younger & Adams 1999). This method has been used in the Northumberland and Durham Coalfields, as well as in East Fife, and has so far proved to be very accurate. Figure 8 shows an example from Easington Colliery in County Durham of a projection compared with the actual recovery curve. A key part of any projection is the source aquifer, and the maximum head of water in the aquifer prior to mining. This information is needed for the project to establish a theoretical maximum recovery level, which gives a more accurate recovery in terms of time.

Monitored mine water recovery curves can also be used to estimate the flow of water into a mine and the likely volume of any surface discharge after recovery. The mining void is calculated by digitizing the area of the mine workings and using recorded extraction thicknesses and roadway sizes. Areas of total extraction and areas of partial extraction are recorded separately to allow the amount of compaction to be varied for the calculation of the residual void. The residual void left in areas of total extraction was generally considered to be 10% of the extraction height (National Coal Board 1972). This figure was based on filling of areas of abandoned workings by known flows. Knowing the rate of recovery and the estimate void space, the volume of water needed to fill this space over a given period can be calculated. Figure 9 shows the calculated inflow for the East Fife Coalfield based on the monitored recovery, and the mine volume derived from the mine plans and recorded

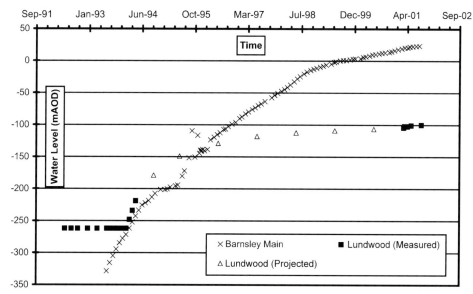

Fig. 7. Mine water recovery at Barnsley Main Colliery and the adjoining Lundwood workings, which were purposely connected by a major roadway at −180 m OD before abandonment. The lack of harmonization of the two recovery curves above this level demonstrates that the roadway has so thoroughly collapsed as to prevent flow from Barnsley Main to Lundwood.

extraction thickness. The graph shows that the water inflow has reduced exponentially, with a fairly rapid initial reduction followed by a more gradual decrease. The later inflow rates can be used to assess the potential surface outflow after recovery. In this case, the modelling was further refined by using the volume of mine water pumped (30 000 m^3 day^{-1}) as the initial inflow. Using this inflow and a calculated residual mine workings void (based on 10% of the total extraction) resulted in a recovery that was much quicker than actually occurred. Using the 30 000 m^3 day^{-1} inflow and matching the recorded recovery, the best-fit figures suggested that at

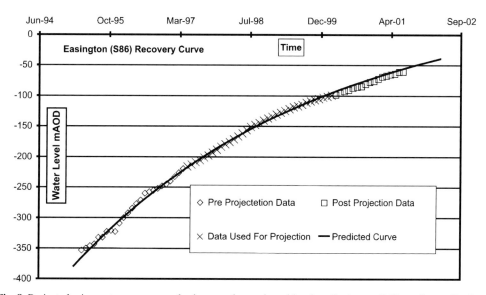

Fig. 8. Projected mine water recovery and subsequently monitored levels at Easington Colliery, County Durham.

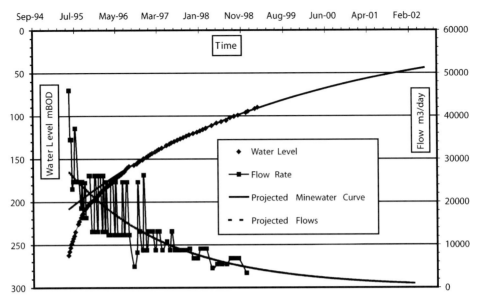

Fig. 9. Measured and modelled mine water recovery and derived values of calculated water inflow over time for the East Fife Coalfield.

300–350 m below OD, the residual void was in the order of 20% of the initial extraction volume. The residual void percentage will vary with lithology and depth, but the East Fife recovery would suggest the British Coal figure of 10% is at the low end of the range.

Monitoring of subsidence above areas of longwall abstraction would also suggest that the residual underground voids are greater than 10% of the original abstraction. At 100 m depth a 200 m wide longwall panel can result in maximum surface subsidence, which is 90% of the extraction thickness. However, at 500 m depth the maximum subsidence would only be 35% of the extraction thickness, with 65% of the extraction void being underground (National Coal Board 1975). The problem then becomes 'how much of this underground void is hydraulically connected with the mine workings'. Sudden hydraulic connection between the mine workings and bed separation (which form a substantial part of the void above the workings) is believed to be the mechanism controlling the large inflows of water noted both in the South Staffordshire and the Selby Coalfields (Whitworth 1982).

Mine water inflow data

Where there is no monitoring of mine water recovery, an alternative method of calculating water inflow is required. In this case the recorded inflow and pumping data from a mine can be used to interpret the steady state situation where inflow generally equals outflow. The various inflows are then fitted onto a conceptual model for each mine to assess how these inflows will vary during recovery. The water quality of the various inflows can also be assessed at this stage using either actual analysis data or water quality based on the depth in source aquifer (e.g. Adams & Younger 2002). The conceptual model for an individual mine or interconnected block is based on a general model, as shown in Fig. 10. The model is based on the general development of mining in the UK and assumes four basic depth controlled mining units, A–D, which may or may not be interconnected. Water inflows into all the units can then be put into three basic categories.

Shaft water. This water may originate from surface superficial deposits, Coal Measures aquifers or major aquifers, such as the Sherwood Sandstone or Magnesian Limestone. Shafts generally form the only major interconnection between a mine and a major aquifer The use of inclined roadways to access the Coal Measures through a major aquifer is rare due to the increased length of drivage in the aquifer and the related increase in costs to seal the roadway. Water qualities are usually good, as most of the aquifers are at relatively shallow depth and, in many cases, water inflows were collected and pumped to surface separately providing accurate data on these flows. During mine water recovery, shaft

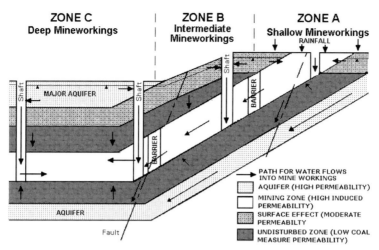

Fig. 10. Schematic diagram of a general conceptual model for water inflow to mine workings.

water inflows from major aquifers will only start to decrease when water levels in the shaft reach the level of the base of the aquifer supplying the water.

Coal measures inflows. Coal Measures inflows are generally small ($<330 \, m^3 \, day^{-1}$) and originate from minor aquifers. Larger flows may occur from major sandstones both above and below workings and from faulted ground. The rate of water inflow to a mine from these sources will decline once the water recovery in the mine has reached the level of inflow, and the head between the mine workings and the source aquifer starts to decrease.

Water quality from these sources is very variable. Chloride levels, for example, are known to decrease logarithmically with depth, the deeper waters can have chloride values in excess of $200\,000 \, mg \, l^{-1}$. (Glover & Chamberlain 1976).

Figure 11 shows a simple conceptual model of inflow to Cronton Colliery in Lancashire based on a small make ($72 \, m^3 \, day^{-1}$) from Type A shallow workings, and a moderate flow ($459 \, m^3 \, day^{-1}$) from Coal Measures aquifers.

Shallow workings water. This water will only affect the older shallow Type A workings but can gravitate to the deeper workings via mining connections. The exact course of water may not be known, but inflows are closely linked to rainfall and will have a more dominant effect on water levels in the later stages of recovery when other inflows have decreased. Figure 12 shows an example of shallow mine workings recovery in Yorkshire with annual fluctuations in water level related to variations in rainfall. The quality of shallow workings water can be very variable, dependent on source, the pathway and the length of the pathway travelled by water. Shallow workings water will include shaft water, Coal Measures aquifer water and surface water.

Modelling recovery using mine water inflow data

The simplest method of calculating mine water recovery time is to assume no reduction in water inflows as the water level in the mine recovers (see Sherwood 1997; Banks 2001, for discussions of this point). This approach yields a minimum recovery period for a mine. Any seasonal variations in water inflow to a mine can be addressed by using an average of the long-term pumping rates for the mine. This method of modelling will only be accurate when all the water entering a mine originates at very shallow depth. Where significant volumes of water enter the mine at depth, either from the shafts or from the Coal Measures, a simple linear recovery is likely to be highly inaccurate. In these cases a reduction in water flow is required to reflect the gradual reduction in head difference between the source aquifers and the water level in mine workings.

Data on the piezometric head in the source aquifers may be readily available in the case of the major aquifers, but in many cases there are no data and so have to be assumed. This can be carried out by looking at the topography and geology of the mining area under consideration

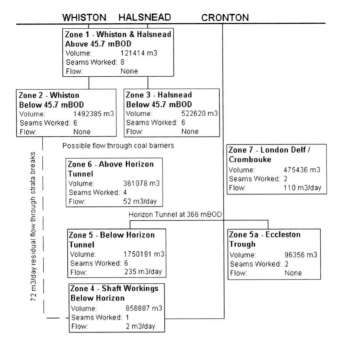

Fig. 11. Simplified conceptual model of mine water inflow to Cronton Colliery.

and estimating pre-mining water levels from springs, water courses and aquifer outcrops. Once the maximum head of the inflow source has been established, several methods of reducing the water inflow between the mine inflow level and the maximum head level can be applied. A straight linear reduction in inflow could be used but the evidence from East Fife suggests that an

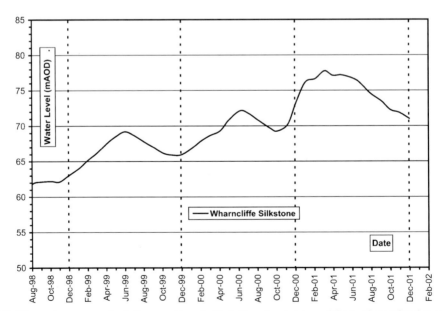

Fig. 12. Mine water recovery in shallow workings in Yorkshire, showing annual fluctuations related to seasonal variations in rainfall.

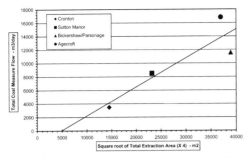

Fig. 13. The relationship between total water inflow from the Coal Measures strata and the area underlain by workings for four mining units in South Lancashire.

exponential reduction in flow, possibly related to the area of the drainage, as in a pumping test, is more appropriate.

Area-related flow model

A simple method for predicting the reduction of the water inflow during recovery was developed by relating the inflow to the area of workings. Plotting of the Coal Measures water inflows against the approximate boundary length of the total extraction area for four mining blocks in South Lancashire shows a general linear relationship (Fig. 13). The boundary is based on the square root of the area of total extraction multiplied by 4, i.e. the working area is assumed to be square. As the mine water recovers the inflows to the workings that originate from the Coal Measures are decreased in proportion to the area or boundary length of workings still to be flooded.

Average permeability model

This method of predicting water inflow reduction is based on an average permeability calculated for a theoretical zone of interaction 30 m above and below the boundary of the total extraction area of workings. The dip of the Coal Measures strata in each mining block is assumed to be constant, therefore, using Darcy's flow equation, the head difference divided by the length of flow path will also be a constant (Darcy's law):

$$V = \frac{K(h_1 - h_2)}{L} \quad (1)$$

where K is the Hydraulic conductivity; h_1 is the maximum hydraulic head (in the source aquifer); h_2 is the minimum hydraulic head (in the mine workings); L is the length of the flow path; and V is the velocity or specific discharge.

Using the total Coal Measures water flow into the mine or mining block and the area of interaction around the total abstraction mining, an average permeability of the Coal Measures strata can be calculated. The Coal Measures flow entering a mine is equal to the flow velocity times the cross-sectional area of interaction:

$$Q = VA \quad (2)$$

where Q is the flow; V is the velocity; and A is the cross-sectional area of interaction.

Using the flow at closure and the interaction area based on a zone 30 m above and below the workings boundary, an average permeability can be calculated:

$$K = \frac{Q}{A} \quad (3)$$

where K is the hydraulic conductivity; Q is the flow from Coal Measures strata at closure; and A is the total area of interaction around the mine workings.

The water inflow to the mine from the Coal Measures is then calculated for various depths using the mining interaction area and the average permeability.

Average Coal Measures permeabilities based on the interaction area around total extraction areas were calculated for several mining blocks in South Lancashire. The permeabilities for each block were generally similar (see Table 2) suggesting that in multiseam workings there is a general relationship between Coal Measures inflow rate and total extraction area.

Table 2. *Permeabilities derived from mine water inflow and interaction areas for five mining blocks in Lancashire*

	Mining block	Average permeability
1.	Bickershaw–Golborn	1.612×10^{-3} m day^{-1} (1.865×10^{-8} m s^{-1})
2.	Cronton	3.92×10^{-3} m day^{-1} (4.537×10^{-8} m s^{-1})
3.	Sutton Manor–Clockface	6.165×10^{-3} m day^{-1} (7.135×10^{-8} m s^{-1})
4.	Ashton Green–Bold	6.55×10^{-3} m day^{-1} (7.581×10^{-8} m s^{-1})
4.	Agecroft	8.82×10^{-3} m day^{-1} (1.021×10^{-7} m s^{-1})

Logarithmic flow model

The area-related and average permeability flow models are methods to reduce Coal Measures flow into the mine. For flows from shallow workings or from aquifers within shafts, the controlling factor is not usually the area of working but the permeability of the mining connection or the shaft lining. Where mine workings extend above the levels of the inflows from shallow workings or shafts, reductions in the flow predictions in the mining area or average Coal Measures permeability can be used once the water in the mine has reached the level of the shallow inflow or the shaft aquifer. However, where there are now shallow mine workings an alternative method of flow reduction is required. In these cases a simple logarithmic scale was applied between the inflow level and the maximum head in the shallow mine workings or the aquifer. A similar technique was also applied to the deeper Coal Measures inflows to give a comparison with area-related and average permeability models.

Results of mine water recovery modelling

The primary aim of mine water recovery modelling has been the prediction of the timing, the site and flow rate of potential future surface mine water discharges. These predictions can then be used to design and correctly place remediation schemes to either prevent or to treat the potential mine water discharge. Where there is no mine water level monitoring and recovery has to be calculated from an estimated inflow and an estimated void, the principal problem is the void-space calculation. The various flow models used (including the simple linear trend) generally give very similar recovery curves except in the late stages of recovery. Figure 14 shows the predicted recovery curves at Cronton, using the different inflow models, based on a residual void-space

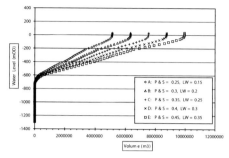

Fig. 15. Variations in predicted mine water recovery times at Cronton Colliery using the linear inflow model for various estimates of residual void volumes.

calculation of 20% for total extraction and 30% for pillar-and-stall workings. However, if the void-space estimation is wrong the recovery period could vary significantly. Figure 15 shows several linear mine water recovery curves based on varying combinations of residual void space for both total extraction and pillar-and-stall workings. There is approximately 30 years difference between the shortest and the longest period.

To date, there has been no monitoring of the actual mine water recovery in a mine or mining block where recovery modelling alone was used to predict recovery. However, of the four major mining units assessed in Lancashire, Agecroft was predicted to have mine water recovery nearest to surface and Bickershaw–Parsonage to have the longest recovery. Recent boreholes have confirmed that mine water levels are still at depth in the Bickershaw–Parsonage block and that water levels in the Agecroft area are close to surface.

Conclusions

The modelling of mine water recovery can be carried out using mine water inflow data and mine plans alone. However, to achieve accurate predictions on timing, this should be coupled with detailed monitoring of the actual water levels especially in the later stages of recovery.

Linking mine water inflow data with void-space calculations and monitored recovery data will give a better estimate of the reduction in water flow during recovery and help in calculating the residual void space in a mine. With this information the timing, site and flow rates of potential mine water discharges can be predicted to reasonable accuracy.

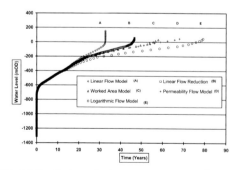

Fig. 14. Predictions of mine water recovery at Cronton obtained using five alternative inflow models.

The author would like to thank the Coal Authority for permission to use the mine water monitoring data and the Environment Agency for allowing publication of the results of the South Lancashire modelling. The author would also like to acknowledge the help of S. Croxford and J. Dannatt of IMC Consulting Engineers in researching the mining data and the development of the Excel spreadsheet models.

References

ADAMS, R. & YOUNGER, P.L. 2001. A strategy for modeling ground water rebound in abandoned deep mine systems. *Ground Water*, **39**, 249–261.

ADAMS, R. & YOUNGER, P.L. 2002. A physically based model of rebound in South Crofty tin mine, Cornwall. In: YOUNGER, P.L. & ROBINS, N.S. (eds), *Mine Water Hydrogeology and Geochemistry*, Geological Society, London, Special Publication, 198, 89–97.

BANKS, D. 2001. A variable-volume, head-dependent mine water filling model. *Ground Water*, **39**, 362–365.

British Coal, 1986. *Hydrogeological Assessment of the North East Leicestershire Prospect*, Golder Associates.

BURKE, S.P. & YOUNGER, P.L. 2000. Groundwater rebound in the South Yorkshire Coalfield: a first approximation using the GRAM model. *Quarterly Journal of Engineering Geology and Hydrogeology*, **33**, 149–160.

GLOVER, H.G. & CHAMBERLAIN, E.A.C. 1976. Water quality systems in Coal Measure Formations. In: *Symposium on Environmental Problems Resulting from Coal Mining Activities*, Katowice.

National Coal Board. 1956. *The Coal and Other Mines (Precautions against Inrushes) Regulations 1956*, National Coal Board, Mining Department.

National Coal Board. 1972. *Design of Mine Layouts. Working Party Report 1972*, National Coal Board, Mining Department.

National Coal Board. 1975. *Subsidence Engineers Handbook 1975*, National Coal Board, Mining Department.

National Coal Board. 1978. *Water Around a Coal Face*. Unpublished Branch Seminar 1978. National Coal Board, Mining Department.

NUTTALL, C.A., ADAMS, R. & YOUNGER, P.L. 2002. Integrated-hydraulic hydrogeochemical assessment of flooded deep mine voids by test pumping at the Deerplay (Lancashire) and Frances (Fife) Collierres. In: YOUNGER, P.L. & ROBINS, N.S. (eds), *Mine Water Hydrogeology and Geochemistry*, Geological Society, London, Special Publication, 198, 315–326.

ROBINS, N.S., DUMPLETON, S. & WALKER, J. 2002. Coalfield closure and environmental consequence – the case in south Nottinghamshire. In: YOUNGER, P.L. & ROBINS N.S. (eds), *Mine Water Hydrogeology and Geochemistry*, Geological Society, London, Special Publication, 198, 99–105.

SHERWOOD, J.M. 1997. *Modelling Minewater Flow and Quality Changes After Coalfield Closure*. Unpublished PhD Thesis. Department of Civil Engineering, University of Newcastle.

SHERWOOD, J.M. & YOUNGER, P.L. 1997. Modelling groundwater rebound after coalfield closure. In: CHILTON, P.J., *et al*. (eds), *Groundwater in the Urban Environment, Volume 1: Problems, Processes and Management. Proceedings of the XXVII Congress of the International Association of Hydrogeologists, Nottingham, UK, 21–27 September 1997*. A. A. Balkema, Rotterdam 165–170.

WHITWORTH, K.R. 1982. Induced changes in the permeability of Coal Measures strata as an indicator of the mechanics of rock deformation above a longwall face. In: FARMER, I.W. (ed.), *Strata Mechanics. Proceedings of the Symposium on Strata Mechanics held in Newcastle Upon Tyne, 5–7 April 1982*. Developments in Geotechnical Engineering No. 32. Elsevier, Amsterdam, 18–24.

YOUNGER, P.L. & ADAMS, R. 1999. *Predicting Mine Water Rebound*. Environment Agency, R&D Technical Report, **W179**.

YOUNGER, P.L. BARBOUR, M.H. & SHERWOOD, J.M. 1995. Predicting the consequences of ceasing pumping from the Frances and Michael Collieries, Fife. In: BLACK, A.R. & JOHNSON, R.C. (eds), *Proceedings of the Fifth National Hydrology Symposium, British Hydrological Society*, Edinburgh, 4–7 September 1995, 2.25–2.33.

Effects of longwall mining in the Selby Coalfield on the piezometry and aquifer properties of the overlying Sherwood Sandstone

STEPHEN DUMPLETON

British Geological Survey, Kingsley Dunham Centre, Keyworth, Nottingham NG12 5GG, UK
(e-mail: sdu@bgs.ac.uk)

Abstract: In the UK, the first longwall faces at Wistow Mine in the Selby Coalfield, with only 80 m depth of cover to the base of the Permian, experienced several inrushes of groundwater derived from the overlying Lower Magnesian Limestone, causing serious disruption to coal production. Subsequent decrease in panel width coupled with increased depth of cover to the Permian reduced the incidence of water problems. However, there has never been any quantitative investigation to determine the effects of mining on the hydraulic properties of the stratigraphically higher Permo-Triassic age Sherwood Sandstone, a major aquifer of regional importance.

This opportunistic study fills that gap. Precautionary observation boreholes had been drilled above and around the margins of two proposed longwall panels prior to working the 2.5-m thick Barnsley seam at a depth of 550–600 m. Data loggers permitted continuous monitoring of the Sherwood Sandstone and Drift piezometric levels over a 2-year period. Widespread drawdown and recovery effects due to intermittent groundwater abstraction from a nearby factory were observed. Standard aquifer pumping test analyses of the hydrographs allowed transmissivity and storativity to be determined before, during and after mining. The results showed apparently permanent post-mining transmissivity increases of up to 149% around the margins of the panels, and up to 234% directly over the first panel. Post-mining storativity remained mostly unchanged. However, the greatest effects were noted during the closest approach by the second longwall panel, which also caused some additional subsidence over the first panel, when peak transmissivity increases of 1979% and storativity increases of 625% occurred. Anomalous, intra-cycle, recovery–drawdown events were also observed during this phase and interpreted as indicating rapid mining-induced dilation and compression of fractures within the aquifer fabric.

The results are consistent with similar investigations carried out at relatively shallow depths (<220 m) in USA coalfields. However, the Selby study shows that mining at much greater depths still has a significant impact on shallow aquifers, with implications for enhanced aquifer recharge, abstraction well yield and possible increased contaminant transport rates.

The impact of deep mining on groundwater has been a problem from the earliest days of the coal mining industry, but is largely one of safety, i.e. how to prevent or minimize the (sometimes disastrous) inflow of water to mine workings. With expansion of the industry from the early years of the nineteenth century, inrushes of water have claimed the lives of hundreds of miners. Nearly all these incidents have been due to the mine workings encountering undocumented flooded old workings, although an inundation of the sea caused by working too close beneath the sea bed was responsible for the deaths of 27 miners at Workington, Cumbria in 1837 (Younger & Adams 1999). The terrible hazard of old workings was highlighted tragically in 1973 at Lofthouse Colliery, West Yorkshire, when a modern, mechanized longwall coal face intersected a flooded shaft whose presence was hitherto unsuspected. The resulting inrush of water and debris killed seven men (Calder 1973).

Mine workings in close proximity to aquifers or other water-bearing formations such as peat (Younger & Adams 1999) can experience severe disruption, as well as posing a potentially life-threatening situation. In the Selby Coalfield, North Yorkshire, production from the Barnsley seam at Wistow Mine commenced in 1983. During the first year, with only 80 m depth of cover to the base of the Permian, several inrushes of water occurred causing coal production to be halted. Initial flow rates reached $10\,368\,m^3\,day^{-1}$ ($120\,l\,s^{-1}$) but gradually reduced to $3197\,m^3\,day^{-1}$ ($37\,l\,s^{-1}$) during the following few weeks (Bigby & Oram 1988; North & Jeffrey 1997). Analyses showed that the water was derived from the overlying Lower Magnesian Limestone, and flowed into the workings along fractures associated with the first mining roof breaks that propagated up to the Permian rocks. Subsequent changes in mining methods and increased depth of cover to the Permian reduced the incidence of water problems, although smaller inflows occurred from time to time. Current mining at Wistow now encounters only 'nuisance' water. The quantity pumped from the entire mine, including water deliberately introduced into the mine, e.g. for dust-suppression sprays, is reported to be only around $655 \text{ m}^3 \text{ day}^{-1}$ (100 gall min^{-1}) (S. Peace, Subsidence Engineer, UK Coal Mining Ltd pers. comm.).

More recently, in times of greater environmental awareness, instances of deep coal mining having a deleterious effect on both surface water courses and groundwater resources have been documented. For example, in the USA, at the Windsor Mine, West Virginia, longwall workings in the 1.45-m thick Pittsburgh No. 8 Coal at depths of 52–117 m caused monitoring wells and surface springs to become dry (Straskraba et al. 1994). At Saline County, Illinois, longwall workings at a depth of 122 m in the 2 m thick Herrin Coal caused water levels in piezometers completed in an overlying sandstone aquifer to decline by up to 30 m. In one piezometer, permanent post-mining increases in the aquifer hydraulic conductivity of up to 1900% of its original value were measured, while, during mining, temporary increases of an order of magnitude were observed. However, other piezometers showed no significant increase, demonstrating the complexity of the situation (Booth et al. 1994; C. J. Booth, Northern Illinois University pers. comm.). Both of these examples deal with mine workings at shallow depth. The present study investigates the effect of coal mining by longwall extraction at greater depths (550–600 m) on the aquifer properties of the Sherwood Sandstone in the Selby Coalfield.

Background to the present study

In 1996, the British Geological Survey (BGS) was commissioned jointly by UK Coal Mining Ltd (formerly RJB Mining (UK) Ltd) and Unitrition International–BOCM Pauls Ltd (Unitrition–BOCM), manufacturers of vegetable oil and animal feedstuffs, to assess the potential effects at the Unitrition–BOCM Selby, North Yorkshire, site (SE 625 329) due to proposed coal extraction by a retreat longwall panel (H93$'$s) at Wistow Mine (Dumpleton 1996, 1999). The panel was planned to be 170 m wide, extracting 2.5 m of Barnsley seam at a depth of around 550–600 m.

One concern to be addressed was the possible effect of the mining on groundwater resources in the overlying Permo-Triassic Sherwood Sandstone. This extensive, major aquifer is of great importance in the UK, second only to the Chalk (Price 1996, p. 83.) for public, industrial and private water supplies. In the area considered in this study, the aquifer is wholly confined by the overlying Vale of York 25-foot Drift. Under Environment Agency Abstraction Licence No. 133(1), Unitrition–BOCM abstracted groundwater from two boreholes drilled into the Sherwood Sandstone. This abstraction forms an essential part of their operations and any interruption or deterioration of the groundwater supply would have severe consequences on production.

Prior to the commencement of coal production, piezometers fitted with automatic data loggers were installed above and around the margin of H93$'$s. As a control, a data logger was also installed in an Environment Agency monitoring borehole completed in the Sherwood Sandstone at Barlby sewage treatment works, a location removed from the effects of mining subsidence (Fig. 1). Monitoring was planned for a year in the first instance but, at the request of UK Coal Mining, was subsequently extended for a further year to observe any effects due to working of the adjacent retreat panel H94$'$s. Initial inspection of the piezometer hydrographs in the first few days of operation showed marked and widespread drawdown and recovery resulting from weekday pumping and weekend shutdown at Unitrition–BOCM. Although the investigation was purely to monitor water level fluctuations, it was thought that the cycles of drawdown–recovery could also be used as a conftinuing series of aquifer pumping tests yielding information about aquifer properties such as transmissivity and storativity. This was a unique opportunity to investigate how these aquifer properties might change with time and, particularly, whether or not there were changes that could be attributed to the effects of mining.

Site description

The site is located about 1 km to the NE of Selby, North Yorkshire, around Bank House Farm (SE 623 330) and Unitrition–BOCM (SE 625 329) (Fig. 1). The River Ouse flows through the Selby area in a general southeasterly direction, with the north–south, 2 km long, Barlby Reach and the east–west, 1 km long, Selby Long Reach being the dominant features in the vicinity of the site. The river is tidal and about 50 m wide. The topography is flat and low-lying, generally 3–4 m

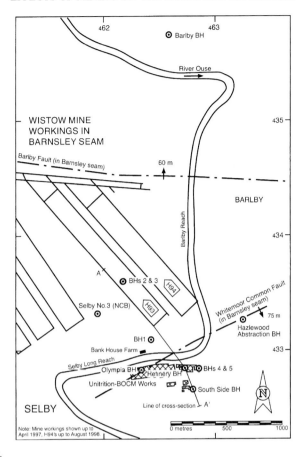

Fig. 1. Location plan.

above Ordnance datum (OD) and prone to flooding. Artificial levees have been built on the river banks to an elevation of 6–7 m OD. Extending northwards from Bank House Farm, a prominent flood defence embankment of similar elevation has been constructed from colliery spoil to offer further protection from flooding exacerbated by mining subsidence (S. Peace pers. comm.). East of Barlby Reach and north of Selby Long Reaches, land use is agricultural, chiefly arable and grazing for cattle and sheep, while south of Selby Long Reach are the factory premises of Unitrition–BOCM and Rank Hovis. Barlby village lies to the east of Barlby Reach and Barlby sewage treatment works (STW) is situated about 200 m north of another east–west reach beyond Barlby Reach.

Annual rainfall during the period 1941–1970 was around 600 mm (Institute of Geological Sciences 1982). This compares favourably with annual rainfall of 480 and 528 mm in 1996 and 1997, respectively, measured at the nearby Selby sewage works rain gauge (SE 634 313) (data supplied by the Environment Agency).

Geology

The geology of the area is covered by the British Geological Survey 1:50 000 Sheet No. 71 (Selby) and 1:10 560 Sheet No. SE 63 SW. There is no memoir available for the Selby sheet, but the area immediately south of the site is included in the memoir for 1:10 000 Sheets Nos 79 and 88 (Goole and Doncaster) (Gaunt 1994). The generalized geological succession is given in Table 1. This is based on: (i) nearby shallow borehole and well logs; and (ii) National Coal Board Selby No. 3 exploration borehole (SE 61945 33323), to which the H93's panel made a closest approach of 300 m. The cross-section (Fig. 2) is largely based on Selby No. 3 borehole and the generalized vertical section of the BGS 1:50 000 Selby sheet.

Table 1. *Generalized geological succession in the vicinity of the site*

	Formation	Typical lithologies	Thickness in site vicinity (m)
Holocene	Made ground Alluvium	Sand, clay and mud	up to 42.1–4.6
Quaternary	Vale of York 25-foot Drift	Interbedded sands and clays, with some peat	10.7–16.2
	Sherwood Sandstone Group	Mainly red sandstones with some marl bands	175
	Upper Marl (Roxby Formation)	Marl. Anhydrite, marl and gypsum	20
			16
	Upper Magnesian Limestone (Brotherton Formation)	Limestone, dolomitic. Marl with anhydrite and gypsum.	23
			16
Permo-Triassic	Middle Marl (Edlington Formation)	Halite and marl. Anhydrite with marl beds.	5
			16
	Lower Magnesian Limestone (Cadeby Formation)	Limestone, dolomitic, with anhydrite in upper part.	74
	Lower Marl	Mudstone with thin limestone at base	1
	Basal Sands	Sandstone, soft	3
Carboniferous	Coal Measures (Westphalian)	Mainly mudstones, siltstones and thin coals; Woolley Edge Rock 19.50 at 429.50; un-named sandstone 7.44 at 535.34; Barnsley seam 2.48 at 561.44	259 proved in Selby No. 3 borehole

Coal Measures, probably totalling some 750 m in thickness, occur in the Selby area. In Selby No. 3 borehole, the 259 m of Measures proved below the base of the Permian consist of a cyclic succession of siltstones, mudstones, seat earths and at least 15 coals. The 2.5-m thick Barnsley seam lies at a depth of around 550 m at the southern limit of the H93's panel. The only sandstones of note are the Woolley Edge Rock, 19.50 m thick, and an unnamed sandstone 7.44 m

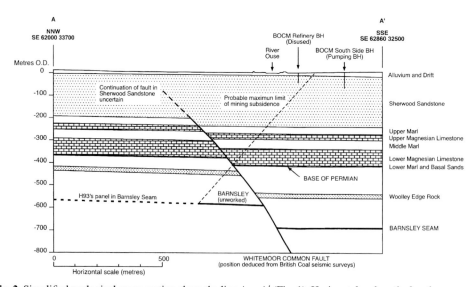

Fig. 2. Simplified geological cross-section along the line A – A′ (Fig. 1). Horizontal and vertical scales are equal.

thick, 129 and 24 m, respectively, above the Barnsley seam roof.

Unconformably overlying the Coal Measures are rocks of Permian and Triassic age. The sequence given in Table 1, based on Selby No. 3 borehole, may be regarded as typical for the vicinity of the Unitrition–BOCM site.

The Vale of York 25-foot Drift (Quaternary) overlies the Sherwood Sandstone at the site. These sediments represent the infilling of the former 'Lake Humber' – a glacially dammed lake that existed in Devensian times in the southern part of what is now the Vale of York. Gaunt (1994) described a silt and clay subdivision, both overlain and underlain by sands. This appears to be broadly reflected at the Unitrition–BOCM site, where borehole and well records indicate an interbedded sequence of sands and clays, with individual units showing considerable variation in thickness and lateral extent. Alluvial sands, clays, and peat and peaty silt overlie the 25-foot Drift and occur over the entire site.

Geological structure

Considerable detailed information is known of the geological structure of the Coal Measures as a result of both extensive coal exploration (including boreholes and seismic surveys) and underground mine workings. This information is summarized in Figs 1 and 2. In the area of interest, the Coal Measures (at the Barnsley seam horizon) dip in an easterly direction at around 1 in 30.

There are two main faults in the study area: (i) the Whitemoor Common Fault striking ENE–WSW; and (ii) the Barlby Fault striking E–W. These faults form the natural boundaries of a block of reserves that was exploited from Wistow Mine, and included the H93's and H94's panels. In the Barnsley horizon the Whitemoor Common Fault and the Barlby Fault, respectively, have downthrows of 75–100 m southeasterly and 30–60 m northerly. British Coal seismic surveys (lines 94-SRL-32 and 90-SRL-01) indicate that these faults continue up into the Permo-Triassic rocks, at least as far as the Middle Marl anhydrite, although with reduced throws. The dips of the fault planes appear to decrease in the uppermost Coal Measures and the lower part of the Permo-Triassic. The Whitemoor Common Fault is indicated on the cross-section in Fig. 2.

The structure of the Permo-Triassic rocks is nearly as well known as the Coal Measures, due to the extensive coal exploration in the area. In particular, the anhydrite in the Middle Marl is a prominent seismic reflector. Above this horizon, however, seismic resolution is poor, and only the broad structure of the Sherwood Sandstone is known. It is not clear from the seismic interpretation whether the Barlby and Whitemoor Common faults are present in this formation. Generally, the dip of the base of the Permian in the vicinity of the site is 1 in 34 to azimuth 076°, very similar to the Coal Measures.

Hydrogeology

The following summary is mainly based on the Institute of Geological Sciences (1982) and Gaunt (1994).

In the Coal Measures the Woolley Edge Rock forms a local minor aquifer at or near outcrop (about 24 km WSW of Unitrition–BOCM), otherwise the sequence in the study area is dominated by siltstones, mudstones and seatearths likely to be of low permeability.

The Basal (Permian) Sands consist of a friable and locally incohesive sandstone 0–30 m thick, which can form an effective aquifer, although significant development is restricted by the limited thickness and recharge area.

The Lower Magnesian Limestone (Cadeby Formation) and Upper Magnesian Limestone (Brotherton Formation) both form locally important aquifers, having been exploited for numerous domestic and farm supplies. In the vicinity of major faults, hydraulic conductivities of 70 m day^{-1} have been measured, and borehole yields of more than $30 \, 1 \, s^{-1}$ (2.6 Ml day^{-1}) can be obtained.

The Middle Marl (Edlington Formation) and Upper Marl (Roxby Formation) both act as aquicludes, the former between the Upper and Lower Magnesian Limestones, the latter separating the Upper Magnesian Limestone from the Sherwood Sandstone.

The Sherwood Sandstone Group consists mainly of a brownish-red, fine- to medium- (less commonly coarse-) grained sandstone, interbedded, especially in the basal 40 m, with thin layers and lenses of brownish-red and greenish-grey mudstone and siltstone. It forms a major aquifer of regional extent and of great importance for public, industrial and private water supplies. In the Vale of York, it is confined by the Mercia Mudstone to the east and partly confined by drift. Infiltration occurs mainly where the drift is clay-free or absent, and it is in these areas that the main public supply boreholes are located.

Permeability in the Sherwood Sandstone is mainly primary, i.e. intergranular, but there is also a significant secondary fracture flow component. Horizontal to vertical permeability ratios may be in the range of 1.2–2.5 (based on core sample studies from elsewhere in the Vale of

York). The presence of low-permeability marl bands suggest that significant horizontal components of groundwater flow may be expected. In the general Selby area, pumping test data indicate that the Sherwood Sandstone has a transmissivity range of between 50 and 400 m^2 day^{-1}. At the nearby Hazlewood Preserves site (SE 632 332), pumping tests in two boreholes in the Sherwood Sandstone gave transmissivities in the range 400–1300 m^2 day^{-1}, although the latter was considered excessively high. A value of 450 m^2 day^{-1} was considered to be representative of the true aquifer transmissivity (Southern Water 1997; Dumpleton et al. 1998).

There are few values for Sherwood Sandstone storativity available in the Selby area. A pumping test at Cowick, to the south of the area, gave two different values of 0.02–0.03 and 0.005–0.006. The latter is thought to be more representative of the confined aquifer storage (Aspinwall & Co. 1996). At the Hazlewood Preserves site, Southern Water (1997) assumed maximum and minimum values of 0.08 and 0.0005, representing unconfined and confined conditions. As the aquifer was suspected to be leaky, the actual value probably lies between these two extremes.

Sands and gravels of the Vale of York 25-foot Drift are recognized as a minor aquifer (Aspinwall & Co. 1996). Although in places overlain by thin clays, these deposits are generally regarded as unconfined and vulnerable to pollution due to a permeable and locally thin unsaturated zone. Elsewhere in the Vale of York, the degree of infiltration to the Drift varies according to the presence or absence of clay cover. Where clay is thin or absent, 90–100% of the potential infiltration is achieved (Spears & Reeves 1975). In the present study area there is little information concerning the hydrogeological characteristics of the Drift. Following the installation of the piezometer in BH3, completed in a Drift sand horizon, a simple falling head test yielded a hydraulic conductivity of 6.2×10^{-2} m day^{-1}.

Mining subsidence

Prior to working, mining subsidence predictions were made for H93's and H94's, concentrating particularly on the area around and adjacent to the Unitrition–BOCM site, which was most likely to be affected. Calculations were made: (a) using the *Subsidence Engineers' Handbook* (SEH) methodology (National Coal Board 1975) (Fig. 3); and (b) using an analogue method based on observed subsidence at Bondgate Lane, Selby, following extraction of earlier panels (Wistow H73's and H90's) in a broadly similar geological and depth setting (Dumpleton 1996; 1997). The analogue method gave generally comparable results, although with a slightly wider (by about 130 m) zone of influence at the surface. The latter was tentatively attributed to modification of the subsidence 'angle of draw' by the Permian evaporites. In Fig. 2 the probable maximum limit of subsidence line shown is based on the analogue method.

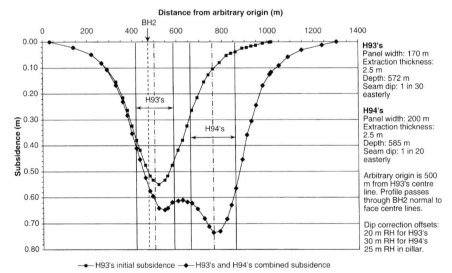

Fig. 3. Calculated subsidence profiles for H93's and H94's using the *Subsidence Engineers' Handbook* method (National Coal Board 1975).

The observed subsidence at Bondgate Lane probably included a degree of subsidence re-activation due to working of adjacent panels. Ferrari (1996) cites examples where working an adjacent panel caused subsidence above the original panel of more than twice the amount predicted by the SEH method. Similar effects have been observed elsewhere in the Selby Coalfield (S. Peace pers. comm.).

The effect of the Whitemoor Common Fault on surface subsidence was not known at the time of making the subsidence predictions. As shown in Figs 1 and 2 the fault in the Barnsley seam passes beneath and just to the north of the Unitrition–BOCM site. However, the fault plane dips in the opposite direction to the angle of draw. If the fault plane had been parallel, or nearly so, to the angle of draw, mining subsidence might have been directed along the fractures associated with the fault, possibly causing a marked subsidence feature along the rockhead intersection of the fault plane (Dumpleton 1996). Subsequent to working of H93's and H94's, no such feature has been observed and, therefore, it is likely that the fault has not had a significant influence on subsidence at the surface.

Piezometer installation

Drilling and piezometer installation was carried out by Soil Mechanics Ltd in July 1996 using a truck-mounted rotary rig. Air flush was specified in order to minimize clogging of the borehole walls with fines. The boreholes were drilled at 150 mm diameter. The piezometers were 50 mm internal diameter proprietary plastic tubing with 1 m long screens set in a gravel pack and isolated from the rest of the borehole by bentonite pellets and a bentonite/cement grout. Borehole surface elevations were determined by levelling carried out by H&H Surveys, Barnsley (contractors for UK Coal Mining Ltd), who also carried out periodic check levelling to determine ground subsidence as mining proceeded during the course of the investigation. Each piezometer was fitted with an automatic data logger to record depth to water at hourly intervals. Data were periodically downloaded to a laptop computer. Table 2 summarizes the details of the boreholes.

BH1 was situated outside the margins of the H93's panel, but within the zone of subsidence. BH2 and BH3 were directly over the panel, about 27 and 22 m, respectively, from the panel centre line. BH4 and BH5 were situated on the Unitrition–BOCM premises in a location where minimal subsidence effects were expected (Fig. 1).

Groundwater abstraction at Unitrition–BOCM

There are three boreholes on the site abstracting groundwater from the Sherwood Sandstone under Environment Agency Abstraction Licence No. 133(1)(A), although only No. 1 (Olympia) and No. 2 (South Side) boreholes are in current use. Pumping at South Side borehole is normally continuous from early Monday morning until the weekend shut-down on Friday night or early Saturday morning. At Olympia borehole, pumping is by airlift pump and is used intermittently to top up a nearby header tank (Table 3).

Table 2. *Summary details of monitoring boreholes*

Borehole	National Grid Ref.	Surface elevation (m OD)	Datum for piezometers and data loggers (m OD)	Total depth (m)	Screen interval (depth below surface) (m)	Formation monitored
BH1	SE 62422 33093	3.75	3.95	44.70	43.40–44.40	Sherwood Sandstone
BH2	SE 62171 33600	2.15	2.35	44.80	43.00–44.00	Sherwood Sandstone
BH3	SE 62174 33601	2.17	2.27	9.00	7.85–8.85	Drift (sand)
BH4	SE 62854 32837	3.76	3.58	44.70	43.45–44.45	Sherwood Sandstone
BH5	SE 62854 32841	3.76	3.58	11.35	10.35–11.35	Drift (sand and gravel)
Barlby BH (control)	SE 62585 35750	7.52	7.04	Approx. 25	Unknown	Sherwood Sandstone

Table 3. *Abstraction boreholes at Unitrition–BOCM*

Borehole	National Grid Ref.	Surface elevation (m OD)	Depth (m)	Typical pumping rates (Ml day^{-1})
No. 1 (Olympia)	SE 62335 32828	approx. 4	77.11	0.15–0.2
No. 2 (South Side) (main abstraction borehole)	SE 62800 32660	approx. 3.7	77.11	1.8–2.4*
No. 3 (Refinery)	SE 62728 32845	approx. 4	48.77	Disused

* These values based on rising main flow meter readings April – September 1998.

At the South Side borehole, times of pump start-up and shut-down, and flowmeter readings were recorded by Unitrition–BOCM staff. These times were checked independently by comparison with the hydrograph from BH4 and, during the latter part of the investigation, from a data logger installed in the South Side borehole as part of a further BGS study (Dumpleton 1998; Dumpleton *et al.* 1998). The pumping water level is typically about 40 m below ground level. During pumping shut-down, the water level recovers to around 11 m below ground level.

Data analysis

Monitoring of water levels commenced on 14 August 1996, except in the case of BH2 where, due to an initially faulty data logger, monitoring did not commence until 30 August 1996. Coal production from H93's started on 19 September 1996, so the time available for establishing baseline conditions was very short. BH2 and BH3 were underworked by the face on 14 November 1996. The face finished production on 29 March 1997. Production from the adjacent panel, H94's, took place from 27 November 1997 to 28 August 1998. The closest approach to BH2 (186 m plan distance) occurred on 25 March 1998.

Observed subsidence at the monitoring boreholes is shown in Fig. 4. It can be seen that the maximum observed subsidence at BH2 due to H93's was 0.31 m. This increased to 0.63 m following working of H94's. This can be compared with the calculated subsidence of BH2 due to H93's and H94's of about 0.5 and 0.56 m, respectively (Fig. 3). Observed maximum subsidence due to H93's at BH1, about 120 m outside the H93's panel margin, was 0.12 m, increasing to 0.14 m after working of H94's. This compares favourably with the calculated subsidence of about 0.11 m for both panels (H94's having little additional effect).

In the Sherwood Sandstone piezometers, BH1, BH2 and BH4, and the control Barlby BH, hydrograph variations were identified that were attributed to features such as seasonal

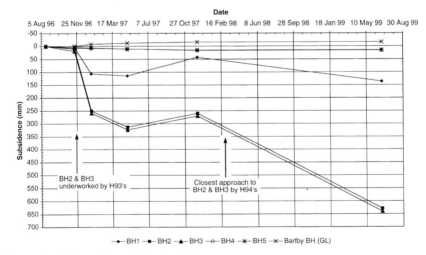

Fig. 4. Observed subsidence at monitoring boreholes.

recharge–discharge, Unitrition–BOCM drawdown–recovery events and loading of the confined Sherwood Sandstone aquifer by diurnal tides on the River Ouse. The latter effects were filtered out as necessary using a 24 h moving mean, to remove the effects of the tidal fluctuations. Hydrographs of piezometers completed in the Drift showed no effects due to Unitrition–BOCM pumping and recovery, nor due to tidal loading response.

A total of 84 individual Unitrition–BOCM pumping cycles were identified that had well-defined start–stop times and flow rate data, and which could be identified in the Sherwood Sandstone piezometers. Each cycle was corrected for rainfall recharge by comparison with Barlby BH. Plots of drawdown versus time for each of the pumping cycles indicated a leaky aquifer response and were analysed manually using the Walton curve matching method (Walton 1962; Kruseman & de Ridder 1990, pp. 81–84) to determine transmissivity and storativity in each of the Sherwood Sandstone piezometers. In addition, the drawdown data for BH4 were corrected for partial penetration effects using Weeks's modification of the Walton method described in Kruseman & de Ridder (1990, p. 169). These parameters were plotted as time series and compared with the timing and position of the mining activity to enable conclusions to be drawn about the effects of the mining on the aquifer hydraulic properties.

Results

Figure 5 shows typical smoothed drawdown–time responses for BH1, BH2 and BH4. Each piezometer hydrograph has two components: (a) unsmoothed; and (b) smoothed by application of a 25-h moving mean to remove the tidal loading effects. Both components were used when performing the manual matching to the Walton-type curves. For each set of data a 'maximum drawdown fit', 'minimum drawdown fit' and 'best-fit' were obtained. Values of transmissivity and storativity were calculated in the usual way (Kruseman & de Ridder 1990). Transmissivity and storativity variations in BH1, BH2 and BH4 are shown in Figs 6–11. The error bars indicate the maximum and minimum fits. A three-point moving mean was applied to the 'best-fit' values to illustrate the changing trends more clearly. Mean values, based on 'best-fit' results, before, during and after mining are shown in Table 4. Changes in the aquifer properties are indicated by the post-mining/pre-mining ratio, expressed as a percentage (i.e. 100% indicating no change).

Pre-mining mean transmissivity values in the range 186–231 $m^2 day^{-1}$ increased significantly in BH1 and BH2 during underworking or close approach by H93's and H94's, when peak values of 945 and 4568 $m^2 day^{-1}$ (increases of 466 and 1979%) were recorded in BH1 and BH2, respectively. After working of both panels had been completed, post-mining transmissivity values

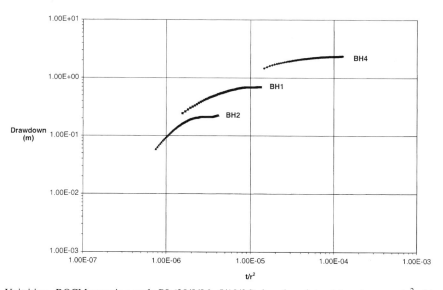

Fig. 5. Unitrition–BOCM pumping cycle P9 (30/9/96–5/10/96): log–log plots of drawdown vs t/r^2 of 'normal' smoothed (25-h moving mean) responses in BH4, BH1 and BH2, where t = time since pumping began and r = distance from Unitrition–BOCM pumping borehole. Compare with Fig. 12.

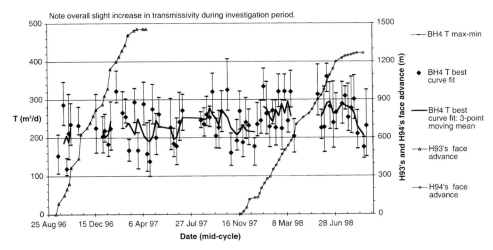

Fig. 6. BH4 transmissivity variations during period of investigation.

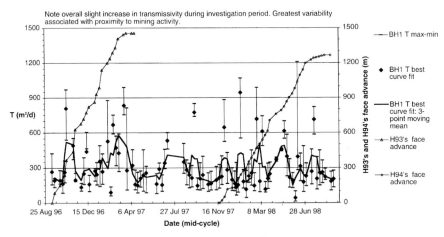

Fig. 7. BH1 transmissivity variations during period of investigation.

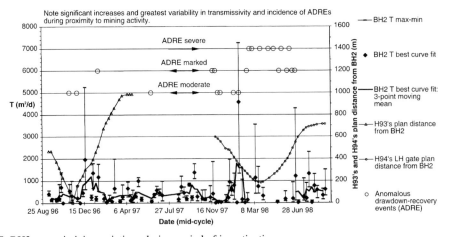

Fig. 8. BH2 transmissivity variations during period of investigation.

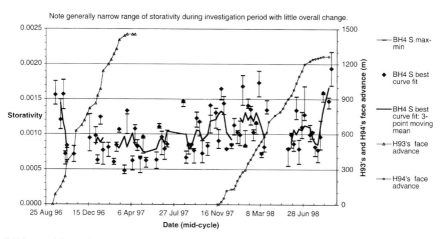

Fig. 9. BH4 storativity variations during period of investigation.

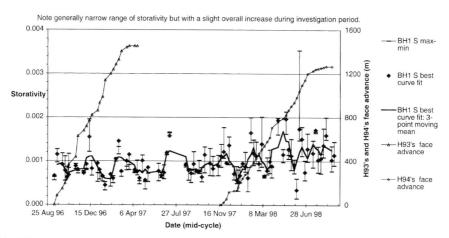

Fig. 10. BH1 storativity variations during period of investigation.

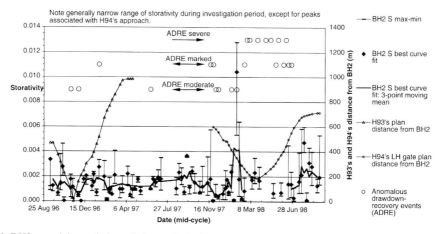

Fig. 11. BH2 storativity variations during period of investigation.

remained higher than the original values at 257–540 m² day⁻¹ (increases of 138–234%). Pre-mining storativity values fell in the range $1.44 \times 10^{-3} – 8.77 \times 10^{-4}$. In BH1 and BH2, during closest approach by H94's, peak increases to 1.97×10^{-3} and 1.04×10^{-4} (224 and 625%) were recorded. Generally, however, storativity appeared to be less affected by the mining activity than transmissivity with overall, post-mining changes falling in the range 79–126%.

Some pumping cycles proved more difficult to fit to the type curves than others. This is reflected by the range between the maximum and minimum error bars. Possible reasons for slight departure from the type curves may include small, undocumented adjustments in pumping rate within the main pumping cycle at Unitrition–BOCM and/or rapid changes in the general aquifer piezometric surface as a result of rainfall recharge. When they occurred, these variations were noted in all three Sherwood Sandstone piezometers, suggesting a general cause. However, more dramatic departures were observed only in BH2, associated with the times of underworking by H93's and the proximity of H94's. Characterized by rapid fluctuations of drawdown and recovery within the main drawdown pumping cycle, they were of a magnitude and frequency that rendered it impossible to match the data points to the type curves (Fig. 12). These anomalous drawdown–recovery events (ADRE) were classified on a visual basis as moderate, marked or severe (Figs 8 and 11) and were interpreted as piezometric head fluctuations responding to rapid mining-induced dilation and compression of fractures within the aquifer fabric. Similar effects were described by Liu & Elsworth (1999) in West Virginia, USA, where the Pittsburgh seam (1.7–1.8 m extraction) was worked by 183 m wide longwall panels at a depth of around 215 m.

Conclusions

Piezometers were installed above and around the margins of two longwall panels prior to working the 2.5 m thick Barnsley seam at a depth of 550–600 m. Automatic data loggers permitted continuous monitoring of Sherwood Sandstone and Drift piezometric levels over a 2-year period. Coincidentally, intermittent but heavy groundwater abstraction from the nearby Unitrition–BOCM factory led to a widespread series of drawdown and recovery cycles observable in the Sherwood Sandstone piezometer hydrographs, but little or no effect in the Drift aquifers. Standard leaky aquifer pumping test analyses of the hydrographs allowed transmissivity and storativity of the Sherwood Sandstone aquifer to be determined before, during and after mining.

The results (Table 4) show apparently permanent post-mining Sherwood Sandstone transmissivity increases of up to 149% around the margins of the panels, and 234% directly over the first panel. Post-mining Sherwood Sandstone storativity around the margins showed a decrease (79%) in BH4 and an increase (126%) in BH1. Directly over H93's in BH2, storativity remained virtually unchanged (101%). The decrease seen in BH4 is interpreted to be minimal, possibly a

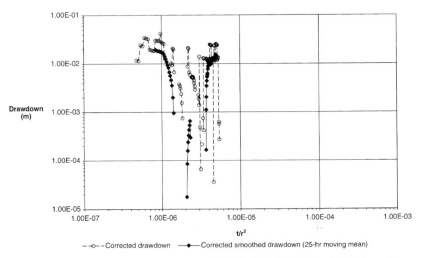

Fig. 12. BH2 anomalous drawdown–recovery events (ADREs) within Unitrition–BOCM pumping cycle P105 (17/5/98–23/5/98). Compare with Fig. 5.

Table 4. Summary of transmissivity and storativity changes before, during and after mining

	Transmissivity ($m^2\, day^{-1}$)			Change in transmissivity (post-mining/pre-mining ratio × 100%)			Storativity			Change in storativity (post-mining/pre-mining ratio × 100%)		
	BH4	BH1	BH2	BH4	BH1	BH2	BH4	BH1	BH2	BH4	BH1	BH2
Pre-mining mean	186	203	231	–	–	–	1.44×10^{-3}	8.77×10^{-4}	1.67×10^{-3}	–	–	–
Peak, H93's	322	834	1970	173	411	854	1.33×10^{-3}	1.55×10^{-3}	2.78×10^{-3}	92	176	167
Post-H93's mean	232	258	406	124	127	176	9.41×10^{-4}	8.72×10^{-4}	1.51×10^{-3}	65	99	91
Peak, H94's	361	945	4568	194	466	1979	1.94×10^{-3}	1.97×10^{-3}	1.04×10^{-2}	134	224	625
Post-H94's mean	257	302	540	138	149	234	1.15×10^{-3}	1.11×10^{-3}	1.69×10^{-3}	79	126	101

function of incorrect pre-mining values due to insufficient time available to establish true baseline conditions.

The greatest effects were noted during closest approaches by the workings to the boreholes, particularly BH2. During the closest approach by the second longwall panel, peak transmissivity increases of 1979% and storativity increases of around 625% occurred. Anomalous, intra-cycle, drawdown–recovery events (ADREs) were also observed during this phase and interpreted as indicating rapid mining-induced dilation and compression of fractures within the aquifer fabric.

The results are broadly consistent with those from similar investigations carried out at relatively shallow depths (<220 m) in the Illinois and Appalachian Coalfields. However, the Selby study shows that mining at much greater depths has a more significant impact on shallow aquifers than had previously been recognized. Within and immediately adjacent to the area affected by mining subsidence, the increased transmissivity and storativity in the 'surface zone' has implications not only for enhanced aquifer recharge and abstraction well yield but also for increased contaminant transport rates in the event of any future pollution incident.

The author gratefully acknowledges assistance from UK Coal Mining Ltd, Unitrition International–BOCM Pauls Ltd and the Environment Agency, without which this investigation could not have taken place. The investigation is the subject of a PhD thesis, submitted to the University of Sheffield, Department of Civil and Structural Engineering, in July 2001. This paper is published here with the permission of the Director of the British Geological Survey (NERC).

References

ASPINWALL & CO. 1996. *Appraisal of Groundwater Resources From the Sherwood Sandstone Aquifer, Selby Region, North Yorkshire*. Report prepared for the National Rivers Authority, **NR1711A**.

BIGBY, D.N. & ORAM, J.S. 1988. Longwall working beneath water. *Coal International* 26–28.

BOOTH, C.J., CURTISS, A.M. & MILLER, J.D. 1994. Groundwater response to longwall mining, Saline County, Illinois, USA. In: *Proceedings of the Fifth, International Mine Water Congress, Nottingham (UK), September 1994* IMWA & The University of Nottingham, 71–81.

CALDER, J.W. *Inrush at Lofthouse Colliery Yorkshire*; Report on the causes of, and circumstances attending, the inrush which occurred at Lofthouse Colliery, Yorkshire, on 21 March 1973. Department of Trade and Industry, HMSO, London.

DUMPLETON, S. 1996. *Problems of Ground Movement, Declining Groundwater Levels, and the Potential Effects of Coal Mining at Unitrition International/*

BOCM Pauls Ltd, Selby, North Yorkshire: A Preliminary Investigation. British Geological Survey Technical Report, **WE/96/13**.

DUMPLETON, S. 1997. *Determination of Subsidence due to Wistow Mine Proposed H94's Panel.* Supplement to British Geological Survey Technical Report, **WE/96/13**.

DUMPLETON, S. 1998. *Groundwater Temperature Monitoring at Unitrition International/BOCM Pauls Ltd, Selby, North Yorkshire.* British Geological Survey Technical Note, **WE/98/16**.

DUMPLETON, S. 1999. *Monitoring of Groundwater Levels in the Vicinity of Wistow Mine H93's and H94's Longwall Panels, Selby, North Yorkshire.* British Geological Survey Technical Report, **WE/99/19**.

DUMPLETON, S., BAKER, S.J. & ROCHELLE, C.A. 1998. *Effects on the Sherwood Sandstone and Drift Aquifers of a Proposed Cooling System and Groundwater Recharge Scheme at Hazlewood Preserves Ltd, Selby, North Yorkshire.* British Geological Survey Technical Report, **WE/98/44**.

FERRARI, C.R. 1996. The case for continuing coal mining subsidence research. *Mining Technology*, **78** (898), 171–176.

GAUNT, G. D. 1994. *Geology of the Country Around Goole, Doncaster and the Isle of Axholme.* Memoir of the British Geological Survey, Sheets 79 and 88 (England and Wales).

Institute of Geological Sciences, *Hydrogeological Map of Southern Yorkshire*, (Sheet 12), scale 1:100 000 Institute of Geological Sciences, London.

KRUSEMAN, G.P. & DE RIDDER, N.A. 1990. *Analysis and Evaluation of Pumping Test Data*, 2nd edn, Publication 47. International Institute for Land Reclamation and Improvement, Wageningen, The Netherlands.

LIU, J. & ELSWORTH, D. 1999. Evaluation of pore water pressure fluctuation around an advancing longwall face. *Advances in Water Resources*, **22**, 633–644.

NATIONAL COAL BOARD, *Subsidence Engineers' Handbook*; National Coal Board, Mining Department, London.

NORTH, M.D. & JEFFREY, R.I. 1997. *Prediction of Water Inflows into Coal Mines From Aquifers*; European Commission Technical Coal Research. Mine Infrastructure and Management. Final Report, **EUR 15198 EN**. European Commission, Luxembourg.

PRICE, M. 1996. *Introducing Groundwater*, 2nd edn. Chapman & Hall, London.

SOUTHERN WATER, 1997. Test Pumping Report, **97/7/1771**. Produced for Dales Water Services Ltd as agents for Hazlewood Preserves Ltd.

SPEARS, D.A. & REEVES, M.J. 1975. The infiltration rate into an aquifer determined from the dissolution of carbonate. *Geological Magazine*, **112**, 585–591.

STRASKRABA, V., FRANK, J., BOSWORTH, W.C. & SWINEHART, T.W. 1994. Study of the impacts of a longwall coal mining operation on surface and ground water resources at the Windsor Mine, West Virginia, USA. In: *Proceedings of the Fifth International Mine Water Congress, Nottingham (UK), September 1994*, IMWA & The University of Nottingham, 125–139.

WALTON, W.C. 1962. Selected analytical methods for well and aquifer evaluation. *Illinois State Water Survey Bulletin*, **49**.

YOUNGER, P.L. & ADAMS, R. *Predicting Mine Water Rebound*, Environment Agency, R&D Technical Report, **W179**.

A physically based model of rebound in South Crofty tin mine, Cornwall

RUSSELL ADAMS & PAUL L. YOUNGER

*Department of Civil Engineering, University of Newcastle,
Newcastle Upon Tyne, NE1 7RU, UK (e-mail:r.adams@ncl.ac.uk)*

Abstract: The recent closure of the South Crofty tin mine, the last working mine of this type in Europe, has raised questions over possible environmental consequences. During several centuries of operation, the mine was dewatered by a series of pumps located at different levels in the mine. Older workings near the ground surface were dewatered by a series of adits that discharge into nearby rivers and streams. The quality of water draining from these shallow workings is generally good and no treatment is required. However, because of bad experiences at the nearby Wheal Jane tin mine, the UK Environment Agency were concerned about the quality and quantity of water which was expected to discharge from the deeper workings when groundwater rebound was completed.

In order to address this problem and make predictions of the timing and volume of the discharge, computer simulations using the SHETRAN/VSS-NET model have been carried out. This model has already been applied to the simulation of groundwater rebound in several UK coalfields. However, the hydrogeological characteristics of coal mines differ considerably from the South Crofty mine. In this mine, the country rocks comprise granite and metamorphic slates, strata that have negligible transmissivity and very low storativity. Most of the groundwater flow is therefore in the 'drives' and stopes from which tin ore was extracted.

The inflows to the mine during its operation were mapped and quantified, and were found to be mainly head-dependent. These inflows usually originated from fault zones in the rock, and also from nearby disused and flooded workings, which surround the modern mine. Predictions of the water level during rebound are compared with the observed water levels in the main shaft which have been measured since the mine closed. The model was then used to predict the dates when surface discharges could be expected to commence. One of the main limitations of the predictions was a 'blank' depth interval in the mine plan records, which could be interpreted in two possible ways: (i) the zone was worked in the mid-nineteenth century but the plans were lost; or (ii) the zone was never worked.

Local professional opinion favoured interpretation (i), and the prediction scenarios considered most probable were based on the assumption that mining-related specific yield values would be similar in this 'blank interval' to those applicable in better-mapped intervals below. In the event, (ii) above appears to have been the true case, resulting in a marked steepening of the rebound curve during the later stages of rebound, with surface discharge commencing in November 2000, as much as a year in advance of most other predictions. A retrospective simulation assuming very low specific yield in the 'blank interval' confirms that the hydrodynamics were otherwise successfully simulated using the SHETRAN/VSS-NET code.

South Crofty mine, located in Cornwall, south-west England, began extracting copper ore in the early 1600s, with production switching to tin ore from the mid-nineteenth century. In October 1985 the price of tin fell dramatically on world markets, with a consequent contraction of the Cornish tin industry. During the 1990s the mine continued to operate at a loss, and closure was announced in 1997, by which time South Crofty was the last remaining tin mine in Europe. Figure 1 shows a plan of the modern mine and its environs.

The possible impacts of the closure of South Crofty have been well documented in a series of reports commissioned by the Environment Agency (Knight Piésold & Partners 1994, 1996, 1997) and also prepared by the mining company (South Crofty plc 1998*a*, *b*). This case study was commissioned by the Environment Agency, who were eager to obtain rebound times from a physically based model rather than from simple void-filling calculations (Adams & Younger 1999).

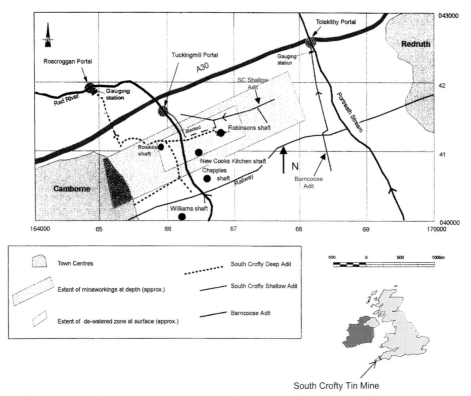

Fig. 1. Location map of South Crofty mine, showing the extent of the modern workings, principal shafts and drainage adits.

The Environment Agency became involved in the contingency planning for closure due to concern over the high risk of polluting discharges affecting the surface water courses (in particular the Red River which flows close to the mine), following a rebound of mine water. Before mine closure, mine water discharging into the Red River from the Dolcoath (Deep) Adit originated from long-abandoned shallow workings and was generally of good quality (Knight Piésold & Parterns 1997). A second adit (Barncoose) also drains shallow workings above the modern mine into the Portreath Stream; however, this adit is at a higher elevation than the Dolcoath Adit. It was anticipated, therefore, that rebounding mine water would reach the Dolcoath Adit before Barncoose Adit. Figure 1 shows the location of the mine and the various adit systems, some of which date back to the eighteenth century. At closure, the adit system was slightly modified to ensure that the mine water would emerge from the circular, brick-lined Roskear Shaft, and flow via a short roadway into the main adit.

The Environment Agency's concern about the pollution risk associated with the South Crofty discharge was partly a consequence of the serious pollution incident caused by the discharge of mine water from the nearby Wheal Jane tin mine in January 1992 (Knight Piésold & Partners 1995). The Agency was anxious to prevent a similar incident following closure of South Crofty, however the quality of the water decanting from South Crofty has actually been of better quality than that of the Wheal Jane mine water due to the lower acid-generating potential of the minerals in South Crofty (Knight Piésold & Partners 1997). The 'first flush' of mine water, however, may result in increased concentrations of copper and zinc in the Red River (which had previously received pumped water from the mine of generally acceptable quality), and some treatment of the discharge has been mooted (Environment Agency pers. comm. 1999) in order to comply with the Environmental Quality Standards (EQS) limits for these metals. EQS are based on the EC Dangerous Substance Directive (Directive 76/464 On Pollution Caused by Certain Dangerous Substances Discharged into the Aquatic Environment, and associated daughter directives).

Mine layout

Workings in South Crofty were divided into a series of near-horizontal levels. Each level formed the base of the stopes (workings), which were driven vertically to extract the tin ore. In traditional Cornish mining terminology, the level is named according to its depth and the shaft used to haul its ore to the surface (e.g. 310 Cooks), with the number referring to the depth in-fathoms (1 fathom = 1.83 m) from adit level in the named shaft to the connection to the workings. The other datum used in more modern references to the mine is the metric mining datum, located 2000 m below sea level (see Fig. 2). Elevations relative to this datum are referred to subsequently as 'm AMD'. Ground surface (at Cooks Kitchen Shaft) is approximately 2111 m AMD, adit level 2051 m AMD. The most modern and deepest workings in South Crofty were at the 470 fathom (1200 m AMD) level, up to 900 m below ground surface. The total plan area of the workings (as implemented in the conceptual model) is around 2 km^2. Figure 2 shows a cross-section north–south through Cooks Kitchen Shaft, showing the workings (principal veins worked in the modern mine are shown by thin black lines) and the dewatered area of the modern mine.

Mine hydrogeology

During the operation of the mine, water was pumped from four major pumping stations from the 420 fathom (1295 m AMD) level (the deepest) to the 195 fathom (1730 m AMD) level (the shallowest), using the Cooks Kitchen Shaft. Between the 225 fathom (1695 m AMD) level and adit level were old workings which were partly dewatered by the 195 fathom level pumps. Estimating the volume of the dewatered workings was extremely important in order to make accurate rebound predictions, therefore, mining engineers with first-hand experience were consulted. The volume was estimated at around 4.5×10^6 m^3.

An assessment of the historical record of pumping was made in order to derive an estimate of the inflow rate of water (water make) during the operation of the mine. Secondly, the dewatered volume of the mine was examined from the data and diagrams in the various reports (South Crofty plc 1998a), in order to assist in the setting up of initial and boundary conditions in the conceptual model described below.

The 195 fathom pumps were used to pump water from the lower workings to the ground surface. The water make from the 195 fathom and higher levels was negligible due to engineering

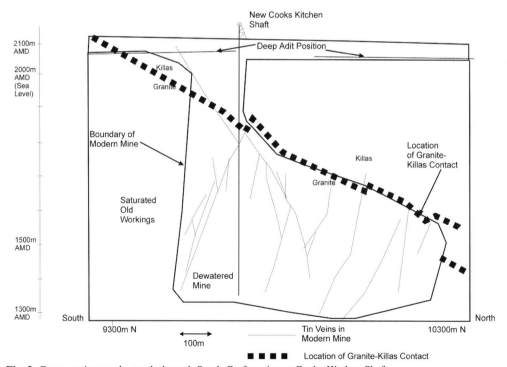

Fig. 2. Cross-section north–south through South Crofty mine at Cooks Kitchen Shaft.

Table 1. *Pumped mine water totals at different levels*

Level (fathoms) (m AMD)	195 (1760)	340 (1495)	380 (1422)	420 (1340)
Average pumped total 1989–1997 ($m^3 h^{-1}$)	273	273	124	48
Average make 1989–1997 ($m^3 h^{-1}$)	0	151	76	48

work carried out in the early 1990s to divert water from the shallow workings into the adit system and away from the mine (Carnon Consolidated 1990). A large flooded area of old workings, called 'the Pool', located between the 195 and 225 fathom (1700–1730 m AMD) levels was used as a reservoir to store the water before it was pumped from the 195 fathom level. From Table 1, assuming that the water pumped at each level migrated downwards to the pump station immediately below, it is clear that the levels between 225 and 340 fathom (1495–1700 m AMD) were contributing the bulk of the water make (55%), with the deeper levels contributing less water make. The majority of the inflows originated from the eastern side of the mine. The largest inflows originated from the 290 and 260 fathom level (1584–1667 m AMD) and were thought to originate from higher flooded workings connected via a major fault zone to the modern mine (M. Owen pers. comm. 1999).

Meteorological data

The meteorological data recorded at stations near the mine were obtained and examined in order to: (i) assess the likely recharge into the mine; and (ii) to examine the relationship between rainfall and the volume of water pumped from the mine. A long record (starting 1950) was obtained from three local rain gauges. The 1961–1990 average annual rainfall at the gauge nearest the mine was 1106 mm. In order to assess the likely timing of rebound in the mine it was necessary to study a long record of rainfall. The current baseline climate period of 1961–1990, extended to include the recent record up to 1998, were analysed. Two extreme scenarios were investigated and 3 year periods of monthly data extracted for input to the model (see below). The aim was to produce an appraisal of the post-closure behaviour of the mine under 'best case' (i.e. extreme dry conditions) and 'worst case' (i.e. extreme wet conditions), and produce an 'envelope' of different times at which flow into the Dolcoath Adit from the Roskear Shaft would commence. The magnitude of the extreme conditions was predicted from the long-term rainfall record for the area. The two scenarios are described below.

Calculation of infiltration into the mine. The infiltration into the mine was calculated from the rainfall and evapotranspiration data using equation (1). No infiltration is possible if actual evapotranspiration exceeds rainfall:

$$I = 0.25 \times (P - E) \qquad I \geq 0 \qquad (1)$$

where I is the infiltration; P is the rainfall; and E is the actual evapotranspiration. All units are in mm.

Values of infiltration were calculated on a daily time series for January 1989 – January 1998. In equation (1) the factor 0.25 is derived by assuming that surface runoff equalled 75% of net rainfall. In an urban district like the Camborne–Redruth conurbation, in which South Crofty mine is situated, infiltration can be as low as 25% of net rainfall. This is due to direct runoff into the local storm drainage and sewer system, from where the rainfall will be discharged directly to local surface watercourses. The high runoff is a consequence of having large paved areas such as industrial estates and car parks in the vicinity of the mine.

Relationship of pumping data with rainfall. The relationship between net rainfall (i.e. infiltration) totals and pumped totals was examined in order to assess: (a) which inflows to the mine were head-dependent and which inflows originated from rainfall infiltrating the shallow workings and shafts; and (b) the time taken for fluctuations observed in the rainfall to be shown in the variation in water pumped from the mine. A series of regression analyses were performed using the weekly pumping totals from the individual pumping stations to determine these relationships. The results indicated that the inflows to the deeper pumps were head-dependent and not from infiltrating rainfall, as the correlation coefficients between rainfall and pumping rate at the 380 fathom and 420 fathom level pumps were very small. The correlation coefficients were much higher between the weekly rainfall totals and pumping totals at the 195 fathom level pumps (0.57). The analyses also identified a lag of approximately 1 month between peaks in the rainfall and peaks in the pumping rate. For the purpose of

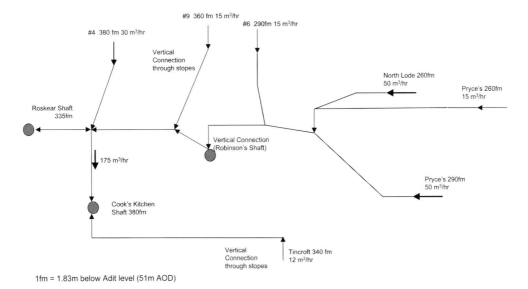

Fig. 3. Schematic diagram of modelled inflows and the pipe network included in the conceptual model, depicting flow rates and location (by mined level).

this study, the inflows were taken as average values, as insufficient data were available for any detailed analysis. Figure 3 shows the locations and magnitudes of the different inflows included in the final model. On average, from analysis of the pumping data, head-dependent inflows accounted for approximately 60% of the total inflow to the mine.

Hydrogeological modelling of rebound

Overview of the SHETRAN/VSS-NET model

The simulation of rebound in South Crofty mine was carried out using the SHETRAN physically based modelling system (Ewen et al. 2000), incorporating the VSS-NET component (Adams & Younger 1997). The VSS-NET component can represent: (i) laminar flow in variably saturated three-dimensional (3-D) porous media (Parkin 1996); and (ii) turbulent flow in open roadways or shafts as a result of the flooding up of old mine workings. The model had been previously applied to UK abandoned coal mines (Younger & Adams 1999).

A 3-D model such as SHETRAN requires large datasets to parameterize the model domain. In the case of South Crofty, the model domain was selected based on the following assumptions:

- The lateral extent of the mine is defined by the location of the modern workings from 195 (1730 m AMD) to 445 fathoms (1295 m AMD), shown in Fig. 2.
- The mine was dewatered at the time of closure to at least the depth of the lowest workings. The initial conditions in the model were set up accordingly (see below), and the bottom boundary of the model is defined by the level of the lowest stopes on the 445 fathom (1340 m AMD) level. Figure 4 shows a cross-section east–west through the SHETRAN conceptual model. The horizontal lines between the vertical columns indicate the lateral flow pathways between model elements.
- Intact rock (granite and slates) is assumed to be of low permeability and assigned a low porosity and hydraulic conductivity ($K = 10^{-5}$ m day^{-1}, porosity = 0.1%). Therefore, outside the dewatered area of the mine, there is negligible groundwater flow even from saturated old workings to neighbouring unsaturated dewatered working, except through faults and fracture zones.
- Areas where the mine plans indicate stoping have a much higher hydraulic conductivity and a porosity calculated from the void volume divided by the total volume of the workings ($K = 1$ m day^{-1}, porosity = 0.14–0.65%). The different porosity values are indicated by different shades of grey in Fig. 4. Where two levels have the same colour (e.g. 340 and 260 fathoms), this

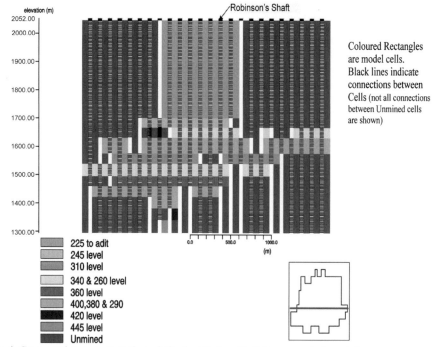

Fig. 4. Cross-section east–west through Cooks Kitchen Shaft (in conceptual model).

grouping has arisen because both layers have similar porosities.

- The head-dependent inflows will be routed to Cooks Kitchen Shaft at the 380 fathom (1422 m AMD) level by a pipe network, according to a schematic diagram (South Crofty plc 1998b) (see Fig. 3). This network also allowed turbulent flow between the Roskear and Cooks Kitchen shafts in either direction. Head-dependent inflows were reduced gradually from an initial maximum value to zero when the water level in the mine rebounded above the inflow level. Inflows from rainfall and surface runoff were applied to each grid element at adit level in the model, forming the top boundary condition.

The model domain comprised 167 125 m^2 finite-difference elements vertically discretized using 200 cells representing 10 layers (Fig. 4). The layers varied in thickness between 5 m (at and just below adit level) with 0.5-m cells and 348 m with 5-m cells.

Simulations

The first simulation carried out by the model was a baseline simulation representing an 'historical' period during which the model results could be compared with the observed water levels measured in the mine during rebound. For this period, the historical rainfall data and actual evapotranspiration data were available allowing daily infiltration values to be calculated using equation (1). After this simulation was run, the two extreme meteorological scenarios, described above, were run. These scenarios comprised the wettest and driest 3 year periods extracted from the 1961 to 1998 monthly rainfall (described above). These simulations are listed below.

- 'Historical' – the model was run from 1 February 1997 to 31 December 1999, and the water level rebound predicted by the model compared with the dips taken in the Cooks Kitchen Shaft. A pumping well (in the model) was located at the corresponding grid element, and the pumping rate was calculated from the historical pumping rates for 1997 and the first 2 months of 1998.
- 'Wet' – infiltration values were calculated from the 1965–1968 rainfall series (3 years commencing July 1965). These data were used to run the model from January 2000 onwards. The average annual rainfall for this period was 1269 mm $year^{-1}$.

- 'Dry' – infiltration values were calculated from the 1990–1993 rainfall series (3 years commencing March 1990). Again, these data were used to run the model from January 2000 onwards. The average annual rainfall for this period was 955 mm year^{-1}.

Results

Historical simulation

The 'historic' simulation was initially run for the final year of mining operations (the year ending March 1998) to establish steady-state conditions by pumping water from the Cooks Kitchen Shaft. Later, the simulation was extended after more recent rainfall data became available. After pumping of the mine ceased in March 1998, water levels in the mine, and in the model, began to rise. Figure 5 shows the predicted water levels in Cooks Kitchen Shaft (solid line), obtained from the model results, against the observed water level dips (shown as black crosses). The time axis begins in May 1998, as before this time there were few measurements of water level taken. In general, the model predictions are within 20 m of the observed water levels, and the general rate of increase of water level has been reproduced. There is a slight underprediction of water levels during the summer of 1999 once the water level reached 1650 m AMD and this could be due to an overestimation of the storage coefficient in this level of the mine (260 fathoms). The rate of rebound of mine water in the upper levels would also be underestimated if the head-dependent inflows were reduced to zero too early during the rebound process, as there were no data on the head elevations in the water-bearing workings producing the inflows. At the start of rebound, head-dependent inflows totalled 182 m^3 h^{-1}, this was approximately 60% of the total inflow to the mine (depending on the net rainfall).

Future scenarios

From January 2000 onwards, the 'wet' monthly time series of recharge values were used in the model. The simulation was run until the water levels in the Roskear and Cooks Kitchen shafts were predicted to reach adit level (2051 m AMD). An estimate of the flow rate discharging from the Roskear Shaft into the Dolcoath Adit was also made, the maximum discharge into the adit was predicted to be approximately 121 s^{-1}. The predicted water level in the Roskear Shaft under the 'wet' scenario is shown in Fig. 6 by the black line, under the 'dry' scenario by the grey line and the black triangles show the observed water levels recorded during 2000 for comparison. Again, the underprediction of the levels was initially around 20 m; however, the model subsequently failed to reproduce the sharp increase in the rate of rise of water level during 2000, which resulted in discharges into the Dolcoath Adit commencing on 8 November 2000. The results of the 'wet' and 'dry' scenarios are summarized in Table 2 with the observed time of rebound for comparison.

In order to ascertain whether the model was capable of reproducing the actual rebound curve a final simulation has recently been carried out (again using the 'wet' monthly time series of recharge values from January 2000). The model was 'back-fitted' to the rebound curve by adjusting the storage parameter in the top layer, which extends from the 225 fathom level to the Dolcoath Adit. Previously it was assumed that this layer was substantially mined (no mine plans clearly indicating workings in this layer were available, the plans showed a 'blank space' here), however, the rapid rise in water levels from the spring of

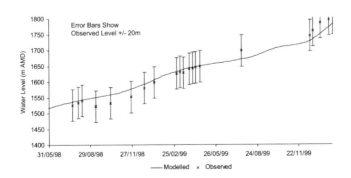

Fig. 5. Observed and modelled water levels in Cooks Kitchen Shaft from June 1998 to January 2000.

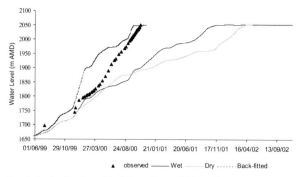

Fig. 6. Modelled water levels in the Roskear Shaft from June 1999 until emergence of mine water in the Dolcoath Adit, under wet, dry recharge and 'back-fitted' scenarios.

Table 2. *Predicted rebound times under different scenarios*

Rainfall scenario	Days to start of discharge	Discharge starts	Month when water reaches 2031 m
Wet	1362	November 2001	September 2001
Dry	1500	April 2002	February 2002
Actual	983	8 November 2000	18 October 2000
Back-fitted	992	17 November 2000	September 2000

2000 indicated that the actual storage in this layer was much smaller than anticipated (back-filling with waste rock could have occurred). Therefore, the porosity of this layer was reduced by 27% and the new prediction is shown in Fig. 6 by the dotted line. The new model simulation predicted that mine water would emerge on 18 November 2000.

Conclusions

The conceptual model developed to simulate rebound in South Crofty tin mine has predicted dates of mine water emergence in the Dolcoath Adit. However, discharges into the adit commenced in November 2000, much earlier than predicted. The main reason for the overestimate of the time to rebound is thought to be an overestimate of the volume of voids in the upper (225 fathoms to adit) levels of the mine, the results of this study indicate the importance of accurate measurements of the storage coefficient in predictions of rebound times. A second reason may be the extremely high rainfall recorded in southwest England in late summer and autumn 2000. Thirdly, it is possible that some of the old workings were backfilled with waste rock. However, with the benefit of hindsight, it was possible to back-fit the rebound curve to match the predicted date within a few weeks, by reducing the porosity in the top layer.

The model has also incorporated head-dependent inflows of water into the mine workings from adjacent flooded mines through faulted and fractured zones. These were thought to be active during the early stages of the water level rise in the mine. Until early 2000 (when the simulations were initially run) predicted water levels in the main shaft were within 20 m of the observed levels. The study also showed a method of distinguishing head-dependent from rainfall-derived inflows by regressing historical pumping rates with rainfall data. Current monitoring will determine if the water discharging into Dolcoath Adit is of poor quality and requires some treatment.

The authors would like to thank the following individuals for their assistance with this research: M. Owen (Mining Geologist formerly with South Crofty plc); J. Wright, C. Marsden and T. Howells of the Environment Agency (South West Region) and B. Sansom (Knight Piésold & Partners Ltd). The modelling work made use of a UNIX workstation funded jointly by United Kingdom Nirex Limited and NERC grant GR3/E0009.

References

ADAMS, R. & YOUNGER, P.L. 1997. Simulation of groundwater rebound in abandoned mines using a physically based modelling approach. *In*: VESELIC,

M. & NORTON, P.J. (eds) *Proceedings of the Sixth International Mine Water Association Congress, 'Minewater and the Environment' 1*. Bled, Slovenia, 8–12 September 1997. **Vol. 2**, 353–362.

ADAMS, R. & YOUNGER, P.L. 1999. *Modelling the Timing of rebound at South Crofty*. Report to the Environment Agency, South West Area, Nuwater Consulting Services Ltd, Newcastle Upon Tyne.

CARNON CONSOLIDATED. 1990. *South Crofty Hydrology Study*. Unpublished Report by BUCKLEY, J.A., Carnon Consolidated, Camborne.

EWEN, J., PARKIN, G. & O'CONNELL, P.E. 2000. SHETRAN: a coupled surface/subsurface modelling system for 3D water flow and sediment and contaminant transport in river basins. *ASCE Journal of Hydrologic Engineering*, **5**, 250–258.

KNIGHT, PIÉSOLD & PARTNERS. 1994. *South Crofty Minewater Study*. Final report to National Rivers Authority, South Western Region, Exeter.

KNIGHT, PIÉSOLD & PARTNERS. 1995. *Wheal Jane Minewater Study, Environment Appraisal and Treatment Strategy*. Report to the National Rivers Authority, South Western Region, Exeter.

KNIGHT, PIÉSOLD & PARTNERS. 1996. *South Crofty Mine Closure Contingency Plan*. Final report to Environment Agency, Bodmin.

KNIGHT, PIÉSOLD & PARTNERS. 1997. *South Crofty Strategic Closure Study*. Provisional report to Environment Agency, Bodmin.

PARKIN, G. 1996. *A Three-dimensional Variably-saturated Subsurface Modelling System for River Basins*. Unpublished PhD Thesis, Department of Civil Engineering, University of Newcastle Upon Tyne.

SOUTH CROFTY PLC. 1998a. *Dolcoath Deep Adit, Report on its Condition and the South Crofty Mine Decant Location*. Report to Environment Agency, Bodmin.

SOUTH CROFTY PLC. 1998b. *South Crofty Mine Closure*. Internal Report, South Crofty plc, Camborne.

YOUNGER, P. L. & ADAMS, R. 1999. *Predicting Mine Water Rebound*. Environment Agency, R&D Technical Report, **W179**.

Coalfield closure and environmental consequence – the case in south Nottinghamshire

N.S. ROBINS[1], S. DUMPLETON[2] & J. WALKER[3]

[1]*British Geological Survey, Maclean Building, Wallingford, Oxfordshire, OX10 8BB, UK (e-mail: N.Robins@bgs.ac.uk)*
[2]*British Geological Survey, Kingsley Dunham Centre, Keyworth, Nottingham, NG12 5GG, UK*
[3]*Scott Wilson Kirkpatrick & Company Limited, Bayheath House, Rose Hill West, Chesterfield, S40 1JF, UK*

Abstract: The strata within and above the South Nottinghamshire Coalfield dip gently towards the east. There are many abandoned shallow workings in the western area where the coalfield is exposed, but to the east the coalfield is concealed beneath Permo-Triassic strata. The coalfield has yet to suffer closure, mine water rebound and the acid mine drainage (AMD) cycle. A very large area has been exploited with complicated internal drainage systems dependent on the maintenance of existing pumping regimes. An evaluation of the AMD threat has been carried out with particular regard to the risk posed to the Sherwood Sandstone aquifer, which overlies the concealed part of the South Nottinghamshire Coalfield. The evaluation has been assisted by three-dimensional (3-D) visualization that has enabled lumping of plentiful mine abandonment data, and predictive runs using the University of Newcastle GRAM model. These studies indicate that the critical spill-over elevation is 41 m above Ordnance datum (aOD), and that the aquifer will be at risk about 20 years after pumping ceases from the Coal Measures.

Wholesale abandonment of deep coal mining took place throughout the UK during the 1990s. This created an urgent and desperate need to understand the physical and chemical processes that accompany mine abandonment in order to predict environmental consequence and, if necessary, to trigger intervention. In fact, the mine closure programme was effected so quickly that need for intervention was almost immediately replaced by need for remediation (Younger 1999). A wide range of investigation and analysis has now been carried out, with field observations used to validate simulations with a satisfactory degree of confidence. As a consequence, the chemical quality of acid mine drainage (AMD) at any given time following mine abandonment and the rate of mine water rebound can be predicted with some certainty for individual collieries and for small groups of interconnected mine workings (Younger 1998). However, the constraints of data handling require that larger and more complicated coalfields be analysed using lumped data and this may have a detrimental effect on the confidence of the results.

The coalfields of Scotland and South Wales are, for the most part, at an advanced stage of the mine water rebound and the AMD cycle. The key issues are remedial action using cost-effective clean-up such as reed bed technology, and the protection of geotechnical structures such as foundations, tunnels or landfill sites from attack by aggressive AMD. Discharge from areas such as the Coventry and Staffordshire Coalfields has yet to peak, but these mine systems are relatively small, particularly when compared with the South Nottinghamshire Coalfield. Similarly, the problems currently facing the Northumbrian and Durham Coalfields are also manageable in terms of their size and data assimilation, analysis and modelling. However, the magnitude of the South Nottinghamshire Coalfield, and the daunting wealth of data pertaining to it, overshadow the prospect of sophisticated predictive modelling, the more so when effective clean-up methodologies are available. More importantly, the deep South Nottinghamshire Coalfield has yet to be abandoned.

The South Nottinghamshire Coalfield is the last remaining large AMD problem area yet to materialize in the UK. The coalfield retains several deep working pits and there are large areas of multiseam workings currently being

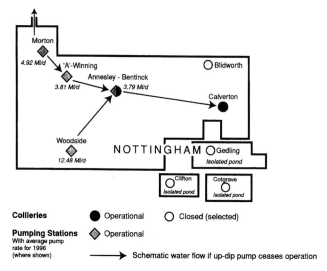

Fig. 1. Schematic hydraulic connections within the South Nottinghamshire Coalfield.

dewatered. The quality of the pumped mine waters to date has been adequate for discharge to streams and rivers without treatment. Up-dip abandoned workings are also dewatered to intercept water percolating down from the exposed coalfield to the west where there are many long-abandoned shallow workings. The interconnections and perceived hydraulic barriers in the coalfield are shown schematically in Fig. 1, and an outline cross-section of the exposed coalfield in the west and the concealed coalfield in the east is shown in Fig. 2. Despite the size and complexity of this coalfield there is an urgent need to evaluate abandonment processes to determine the economics of whether the pumps could be switched off or if the consequential environmental damage would be too great.

In this area not only is the shallow and surface aqueous environment at the exposed coalfield under threat from the emergence of AMD, just as it is or has been in all the other UK coalfields, but the important Sherwood Sandstone aquifer above the concealed part of the coalfield might also be at risk. Although the quality of mine dewatering discharge is generally favourable, flooding of dry workings, containing minerals such as soluble oxidized products of pyrite, creates poor quality, metal-rich water or AMD.

In Nottinghamshire, the Sherwood Sandstone Group forms the principal regional aquifer responsible for meeting potable demand of nearly 400 Ml day^{-1}. Much of the abstraction is used for public water supply, but private agricultural and industrial use is also significant. Hydrographs for the aquifer north of Nottingham indicate a

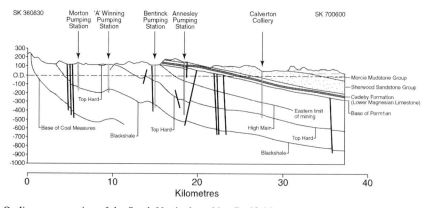

Fig. 2. Outline cross-section of the South Nottinghamshire Coalfield.

slight downward trend in piezometric level over recent years, and there is a concern from both the Environment Agency and local environmental groups for the sustainability of current resource exploitation and for sensitive surface water and wetland sites.

There is already one case of upward migration of mine water polluting an aquifer in the UK. In the southern part of the Durham Coalfield, Coal Measures are covered by the Permian Magnesian Limestone aquifer, which is heavily exploited for public supply. One mine, the Mainsforth Colliery, is situated close to the feather edge of the Permian outcrop, and there are numerous subsidence fractures that locally propagate up into the aquifer. Frost (1979) reported sustained inflow of Magnesian Limestone water into the former colliery workings. Once pumping had ceased, these same fractures allowed flow to occur in the reverse direction rapidly creating a 10 m rise in the piezometric head in the limestone aquifer. This was accompanied by a significant deterioration in water quality in the aquifer.

The South Nottinghamshire Coalfield

The Carboniferous strata within the coalfield have an easterly dip away from the Pennine anticline. The Millstone Grit Series lies below the Lower and Middle Coal Measures, and these strata collectively dip gently eastwards beneath Permian marls, which are in places locally absent but are generally thickest in the east, and the overlying Triassic Sherwood Sandstone aquifer (Fig. 2). The Coal Measures have been extensively exploited in the South Nottinghamshire and Derbyshire Coalfields. The Coal Measures comprise alternations of grey sandstones, which are typically fine grained, with grey siltstones, mudstones, coals and seat-earths, and ironstones. The sandstones in this area generally constitute less than a quarter of the succession, but there are several large channel deposits that are up to 8 m thick.

The Coal Measures Group constitute a complex multilayered aquifer. The predominantly argillaceous strata form aquitards that isolate subordinate sandstone horizons, which act as discrete groundwater bodies. The strata are extensively faulted and vertical hydraulic connections are common across otherwise weakly permeable mudrocks and coals. However, some faulted blocks of interbedded sandstone may be isolated from any source of recharge, and once dewatered may remain largely dry. Little, if any, natural water transport occurs beneath depths of about 200 m in unmined areas because of the poor interconnectivity of sandstone horizons (Rae 1978). However, collieries that were situated in local synclinal areas were far wetter than those on anticlinal ridges because groundwater tended to migrate down-dip in the Coal Measures (Downing et al. 1970).

The widespread extraction of coal in south Nottinghamshire has greatly enhanced the transmissive properties of the Coal Measures. The creation of open shafts, roadways and galleries, together with collapsed, goaf-filled, longwall panel workings, has altered the hydrogeological properties of the remaining strata to create a complex of worked areas separated by in situ, but disturbed, country rock. Aided by mining-induced subsidence, the hydraulic conductivity of the whole sequence is now greatly enhanced over the natural state. Nearly all the deep mining has taken place within a sequence approximately 350–500 m thick between the Blackshale seam in Westphalian A to the High Main seam in Westphalian B (Fig. 2). At the centre of the sequence is the Top Hard, which has been worked to exhaustion in many collieries, many of which were linked underground through this seam.

It has generally been considered that the Permian Marl sequence overlying the Coal Measures was sufficient hydraulic seal to prevent ingress of mine water contamination into the overlying Sherwood Sandstone aquifer should significant mine water rebound be allowed to take place. However, the marl is not universally present and in places the Triassic sandstones come into close proximity with, or rest directly on top of, the Coal Measures strata (Bishop & Rushton 1993). In addition, inadequately sealed boreholes and shafts may provide a hydraulic connection across the marl (Fig. 3). Finch (1979) reported small areas of saline water in the Sherwood Sandstone aquifer associated with vertical fracturing. Rather than upward move-

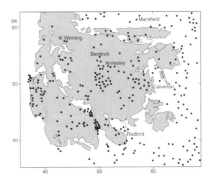

Fig. 3. Boreholes and shafts penetrating the base of the Permo-Triassic, and pumping stations. The extent of workings in the Top Hard, Blackshale and Kilburn is stippled.

ment of mine water, however, he concluded that the pockets of salinity derived from downward percolation of pumped mine water that was discharged to surface waters, which were losing to the sandstone aquifer.

Water make in the concealed coalfield from downward transport of groundwater from the Permian and Triassic strata is small to negligible compared with down-dip flow of water recharged directly to the exposed part of the coalfield. However, once the concealed coalfield is flooded and the base of the Permo-Triassic sequence is subject to a reversal in head gradient, fractures and other flow paths may open up allowing mine water to migrate towards and into the Sherwood Sandstone aquifer. Besides, there are many shafts and boreholes penetrating through to the Coal Measures (Fig. 3), and some of these will inevitably contain imperfect hydraulic seals.

Mine water risk evaluation

The present-day dewatering scheme (as at January 2001) protects the remaining working pits. It can be visualized as a set of interlinked and partially overflowing underground ponds. The shallow ponds are recharged locally and flow down-dip to be intercepted by pumping stations and working dewatered collieries. If the pumps were to be switched off, the ponds would slowly fill and mine water would rise towards the Sherwood Sandstone aquifer, as well as emerge at surface in the exposed coalfield.

Quantification of this risk and prediction of both breakout time after pumping ceases and of water quality in the future is both difficult and fraught. This is because the coalfield is large and contains a complex interlinked system of mine workings penetrating a variety of different coal seams. A key to the evaluation is the identification of the main recharge process to the system in order to quantify the volume of incoming water. In broad terms, the volume of recharge equates to the dewatering volume currently taking place. A second important parameter is the actual void space that needs to be filled to saturate the system of mine voids, high porosity goaf and fractured rock areas. Neither are easy to establish with any degree of certainty.

The quality of the generated AMD depends on the pyritic sulphur content of the strata being flooded. The distribution of sulphur in the South Nottinghamshire Coalfield has not been mapped and quality prediction can only sensibly be carried out by analogy to abandoned coalfields elsewhere and by the simple estimation techniques such as those advocated by Younger (2000). As a general rule the sulphur content is greatest wherever marine bands occur.

A number of different modelling approaches have been made to address the problem of AMD breakthrough prediction. The University of Newcastle program GRAM, Groundwater Rebound in Abandoned Mineworkings, has been successfully deployed to predict mine water arisings at the surface in a number of different colliery settings (Sherwood & Younger 1997). This model is based on planar recovery of levels in discrete ponds, each overflowing into the next lower one at a critical lip, probably a roadway or interconnecting adit between collieries. More recently, the surface and groundwater modelling package SHETRAN has been used to evaluate mine water rebound (Adams & Younger 2000), as well as a number of other packages that have also been successfully employed.

One of the basic problems in the larger-scale evaluations is the bringing together of diverse sets of data to create a conceptual model. This, usually difficult and often subjective, process can be formalized by adopting a three-dimensional (3-D) geographical information system (GIS) and visualization package for data assimilation. There are a number of commercial packages available including EARTHVISION and VULCAN. It was the latter that was adopted for analysis of the South Nottinghamshire Coalfield (Dumpleton et al. 2001). However, even using this 3-D package, it was found that not all the available data could usefully and cost-effectively be incorporated into the model, and data lumping was an essential part of the model design. It was considered that the implied approximation of mine void data caused by data lumping was acceptable given the already coarse nature of the water balance estimation (Table 1).

The visualization of the southern portion of the South Nottinghamshire Coalfield concentrated on bringing together the mapping and borehole geological data beneath a digital terrain model. Mine abandonment plans were more difficult to deal with because of the volume of data that needed to be digitized, the range of different types of plan and the accuracy associated with them. Three key horizons were selected for digitization: the Top Hard seam was adopted as the central important coal seam; the High Main the upper seam; and the Kilburn the lower seam of interest. Of the other 30 or so seams that have been worked in the area, data were selected for critical areas only, such as mine pumping stations and reported wet areas where mine water makes have been particularly high. Features that were incorporated into the model included:

Table 1. *Outline water balance* for the catchment area of the south Nottinghamshire exposed coalfield (after Dumpleton et al. 2001)*

Inputs	Volume (Mm^3 year^{-1})	Outputs	Volume (Mm^3 year^{-1})
River flow onto catchment area (Rivers Amber and Erewash)	Not known	River flow out of the catchment	107
Sewer water/main leakage	Small	Mine water pumping	9
Effective precipitation	77, of which stream flow separation and distribution of sandstones suggests recharge may amount to 35	Groundwater abstraction	0.5

* The water balance assumes that the rivers gain as baseflow the difference between estimated recharge and mine water plus groundwater pumping.

- areas of goaf;
- principal in-seam mine roadways linking areas of goaf;
- cross-measures drifts;
- seam contours and spot levels;
- selected shafts and boreholes.

Virtual flooding of the visualization model identified a set of 'hotspots' at which AMD arisings may potentially cause problems. In the exposed coalfield, a critical low-elevation area occurs near Radford (SK5498 4101), where the Top Hard workings area are shallow and old shafts exist at the former Radford and Newcastle collieries. The critical spill-over elevation to ground surface is estimated to be 41 m above Ordnance datum (aOD). This becomes the key control to mine water recovery levels in the whole coalfield, and it is this elevation that prescribes the risk of mine water rising into the Sherwood Sandstone aquifer in the concealed part of the coalfield (Fig. 4).

After the initial flooding of the mine complex, small volume laminar flow will prevail through roadways and other effective pipe-flow voids. The head difference between the concealed coalfield and down-dip beneath the concealed coalfield is, therefore, likely to be small. It is assumed that as near steady state is approached the head on the Coal Measures in the concealed coalfield will attain an elevation approaching 41 m aOD, perhaps locally constrained in relatively undisturbed areas to a slightly lower head. The regional head in the Sherwood Sandstone aquifer over the concealed coalfield ranges from only 20 m aOD in the eastern down-dip part of the aquifer, to between 40 and > 100 m aOD along the western feather edge of the aquifer. It is, therefore, the critical central and eastern area of the Sherwood Sandstone aquifer above the worked concealed coalfield, where there is both an upward groundwater (mine water) flux from the Coal Measures and where the aquifer is undermined, that is potentially vulnerable to migration of AMD.

Borehole-specific capacites in the Sherwood Sandstone aquifer range typically between 1 and

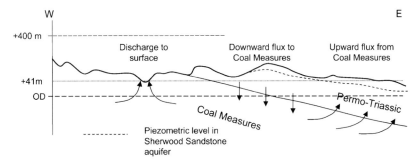

Fig. 4. Schematic cross-section showing likely surface discharge and groundwater fluxes to and from the Sherwood Sandstone aquifer at the critical mine water recovery level of 41 m aOD.

$2 \, \text{l s}^{-1} \, \text{m}^{-1}$, so that a high yielding public supply borehole, yielding $40 \, \text{l s}^{-1}$, would produce a drawdown of between 20 and 40 m. Test pumping data for boreholes at Edwinstowe (SK 6267) indicate a transmissivity of $1500 \, \text{m}^2 \, \text{day}^{-1}$ and a storage coefficient of 0.005, typical for this aquifer. The 4-day specific capacity for this site derived from a drawdown of 16 m, created by a pumping rate of $28 \, \text{l s}^{-1}$, is $1.8 \, \text{l s}^{-1} \, \text{m}^{-1}$. Drawdowns of this order have obvious implications for potential upconing of AMD leaking up from the base of the aquifer, given the small difference in head between the aquifer and the recovered Coal Measures water below.

Interrogation of the visualization model rapidly identifies critical areas in the concealed coalfield. These occur where worked seams approach the base of the Permo-Triassic strata, where the Permian marls are locally thin or absent, or where shafts and boreholes occur that may be inadequately sealed against the rising head in the Coal Measures.

Dumpleton et al. (2001) predicted the time taken to attain these critical elevations after pumping ceases using GRAM and a variety of different pumping scenarios. These were based on either a single 'pond' system or a system of two ponds. Inflow was derived from a simple, but essentially crude, water balance estimate (Table 1), with vertical inflow to the concealed coalfield considered to be relatively insignificant. The lapsed time between cessation of pumping or shut-down of selected pumps to AMD discharge at the critical elevation of 41 m aOD ranges from only 20 years if the system is modelled as a single pond to over 50 years if it is considered as two separate ponds. This model indicates that criticality is possible, as a worst case, from about 20 years onwards.

Environmental consequence

Loss of present-day good quality mine water discharges to surface waters may have an adverse effect on low-flow quality, for example through the dilution of sewage treatment effluents in some rivers and streams. AMD arising at surface in the Newcastle and Radford areas of the exposed coalfield can be dealt with by remediation, possibly also by interception pumping, but is not likely to cause significant environmental damage. AMD percolating into the base of the Sherwood Sandstone aquifer may have more serious consequences such that a programme of scavenger pumping could be required to safeguard the quality of the aquifer. The need to prevent pollution in the Sherwood Sandstone aquifer is underlined in the EC Water Framework Directive. This ensures the prevention of mine water ingress to the aquifer (even if the public water supply could be maintained with the loss of this resource). However, the precise quality of the AMD generated in the system is not known, but the presumption is that it will be net alkaline and aggressive during the initial phases of flooding.

Interrogation of the VULCAN visualization model reveals areas of potential risk to the Sherwood Sandstone aquifer. The controlling elevation of mine water discharge is at 41 m aOD in the main, fully interconnected, part of the South Nottinghamshire Coalfield. In the eastern part of the Coalfield, the base of the Sherwood Sandstone generally lies below this level. The continuing easterly dip brings the piezometric level in the Sherwood Sandstone Group below this critical elevation in the eastern extremity of the South Nottinghamshire Coalfield above much of the Bestwood, Calverton and Blidworth mine workings, i.e. there is an upward flux from the Coal Measures. The stacking effects of multiple seam workings, particularly in the High Main, Abdy-Brinsley, High Hazles and Top Hard seams, on the overlying Permo-Triassic strata have not been fully investigated. It is possible that such effects may create new pathways for future upward flow of mine water. Deteriorating seals in mine shafts and exploration boreholes may also provide similar flow paths.

Conclusions

Mine water recovery in most of the UK coalfields is at an advanced stage and requires remediation rather than prevention. Mine dewatering is still occurring in a few coalfields. Most of the coalfields are relatively small and limited to few underground interconnections so that data assembly, handling and modelling is a manageable task. The South Nottinghamshire Coalfield is still pumped and the mine water rebound cycle will only commence when pumping ceases. Predictive modelling of mine water levels, given shut-down of all or just of selected pumps, has been successfully carried out using a combination of 3-D visualization of the data and the University of Newcastle GRAM mine water rebound estimation model. This modelling combination was best suited to the large area of the South Nottinghamshire Coalfield and the overwhelming volume of mine plan data, albeit of variable quality, available for this area.

Given that up to 30 different seams have been worked in the South Nottinghamshire Coalfield, effort was concentrated on digitizing three key horizons into the 3-D model. The remainder of the mine plan data (i.e. for the thinner and lesser worked seams) was lumped and generalized for the construction of the GRAM model, which was used to predict various rebound scenarios.

Although mine plan data are abundant, the total potential recharge to the coalfield could not be determined other than by equating it to coalfield pumping rates. The coarse nature of this vital input offered justification for the lumping of other data.

The main outcome from the modelling is the prediction that the coalfield will overflow to surface at an elevation of 41 m aOD in the vicinity of the former Radford and Newcastle collieries. This will take at least 20 years to occur, but could be longer depending on the complexity of the flow system. At that time mine water may percolate upwards within the concealed coalfield to emerge at the base of the Sherwood Sandstone aquifer. Interception or scavenger pumping may then be required in order to satisfy the requirement of the EC Water Framework Directive and to retain groundwater quality sufficient for public supply.

Analysis of the adjacent coalfields to the north of the current investigation have not yet been attempted with any degree of detail. However, it is suspected that these areas offer a greater degree of hazard to groundwater in the overlying Permo-Triassic strata because of increased fracture density and the relative geometry of aquifer and Coal Measures in this area.

The work described in this paper was jointly funded by the Environment Agency and NERC. Published by permission of the Director, British Geological Survey.

References

ADAMS, R. & YOUNGER, P.L. 2000. Simulating groundwater rebound in a recently closed tin mine. In: *Proceedings of the Seventh International Mine Water Association Congress, Mine Water and the Environment*, IMWA & Universytet Slaski, Katowice, Poland, 218–228.

BISHOP, T.J. & RUSHTON, K.R. 1993. *Water Resource Study of the Nottinghamshire Sherwood Sandstone Aquifer System of Eastern England: Mathematical Model of the Sherwood Sandstone Aquifer*. Technical Report Department of Civil Engineering, University of Birmingham for the National Rivers Authority.

DOWNING, R.A., LAND, D.H., ALLENDER, R., LOVELOCK, P.E.R. & BRIDGE, L.R. 1970. *The Hydrogeology of the Trent River Basin*; Institute of Geological Sciences, London, Hydrogeological Report **5**.

DUMPLETON, S., ROBINS, N.S., WALKER, J.A. & MERRIN, P.D. 2001. Mine water rebound in South Nottinghamshire: risk evaluation using 3-D visualisation and predictive modelling. *Quarterly Journal of Engineering Geology and Hydrogeology*, **34**, 307–319.

FINCH, J.W. 1979. *The Further Development of Electrical Resistivity Techniques for Determining Water Quality*. PhD Thesis, University of Birmingham.

FROST, R.C. 1979. Evaluation of the rate of decrease in the iron content of water pumped from a flooded shaft mine in County Durham, England. *Journal of Hydrology*, **40**, 101–111.

RAE G.W. 1978. *Groundwater Resources in the Coalfields of England and Wales; the Yorkshire Coalfield*. Technical Note, Central Water Planning Unit, Reading.

SHERWOOD, J.M. & YOUNGER, P.L. 1997. Modelling groundwater rebound after coalfield closure. *In*: CHILTON, P.J. *et al.* (eds) *Groundwater in the Urban Environment: Problems, Processes and Management*. Balkema, Rottedam, 165–170.

YOUNGER, P.L. 1998. Coalfield abandonment: geochemical processes and hydrochemical products. *In*: NICHOLSON, K. (ed.) *Energy and the Environment: Geochemistry of Fossil, Nuclear and Renewable Resources*. Macregor Science, Abedeenshire, 1–29.

YOUNGER, P.L. 1999. Restless waters of the Durham Coalfield: pollution risks and popular resistance. *Bands and Banners*, **1**, 19–21.

YOUNGER, P.L. 2000. Predicting temporal changes in total iron concentrations in groundwaters flowing from abandoned deep mines: a first approximation. *Journal of Contaminant Hydrology*, **44**, 47–69.

Depressurization of the north wall at the Escondida Copper Mine, Chile

PATRICK MCKELVEY[1], GEOFF BEALE[1], ADAM TAYLOR[1], SIMON MANSELL[2], BENJAMIN MIRA[2], CRISTIAN VALDIVIA[2] & WARREN HITCHCOCK[2]

[1]*Water Management Consultants, 23 Swan Hill, Shrewsbury, UK*
(e-mail:pmckelvey@watermc.com)
[2]*Minera Escondida Ltda, 501 Avda de la Minera, Antofagasta, Chile*

Abstract: Concurrent dewatering and slope depressurization operations have been underway at the Escondida open pit since 1996. A hydrogeological investigation has been undertaken as part of the depressurization operation on the north wall. It has shown that there are two distinct hydrogeological units in the slope, the altered and unaltered Escondida Porphyry. The hydrothermally altered Escondida Porphyry is clay rich and has relatively high matrix permeability. The underlying silicified, or unaltered porphyry, has very low matrix permeability and groundwater flow is in steeply dipping NW-trending fracture systems. Pore pressures in the silicified porphyry are higher than the altered material.

A system of horizontal drains and vertical wells has been in operation to reduce pore pressures in the altered porphyry. To investigate effective drainage measures for the silicified porphyry, a groundwater model of the slope was constructed. This analysis showed that a drainage tunnel with horizontal drains was the most effective method of draining significant areas of the slope to the required pore pressure targets.

The Escondida copper mine, owned and operated by Minera Escondida Limitada (MEL), is located 160 km SE of Antofagasta in the Atacama Desert of northern Chile (Fig. 1). The mine has been operating since 1990 and currently 900 000 t of copper are extracted per year from a 400 m deep open pit.

The copper mineralization is associated with a quartz–monzonite intrusive of Eocene–Oligocene age, known as the Escondida Porphyry. Younger rhyolites cover and truncate the porphyry, while andesite is found mainly in the south and west of the pit. In the Hamburgo Basin, 700–1200 m to the east of the mine, alluvial and colluvial deposits of gravel and sands form a groundwater storage reservoir of up to 100 m thickness (Water Management Consultants WMC 1999).

The porphyry copper system of concentric alteration haloes can be generally applied to the deposit, but is modified due to overprinting from multiple volcanic and intrusive events. There are at least three alteration types: (1) potassic alteration seen as pervasive biotite and feldspar; (2) phyllic alteration comprising chlorite–sericite and quartz–sericite; and (3) advanced argillic alteration, which is present as pervasive clay assemblages infilling the fracture networks. The zone of advanced argillic alteration reaches its greatest extent on the north and northeast walls, as shown in Fig. 2. The main structural features are NNW oriented, and define boundaries of hardness, mineralization type, alteration and copper grades. The Panadero Fault Zone on the east of the mine is one of these features (see Fig. 2). It forms a barrier to flow perpendicular to the zone but allows enhanced flow along the strike of the structure (Water Management Consultants 1999).

In 1996, discrete seepage faces formed at high levels on the east wall of the pit, to the west of the Panadero. In addition, water level monitoring showed that groundwater levels in the rhyolites to the east of the pit had risen by 20–30 m from the pre-mining water level. An evaluation confirmed that groundwater was flowing from the Hamburgo Basin to the pit. This recharge flow is partially blocked by the Panadero Zone but enters the pit through specific structural offsets, or leakage points, in the Panadero. Fig. 2 demonstrates the conceptual model of flow to the pit. It is calculated that a maximum of $70 \, l \, s^{-1}$ is flowing to the mine from the Hamburgo Basin.

Since 1996, a major programme of interceptor well drilling has been implemented to cut off

Fig. 1. Location plan.

flow from the basin and to reduce water levels in the eastern rhyolites. Currently, the east wall interceptor system abstracts up to $150 \, \text{l s}^{-1}$ from the rhyolites on the margin of the pit. In association with this interception operation, depressurization work is underway to reduce pore pressures within the pit. The pore pressures within the in-pit rocks are the result of slow dissipation of pore pressure within low permeability material rather than the recharge flow from Hamburgo. Geotechnical studies have been carried out to optimize the stability of the slopes. These studies have shown that pore pressures need to be reduced on the north wall. Therefore, specific hydrogeological investigations were implemented to identify the most effective depressurization methods for that sector of the mine.

Field programme and database

Figure 3 shows the layout of the north wall, with the locations of boreholes and horizontal drains. The rim of the wall is at 3100 m above sea level (m asl) while the base is at 2700 m asl. Approximately 100 boreholes have been constructed on the wall. Initially, single standpipe boreholes were constructed to gain basic information on the phreatic surface distribution. Subsequently, single and multiple piezometers were constructed to measure pore pressures at specific intervals. The dataset for each borehole comprises drilling information (airlift flows and penetration rate), geology (alteration pattern and lithology) and water levels. On most boreholes, a downhole geophysical suite was run comprising caliper, resistivity, gamma, density, neutron and acoustic televiewer.

In addition, there was a targeted investigation programme carried out on the north wall between September 1999 and February 2000. In this

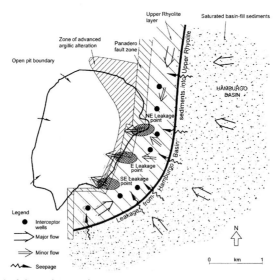

Fig. 2. Conceptual model of the recharge mine.

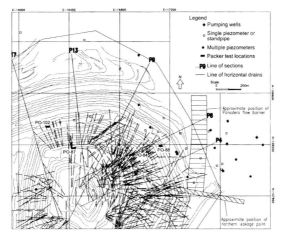

Fig. 3. North wall layout.

hydrogeological investigation the work programme consisted of:

- The construction of multiple piezometer groups – to investigate pore pressures in the various lithological and alteration units throughout the slope;
- The drilling of four 300 m deep holes with packer permeability testing – to assess the drainability of each hydrogeological unit.

Interpretation of the results of the north wall investigation programme, together with the data from existing boreholes, has shown that (Water Management Consultants 2000):

- There is a multilayered rhyolite at the top of the north wall. Groundwater heads in the rhyolite are up to 160 m below ground level (m bgl). Potential recharge flows to this unit are partially blocked by the Panadero Fault Zone or have been cut off by the interceptor wellfield.
- Below the crest of the slope, the geology changes to Escondida Porphyry. The thickness of argillic altered material is around 250 m near the top of the slope. Towards the toe of the slope, the thickness of altered material reduces. Underlying the argillic altered porphyry is silicified porphyry. These units are termed the altered and silicified porphyry, respectively.
- During drilling, airlift flows are not recorded above the altered–silicified rock interface. Airlift flows are variably recorded from transmissive structures below the altered–silicified rock interface. However, significant transmissive structures are not evident in any of the four packer test holes.
- There is a difference in heads between the altered and silicified rock. Silicified rock heads are generally up to 30 m higher than heads in the altered material.
- There is a reduction in permeability between altered material and silicified rock. This is seen in the packer test results, which are displayed in Fig. 4. Altered material permeabilities range from 10^{-4} to 10^{-6} cm s^{-1}. Unfractured silicified rock matrix permeabilities are measured in the range 10^{-7}–10^{-8} cm s^{-1}. However, 10^{-8} cm s^{-1} is about as low a permeability as can be measured using packer test equipment, and the matrix permeability of some sections of the unaltered material is expected to be lower.
- These permeability values suggest that the altered material can be globally drained, but that the silicified rock can only be drained if a sufficient number of transmissive structures are intercepted.

Approximately 100 drains have been drilled into the north wall. Data available for these drains include position, orientation, length and weekly flow measurements. The majority of the drains are drilled in the middle sections of the slope. Generally, they are drilled horizontally to a length of 300 m. Drain flow is piped to sumps and pumped out of the pit.

Regular monitoring of the drain programme shows that there is a relatively rapid decline in total drain flow as the drains remove water stored within the rock in their area of influence. High combined flows are only sustained where new drains are continually being drilled into zones that have not yet been depressurized. When drains

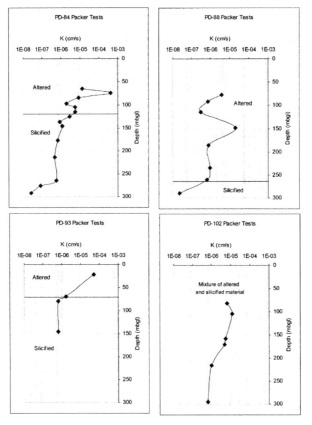

Fig. 4. Packer test results.

are not replaced, then there is a steady drop off in total flow. This is demonstrated in Fig. 5, which shows combined flow from the north wall drains from 1996 to 2000. The drain data infer that the amount of active recharge to the slope is minimal. In other sectors of the pit where recharge from the east is occurring, the initial high flows from each drain are typically more sustained.

Current pore pressures

A selection of hydrographs are displayed in Fig. 6. The graphs show that the east wall interceptor

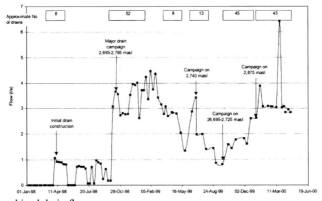

Fig. 5. North wall combined drain flow.

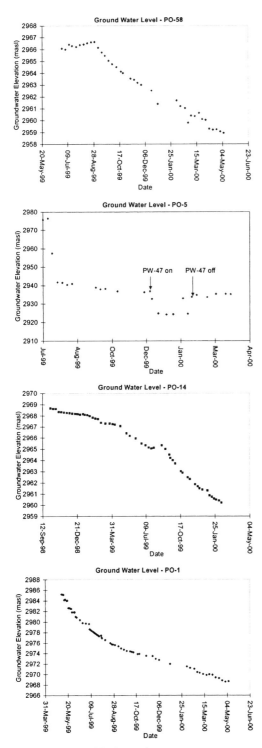

Fig. 6. Water level hydrographs.

wellfield is starting to cut off recharge flow to the slope. As PO-58 and PO-14 show, heads at the top of the wall were static or rising but are now dropping. PO-51 confirms the impact of the adjacent pumping well, PW-47. In December 1999, there was a head reduction of up to 20 m when PW-47 commenced pumping. When the well is shut off, there is a very quick recovery. In general, over most of the slope, there is a background pore pressure reduction rate of about 0.5 m month^{-1}. Therefore, under current conditions, a reduction in pore pressure of about 6 m year^{-1} would be expected.

Data from the northeast section of the wall confirm that the interceptor system is beginning to cut off recharge to the north wall, with an accompanying fall in heads. Boreholes on the central and western sections of the wall show that there is a steady background reduction in pore pressure caused by the existing drains and by passive seepage. Higher drawdowns are created in the hydraulic compartments around pumping wells.

A series of pore pressure profiles were drawn over the wall, locations are as displayed in Fig. 3. The profiles were drawn to provide pore pressure data for the geotechnical analysis of the north wall. Three of the sections are displayed in Fig. 7 and described below. For consistency, heads in the altered material are described as the 'phreatic surface'; the heads in the unaltered or silicified rock are described as the 'piezometric surface'.

P4

This section displays the hydrogeology from just to the east of the Panadero Fault. The section shows that to the west of the Panadero Zone, altered Escondida Porphyry is underlain by silicified porphyry. The piezometric surface in the unaltered material at the top of the slope is approximately 80 m bgl. The phreatic surface in the altered material is 20 m lower than this. Therefore, there is an upward head gradient from silicified rock to altered porphyry. This trend continues towards the bottom of the slope. At the toe of the slope, the head in the silicified rock is at ground level. The altered material is very thin or absent. Significant flows are encountered in silicified material at shallower depths. The flows may be related to fracture zones that have been enhanced by stress relief and rock mass unloading.

P6

The pattern in this section is broadly similar to that seen in the previous section. The section

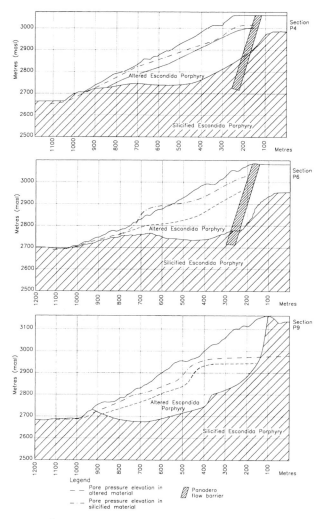

Fig. 7. Pore pressure cross-sections.

shows that the north wall is separated from the rhyolites to the east by the hydraulic barrier of the Panadero system. To the west of the Panadero, altered material overlies silicified porphyry. The altered material is thickest (250 m) at the top of the slope and thins towards the bottom.

In the top of the slope, the piezometric level in the silicified rock is approximately 50 m above the phreatic surface in the altered material. This pattern is also seen in the middle slope section, and is confirmed by packer tests and multilevel piezometers. The altered material in this sector is approximately 250 m thick. The phreatic surface in this unit is at 80 m bgl. However, the piezometric surface in the silicified rock is 40 m higher than this. Further down the slope, the phreatic surface in the altered material is closer to ground level at 42 m bgl. However, artesian pressures were encountered in the silicified rock, and the piezometric surface in this unit is approximately 30 m above ground level. Therefore, in this sector of the slope there is an upward head difference of at least 70 m.

The influence of the existing horizontal drains in the altered material can be seen in this section. In the bottom section of the slope, the phreatic surface is relatively flat because of depressurization caused by the horizontal drains. Towards the upper middle section of the slope, beyond the zone of influence of 300 m drains, the phreatic surface becomes steeper. At the toe of the slope, the piezometric surface is at ground level or slightly above.

P9

In this cross-section there is only minimal active recharge in the rhyolites behind the crest of the pit. Consequently, heads within the rhyolites north of the pit are low at around 160 m bgl. Heads in the deeper rhyolite layer are approximately 25 m higher than measured in the shallow rhyolite layer.

At the top of the slope, in the Escondida Porphyry, the phreatic surface in the altered material is approximately 100 m bgl. The piezometric surface in the silicified rock is 25 m higher than this. This upwards head gradient is maintained through the middle slope and to the toe, with the pore pressure in both units tending to become closer to ground level towards the bottom of the pit. At the bottom of the pit, the piezometric surface in the silicified rock is close to ground level.

Conceptual model of groundwater flow

A conceptual model was prepared to assist in numerical modelling of the slope; the components comprise the rhyolites outside the pit and the Escondida Porphyry within the pit as follows (Water Management Consultants 2000).

Rhyolites

On the northern margin of the pit, rhyolite of intermediate permeability overlies very low permeability rhyolite and andesite. The hydrogeology within the rhyolite is compartmentalized by the steeply dipping NW–SE faults. These create barriers to southward flow within the rhyolite towards the pit, so that continuous major seepage within the rhyolite along the north wall is not observed.

Groundwater heads are significantly below ground level, up to 160 m bgl. Active recharge from the Hamburgo Basin to this zone is partially blocked by the Panadero flow barrier. Therefore, depressurization is not required in the rhyolites to the north of the north wall.

Silicified Escondida porphyry

Heads are typically 100 m below topographic level beneath the crest of the slope, up to 40 m above, at mid-slope, and at or near the topographic level near the toe of the slopes.

The matrix permeability of the silicified porphyry is less than 10^{-8} cm s^{-1}. However, the unaltered porphyry does contain discrete open fractures, which will allow it to be drained and depressurized. The fracture flow groundwater system in the unaltered porphyry is highly compartmentalized by faulting, both laterally and vertically. Intersection of any given fracture will only allow drainage of the discrete hydraulic block that contains the fracture. Therefore, to achieve reasonable drainage of the overall pit slopes, fractures at multiple locations will have to be encountered and drained.

The unfractured rock mass has very low drainable porosity (less than 0.1%). Virtually all of the water is contained in fractures (the main fracture zones and interconnected microfracture systems). Main fracture zones may typically yield $0.1-3.0$ l s^{-1}, sufficient to sustain flows in vertical wells and horizontal drains. Where drains in the silicified rock intersect such transmissive fractures, high yields can be obtained. An example of this is the recent pilot hole for a drainage tunnel. The yield of the drain was 2 l s^{-1} at a distance of 200 m into the slope.

Water in the silicified porphyry is currently leaking upward into the altered porphyry. The water is being removed from the altered material by horizontal drains and seepage to the face.

Vertical wells, constructed in the silicified porphyry, drain storage within relatively small hydraulic compartments. Nearby piezometers react quickly to the onset of pumping. Observation boreholes at greater distances do not record any obvious effects from well pumping.

Pilot hole drilling on the north ramp has not yielded any sites with the required fractures and airlift flows to justify the construction of additional production wells. The reasons for the lack of productive fractures may be that the wells only intersect approximately 40 m of unaltered rock below the altered material. Therefore, not enough of the silicified porphyry has been 'sampled'. In addition, the vertical wells may not be intersecting the steeply dipping structures at the head of the slope.

Altered (argillic) porphyry

Current piezometric heads in the altered porphyry are typically 20–60 m lower than in the underlying silicified porphyry. Therefore, over most of the slope, there is a strong upward hydraulic gradient. Heads are typically 50–150 m below topographic level beneath the crest of the slope and are typically 50–60 m below topographic level at mid-slope. The altered porphyry is typically thin to absent near the current toe of the slopes, but, where present, the phreatic surface is less than 10 m bgl.

The matrix permeability of the altered porphyry is typically between 10^{-4} and 10^{-5} cm s^{-1}. Although it is still controlled to some extent by relict structures, groundwater flow in the altered porphyry is much more homogenous than flow in the unaltered porphyry. The altered porphyry does not contain any discrete high-permeability fracture zones. These have become 'healed' as part of the alteration and weathering process. The unit behaves more like a 'sponge'. Compartmentalization is much less important than in the unaltered Escondida Porphyry.

The altered porphyry has a drainable porosity typically within the range 0.1–0.5%. Water in the upper layers of altered porphyry is draining mainly to horizontal drains. Drains can yield an average of $0.1 \, \mathrm{l\,s^{-1}}$ in the altered material. Combined flows amount to $2–3 \, \mathrm{l\,s^{-1}}$ when there is active drain construction. However, flows of up to $1 \, \mathrm{l\,s^{-1}}$, necessary for an effective vertical well, are unlikely to be sustained. In addition, there is slow lateral seepage to the pit face.

Numerical modelling

The purpose of the model was to allow the simulation of a number of depressurization options and, therefore, to aid in the design of the most effective depressurization system for the north wall. The construction of the model is summarized as follows:

- The flow system was modelled using the MODFLOW finite-difference code (McDonald & Harbaugh 1988), via the Groundwater Vistas pre/post-processing package, and a specially constructed spreadsheet.
- A two-dimensional, axi-symmetric, vertical slice model was constructed based on hydrogeological section P6. The model simulates the flow conditions in a sector of the northwest wall of the pit, centred on section P6 (Fig. 3).
- In modelling the flow system as axi-symmetric, it is assumed that the dominant flow directions are radial flow towards the centre of the pit, and vertical flow between the silicified and altered Escondida Porphyry. Flow parallel to the pit slope is assumed to be negligible.
- The fracture network in the porphyry, through which groundwater flows, has been represented by an 'equivalent porous medium'. This is a simplification of the field situation, which is based on the assumption that the occurrence of water-bearing fractures does not vary greatly from one model cell to another.
- The groundwater model consists of a single layer with 48 rows and 96 columns. Each model cell represents a volume 12.5 m high by 12.5 m long. The width of each cell (tangential to the radial direction) depends on the distance of that cell from the origin of the axi-symmetric model.
- As MODFLOW is designed for modelling flow in systems with Euclidean geometry (using an x–y–z co-ordinate system), the model parameters had to be calculated in such a way that would enable MODFLOW to represent a system with axi-symmetric geometry (an r–θ–z co-ordinate system).

Starting from initial conditions with groundwater heads set uniformly to 3000 m asl, the model simulates 2 years during which water is removed solely by evaporation from the pit wall surface. This is followed by a further 2 years, during which additional water is removed via the horizontal drains. At this point, the modelled groundwater heads were compared to the observed head profile along the cross-sections. The match between the modelled and observed head profiles on section P6 is shown in Fig. 8. This figure demonstrates that there is a good fit; in particular the upward head difference throughout the slope is replicated.

The water balance for the calibration run shows that:

- Recharge to the slope from the constant heads behind the crest is of the order of $0.1 \, \mathrm{l\,s^{-1}}$. This replicates the situation on the north wall west of the Panadero where there is very little recharge from the rhyolite to the north.
- The initial high evaporation (EVT) flows of $10 \, \mathrm{l\,s^{-1}}$ reduce to $2 \, \mathrm{l\,s^{-1}}$ as the phreatic surface is drawn back from the face and evaporation is reduced. There is a further reduction in EVT when the drains are constructed and the phreatic surface drops further.
- When the drains are constructed, there is an initial high drain flow of up to $10 \, \mathrm{l\,s^{-1}}$. After 2 years of operation, combined drain flow is of the order of $2.5 \, \mathrm{l\,s^{-1}}$. This compares well to the current combined drain flow of $2–3 \, \mathrm{l\,s^{-1}}$.
- Average flow from the silicified porphyry to the overlying altered porphyry is of the order of $1–2 \, \mathrm{l\,s^{-1}}$. This flow is driven by the upward head difference between the two units. The current drains are constructed mainly in the altered porphyry so the phreatic surface drops faster than the piezometric surface. Therefore, it can be inferred that the upward head difference has reduced over the last few years and the upward flow has diminished.

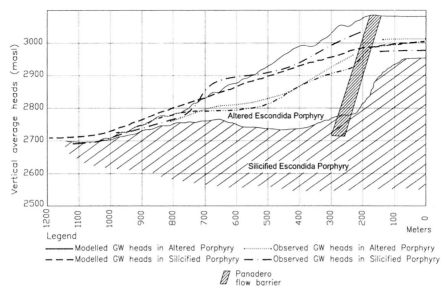

Fig. 8. Model calibration.

- It can be concluded that any depressurization system needs to stop $1–2\,l\,s^{-1}$ of upward recharge flow from the unaltered to altered porphyry, as well as removing stored water.

The numerical model was used to assess a number of depressurization options. At this stage, each option has been simulated for a two-year period. The model results represent conditions at the end of two years. Except for the first case, the prediction runs assume that the current horizontal drain network remains in operation, in addition to the simulated new drainage measures. The depressurization options that were modelled in the predictive runs include:

- conditions without the existing north wall horizontal drains in the altered material;
- wells in the silicified rock at the base of the pit;
- wells in the silicified rock towards the top of the slope;
- drains in the silicified rock;
- a drainage tunnel in the silicified rock with various combinations and spacings of horizontal, subvertical and vertical drains.

The cross-sections in Figs 9 and 10 present a selection of the modelled piezometric and phreatic surfaces. These are presented in relation to the target pressures for the slope, produced by a geotechnical analysis. The modelling results show the following:

- If the existing drains had not been installed, the phreatic surface would have been close to the face of the slope with discharge via evaporation and the piezometric surface would have been up to 30 m above ground level.
- Currently the upward head difference drives a flow of approximately $1–2\,l\,s^{-1}$ from the silicified rock into the altered material.
- Wells at the base of the pit initially produce significant local drawdown at the toe of the slope but no significant change to the pressures in the top and middle sectors of the slope.
- Wells at the top of the slope have a local effect on the piezometric level but the target is only achieved up to a distance of approximately 100 m either side of the wells.
- When drains are drilled into the silicified rock there is significant additional reduction of the piezometric and phreatic surfaces in the middle sector of the slope. The piezometric surface target is achieved over the area of coverage of the drains, i.e. 500 m into the slope.
- A drainage tunnel with horizontal and subvertical drains produces drawdowns at the top and middle of the slope, of 50–100 m, which are significantly higher than all other cases. Upward flow from the unaltered to altered porphyry is stopped and

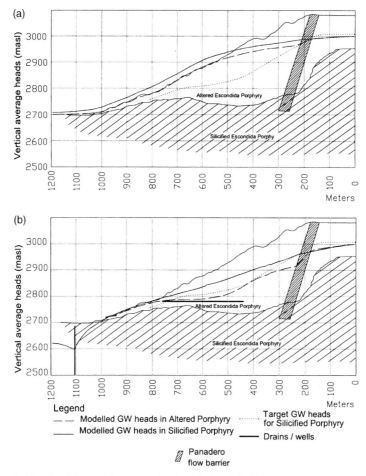

Fig. 9. (a) Modelled heads with no active pressurization. (b) Modelled heads with vertical wells.

the targets are achieved at distances of up to 800 m from the toe of the slope, i.e. over the length of coverage of the tunnel system.

The depressurization system that has the most significant effect on all sectors of the slope is the tunnel option with horizontal and subvertical drains. The model indicates that this layout produces pressure reductions that are up to 15 times those currently measured. The targets for the slopes are effectively achieved with this option.

An examination of the water balance from the tunnel prediction shows that the tunnel stops the upward flow of $1-2\,\mathrm{l\,s^{-1}}$ from the silicified rock to the altered material. However, the pore pressure reduction in the silicified porphyry is not sufficient to induce flow downwards from the altered material. Therefore, active drainage measures will still be required to remove stored water in the altered porphyry. Although horizontal drains in this unit will become more inefficient as the phreatic surface drops, they are still the most effective method of reducing the phreatic surface. Drain drilling in the altered material needs to be implemented on a continual basis, so that $2-3\,\mathrm{l\,s^{-1}}$ is discharged from that section of the slope.

Depressurization system design

The analysis and modelling demonstrates that an effective depressurization system for the north and NE walls should consist of a combination of elements, as follows.

Elements

Recharge interception. The modelling assumes that there is no significant recharge flow from the Hamburgo Basin. The constant heads on the left-

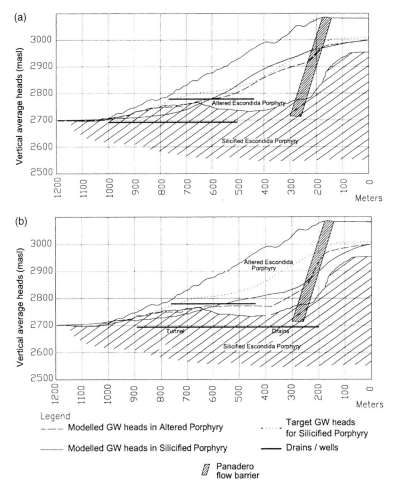

Fig. 10. (a) Modelled heads with drains in silicified porphyry. (b) Modelled heads with drainage tunnel.

hand boundary of the model contribute approximately $0.1\,l\,s^{-1}$ to the model domain. This represents the conditions west of the Panadero, where there is minimal recharge in the rhyolites to the north of the pit.

The planned depressurization methods will remove a long-term maximum of $10-15\,l\,s^{-1}$ from the slope. If there is significant recharge flow from the east, then it is unlikely that pore pressure targets will be achieved. It is, therefore, important that the interceptor wellfield along the east wall of the pit will continue to be operated as planned.

Vertical wells. Vertical wells will be required at the base of the pit. These will reduce the phreatic surface at the bottom of the pit below the working pit floor, and have significant local effects on heads at the toe of the slope.

Drainage tunnel. A 500 m tunnel with an extensive lateral drain network will produce the required depressurization of the silicified porphyry.

Horizontal drains from the face. Horizontal drains, as currently constructed in the altered material, will be a part of the future depressurization system. This is because the current drain network has substantially dewatered the altered material, and horizontal drains are the most effective method of reducing the phreatic level. The tunnel will reduce the piezometric surface in the silicified rock and stop upward leakage to the altered material, but will not cause downward flow from the altered material. Active drainage will still be required in the altered porphyry to remove stored water. Drain drilling should be continued so that, on average, $2-3\,l\,s^{-1}$ is

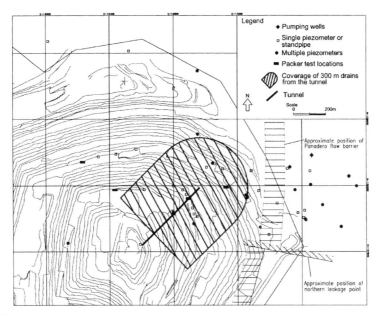

Fig. 11. Tunnel location.

discharged from the altered material. There are sectors of the slope, for example near PO-93, where the phreatic surface is close to ground level and where new drains will have a significant impact.

The depressurization programme for the north wall will be an interactive combination of the above elements. The implementation of each element will need to be flexible and based on the results of the monitoring, and on continual review and re-evaluation.

Tunnel design

The tunnel location is shown in Fig. 11; it will be driven upwards at an angle of 2° from a portal near the base of the slope. Its dimensions will be 3.8 × 3.2 m, with stations for drain drilling every 50 m.

An intensive monitoring programme will be carried out to allow interactive decisions to be made on each subsequent drilling phase. Therefore, in addition to the existing vertical piezometers, a number of piezometers will be constructed from the tunnel to measure pore pressures in areas which are inaccessible from the surface.

A preliminary test phase is also included, using drains drilled from the first two drilling stations, as the tunnel is being constructed. The test will investigate the effect of varying the length and orientation of drains, and construction and completion techniques for drains and piezometers. Once the preliminary testing phase has been evaluated, the optimal procedures for the drain programme will be determined.

The total meterage of tunnel drains has been optimized by using the model. It is estimated that a maximum of 7000–10 000 m of drains will be required; with each drain up to 300 m long. In general, there will be two types of drain: (1) horizontal drains drilled laterally from all drilling stations; and (2) subvertical drains (up to 45° from horizontal) drilled from the last 200 m of the tunnel to drain the increased thickness of saturated material above the tunnel.

Conclusions

The groundwater model of the north wall, plus practical field experience, has shown that the most effective means of reducing pore pressures in the deep silicified Escondida Porphyry is a drainage tunnel. By removing small amounts of water over a large area, the tunnel and associated drains will bring pore pressures below the specified targets and will optimize the stability of the wall. The depressurization of the slope will be an interactive process with additional elements, such as targeted vertical wells and drains from the face, used where appropriate.

We are grateful to Minera Escondida Limitada (MEL) for permission to publish this work. In addition, we would like to thank the staff of MEL and WMC who helped in the study.

References

MCDONALD, M.G & HARBAUGH, A.W. 1988. *A Modular Three Dimensional Finite-difference Ground-water Flow Model*. US Geological Survey, Techniques of Water Resource Investigations, 06-A1.

Water Management Consultants, 1999. *Escondida Mine Interim Dewatering Status Report*, Water Management Consultants, Shrewsbury, UK.

Water Management Consultants, 2000. *Escondida Mine, North Wall Hydrogeology and Design of a Depressurisation System*, Water Management Consultants, Shrewsbury, UK.

Hydrogeological framework for assessing the possible environmental impacts of large-scale gold mines

J. S. KUMA[1,2], P. L. YOUNGER[2] & R. J. BOWELL[3]

[1] KNUST School of Mines, PO Box 237, Tarkwa, Ghana (e-mail: j.s.y.kuma@ncl.ac.uk)
[2] Department of Civil Engineering, University of Newcastle, Newcastle upon Tyne NE1 7RU, UK
[3] SRK, Summit House, 9-10 Windsor Place, Cardiff CF10 3RS, UK

Abstract: Hydrogeological information is crucial to the development of a sound environmental impact assessment (EIA) for a proposed mine, as well as the management of potential environmental impacts during and after exploitation. However, the determination of hydrogeological parameters is not customarily included in mineral exploration surveys, with the result that many EIAs end up being rather light in hydrogeological content. Examples from the Tarkwa gold mining district of Ghana illustrate this point. Consequences of such an inadequate hydrogeological understanding are potentially serious, ranging from an inability to predict future problems in water quality after the cessation of mining, to a lack of understanding of hydrogeological controls on slope stability, which is arguably manifest in the catastrophic spill of cyanide-rich processing effluents from a breached tailings dam at Wassa West, near Tarkwa, on 16 October 2001. To redress this deficiency, we propose that a hydrogeological database be assembled during the mineral exploration phase, according to a specified protocol ('check-list'). Using these data, a rational conceptual hydrogeological model for the mine site and its surrounding area can be developed, providing the basis for a thorough consideration of groundwater aspects within the statutory Environmental Impact Assessment, which is (as in most other countries) required by Ghanaian government statute before a mining lease is approved. The resources required to set-up such a database are small compared to the benefits.

Mineral exploration and exploitation is a multimillion dollar business. Even though the risks are very high, the quest for improved living standards and developments in mineral beneficiation technology both stimulate the exploration and development of new sources of mineral wealth (Woodall 1984). The mining industry is both an important source of employment and a major foreign exchange earner in many countries, and this is certainly the case in Ghana, the former 'Gold Coast' of Africa. Although environmental controls during active mining are increasingly stringent, former mining operations have left a long-term legacy of environmental problems on almost all continents (e.g. Hedin *et al.* 1994; Younger 2001; Dzigbodi-Adzimah 1996). Many of these long-term problems arise because of a lack of appreciation of the hydrogeological setting of the mine, resulting in the pursuit of inadequate closure strategies. This lack of hydrogeological appreciation belies the existence of well-established techniques for relatively low-cost development of conceptual hydrogeological models to underpin the process of environmental impact assessment (EIA) (e.g. Pettyjohn 1985; Kolm 1996; Stone 1999).

This paper illustrates the hydrogeological shortcomings of EIA procedures using examples from the Tarwka gold mining district of Ghana. A protocol for hydrogeological data collection to improve the insertion of groundwater system concepts into future site appraisal and management activities is proposed.

Gold mining and EIA requirements in Ghana

Ghanaian gold mining and EIA requirements

In Ghana, gold mining contributes approximately 30% of the annual foreign exchange earnings and more than 2.3 million ounces of gold is produced by the country annually. Figure 1 shows a simplified geological map of southwest Ghana and the location of large-scale gold mines. A new mining permit can only be issued after the Environmental Protection Agency (EPA) of Ghana has approved an EIA report. The EIA report

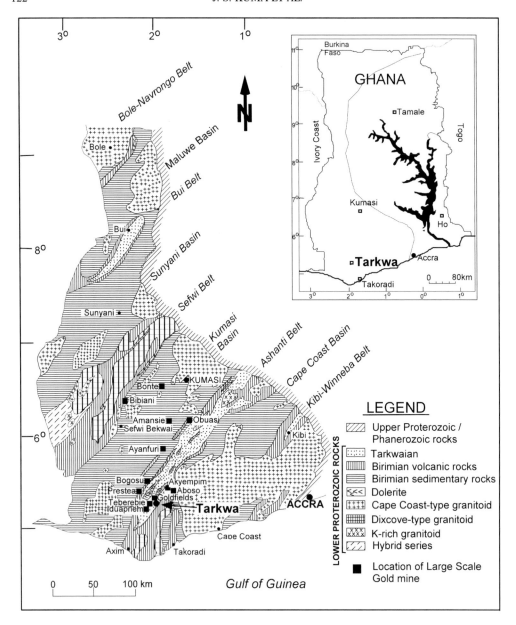

Fig. 1. Simplified geological map of Ghana showing the location of large-scale gold mines (adapted from Eisenlohr 1992).

includes, among other requirements, an Environmental Baseline Study, Environmental Impact Statement and an Environmental Management Plan. On-going mining activities are required to submit Environmental Action Plans (EAP) with an annual report (Anon. 1994). During mine operation the EPA also administers a random monitoring programme.

Examples of hydrogeological shortcomings in Ghanaian mining EIAs

Even though the EPA requires water sections in EIA studies, many such studies lack sufficient hydrogeological detail to permit full scrutiny of the intentions of the mining lease permit applicants. Without such detail, the hydrodynamics of the mining areas cannot be conceptualized to the degree necessary to allow adequate safeguards to be put in place to forestall damage of water resources. There are even times when little or no hydrogeological information is presented. For example, one recent EIA reported that 'very little hydrogeological information was available. Groundwater seems to be abundant as there was the need for dewatering before blasting could be conducted by the mine' (Amegbey 1996). Even in situations where some hydrogeological information is presented, important aspects are normally absent from the document. For instance an internal report intimated that 'no groundwater recharge information for the area exists' (Anon. 1997).

The ultimate consequence of a lack of hydrogeological data is an inability to manage environmental and geotechnical processes that influence/are influenced by groundwater occurrence, movement and quality. Hence, there still exists no regional groundwater monitoring network in the Tarkwa area that would be needed to independently assess the likely rates of filling of post-mining pit lakes, and/or the positions and quality of mine water decants to the surface environment. Without such information, preventative measures cannot be devised (cf. Younger & Adams 1999; Shevenell 2000). Even before mine closure, groundwater data are critically important in the design of safe pit wall angles, and secure bunds for tailings dams and other structures (see Younger *et al.* 2002). With regard to the latter, inadequate control of sub-dam pore water pressures is suspected to have played a role in the partial failure of a tailings dam, which gave rise to a release of cyanide-rich water in June 1997 (spill from the Teberebie gold mine into the Awunabeng stream) and perhaps also contributed to a similar spill (from Goldfields of Ghana gold mine) on 16 October 2001 that introduced cyanide into the River Essuman, which provides drinking water to the villages of Abekroase and Huniso (Fig. 2).

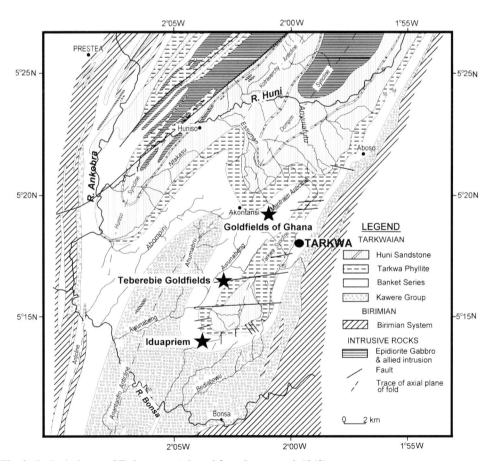

Fig. 2. Geological map of Tarkwa area (adapted from Junner *et al.* 1942).

Hence, both short- and long-term environmental management requries hydrogeological understanding. Such an understanding does not necessarily require extensive groundwater mapping to 'First World' standards and it can, in fact, be achieved using an array of relatively simple methods.

Protocol for low-cost hydrogeological data collection for inclusion in EIA

Groundwater information can be gathered during the phase of mineral exploration, so that it is readily available at the time the EIA is undertaken. Although the following items are described in a segmented manner, it should be borne in mind that they are in reality interconnected. Hence, some of the items covered in one section below are also highlighted in subsequent sections.

Physiography

Historical information on the physiography of the area may be available, but data on current land use, vegetation, relief and drainage information are normally collected early in the preliminary exploration phase by conventional surveying. A map of the physiography at a scale of 1:20 000 or less is normally prepared for the concession, and this scale is reduced as detailed phases of exploration are pursued. Surface water systems provide valuable evidence of groundwater movement (where the two systems are hydraulically connected) and analyses of the surface water system will enable information to be obtained about groundwater flow patterns and water quality (Fetter 1994; Winter et al. 1998). In addition, the drainage pattern of an area is normally a reflection of its geological history and structure (Stone 1999). Thus, due consideration given to the study of the drainage pattern will improve an early conception of the groundwater system in the area, for example, identification of recharge and discharge areas.

Pedology

The physical and chemical characteristics of a soil play a vital role in the recharge and chemical evolution of groundwater. This is because soil texture and structure control hydraulic properties, whereas soil mineralogy affects water quality (Hem 1992; Stone 1999). It has, therefore, been advocated that the nature of the soil and the drainage characteristics of an area be closely considered during hydrogeochemical studies (Edmunds 1981). During these studies the type of clay present should be determined by X-ray diffraction (XRD) if possible, as this information can improve knowledge of the possible chemical reactions likely to occur in the shallow subsurface (Appelo & Postma 1993; Head 1997). In addition, the rainfall characteristics, vegetation and land use all influence infiltration of water through soil, and form a critical component of the hydrological cycle (Dingman 1994; Dunne & Leopold 1998).

In undertaking a pedological survey the soil thickness encountered during drilling should be recorded, to delineate the weathered–unweathered boundary of the area. Furthermore, the soil profile and any peculiar textural and structural characteristics in all excavations should be logged. Soil should be sampled in the B-horizon because this is the most critical horizon on which infiltration of precipitation to the groundwater zone depends (Davis & DeWiest 1991). An optimum of 40 soil samples is recommended for determination of particle size distribution (PSD) analysis, bulk and particle densities, and moisture content tests. Soil texture, sorting, changes in lithology, volume changes as a result of compaction and subsidence, and porosity are determined from these tests. Moreover parameters such as porosity, permeability and heterogeneity, together with shape and size of voids, are likely to be of considerable importance in controlling the movement of pollutants through the unsaturated zone (Hounslow 1985). Assessment of groundwater recharge is also aided by using this information.

Geology. Knowledge of the lithology, stratigraphy and structure of the rocks in a region is essential to understanding the nature and distribution of their water-bearing properties (Fetter 1994). In a mining environment the presence of lithological logs will greatly enhance knowledge of the subsurface geology, which in turn helps to improve conceptions of the hydrogeology of the area.

Lithological logs are collected during the exploration phase. Normally during the stage of detailed exploration, these logs are acquired at spacings as small as 25 m and in a regular pattern, at least in and around the ore body. These logs should be scrutinized for all the hydrogeologically relevant information that they may contain by means of logging and laboratory tests, which can then form the basis of a hydrogeological model of the concession. Younger (1992) has demonstrated the use of a petrological microscopy in the direct measurement of grain size, pore size and porosity, from which estimates of hydraulic conductivity, specific yield and solute retardation factors have been made. It has also been shown that the storativity of an aquifer can be estimated as function of aquifer lithology and thickness alone (Younger 1993). A probe permeameter could also be employed for making inexpensive permeability determinations on borehole core samples (Eijpe & Weber 1971).

Geological logging and petrological studies of cores enable the prediction of water–rock interactions likely to explain the chemistry of groundwater discharges to streams (Sharp & McBride 1989; Appelo & Postma 1993). The presence of sulphide minerals, which are prevalent in some gold ore fields, deserves particular attention due to their propensity to release acidic, metalliferous leachates (acid mine drainage) after they have been aerated. Early knowledge of their occurrence is useful during planning.

There is a paucity of information regarding the aquifer properties of many mine concessions, even though water boreholes are drilled for processing and other needs (e.g. for local communities on mine concessions and for those who have been resettled) during the feasibility stage of exploration. A little time and effort invested in pumping tests of water boreholes drilled for such purpose can be an invaluable addition to the hydrogeological database. Some of the exploration boreholes can also be adapted as monitoring boreholes to establish pre-mining groundwater trends and background hydrochemical patterns.

Hydrology

Information on the climate may be available for the region as a whole. However part of the concession may have been excluded from further consideration (typically 50% or more). Rainfall, evaporation, temperature and other meteorological data are normally a requirement for site design. Therefore, a simple meteorological station is often installed to aid the estimation of evapotranspiration. Evaporation measurements are also necessary for the design and operation of heap-leach pads, which are employed by most of the gold mines. It is advisable to start routine gauging of major streams in the area (for run-off and recharge estimates) at this time so that this information can be synchronized with meteorological data for future analyses and interpretation. In addition, piezometers need to be economically installed at this time at well-chosen sites in the concession so that the flow regime of the ground water can be determined. Recharge can also be estimated and compared with the value from the stream hydrographs. These piezometers must be kept in working order throughout the life of the mine so that future assessments can be conducted without difficulty.

Hydrogeochemistry

Surveys of streams and springs in and around the concession area, particularly during dry periods when there is no surface run-off to mask groundwater contributions, can provide invaluable information on groundwater chemistry (e.g. Pettyjohn 1985). The data also provide a useful input to the baseline information required in EIA reports. However, where such surveys are undertaken, they often include analysis of only the major ions and/or samples prove to have been poorly collected, so that incomplete and/or contradictory information is provided. We propose the following list of determinands: temperature, pH, Eh, Na, Ca, Mg, K, HCO_3, SO_4, Cl, NO_3, SiO_2, Fe, Mn, Al, Co, Cr, Pb, Zn, Cu, CN, Hg, As, DO, TDS colour and turbidity. Further, a major omission is often the assessment of pit-wall or waste rock contribution to discharge water quality and changes that will occur over time (Bowell 2002). This information is normally not collected because an experienced hydrogeochemist may not be included in the exploration team.

At least 5 years (some times up to 20 years) elapse before an exploration prospect actually becomes a mine (Woodall 1984). Therefore, if all the aspects above are conscientiously followed during the exploration stage, a sufficient run of data will have been accumulated to support development of a robust conceptual hydrogeological model of the mine.

Example application of the protocol to the Tarkwa gold mining district

To test the feasibility of applying the above protocol to a real system, it was applied to the Tarkwa mining district in Ghana by means of field surveys between January 2000 and January 2001. The following sections document the findings.

Physiography

The Tarkwa area has a humid tropical climate with an average annual rainfall of over 1750 mm. Rain falls in two main periods in the year: a major wet season from April to July (with a peak in June) and a second minor season from September to November. Air temperature for the area varies in a narrow range (between 28 and 33°C) and relative humidity varies from 83 to 91%. Located in an area transitional between the rain forest and moist semi-deciduous forest, zones of the original vegetation are locally present, while other areas have been cleared for farms, mines and communities (Dickson & Benneh 1988).

The Tarkwa district is located within the Ankobra River basin and is bordered to the west by the southerly flowing Ankobra River (Fig. 2). Both the Huni and Bonsa rivers are major tributaries to the Ankobra, and border the area to the north and south, respectively. The area is highly dissected and of moderate relief with a general decrease in hilltop altitudes towards the south. A series of parallel ridges and valleys oriented along the general NE–SW strike of the rocks define the landscape. This geomorphology is due to pitching fold structures and dip-and-scarp slopes of the Banket Series and Tarkwa Phyllites, which form part of the bedrock (see below). Transverse to the ridges and valleys are smaller tributary valleys and gaps controlled by faulting and jointing (Whitelaw 1929).

Pedological characterization

Infiltration and particle size distribution (PSD) tests were conducted to determine saturated hydraulic conductivities (K_s) and textural characteristics, respectively, of soil at 56 sites in the Tarkwa area (Kuma & Younger 2001). These soil tests were conducted in the B-horizon (Table 1). It was observed, in general, that extremely poorly sorted soils exhibit low porosity and relatively higher K_s while relatively better-sorted soils reveal high porosity and low K_s values (Kuma & Younger 2001). The soils are mainly silty-sands, which exhibit K_s values in the 10^{-5}–10^{-8} m s^{-1} range, although this is dominated by those in the narrower band of 10^{-6}–10^{-7} m s^{-1}. Minor lateritic patches are located on hilly terrain underlain by Banket Series and Tarkwa Phyllite rocks. Banket soils exhibit the most favourable characteristics for infiltration in the area, in terms of both PSD and K_s values. These soils are located on hills and, therefore, act as the main areas for groundwater recharge. Huni and Kawere soils, located in low-lying areas, display characteristics suggesting that the Huni, with a much better sorting coefficient and K_s value, will admit more precipitation for recharge compared to the Kawere. Soil pH varies from very acidic to moderately acidic, that is 1.72–5.01. The lowest pH values are associated with weathered felsic dykes and are caused by the presence of well-disseminated fine pyrite crystals, which are expected to be a potent source for acid rock drainage when found in the unsaturated zone.

Geology

Sediments of the Tarkwaian System are predominantly arenaceous and were 'deposited by high-energy alluvial fans entering a steep sided basin filled with fresh water' (Kesse 1985). They consist, in general, of coarse, poorly sorted, immature sediments with low roundness typical of a braided stream environment. They have been metamorphosed to low-grade green-schist facies (Kesse 1985). Rocks of the Tarkwaian System, in the direction of younging, comprise the Kawere Group, the Banket Series, the Tarkwa

Table 1. *Summary of soil tests conducted in the Tarkwa area*

Soil type	Texture	Percentage				K_s (10^{-6} m s^{-1})			S_o	n	$W\%$	G_s	pH
		Gravel	Sand	Silt	Clay	Min.	Max.	Mean					
Banket	Silty-sand	2	59	29	10	0.21	10.56	4.45	3.74	0.38	17.5	2.66	4.92
	Laterite	69	14	10	7	0.01	20.75	3.46	5.85	0.22	9.8	2.67	4.86
Huni	Silty-sand	2	55	33	10	0.05	7.28	1.57	2.28	0.42	12.8	2.65	5.01
Kawere	Silt sand	0	47	40	13	0.10	1.87	0.72	4.34	0.38	12.4	2.65	4.65
Tarkwa Phyllite	Laterite	62	9	13	16	–	–	–	17.8	–	14.2	2.74	5.01
W. Dyke	Silt	3	20	64	13	0.31	2.57	0.88	2.75	0.40	22.2	2.62	5.22* 1.96†

K_s is saturated hydraulic conductivity, S_o sorting; n is porosity; $W\%$ is moisture content, Gs is specific gravity; W. Dyke is weathered dyke.
*Mafic.
†Felsic.

Table 2. *Subdivision of the Tarkwaian System in the Tarkwa area (modified after Junner et al. 1942). The bold face characters under 'composite lithology' signify the most abundant and important rocks forming each division*

System	Series	Thickness (m)	Composite lithology
Tarkwaian System	Huni Sandstone	1370	Sandstones, grits and quartzites with bands of phyllite
	Tarkwa Phyllite	120–400	Huni sandstone transitional beds and greenish-grey phyllites and schists
	Banket Series	120–600	Tarkwa Phyllite transitional beds and sandstones, quartzites, grits breccias and conglomerates
	Kawere Group	250–700	Quartzites, grits, phyllites and conglomerates

Phyllite and the Huni Sandstone (Fig. 2 and Table 2).

Intrusive igneous rocks make up about 20% of the Tarkwaian System in the Tarkwa area. These rocks range from hypabyssal felsic to basic igneous rocks, principally in the form of conformable to slightly transgressive sills, with a small number of dykes. Faults and joints are common in the area and the most prominent joint sets trend ESE–WNW (although NW–SE and N–S trends are also present). The faults are either strike-parallel and closely associated with folding (and may occur as upthrusts) or dip-parallel, and are most often recognized as breaks in the topography (Hirdes & Nunoo 1994).

With shallow-water beds in a braided environment, the thickness of individual members varies considerably. In addition, a fractured and metamorphosed lithology confers aquifers with dual porosity, limited areal extent and storage properties.

Hydrology

Direct recharge (*sensu* Lerner 1990) was identified as the dominant recharge process requiring estimation because precipitation (1803 mm for the year 2000) is the primary hydrological input. In the development of the water balance model, it was assumed that the area is almost hydrologically closed with respect to both surface and groundwater. This is because on three sides of the area, three rivers, i.e. the Bonsa, Huni and Ankobra, effectively act as surface and groundwater divides with no possible intrusion from beyond the river boundaries. On the eastern boundary, the Banket ridge acts as a water divide. The equation for precipitation recharge for the study area is written as:

$$\text{Recharge } (R) = P - RO - AE \pm SMS - SWS \qquad (1)$$

where: P is the precipitation (mm year^{-1}), AE is the actual evapotranspiration (mm year^{-1}), RO is the catchment surface run-off (mm year^{-1}), SMS is the soil moisture storage (mm year^{-1}) and SWS is the surface water storage (shallow lakes, mm year^{-1}).

The Penman–Pan evaporation relationship is strongest in the more humid areas of Ghana and in the wet seasons, because the potential evapotranspiration and pan evaporation values are similar to those yielded by the Thornthwaite and Papadalus formulae (as described by Acheampong 1986). Dunne & Leopold (1998, p. 136) came to a similar conclusion working in Kenya (which is in the tropics, on about the same latitude as Ghana, although higher in altitude). Considering the results of these studies, pan evaporation data were adopted as suitable for describing the potential evapotranspiration (PE) of the study area.

A method adopted by the British Meteorological Office (BMO) for estimating actual evapotranspiration (AE) of a water basin is to classify the various types of vegetation according to their root constants (RC) and proportionately determine the AE from these after calculating the water balance (Shaw 1994). Based on the BMO model, vegetation in the Tarkwa area was classified as having about 65% mature forest (of which 25% was riparian and always transpires at the potential rate), 10% shrubs, 15% crops, and 10% urban and mine area. The mature forest is expected to have the maximum possible root constant of 250 mm, based on both the soil texture and its vegetation (Grindley 1970). Crops farmed are roots and cereals with approximate RC values of 100 and 150 mm, respectively; shrub RC is about 200 mm, and in urban and mining areas it is 25 mm.

The Thornthwaite & Mather (1957) accounting procedure for the water balance method was adopted, and Table 3 shows an example calculation using a root constant of 250 mm. The average actual evapotranspiration obtained

Table 3. *Monthly water balance at the Tarkwa area with a root constant of 250 mm**

Month	Rainfall (mm)	Pan Evap. (mm)	R-PET (mm)	Acc. PL (mm)	ΔST (mm)	ST (mm)	AE (mm)	Deficit (mm)	Surplus (mm)
January	60.97	79.20	−18.23	−34.00	−16	218	77	2	0
February	59.12	118.90	−59.78	−94.00	−47	171	106	13	0
March	37.33	106.00	−68.67	−163.00	−42	129	79	27	0
April	166.33	115.60	50.73		50	179	116	0	0
May	293.92	102.30	191.62		71	250	102	0	121
June	480.01	65.00	415.01		0	250	65	0	415
July	74.92	85.80	−10.88	−11.00	−11	239	86	0	0
August	91.08	73.70	17.38		11	250	74	0	6
September	123.46	73.20	50.26		0	250	73	0	51
October	187.18	89.90	97.28		0	250	90	0	97
November	160.90	95.80	65.10		0	250	96	0	65
December	68.22	84.40	−16.18	−16.00	−16	234	84	0	0
Total	1803.43	1089.80		−318.00	0		1048	42	755

*Acc. PL is the accumulated potential water loss and is defined as the cumulation of negative values of (R-PET). ΔST is the change in soil moisture. The amount by which PE and AE differ each month is the deficit (SMD). Surplus is the moisture surplus.

for the Tarkwa area using these procedures is 1035.18 mm year^{-1}.

Three of five major streams and one smaller stream, all in the study area, were gauged in order to estimate their surface run-off and base flows (Fig. 3). The computer program HYSEP (Sloto & Crouse 1996) was employed to separate the stream hydrographs. This program separates the hydrograph into base flow (associated with groundwater discharge) and the surface run-off components (associated with precipitation) (see Pettyjohn & Henning 1979). Total surface run-off for the whole study area for the year 2000 was estimated at 449 mm. Figure 4a–d shows the graph of monthly surface run-off and base flow for the four gauged catchments.

Soil moisture (SM) is held in the soil pores by capillary forces, and it is highly sensitive to changes in precipitation and evaporation, so that changes in soil moisture content occur throughout the year (Chidley 1981). It is generally accepted that the net change in water storage on an annual basis is zero, even though significant differences may occur from one year to the next (Shaw 1994).

Shallow basins under water or marshy conditions during the dry season are significant. Some of these are the result of recent distortions in the landscape due to surface mining while others are natural. Those due to surface mining are either abandoned pits created in the process of mining or else have become unintentionally

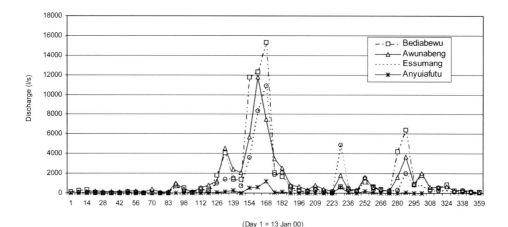

Fig. 3. Plot of discharge of four streams in the Tarkwa area.

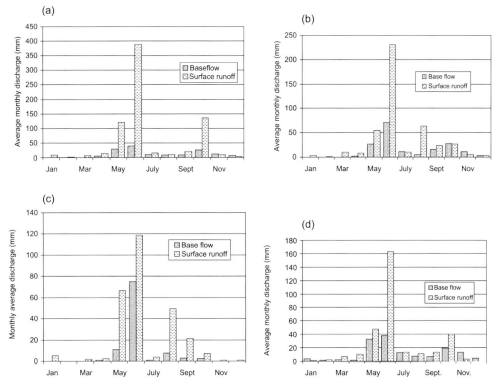

Fig. 4. (**a**) Graph of base flow and surface run-off of the Bediabewu catchment. (**b**) Graph of base flow and surface run-off of the Essuman catchment. (**c**) Graph of base flow and surface run-off of the Anyuiafutu catchment. (**d**) Graph of base flow and surface run-off of the Awunabeng catchment.

impounded by earth moved and tipped during mining. Water stored in this way would previously have formed part of local stream flow. The natural pools are surface manifestations of shallow regolith aquifers. The volume of water present in the depressions depends on precipitation and PE. The presence of water at the end of the dry season implies that PE is less than precipitation. A rough estimate of surface water storage (SWS) was made, i.e. about 20 mm year^{-1}.

Precipitation recharge (RE) from the above considerations is:

$$(1803.43 - 1035.18 - 449 - 20) \text{ mm year}^{-1}$$

$$= 299.25 \text{ mm year}^{-1}$$

i.e. about 17% of annual precipitation.

Stream hydrograph analysis, from which baseflow is separated, may also be used for the estimation of groundwater recharge (e.g. Meyboom 1961; Rorabough 1964; Bevans 1986; Rutledge 1992). The baseflow estimated from stream hydrographs (using HYSEP; Sloto & Crouse 1996) for the area amounted to 147 mm year^{-1}, i.e. about 8% of annual precipitation. For mountainous regions, Mau & Winter (1997) observe that baseflow estimates could provide lower estimates of recharge by about 25%. The moderate relief of the Tarkwa area suggests that, although baseflow is less than recharge, the difference is not as great as 25%. An important reason for the low value of recharge estimated using stream hydrographs compared to the water balance method is that surface mining has distorted the topography such that some of the rainfall is held in mined-out pits and other man-made impoundments, thereby reducing the volume of stream discharge (as noted above). In addition, all three surface mines in the district win gold from the Banket Series, which incidentally has the highest relief in the area. PSD and infiltration tests for the Banket soils revealed they possess the best characteristics for infiltration and provide many recharge areas. Thus, gold mining is inadvertently destroying recharge areas for groundwater. It might be argued that if, for example, recharge estimates were determined ahead of the three surface gold mines commencing production, a better understanding of the groundwater system may have been obtained.

Baseflow index (BFI) is a catchment characteristic and is defined as the percentage of run-off derived from groundwater (Anon. 1980). The BFI determined for four catchments reveal that their storage characteristics are low: Awunabeng (33%), Bediabewu (17%), Essumang (28%) and Anyuiafutu (27%) (Fig. 4a–d). (Stream names are also used to define the catchments in Fig. 2.) Run-off varies with the depth and texture of the soil, and physiography of the catchment (Dunne & Leopold 1998). Therefore, these values have both lithological and topographic significance. The Bediabewu flows more or less between the Banket and Tarkwa Phyllites before going over Kawere rocks. Both Banket and Tarkwa Phyllites form narrow ridges in the area and run-off is swift, resulting in a very low baseflow (Fig. 4a). The Essumang and Anyuiafutu streams, with virtually equal BFI values, are both largely confined to the Huni Sandstone, which has a very gentle topography and, therefore, these values probably reflect a strong soil influence (Fig. 4b & c). The Awunabeng catchment includes the Awunabeng and Ahumabru streams, and is underlain by Banket and Kawere rocks. The Banket terrain is large, and is drained by numerous tributaries of the Awunabeng stream. Despite the fact that mining has destroyed part of the recharge area of the Banket, the Awunabeng catchment still exhibited the highest BFI of the streams gauged (Fig. 4d), confirming the status of the Banket as the best yielding aquifer lithology in the Tarkwaian System.

Hydrogeochemistry

The following water features were identified for investigation:

- Dry weather water samples were collected from all streams in the area at reach intervals of about 3 km. The dry weather (or 'low-flow') period allowed any spatial variations in in-coming groundwater chemistry to be identified (Pettyjohn 1985).
- Temporal variability in stream chemistry was investigated by means of periodic gauging and sampling. All the four streams were sampled over a 12 month period on a weekly basis for pH, Eh, temperature, bicarbonate alkalinity, Cl, Ca, SO_4 and SiO_2. The suite of major and some minor ions was analysed on a monthly basis.
- Springs and shallow wells give direct samples of groundwater, and boreholes provide samples of bedrock groundwater. (The average depth of the boreholes is 60 m.) During these studies, the lack of depth sampling equipment meant that 'bulk' groundwater samples were taken.

Table 4 summarizes results of the water properties investigated. Spatial and temporal variations are observed, and are attributed to mine water discharging into receiving streams and local community use of the water resources. Some of the important conclusions arrived at are as follows.

Different water types are observed for the different water regimes sampled. The hydrochemical facies identified with water samples from springs, shallow wells and hand-dug wells are dominated by Na–Cl–HCO_3–Cl. Streams which were perceived not to be directly or indirectly linked with mine areas are grouped as 'pristine'. Pristine water samples fall into two hydrochemical zones, namely, water with an intermediate chemical character, i.e. no cation–anion pair exceeds 50% and the fresh water facies. These water types are described by Na–Ca–Mg–HCO_3–SO_4 and Ca–Na–HCO_3 chemical facies, respectively.

Streams that have passed through and/or received mine water display Na–SO_4–HCO_3 and Na–Ca–HCO_3–SO_4 hydrochemical facies and are identified, respectively, with the Awunabeng and Bediabewu streams. Their water chemistry ranges from slightly saline to saline, with the highest Na sample of 373.6 mg l^{-1} from a tributary of Bediabewu flowing from Teberebie Goldfields concession.

The Essuman stream exhibits a Ca–Na–HCO_3 hydrochemical facies, while Anyuiafutu shows equal concentrations of the major cations and almost equal concentrations of SO_4 and Cl, i.e. Na–Ca–Mg–Cl–SO_4. Low HCO_3 water concentration associated with Anyuiafutu is likely due to microbial respiration in the soil resulting in a low annual pH of 6.27. Considering that both Essuman and Anyuiafutu traverse the same lithological unit, the expectation is that both will exhibit broadly similar hydrochemical facies. Differences in their water types are attributed to both artificial and natural influences:

- most of the water in Essuman flows through mine territory and some tributaries receive mine water, this is not so with Anyuiafutu;
- Anyuiafutu, because of its small catchment may be receiving only near-surface

Table 4. *Summary of mean chemical characteristics of water in the Tarkwa area*

	Dry weather water chemistry (spatial variation)			Stream gauged water chemistry (temporal variation)			
	Spring and borehole	Pristine areas	Mine water influenced areas	Awunabeng	Bediabewu	Essuman	Anyuiafutu
Facies type	Na–Ca–HCO$_3$–Cl	(a) Na–Ca–Mg–HCO$_3$–SO$_4$ (b) Ca–Na–HCO$_3$	Na–SO$_4$–HCO$_3$	Na–SO$_4$–HCO$_3$	Na–Ca–HCO$_3$–SO$_4$	Ca–Na–HCO$_3$	Na–Ca–Mg–Cl–SO$_4$
HCO$_3$/SiO$_2$	1.74	2.85	11.42	7.53/14.85	5.29/7.33	2.40/3.15	0.90/1.48
pH	5.76	6.48	7.74	7.09/7.68	7.04/7.46	6.75/6.77	6.27/6.58
TDS	141	110	536	192/348	162/227	93/97	93/116
Remark	(i) Units of TDS is mg l^{-1}.			(ii) 7.53/14.85 refers to annual/low-flow figures.			

groundwater, explaining its low pH and HCO$_3$ values. However, the pH of Essuman is higher than the average and may be due to mine water discharge.

Another important factor considered during this investigation is the ratio HCO$_3$/SiO$_2$. A terrain is likely to have undergone silicate weathering if the HCO$_3$/SiO$_2$ ratio is less than 5, and carbonate weathering if the ratio is greater than 10 (Hounslow 1995). Also, a terrain that has undergone silicate weathering exhibits a low total dissolved solids (TDS) value (in the range 100–200 mg l^{-1}), but in carbonate weathered areas, TDS is typically greater than 500 mg l^{-1} (Hounslow 1995).

(a) Dry weather water analyses of perceived pristine areas show HCO$_3$/SiO$_2$ and TDS values of 2.85 and 110 mg l^{-1}, respectively. These values imply that the water in this area has participated in silicate weathering. However, streams passing through mining territory and those that receive mine water show HCO$_3$/SiO$_2$ ratios of 11.42 and TDS of 536 mg l^{-1}, suggestive of carbonate weathering. As the lithology in the mines is not different from that outside of them, this apparent inconsistency can only be due to a direct influence of mine water chemistry on receiving streams, probably by disposal of alkali-dosed waters and/or leakage of spent heap leach pad liquors (de-cyanidized) and/or tailings dam supernatants entering the groundwater system.

(b) The results of (a) apply to streams flowing through and receiving mine water from Ghana Australian Goldfields (GAG) and Teberebie Goldfields Limited concessions. The Essuman stream receives discharges of mine water from Goldfields Ghana Limited, but this stream does not show any significant major ion effect and its chemical facies plot in the fresh water domain, i.e. Ca–Na–HCO$_3$. It is possible to speculate that little or no leakage of unprocessed mine water is released into the Essuman. However, minor and trace ion chemical analysis should be performed on this and all the other streams to ascertain their complete water chemistry (local resources precluded this during the present study).

(c) Mean pH of water in the pristine areas is 6.48, but above 7.00 in streams carrying and receiving mine water from GAG and Teberebie Goldfields (it is 6.77 for Essuman). The higher than average pH values appear to be due to the use of lime in processing gold ore, which seep into streams.

A mean TDS value of $110\,\text{mg\,l}^{-1}$ for pristine areas suggests that the residence time of groundwater in the subsurface prior to their emergence in streams is brief, and hence source concentrations are low. In addition, the low TDS value also suggests a region dominated by local-scale flow systems. No change in anionic composition from recharge to discharge points was observed (for streams in pristine areas), which further corroborates small-scale, localized water–rock interactions.

This short study of the major ion chemistry of water from the Tarkwa gold mining district has shown that mining has affected stream water. A long-term monitoring programme undertaken with an expanded analysis including trace and other metal ions would enable the full impact of mining to be stated.

Using hydrogeological data to assess possible environmental impacts

If all the issues raised are pursued during the exploration and planning phases of mine development, hydrogeological assessment for possible environmental impacts becomes relatively simple and may be conducted by:

- using the hydrogeological model in the EIA report as a baseline for assessment;
- making regular updates to the hydrogeological database during the course of mine operation and production. More monitoring boreholes may then be in operation and can be sampled and gauged for water quality and groundwater head studies, respectively;
- a low-flow water quality and discharge survey conducted in the mine concession leading to hydrogeochemical interpretation to support the baseline report;
- field monitoring of identified poor discharges or problem areas;
- hydrogeochemical studies for possible amelioration.

A summary of the above recommendations is presented in Table 5 and as a flow chart in Fig. 5.

Discussion and conclusion

This study has outlined the importance of early gathering of hydrogeological information during the mineral exploration phase for new mines. It has identified which datasets should be considered and how they can be obtained at relatively low cost, thus easing the data requirements during hydrogeological assessment for environmental impact assessment purposes later in the development of the mine prospect. All these data are collected one way or the other during groundwater studies for other purposes. In the case considered, i.e. during mineral exploration, the problem is that emphasis is almost entirely placed on "what grade and ore body magnitude are we looking at?'. Therefore, the people involved are either not aware that hydrogeological data are available or, if they are aware, the value of data collection is not realized and it is seen as a waste of time and money.

It might also be argued that at an early stage when the geological resource is not yet fully a reserve, it would be a waste of financial resources and time to collect data that will not directly improve the reserve estimation. However, spending at this time on good quality data collection (which should mean employment of a qualified hydrogeologist) could offset:

- the amount to be spent for employing a hydrogeologist during the preparation of that part of the EIA report. This also means that a more accurate report is produced;
- the cost of hiring a consulting hydrogeologist later on when a hydrogeological problem arises at the mine. This is because the solution can be obtained by analysing the detailed report prepared earlier, or the problem will not arise anyway because the information available has been utilized, preventing its occurrence;
- future problems (environmental, monetary or both) due to ill-conceived planning and construction;
- some of the information is used as input to dewatering design and water management.

It is now the practice to introduce environmental studies into many academic programmes. Therefore, as a permanent solution in the future, and to cut costs, it is suggested that relevant programmes for undergraduates and graduates in the field of earth sciences should incorporate fundamentals of hydrogeology into their curriculum. This should familiarize geoscientists with groundwater hydraulics and the groundwater cycle. In addition to the above, an understanding of hydrogeological soil and rock sampling and logging procedures are essential during lithological core logging. This information can be

Table 5. *The recommended phases of mineral exploration most appropriate for hydrogeological data collection and the expected outcomes*

Hydrogeological aspect	Mineral exploration phase	Hydrogeological outcomes
Physiography • Relief • Drainage • Land use • Vegetation	Desk studies and during regional reconnaissance. As more detailed surveys are executed, larger-scale maps are obtained.	Maps at a scale of 1:20 000 – 1:5000 are prepared during coventional surveying. Recharge and discharge areas are determined. Preliminary groundwater flow directions and groundwater boundaries are conceptualized.
Pedology • Soil logging and sampling • PSD • Moisture content, bulk and particle densities • Infiltration tests	Exploratory, outline and evaluation drilling phases. Pitting and trenching are useful if weathering profile is thin.	Soil thickness map constructed. Soil texture, sorting, porosity and changes in lithology determined. Results are also useful during assessment of groundwater recharge.
Geology • Structural • Lithological • Stratigraphic • Thin section and microscopy • Probe permeameter	From the desk study phase through to evaluation drilling, surface mapping and core logging.	From thin section analysis of rock samples, determine: mineralogical composition, grain size, sorting and porosity (by point counting). Use porosity to estimate specific yield and storativity (Younger 1993). Permeability determined from probe permeameter. Prediction of rock–water interactions for assessment of possible discharges to streams. Three-dimensional (3-D) conceptual hydrogeological model from above information coupled with surface geological mapping and core logging.
Hydrology • Meteorological station installed for daily recording • Daily gauging of major streams • Piezometers installed	Some information is available at the desk study phase, but install a meteorological station, install piezometers and start stream gauging – all during evaluation drilling.	Evaporation and evapotranspiration estimation. Use information from soil survey with rainfall, run-off and evapotranspiration data to estimate recharge. Use stream hydrograph to also estimate recharge. Gauging and sampling of piezometers to determine groundwater movement and changes in groundwater quality.
Hydrochemistry • Low-flow stream surveys	Regional reconnaissance stream surveys but mainly during feasibility studies.	Hydrochemistry of the area determined. Weekly or monthly hydrochemical sampling of major stream(s) to determine temporal changes in chemistry of groundwater.

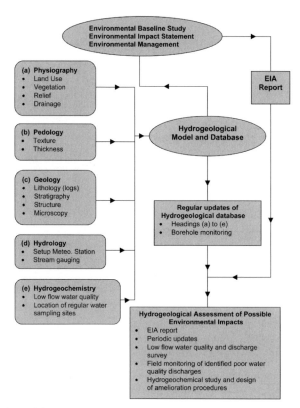

Fig. 5. Proposed flow chart of the hydrogeological framework for assessing environmental impacts of large-scale gold mines.

gathered at the same time and consciously built into a database for later interpretation.

The fact that hydrogeological data are available at such an early stage means that it can be included in the EIA report when presented to the EPA for evaluation before a lease for mining is granted. Interpretation of these data could be a useful input to dewatering design and groundwater management. Any water borehole drilled by a mine should be seen as an opportunity for carrying out test pumping so that aquifer properties can be determined. Piezometers installed for water monitoring need to be kept in working order and measurements taken regularly for valuable data to be available for later assessment.

The onus is, therefore, on the regulation agency to make it a policy for all prospective mines to include vital hydrogeological information in their EIAs to enhance knowledge of the hydrogeology of the area in question. The argument for this procedure is that once the decision is made for detailed drilling to commence, it implicitly means that the prospect is of interest and that it also warrants a hydrogeological assessment. The hydrogeological information can be integrated with the EIA document. This actually eliminates the tendency of looking at time as a constraint and, therefore, precludes a half-hearted (or absent) hydrogeological section in the report. This procedure may also be employed in areas where 'incomplete' EIA reports are a common occurrence.

References

ACHEAMPONG, P.K. 1986. Evaluation of potential evapotranspiration methods for Ghana. *GeoJournal*, **12**, 409–415.

AMEGBEY, N. 1996. *West Coast Explosives (Ltd) EIA on the Bulk Emulsion Explosive Plant at Nsuta-Wassaw*. West Coast Explosive Ltd., Tarkwa, Ghana.

ANON. 1980. *Low Flow studies*. Institute of Hydrology, Wallingford, Research Report, **1**.

ANON. 1994. In: LAING, E. *Ghana Environmental Action Plan (Vol. 2), Technical Background Papers by the Six Working Groups*. Environmental Protection Council (EPC), Ghana.

ANON. 1997. *Drilling of Production Boreholes and Monitor Boreholes at the Tarkwa Mine*. Internal SRK Report, **229708**.

APPELO, C.A.J. & POSTMA, D. 1993. *Geochemistry, Goundwater and Pollution.* A.A. Balkema, Rotterdam.

BEVANS, H.E. 1986. Estimating stream–aquifer interactions in coal areas of eastern Kansas by using streamflow records. In: SUBITZKY, S. (ed.) *Selected Papers in the Hydrologic Sciences*, US Geological Survey, Water Supply Paper. Vol. **2290**, 51–64.

BOWELL, R.J. 2002. The hydrogeochemical dynamics of mine pit lakes. In: YOUNGER, P.L. & ROBINS, N.S. (eds) *Mine Water Hydrologogy and Geochemistry*; Special Publications. Geological Society, London Vol. **198**, 59–185.

CHIDLEY, T.R.E. 1981. Assessment of groundwater recharge. In: LLOYD, J.W. (ed.) *Case Studies in Ground Water Resources Evaluation*. Butler & Tanner, London, 133–149.

DAVIS, S.N & DEWEIST, R.J.M. 1991. *Hydrogeology*, reprint edn. Krieger, Florida.

DICKSON, K.B. & BENNEH, G. 1988. *A New Geography of Ghana*. Longman, London.

DINGMAN, S.L. 1994. *Physical Hydrology*. Prentice-Hall, Englewood Cliffs, New Jersey.

DUNNE, T. & LEOPOLD, L.B. 1998. *Water in Environmental Planning*. W.H. Freeman, New York.

DZIGBODI-ADZIMAH, K. 1996. Environmental concerns of Ghana's gold booms: past present and future. *Ghana Mining Journal*, **2**, 21–26.

EDMUNDS, W.M. 1981. Hydrogeochemical investigations. In: LLOYD, J.W. (ed.) *Case Studies in Ground Water Resources Evaluation*. Butler & Tanner, London, 87–111.

EIJPE, R. & WEBER, K.J. 1971. Mini permeameter for consolidated rock and unconsolidated sand. *American Association of Petroleum Geologists Bulletin*, **55**, 307.

EISENLOHR, B.N. 1992. Conflicting evidence on the timing of mesothermal and paleoplacer gold mineralisation in Early Proterozoic rocks from southwest Ghana, West Africa. *Mineralium Deposita*, **27**, 23–29.

Fetter, C.W. 1994. *Applied Hydrogeology*. Prentice-Hall, Upper Saddle River, New Jersey.

Grindley, J. 1970. *Estimation and Mapping of Evaporation*. IASH publication No. 92. *World Water Balance*, **1**, 200–213.

HEAD, K.H. *Manual of Soil laboratory testing. Soil Classification and Compaction Tests*; Pentech Press, London, Volume 1.

HEDIN, R.S., NAIRN, R.W. & KLEINMANN, R.L.P. 1994. *Passive Treatment of Coal Mine Drainage*, US Bureau of Mines Information Circular, Vol. **9389**, 35.

HEM, J.D. *Study and Interpretation of the Chemical Characteristics of Natural Water*, United States Geological Survey, Water Supply Paper, Vol. **2254**, 263.

HIRDES, W. & NUNOO, B. 1994. The Proterozoic paleoplacers at Tarkwa Gold Mine, SW Ghana: sedimentology, mineralogy and precise age dating of the main reef and west reef, and bearing of the investigations on source area aspects. *BGR Geol Jb*, **D100**, 247–311.

HOUNSLOW, A.W. 1985. Strategy for subsurface characterisation research. In: WARD, C.H., GIGER, W. & MCCARTY, P.L. (eds) *Ground Water Quality*. John Wiley, New York, 356–369.

HOUNSLOW, A.W. *Water Quality Data: Analysis and Interpretation* CRC Lewis, New York.

JUNNER, N.R., HIRST, T. & SERVICE, H. 1942. *The Tarkwa Goldfield*, Gold Coast Geological Survey, Memoir, **6**, 75.

KESSE, G.O. *The Mineral and Rock Resources of Ghana*; Balkema, Rotterdam.

KOLM, K.E. 1996. Conceptualization and characterization of ground water systems using Geographic Information Systems. *Engineering. Geology*, **42**, 111–118.

KUMA, J.S. & YOUNGER, P.L. 2001. Pedological Characteristics Related to Ground Water Occurrence in the Tarkwa Area, Ghana. *Journal of African Earth Sciences*, **33**, 363–376.

LERNER, D.N. 1990. Techniques (Part III). In: LERNER, D.N., ISSAR, A.S. & SIMMERS, I. (eds) *Groundwater Recharge; A Guide to Understanding and Estimating Natural Recharge*, International Association of Hydrogeologists. Vol. **8**, 99–228.

MAU, D.P. & WINTER, T.C. 1997. Estimating ground water recharge from streamflow hydrographs for a small mountain watershed in a temperate humid climate, New Hampshire, USA. *Ground Water*, **35**, 291–304.

MEYBOOM, P. 1961. Estimating groundwater recharge from stream hydrographs. *Journal of Geophysical Research*, **66**, 1203–1214.

PETTYJOHN, W.A. 1985. Regional approach to ground water investigations. In: WARD, C.H., GIGER, W. & MCCARTY, P.L. (eds) *Ground Water Quality*. John Wiley, New York, 402–417.

PETTYJOHN, W.A. & HENNING, R. 1979. *Preliminary Estimate of Ground Water Recharge Rates Related Streamflow and Water Quality in Ohio*. Water Resources Center, Ohio State University, Project Completion Report, Vol. **552**, 333.

RORABAUGH, M.I. 1964. Estimating changes in bank storage and ground water contribution to streamflow. *International Association of Science & Hydrology, Publication*, **63**, 432–441.

RUTLEDGE, A.T. 1992. Methods of using streamflow records for estimating total and effective recharge in the Appalachian Valley and Ridge, Piedmont, and Blue Ridge physiographic provinces. In: HOTCHKISS, W.R. & JOHNSON, A.I. (eds) *Regional Aquifer Systems of the United States, Aquifers of the Southern and Eastern States*, American Water Research Association Monograph Series. Vol. **17**, 59–73.

SHARP, J.M. & MCBRIDE, E.F. 1989. Sedimentary petrology – a guide to paleohydrogeologic analyses, example of sandstones from the northwest Gulf of Mexico. *Journal of Hydrology*, **108**, 367–386.

SHAW, E.M. *Hydrology in Practice*, Chapman and Hall, London.

SHEVENELL, L. 2000. Analytical method for predicting filling rates of mining pit lakes: example from the Getchell Mine, Nevada. *Mining Engineering*, **52**, 53–60.

SLOTO, R.A. & CROUSE, M.Y. 1996. *HYSEP: A Computer Program for Streamflow Hydrograph Separation and Analysis*, United States Geological Survey, Water Resources Investigations Report, Vol. **96–4040**, 46.

STONE, W.J. *Hydrogeology in Practice. A Guide to Characterizing Ground-water Systems*, Prentice-Hall, Englewood Cliffs, New Jersey.

THORNTHWAITE, C.W. & MATHER, J.R. 1957. *Instructions and Tables for Computing Potential Evapotranspiration and the Water Balance publications in Climatology*, (3) Centerton, New Jersey Vol. 5, 185–311.

WHITELAW, O.A.L. 1929. *Geological and Mining Features of the Tarkwa–Aboso Goldfield*, Gold Coast Geological Survey, Memoir, Vol. 1, 46.

WINTER, T.C., HARVEY, J.W., FRANKE, O.L. & ALLEY, W.M. 1998. *Ground Water and Surface Water: A Single Resource*, United States Geological Survey, Circular, Vol. 1139, 79.

WOODALL, R. 1984. Success in mineral exploration: a matter of confidence. *Geoscience Canada*, **11**, 41–46.

YOUNGER, P.L. 1992. The hydrogeological use of thin sections: inexpensive estimates of groundwater flow and transport parameters. *Quarterly Journal of Engineering Geology*, **25**, 159–164.

YOUNGER, P.L. 1993. Simple generalized methods for estimating aquifer storage parameters. *Quarterly Journal of Engineering Geology*, **26**, 127–135.

YOUNGER, P.L. 2001. Mine water pollution in Scotland; nature, extent and preventive strategies. *Science of the Total Environment*, **265**, 309–326.

YOUNGER, P.L. & ADAMS, R. 1999. *Predicting Mine Water Rebound*; Environment Agency, R&D Technical Report, **W179**.

YOUNGER, P.L., BANWART, S.A. & HEDIN, R.S. 2002. *Mine Water: Hydrology, Pollution, Remediation*. Kluwer, Dordrecht.

Predicting mineral weathering rates at field scale for mine water risk assessment

STEVEN A. BANWART, KATHERINE A. EVANS & STEPHANIE CROXFORD

Groundwater Protection and Restoration Group, Department of Civil and Structural Engineering, University of Sheffield, Sheffield S1 3JD, UK
(e-mail: s.a.banwart@sheffield.ac.uk)

Abstract: A general challenge to the environmental management of mine sites is the relatively high costs of site investigation and the associated development of conceptual site models and parameterization of reactive transport models. A particular problem is the inability to predict at field scale the rates of processes that give rise to long-term dissolved contamination due to active sulphide mineral weathering. Mineral weathering rates determined from laboratory and field observations generally do not agree, often exhibiting a discrepancy of two–three orders of magnitude.

Recent work on mine waste deposits has demonstrated that this discrepancy can be explained by considering a small number of bulk physical and chemical properties of mine rock at field sites. The apparent decrease in mass-normalized rates between bench-scale batch reactors and pilot-scale column reactors is predicted by accounting for differences in temperature, where lower temperatures in the column reactors reduced reaction rates due to activation energy effects. The columns also contain a significantly greater mass fraction of larger particles that have lower specific surface area and, thus, exhibit lower weathering rates.

The further apparent decrease in rates between the column reactors and field scale is predicted by additionally accounting for the spatial variability of sulphide-bearing rock at the site, which gives rise to only localized weathering. Localized zones of sulphide weathering are also associated with locally active weathering of silicate minerals due to lower pH. Hydrological factors are also important due to preferential flow within the field site, whereby a fraction of dissolved weathering products are retained within immobile water and do not reach the effluent stream, where ion mass flows resulting from weathering reactions are determined. These results suggest that application of compiled laboratory data to prediction of weathering rates at mine sites may be feasible. This is potentially valuable for application to Tier 1 risk assessment of mine sites, where reliable prediction of weathering rates from tabulated laboratory data would provide significant information to support the generally sparse datasets that are available, particularly for orphan mine sites.

Mining activities worldwide contribute significantly to the solute loads of receiving streams and aquifers. The contribution of sulphide mineral weathering associated with mine sites to sulphate ion loads is currently estimated at 12% of the global fluvial sulphate flux to the world's oceans (Nordstrom & Southam, 1997). Associated with this weathering flux are dissolved metal ions and acidity as potential environmental hazards. Assessment of current environmental risk associated with developed mine sites is necessary in order to make appropriate management decisions for monitoring of site emissions, and, if necessary, treatment or restoration of environmental quality. Assessment of future risk is critical when developing licensing plans for new mining activity in order for owners, operators and investors to have a clear assessment of potential environmental liabilities on their balance sheets.

In the following sections we first present an introduction to mine water risk assessment methods and the problems associated with predicting the rates of chemical weathering at mine sites for a range of minerals that give rise to contamination, and that also help neutralize acidity and thus attenuate contamination loads. The second section outlines hydrochemical methods for assessing rates of mineral weathering in the laboratory and in the field, and outlines possible sources for the significant discrepancy between rates that is generally observed at the two different physical scales. The third section presents results from a site-specific study where this discrepancy was successfully explained by accounting for differences in a few bulk physical–chemical parameters between the two scales. These parameters are temperature, mineralogical composition, ratio of mobile to immobile water content, particle size distribution and related physical surface area, and differences in pore water pH. The concluding section critically discusses the potential to extend this methodology for predicting weathering rates at field scale to other sites and other environmental problems where mineral weathering kinetics play an important role.

Mine water pollution and environmental risk

Within in the UK, implementation of Part IIA of the Environmental Act of 1990 now requires a risk-based approach to the assessment of potential harm or damage to receiving waters, arising from contaminated land. Implications of the new legislation for mines and quarries has recently been outlined by the Environment Agency of England and Wales (Bone 2001). The long history of mining activities, from the Bronze Age forward, has resulted in a large number of now abandoned sites that represent potential environmental risk. These *orphan sites* have no legally responsible party attached to them. Assessment for potential site re-use, including any necessary restoration work, requires either public money or private investment that is to be recouped with the redevelopment of the site. In addition, there are exceptional commercial pressures on current or potentially new mining activity. In both cases, this means that decisions on environmental liability and risk must be made with the absolute minimum of cost.

An important consequence of minimal investment in risk assessment is relatively high uncertainty in the site conceptual model and any associated quantitative estimates of environmental risk. This uncertainty results in conservative decisions on further action. If serious contamination is thought to be present, one consequence may be that inordinately exhaustive and, therefore, expensive restoration schemes may be commissioned in order to make sure that any future potential liability is removed. A more likely outcome of this uncertainty is that many orphan sites will not be redeveloped and further mining at potentially lucrative sites may be unnecessarily curtailed.

Figure 1 illustrates a general conceptual model for sources of mine water pollution, transport pathways for soluble contaminants and potentially sensitive receiving waters that pose targets for contaminant risk. Within this source–pathway–target risk assessment framework, there is a common conceptual approach to mine sites that arises from considering the oxidative weathering of sulphide minerals to release soluble and, therefore, mobile metal ions and acidity to the environment. Weathering of the contaminant source minerals pyrite and sphalerite are shown here as examples:

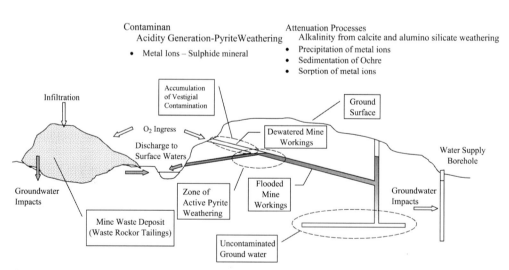

Fig. 1. A schematic representation of contaminant sources, transport pathways and sensitive risk targets in mining environments. Attenuation processes occur during transport from source areas to targets, thus reducing contamination levels and risk. The Tier 1 risk assessment methodology presented in Table 1 considers long-term juvenile contamination arising from active weathering of sulphide minerals above the water table where $O_2(g)$ ingress occurs.

pyrite weathering

$$FeS_2(s) + 3\frac{1}{2}O_2(aq) + H_2O \rightarrow Fe^{2+} \quad (1)$$
$$+ 2SO_4^{2-} + 2H^+$$

sphalerite weathering

$$ZnS(s) + 2O_2(aq) \rightarrow Zn^{2+} + SO_4^{2-}. \quad (2)$$

Further consideration is given to the fact that solubility-limiting oxide and hydroxide mineral phases, such as $Zn(OH)_2(s)$, are generally much more soluble at low pH. Acidity release due to pyrite weathering poses an environmental hazard itself. However, it also can lead to a drop in pH, which generally increases the solubility and, thus, transport of hazardous metal contaminants to potentially sensitive targets such as aquifers or receiving streams.

A particular distinction is made in Fig. 1 between *vestigial contamination* and *juvenile contamination*. Vestigial contamination accumulates as secondary mineral precipitates of the metal ions and sulphate that arise from sulphide mineral weathering in dewatered void spaces in mine environments. A particular relatively short-term problem arises when subsurface workings are flooded, causing these secondary precipitates to dissolve. The time scale for attenuation of the resulting contamination is controlled by the hydraulic flushing time of the mine.

Juvenile contamination arises due to ongoing release of soluble contaminants due to active sulphide mineral weathering that is driven by $O_2(g)$ ingress above the water table. Contamination is subsequently transported with subsurface flow to potential receptors such as receiving streams. Juvenile weathering thus represents a relatively much longer-lived contamination problem. The longevity of contamination is controlled by the rate of sulphide mineral weathering and the mass of reactive minerals in the zone of active weathering. The risk assessment framework described below is relevant to the long-term risk posed by juvenile contamination where mineral weathering rates, but not hydraulic flushing times, are the critical control on the contamination source strength and longevity. Vestigial contamination and the linked hydrogeological problem of mine water rebound are reviewed in detail elsewhere (Younger *et al.* 2002).

Figure 2 shows this general conceptual model for the relative risk from long-term acidity generation to receiving waters from various subsurface mine environments (Younger & Banwart 2002). Key factors affecting the relative risk from a site include:

- the extent of unsaturated flow processes where atmospheric $O_2(g)$ ingress occurs;
- the extent to which sulphide minerals are present within disaggregated, crushed or milled particles allowing greater reactive surface area for reaction;
- the proximity of oxic source zones for sulphide mineral weathering to discharge points whereby shorter flow paths and, therefore, lower residence times provide less opportunity for attenuation processes (see Fig. 1) to occur.

Risk-based methods for assessing mine water pollution

Figure 3 shows a schematic representation of the 'tiered risk assessment methodology'. The diagram is based on the 'risk-based corrective action' (RBCA) guidelines published by the ASTM (1995) for petroleum hydrocarbon pollution in the subsurface. The aim of such methodology is to guide decisions on the use of risk assessment and site investigation resources. Tier 1 represents the preliminary assessment of a site. If the site is relatively easy to understand

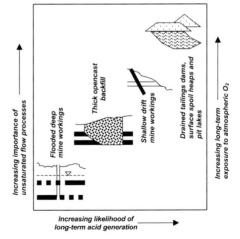

Fig. 2. Qualitative representation of risk arising from the oxidative weathering of pyrite and other sulphide minerals (Younger & Banwart 2002). Ingress of $O_2(g)$ is enhanced by shallow source zones and those above the water table in proximity to the atmosphere. Proximity of source areas to discharge points also increases contamination through shorter transport paths and residence times, thus providing less opportunity for attenuation processes to occur.

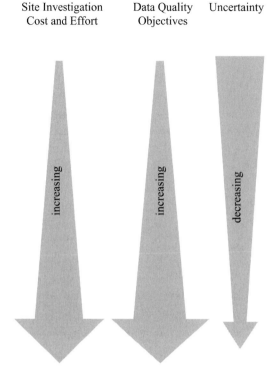

Fig. 3. A diagram of the 'Tiered risk assessment framework' based upon 'risk based corrective action' for subsurface petroleum hydrocarbon contamination (ASTM 1995). Tier 1 represents initial site assessment. Sound understanding of site contamination and risk scenarios allows robust decisions to be made regarding site use and/or remediation at lower Tiers. Poor understanding can result in a 'tier upgrade', which entails greater effort and expense for analysis of site data, and often further site investigation.

based on existing information, then a reliable decision may be taken to use the site in its present state or invest in site remediation. If the nature and type of data available are unsuitable for making such decisions, i.e. the conceptual model for the site is too uncertain, then a decision can be made to abandon the effort or upgrade to a higher tier. Additional site investigation, as well as data analysis, generally accompany a tier upgrade in order to improve the basis for making the best management decision.

Each tier is assigned appropriate 'data quality objectives', i.e. the type and quality of information required to carry out the particular level of assessment and reach a decision. Banwart & Malmström (2001) recently proposed 'Tier 1 data quality objectives' and assessment methodology for risk-based assessment of mine water discharges. The methodology is particularly suitable for spoil deposits and shallow, relatively well-characterized mine environments. Table 1 summarizes the proposed analysis method and the site data that are required for such preliminary assessment.

Younger (2000) recently provided a decision flow sheet for preliminary assessment of the longevity and future water quality (relative acidity and dissolved iron load) of discharges from flooded coal mine workings. The strength of the approach is that it relies on a very general conceptual model and, therefore, requires very little site-specific information. The model was calibrated against data from a large number of UK coal mine discharges and the resulting uncertainty in the model predictions is quantified.

A common philosophy that is evident in these two approaches is the acknowledgement that site data are extremely precious. Orphan sites can be expected to have very sparse site data and in many cases few mine records. Currently, operating or potentially new sites will probably have a very limited budget to carry out new site investigations targeted specifically at environmental

Table 1. *Data quality objectives and site assessment methodology appropriate for Tier 1 (Banwart & Malmström 2001).*

Assessment step	Data required	Corresponding data sources
1. Identify • Possible pollutants • Source areas • Transport pathways • Sensitive targets	1. Discharge composition 2. Mine petrology and mineralogy 3. Stratigraphy/structural geology 4. Groundwater flow paths 5. Location of site boundaries, discharge points, aquifers, streams, boreholes, etc.	1. Water quality analyses 2. Mine records and geological memoirs 3. Mine records and geological memoirs 4. Mine records and geological maps 5. Site maps and plans, regulatory authority records
2. Site water balance	1. Recharge to subsurface 2. Evapotranspiration 3. Number of site discharges 4. Discharge flow rates 5. Groundwater flows	1. Hydrological atlas 2. Hydrological atlas and knowledge of mine ventilation 3. Site maps and plans 4. Pumping records and environmental monitoring records 5. Borehole monitoring records, geological maps, hydrogeology handbooks
3. Discharge solute mass flows	1. Discharge flow rates 2. Discharge solute concentrations 3. Temporal variability in same	1. Environmental monitoring records 2. Environmental monitoring records 3. Environmental monitoring records
4. Source mineral weathering rates	1. Identify source minerals at site 2. Discharge solute mass flows 3. Weathering reaction stoichiometry	1. Geology reports and mine records 2. Step 2. Results 3. Mineralogy handbook
5. Source mineral abundance	1. Spatial extent and porosity of workings and deposits 2. Volume and mass of reactive rock 3. Mineral composition of rock	1. Site maps and mine plans 2. Site maps and mine plans 3. Mine records, geological memoirs, mineralogy handbooks
6. Source mineral longevity	1. Source mineral weathering rates 2. Total reactive source mineral abundance	1. Results from Step 4 2. Results from Step 5
7. Exposure scenario evaluation	1. Longevity of acidity generating minerals 2. Longevity of metals generating minerals 3. Longevity of alkalinity generating minerals	1. Results from Step 6 2. Results from Step 6 3. Results from Step 6

assessment. Even for sites with extensive mining records and a good understanding of the existing or future mining operations, it can be difficult for owners and operators to meet data quality objectives beyond those for Tier 1 assessment. This is largely because of the specialized knowledge and data that are required for quantifying water flow, geochemical reaction rates and contaminant transport pathways in complex hydrogeological systems.

An important strategy to reduce the cost and effort involved in Tier 1 risk assessment is to develop more general conceptual models for such complex hydrogeological systems. This has two immediate benefits. First, this reduces the amount of effort spent in developing a conceptual site model. Second, by starting from a proven conceptual model, effort can then be targeted towards reducing uncertainty in the site-specific conceptual model and improving quantitative analysis of site data.

There is also a third benefit that is explicitly linked to current research programmes in the area of mine water pollution. By developing more general understanding of the processes that give rise to contamination, researchers are able to target underpinning science towards development of more general mathematical descriptions of the physical, chemical and microbiological processes that determine contaminant loads. This allows more fundamental and, therefore, more generally applicable parameter sets to be developed for application in reactive transport modelling. Fundamental parameters do not change with site conditions and can, therefore, be tabulated as handbook values.

The associated science must, of course, provide quantitative methods to scale such fundamental parameters to site-specific conditions. A relatively familiar example in geochemical applications is the use of general thermodynamic databases with computer codes for equilibrium modelling of aqueous chemical speciation. Underpinning science has established that such data will vary in a known way with aqueous solution temperature and ionic strength. The modelling codes thus provide general algorithms to scale the fundamental thermodynamic data to account for these site-specific effects.

A particular problem that will be addressed in some detail in the following sections is the lack of a general conceptual model for the chemical weathering of minerals that act as sources of soluble contamination (source minerals). Associated with this problem is a complete lack of a general scaling procedure to predict mineral weathering rates at mine sites, in spite of extensive compilations of detailed weathering kinetic data from laboratory studies.

Studies of mineral weathering at field scale and associated application of weathering rates determined in laboratory studies have been widely used in the assessment of watershed acidification impacts on surface waters (reviewed by Schnoor 1990). White & Petersen (1990) reviewed a number of datasets arising from catchment and aquifer studies that demonstrate a general discrepancy where weathering rates from field studies are generally two–three orders of magnitude slower than those observed in laboratory studies. With respect to mine water pollution, the ability to predict weathering rates at field scale, based on compiled laboratory data and basic site investigation results, would constitute a major improvement in data quality objectives for contaminant source term modelling at Tier 1 risk assessment (Table 1, steps 4–7). This would also find important application for parameterization of more advanced reactive transport models corresponding to Tier 2 risk assessment (Fig. 3). Successful prediction of weathering rates at field scale from laboratory data would also constitute a major step forward in the understanding of other environmental problems, such as watershed acidification where mineral weathering plays an important role.

Scaling parameters for mineral weathering rates

Chemical weathering is controlled by the kinetics of surface chemical reactions, namely the change in the metal ion co-ordination environment from that within the crystal lattice to that of the dissolved metal ion or complex in aqueous solution. The reaction occurs at the mineral–aqueous solution interface, where the ordering of the crystal lattice terminates with some metal bonds maintained to the lattice, and others to aqueous species such as chemically bound water molecules. An important consequence of these reactions is that reaction rates are generally proportional to the amount of reacting mineral surface area in contact with aqueous solution. An excellent overview of the theory of chemical weathering is provided by Stumm & Morgan (1996, chap. 13). White & Brantley (1995) and Younger et al. (2002, chap. 2 and included references) provide additional reviews of silicate mineral weathering and sulphide mineral weathering, respectively.

Laboratory derived rates are usually normalized to the mass of the reacting mineral or rock sample. These rates are reported in units such as $mol\, g^{-1}\, day^{-1}$. The scaling parameter is therefore the mineral mass, m (g). The specific surface area (a, $m^2\, g^{-1}$) relates the physical surface area to the mass of reacting sample and allows laboratory rates to subsequently be normalized to the reacting surface area (r_{lab}, $mol\, m^{-2}\, day^{-1}$). In this case surface area becomes the scaling parameter, which in turn depends on both reacting mineral mass and the corresponding specific surface area.

Equation (3) shows the relationship between field weathering rate, laboratory rates and these scaling parameters, where the subscript 'py' refers by way of example to the mineral pyrite:

$$R_{py} = r_{lab,py} m_{py} a_{py} \quad (mol\, day^{-1}). \quad (3)$$

This relationship implies that if the values of m_{py} (g) and a_{py} ($m^2\, g^{-1}$) are known for a mine site, then the laboratory rates could be used to predict the weathering rate (R_{py}, $mol\, day^{-1}$) at the site. In practice it may be possible to estimate mineral mass within a mine workings based on mine plans and mineralogical assay information on the ore body and surrounding rock. However, values for specific surface are generally not known. White & Petersen (1990) reviewed the role of the reacting surface area as a scaling parameter for mineral weathering rates in detail.

Although there is currently no general method for reliably extrapolating laboratory test results to the complexity and physical scale of mine sites, it is still possible in some cases to get order of magnitude estimates of contaminant loadings. As explained below, this is achieved by transferring information about weathering rates and solute fluxes from another site where measurements have been made. This methodology is based on the premise that sites with similar geology and mining history will have a common conceptual description and similar site characteristics.

Scaling weathering rates between similar field sites. The specific surface area of an individual mineral at a field site presumably varies considerably with the mineral composition, morphology and size distribution of weathering particles in the rock mass, and therefore on the physical structure of geological material associated with a mine workings. Because of the spatial heterogeneity of field sites, this information is extremely uncertain. White & Petersen (1990) also present considerable discussion on what constitutes 'reactive' surface area (A_r, m^2 g^{-1}) and how this may relate to the physical surface area of weathering rock. There is not a simple relationship between the two parameters.

An alternative approach to this *scaling problem* is to estimate solute fluxes at the field site, and compare these with laboratory rates in order to obtain an empirical measure of A_r for rock that is weathering at the site. The value for A_r is obtained by dividing the weathering rate at the site, normalized against rock mass (mol g^{-1} day^{-1}), by the laboratory weathering rate (r_{lab}, mol m^{-2} day^{-1}) yielding a value with units of m^2 g^{-1}. This 'calibrated' value could then be used to scale laboratory weathering rates to similar types of field sites. Table 2 shows calibrated values for A_r determined from a variety of solutes in the discharge from a field site. The range of values spans a five-fold difference, which is remarkably small given the uncertainties involved.

This approach assumes, using pyrite as an example, that sulphate ions are conserved in the discharge, i.e. not accumulated within the workings as secondary precipitates or in immobile water. Banwart & Malmström (2001) discuss the limitations of such assumptions, and how to test them, in considerable detail. As an example here, the mass flow of sulphate normalized to total rock mass, F_S, (mol Kg^{-1} day^{-1}) is treated as a direct measure of pyrite weathering at field scale, but it should be borne in mind that sorption of sulphate or formation of sulphate-bearing secondary precipitates is a potential source of error when applying such methods to analysis of field data.

Equation (4) relates reactive surface area, A_r, to the sulphate mass flow at a field site and to the laboratory weathering rate. This relationship implies that solute mass flows originating from a single mineral, e.g. pyrite, must be corrected for the amount of pyrite in the rock. The simplest approach is to assume that the fraction of reactive surface area corresponding to pyrite is similar to the mass fraction of pyrite in the rock, X_{py}. Sulphur mass flow must also be divided by a factor of 2 to reflect the weathering stoichiometry of pyrite (FeS$_2$(s)), where 2 mol of dissolved sulphur are released for each mol of pyrite dissolved:

$$A_r = \frac{F_s}{2X_{py}r_{py}} \quad \text{(m}^2 \text{ of reactive rock)}. \quad (4)$$

Values for A_r, such as those in Table 2, can be used in equation (4) along with tabulated laboratory data on r_{py} and estimates of pyrite

Table 2. *Reactive surface area estimated at field scale*

Mineral Abundance in Field[a] (vol.%)	Tracer in discharge	Weathering rate from tracer flux in field[b] (mol kg^{-1} s^{-1})	Laboratory rate, pH 4 (mol m^{-2} s^{-1})	Reactive surface area at field Scale[c] (m^2 kg^{-1})
Albite 13%	Na$^+$	3.7×10^{-12}	3.1×10^{-11}[d]	0.9
Anorthite 6%	Ca^{2+}	8.5×10^{-12}	3.1×10^{-11}[d]	4.6
Biotite 8%	Mg^{2+}	2.8×10^{-12}	1.9×10^{-11}[d]	1.8
Pyrite 0.6%	SO$_4^{2-}$	1.2×10^{-11}	5×10^{-10}[e]	4.0
Chalcopyrite 0.1%	Cu^{2+}	5.3×10^{-13}	2.5×10^{-10}[e]	2.1

[a] Biotite gneiss with mica schist reported in Strömberg & Banwart 1994.
[b] From Banwart *et al.* (1998). determined for field site with mineralogy given in column 1.
[c] Determined by dividing weathering rate from tracer flux in field by the laboratory rate.
[d] From Stumm & Morgan (1996, Table 13.3). Albite (Al) and anorthite (An) dissolution rates correspond to weathering of a single plagioclase with a composition 70%Al–30% An.
[e] From Strömberg & Banwart 1994.

Table 3. *Assessment methodologies for source mineral weathering in mine water risk assessment*

Assessment method	Description of methodology	Relation to data quality objectives
1. Acid–base accounting (ABA)[a]	• Compares neutralization potential (NP) and acidity potential (AP) • NP determined by reacting crushed rock with hot HCl with back-titration to determine amount of HCl neutralized • AP determined by relating chemical analysis of total sulphur content of rock to equivalents of H_2SO_4	• Provides pre-mining assessment of contamination potential • If AP > NP, contamination is anticipated • Does not account for rates of weathering • Does not react at field conditions • Assumes all sulphur is acid-generating
2. Net acid generation[b]	• Reacts crushed rock with hydrogen peroxide to oxidize sulphide minerals • Measures the amount of oxidizable sulphur in rock	• Similar objectives as for ABA • Accounts specifically for acidity due to oxidation of sulphide minerals • Does not account for rates of weathering
3. Humidity cells experiments[c]	• Reacts mill tailings under hydraulically-unsaturated conditions • Dry air blown through tailings for fixed time followed by • Moist air blown through for same amount of time • Then flushed to capture oxidation products • Cycle repeated until source minerals are exhausted or pH drops	• Assesses time to exhaustion for source minerals • Realistic representation of tailings deposit conditions • Tells if alkalinity generating minerals have greater or less longevity than sulphide minerals
4. Batch experiments[d]	• Kinetic test of weathering rates • Solute accumulation rate related to individual source minerals	• Provides detailed weathering rate data • Does not account for flushing in mine environments
5. Column experiments[e]	• Solute fluxes measured in column effluents • Fluxes related to individual weathering rates of source minerals	• Provides detailed weathering rate data • Accounts for flushing of reaction products • Requires significant lab prowess

Table 3 – continued

Assessment method	Description of methodology	Relation to data quality objectives
6. Field hydrochemical assessment[f]	• Solute fluxes measured in site discharges • Fluxes related to individual weathering rates of source minerals	• Requires field monitoring budget • Does not account for sorption/precipitation
7. $O_2(g)$ consumption[g]	• Measures $O_2(g)$ decay with time in packed-off boreholes • Relates $O_2(g)$ vs time data to yield rate laws and rate constants for weathering of sulphide minerals	• Provides spatial resolution of weathering rates • Does not distinguish between sulphide minerals • Does not distinguish sulphide oxidation from carbon and Fe(II) mineral oxidation

[a]Sobek et al. 1978; [b]Finkelman & Griffen 1986, O'Shay et al. 1990; [c]ASTM 1996, Price 1997; [d]Strömberg & Banwart (1999a); [e]Strömberg & Banwart (1999b); [f]Strömberg & Banwart 1994; [g]Blowes et al. 1991, Harries & Ritchie 1985.

content as mass per cent in order to obtain a very crude estimate of solute mass flows in the absence of water quality data; for example, when assessing possible impacts of a *planned* mining operation for sites with similar geology and hydrology. Calculations of this type thus help to answer the question 'how bad will the contamination be'. Better site information such as discharge rates, water quality data, geometry of workings and mineralogical information are all important pieces of information to improve estimates of (or to help interpret) contamination loads.

Implications for extrapolating weathering rates from the laboratory to field scale. A common feature of these studies is that empirical values for reactive surface areas, derived from A_r and mass of rock at the field sites, are always two–three orders of magnitude lower than the specific surface area determined in laboratory studies (White & Petersen 1990). The range of values in Table 2 suggest that this scaling discrepancy between A_r and a can be similar for many minerals at a given field site (Strömberg & Banwart 1994). This implies that the factors responsible for the discrepancy are largely independent of the various dissolution mechanisms for different minerals.

Although the exact weathering rate for any individual mineral at a field site will probably not be possible to predict from laboratory rates alone, at least the *relative* rates of weathering for different minerals at a field site may be similar to those determined from laboratory studies. Because the development of acidic versus alkaline mine water discharges depends on the relative rates of pyrite, carbonate and alumino-silicate mineral weathering (see Banwart & Malmström 2001), the corresponding laboratory data are still useful for understanding development of water quality and its temporal trend. The main drawback is that even if relative rates of weathering are known, the longevity of source minerals depends on the *absolute* rates of weathering at a site. In the absence of these rates, the contamination lifetime cannot be predicted with any certainty.

Determination of weathering rates by aqueous chemical methods

There are a number of assessment methodologies for source mineral weathering rates that range from relatively simple laboratory leaching tests to much more sophisticated field-based methods that resolve spatial variability in weathering rates. Table 3 outlines these methods and the assumptions involved in interpreting the measurement results. These methods can be outlined in broad terms as *static tests* (Table 3, methods 1 and 2) and *kinetic tests* (Table 3, methods 3–7). Static tests are based on analysis of total amount of weathering products in order to establish chemical capacities for acidity and alkalinity production, while kinetic tests follow reactants and/or products with time in order to obtain reaction rates.

Although field methods capture processes that are actually occurring at a site, interpretation of the data is generally fraught with difficulty due to the complexity of real environments. Alternatively, laboratory tests are carried out under controlled conditions where source mineral weathering rates can be reliably determined. However, the problem remains of how to apply such laboratory data to reliably predict what the corresponding rate of weathering might be at a field site. In the subsequent section, a potentially general scaling procedure is presented and tested whereby laboratory rates may predict field rates. Two particular types of laboratory methods to determine mineral weathering rates have been used to obtain the data for that study. These methods, namely batch and column experiments, are described below.

Batch reactor methods

Batch reactor experiments are a relatively simple approach to determining the potential for a given geological sample to generate contamination under hydraulically saturated conditions. This method can also be applied in order to ascertain release rates of contamination under controlled laboratory conditions, and to relate these release rates to the weathering rates of individual minerals within the sample. It can also be used to establish the extent and rate of acid consumption by weathering of oxide, carbonate and silicate minerals in relation to pH buffering reactions. Figure 4a shows a drawing of a typical reactor configuration.

The reactors are usually set up using a known weight of solid sample and a known volume of aqueous solution, often with the solution ionic strength fixed by a swamping electrolyte of known composition in order to avoid changes in speciation due to changes in ionic strength during the experiment. Oxic or anoxic conditions can be controlled by maintaining the reacting solution in equilibrium with a specified gas such as $N_2(g)$ to maintain anoxia and $N_2(g)$–$O_2(g)$ mixtures of known composition to maintain fixed $O_2(aq)$ concentrations. The batch reactor solutions are stirred sufficiently rapidly in order to maintain well-mixed conditions, i.e. to prevent diffusion-controlled mass transfer in the fluid film surrounding the reacting solid sample. Reactors are often maintained at constant temperature conditions with typical experimental results reported for a temperature of 25°C as a standard condition.

Batch reactors can be set up as a large number of relatively low-maintenance experiments that can be left to run over a period of months. Continuous monitoring of the temperature and pH can give a general indication of the reactor performance and sample reactivity. Periodic sampling and chemical analysis of the sample solution composition allows the reaction progress to be followed. Time series of solute concentration data are then used to determine ion release rates from the reacting material. If solutes can be reliably related to the source minerals present, then weathering reaction rates for source minerals can be determined. Figure 4b shows a time series for copper ion release with time for a sample of mine waste rock. In this case, Cu release rates can be related to the source mineral chalcopyrite ($CuFeS_2(s)$). The effect of reacting surface area is demonstrated by the fact that small particle sizes release ions much more quickly than larger particle sizes. In fact, an important conclusion from the cited work shown in Fig. 4b is that particles with a nominal diameter greater than approximately 4 mm contribute negligibly to mass of solute released by the weathering reactions. This greatly simplifies a possible scaling procedure as a large mass of rock at a field site, i.e. above the 'particle-size cut-off' of approximately 4 mm (specific to the sited study), can be effectively neglected.

Batch reactors, which have the advantage of being a relatively simple laboratory procedure, are restricted in their ability to provide mineral weathering rates that can be extrapolated to the field. This is for two main reasons. First, the water to rock ratio within a batch reactor is far higher than it would be in any natural rock system. This leads to problems with gas access and the concentration of pore water solutes with respect to comparisons of calculated mineral solubility equilibria, i.e. much greater dilution of released ions occurs in batch reactors than would normally occur in pore waters of geological formations. Secondly, the solution composition is constantly changing as a result of mineral dissolution, so that steady-state conditions, that might be approached for flushed voids in subsurface mine environments, is not achieved.

Column reactor methods

Column experiments may be run under hydraulically saturated or unsaturated conditions, and thus offer the opportunity for realistic water to rock ratios and gas access. If they are run for long enough, they may reach steady-state conditions, where the percolating solution evolves steadily from the chosen influent to an effluent of temporally constant composition. In addition, column experiments provide the opportunity to

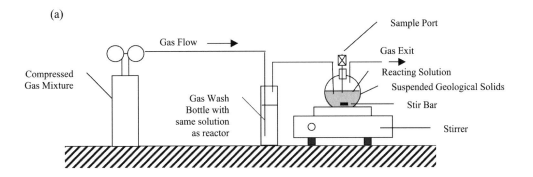

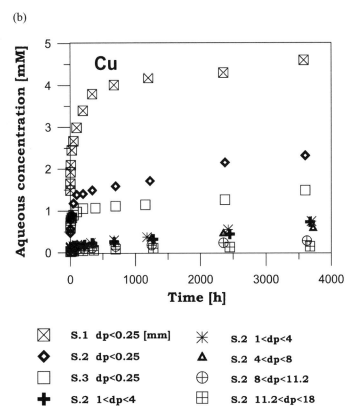

Fig. 4. (**a**) A schematic diagram of a batch reactor for assessing ion release and the related rates of mineral weathering in geological solids. Raes are evaluated by rate of solute accumulation in solution with time. (**b**) Batch reactor data showing copper ion release from the weathering of chalcopyrite, $(Cu,FeS_2(s))$,-bearing mine rock. Rapid and extensive copper ion release results from the smaller size fractions, while larger size fractions show negligible extent and rate of weathering (Strömberg & Banwart 1999a). S.1–S.3 refer to three replicate rock samples and d_p refers to the nominal diameter denoting particle size.

study element release as a function of flow rate, providing valuable information on weathering mechanisms. This includes assessing whether solubility equilibria between effluent and source or secondary minerals occurs (solute flux is proportional to flow rate), whether ion release is controlled purely by chemical kinetics (solute flux is independent of flow rate) or whether control of ion release by diffusion-controlled mass transfer limitation occurs (solute flux increases less than proportionally to flow rate).

Determination of weathering rates from experimental data, and extrapolation to field conditions, is still complicated by issues such as heterogeneity within the columns, representative sampling for column materials, sample preparation and length scaling. This last is particularly challenging, as a 1 m column can provide a reasonable analogue of the top metre of spoil in a heap but may be quite incapable of providing information on greater depths.

There are four essential components to any column experiment (Fig. 5a): reservoir, influent application, the column itself and sample collection. The reservoir holds a source of influent. In the simplest case this may be deionized water, otherwise the influent may be either a natural solution (e.g. mine water, rainwater) or solutions made up in the laboratory (e.g. synthetic groundwater). It may be necessary to pressurize the influent reservoir with a gas of known composition to preserve its composition, i.e. 100% N_2 to preserve anoxia, fix $PCO_2(g)$ for the carbonate buffer system, fix $PO_2(g)$ for specific oxic conditions, etc. Influent application is generally achieved using a peristaltic pump to feed the column.

In a saturated column it is advantageous for the flow to be pumped upwards to avoid problems with air trapping, while unsaturated columns require irrigation by downwards flow driven by gravity plus optional enhancement by additional pumps. Homogeneous distribution of influent is critical as heterogeneities can lead to preferential flow within columns and concomitant problems with data interpretation. Unsaturated columns frequently use sprinkler systems, although an alternative option is to use a disc of higher matrix suction (e.g. glass sinter or porcelain) than the underlying material so that the disk will transmit water only when saturated.

The details of column design depend entirely on the desired outcome of the experiment, although common elements include the need to prevent flow down the sidewalls, the possibility for sampling ports along the length of the column, and the problem of representative and reproducible packing procedures. Sample collection involves the flow of effluent, via in-line flow cells to measure parameters such as pH and Eh (redox potential, V), if required, into a receiving vessel, which may be a simple flask or an automated fraction collection system. Care must be taken to avoid evaporation at this stage, and the question of averaging element fluxes over periods that may be considerable at low flow rates must be addressed.

Figure 5a shows a schematic diagram of a column reactor. The associated results in Fig. 5b were obtained from a reactor with the following specifications. Unsaturated columns were constructed from 10 cm diameter uPVC drainpipe with high-suction glass sinters top and bottom for influent distribution and effluent collection. Column lengths were varied to investigate the effect of varying residence time on effluent chemistries, as it was felt that multiple sampling ports at intervals down the columns would induce preferential flow. Influent was deionized water and effluent passed through low-volume ($<1\,cm^3$) Teflon flow cells in which pH, Eh and dissolved oxygen content were continuously monitored.

The columns were filled with material crushed to a median grain size of 1 mm. A variety of materials, ranging from lead and zinc ore-bearing limestone to Coal Measures mudstone, sandstone and coal, were run in the columns for periods of approximately 3 months. Typical runs incorporated hydrological and geochemical characterization procedures, designed to extract information regarding water and gas distribution, and element release rates. Samples were taken every weekday and analysed by ICP-AES (inductively coupled plasma-atomic emission spectrometry) for cations and by Dionex ion chromatography for anions.

Figure 5b illustrates steady state results for sulphur and iron release from a 50 cm long column containing approximately 7 kg of Coal Measures sandstone material. Variation of element concentrations as a function of flow rate may be used to infer mechanisms for element release. Iron concentrations are effectively independent of flow rate up to 0.18 ml min^{-1}, consistent in this range of flow with control of release by equilibrium with primary or secondary minerals. Sulphur concentration is inversely proportional to flow rate, indicative of element release by kinetically controlled reactions. Release rates may be combined with bulk and mineralogical composition data to calculate dissolution rates of the minerals concerned, although care must be taken with the identity of secondary minerals, which, although they may be present in quantities too

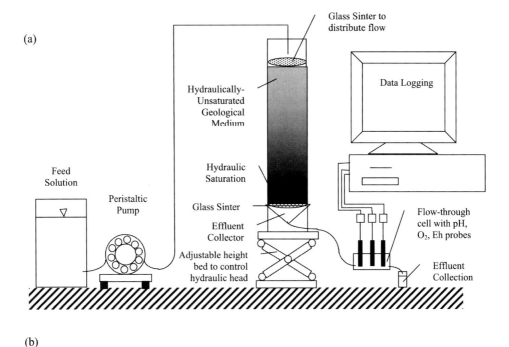

Fig. 5. (**a**) A schematic diagram of a column reactor for assessing ion release and the related rates of mineral weathering in geological solids. Solute mass flows in the column effluent, and associated weathering rates of the source minerals, are determined from effluent discharge rates and the ionic composition of the effluent. (**b**) Effluent sulphate ion concentration (■) as dissolved sulphur and dissolved iron concentration (○) are plotted against flow rate for a hydraulically unsaturated column. Kinetic control of pyrite mineral dissolution results in the observed inversely proportional dependence of sulphur (S) concentration on flow rate. Iron concentrations are independent of flow rate in the range flow rate $<0.18\,\mathrm{ml\,min^{-1}}$, which is consistent with thermodynamic solubility equilibrium control. In this case, it is likely that dissolved iron species are maintained in equilibrium with secondary iron oxyhyroxide mineral phases.

small for identification by bulk techniques such as X-ray diffraction, can exert considerable control on element concentrations.

Solute mass flows from field sites

Within the Tier 1 risk assessment methodology outlined in Fig. 2, weathering rates for source minerals are determined in Step 4 by solute mass flow measurements at a field site. Such data require measurements of discharge flow rates and the associated solute concentrations. Flow rates are typically determined by either flow gauging with a weir installed in the effluent channel or by bucket and stopwatch. Grab samples of effluent at the time of flow measurement allows subsequent chemical analysis to be carried out to determine solute concentrations and the associate solute mass flows in the effluent.

As for batch and column experiments, the field results are also dependent on a number of assumptions such as: (1) the solutes arise solely from mineral weathering; (2) that solutes can be reliably assigned to the specific source minerals; and (3) that solutes are conservative and are not lost to sorption or secondary mineral precipitation along transport pathways from the source zone to the point of sampling.

Resolving the scale dependence of weathering rates between laboratory and field

As discussed above, the very large differences in the physical scale and complexity between laboratory tests and field sites makes extrapolation of laboratory results extremely uncertain. Scale effects that are commonly cited as contributing to discrepancies between laboratory and field observations of mineral weathering rates are:

- Particle size effects where large rock aggregates at field sites contribute significant mass but little reactive surface area for weathering. Laboratory tests usually focus on smaller particle sizes that have greater specific surface area and thus weather more quickly.
- Temperature effects can be significant for field sites in cold weather regions. Laboratory tests at room temperature are expected to yield weathering rates that are significantly faster than those at field sites with low average temperature. Owing to the activation energy effects (Equation 10), mineral weathering is faster at higher temperatures.
- Spatial variation in mineralogy at field sites can have a large impact on contaminant generation in discharges. Material that is suspected to be reactive is usually used for the types of laboratory tests outlined above. At field sites, the extent of such material may be limited. Contaminant solutes originating from such 'hotspots' may be diluted in the discharge.
- Preferential flow through mine workings results in some contamination being retarded in immobile water, rather than contributing to contamination in the discharge.
- The availability of O_2 may be much more restricted in the interior of mine sites, particularly below the water table, where O_2 diffusion is much slower. This means that the actual extent of reacting rock may only be a small fraction of that present, due to large parts of a site being anoxic and therefore unreactive.

Elucidating these five factors has resulted in a general conceptual model for weathering processes in mine sites (Malmström *et al.* 2000), which is reviewed below. It seems likely that these factors will be common across a wide range of mine sites. If the impact of these factors on weathering rates at field scale can be generally quantified, it would be possible to use the large compilation of laboratory data on mineral weathering rates to account for source mineral weathering at field sites.

A large body of data is available from the well-characterized deposits of mining waste rock at the Aitik site in northern Sweden, for which weathering rates have been previously published. The site is suitable as a model system for investigating the apparent scale dependence of mineral weathering rates, showing a two orders of magnitude discrepancy between mass-normalized weathering determined in the laboratory and the field site. The magnitude of this discrepancy is similar to those observed for weathering studies in natural catchments and aquifers. The scale dependence exhibited by the Aitik data is, to a large degree, predictable by quantification of the effects of a few critical and readily available, bulk-averaged physico-chemical characteristics. The fact that the scale dependence exhibited by the Aitik data is consistent with other laboratory, watershed and aquifer studies suggests that at least some of the quantified effects are of general applicability and importance when extrapolating weathering rates from the laboratory to the field.

Resolving the scale dependence of weathering rates at the Aitik mine, Sweden

Site description

The Aitik mine, located close to Gällivare in the northern part of Sweden, is Europe's largest operating copper mine, producing also waste rock that is deposited in heaps at the site (Fig. 6). The Aitik waste rock heaps currently extend over an area of about 260 ha, with a depth of 15–20 m, and have been well characterized in terms of physico-chemical parameters, mineralogy, chemistry and hydrology (Strömberg & Banwart 1994, 1999a, b; Eriksson et al. 1997). The owner, Boliden Mineral AB, has applied dry covers on the heaps following the outcome of a programme for assessing protective measures after mining activities.

This case study assesses the situation prior to remediation at which point the heaps were unsaturated with respect to water, with an average water content of about 10 vol.%. The heaps, which contain rock material with fragments up to the order of 1 m diameter, remain predominantly oxic over their full depth, with measured oxygen concentrations in the pore gas ranging from 3 to 21 vol.%. The surface temperature at the site varies considerably, with average winter and summer temperatures of -15 and 15°C, respectively, whereas the temperature in the major part of the heaps was fairly constant around 4°C. The drainage water from the heaps, which is collected in two ditches and used as process water in the enrichment plant, has a pH of 3.8–4. The low pH in the drainage water indicates that available carbonates (mainly calcite) have been depleted to a large degree.

Site assessment

The dominant acidity producing processes in the Aitik waste rock heaps have been identified to be oxidative weathering of pyrite and subsequent precipitation of ferric oxyhydroxide (Strömberg & Banwart 1994). The dominant reaction for copper mobilization at the Aitik site has been identified to be oxidative weathering of chalcopyrite (Strömberg & Banwart 1994). The low pH of the drainage water indicated that available calcite has been depleted, or is no longer reacting sufficiently rapidly to release significant amounts of Ca^{2+}, as has been demonstrated for larger rock particles at the site (Strömberg & Banwart 1999a). Hence, we consider plagioclase as the source of Ca^{2+}.

The weathering of biotite, $K(Fe_{1.5}Mg_{1.5})(AlSi_3O_{10})(OH)_2$, and of plagioclase (with a composition of 30% anorthite, $CaAl_2Si_2O_8$, and 70% albite, $NaAlSi_3O_8$) then remain as the major acidity consuming processes. The weathering of, for example, muscovite and K-feldspar, which are also dominant silicates in the Aitik waste rock, is much slower and contributes less to the proton consumption at the site (Strömberg & Banwart 1994).

To estimate the solute mass flows from the site, average concentrations of solutes were multiplied by average flow rates in the ditches draining the waste rock heap. The total solute flows from the site were obtained by summing flows in ditches 1 and 2 (Fig. 6). Hydrological investigations and groundwater monitoring indicate that only a small fraction of the water flow infiltrates the subsurface below the heaps (Axelsson et al. 1992). Therefore, solutes originating from weathering processes in the heaps are transported from the site via the ditches. Hence, solute flows in these ditches represent integrated weathering rates at the site.

Weathering rates for chalcopyrite, R_{chp}, and pyrite, R_{pyr}, were determined from the mass flow of Cu^{2+} and sulphate, respectively. The sulphate arising from chalcopyrite weathering was accounted for using the mineral stoichiometry. The equivalent flow of sulphate was subtracted from the total in order to obtain the pyrite weathering rate:

$$R_{chp} = F_{Cu^{2+}} \quad (mol\ s^{-1}) \qquad (5)$$

$$R_{pyr}^{SO_4^{2-}} = \frac{(F_{SO_4^{2-}} - 2R_{chp})}{2}$$
$$= \frac{(F_{SO_4^{2-}} - 2F_{Cu^{2+}})}{2} \quad (mol\ s^{-1}). \qquad (6)$$

Silicate weathering rates were calculated from Ca^{2+} and Mg^{2+} mass flows, assuming that Ca^{2+} originates from plagioclase and Mg^{2+} from biotite. Hence, when accounting for the mineral stoichiometries, we obtained the weathering rates of biotite, R_{bio}, and plagioclase, R_{pla}, from:

$$R_{bio} = \frac{F_{Mg^{2+}}}{1.5} \quad (mol\ s^{-1}) \qquad (7)$$

$$R_{pla} = \frac{F_{Ca^{2+}}}{0.3} \quad (mol\ s^{-1}) \qquad (8)$$

Mineral abundance was quantitatively characterized from drill cores by thin section and mineral chemical analysis. The reported areal extent and height of the heap, porosity and volume fractions of the minerals, allowed calculation of the volume of the individual

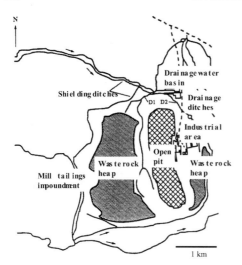

Fig. 6. Diagram of the Aitik mine site showing the extent of the open-pit mine and waste rock heaps, and the locations of the shielding ditches that capture the run-off from the heaps. Weathering rates at field scale were estimated from solute mass flows in the ditches, obtained by simultaneously measuring water flow rates and solute concentrations.

minerals in the deposit. These volumes were converted to the molar amounts for each mineral using the respective mineral densities and molecular weights. Mineral life times were obtained by dividing moles of each mineral by the respective weathering rate.

Comparison between laboratory and field conditions and data

Table 4 compares the physical and geochemical conditions encountered between bench-scale batch reactors, mesocale column reactors and the field site. Systematic differences are noted for rock mass, temperature and water flow rates. The relative abundance of contaminant generating minerals, pyrite and chalcopyrite, when spatially averaged over the field site are approximately half the value determined for the rock used in the batch and column experiments.

Table 5 compares the resulting weathering rates determined from solute release rates in batch reactors and the solute mass flows from the columns and from the field site. The column experiments demonstrate an approximately one order of magnitude decrease in weathering rates, compared to the batch experiments. The field demonstrates a further approximately one order of magnitude decrease in weathering rates, compared to the results from the column reactors.

Scaling procedure for predicting column and field rates from laboratory rates

Equation (9) relates the relative weathering rate in the column or at the field site ($R^{C/F}$) to the rate determined in the laboratory batch reactors (k^B):

$$\frac{R^{C/F}}{a_{sm}^{C/F}} = \alpha^{C/F} \beta_T^{C/F} \beta_{pH}^{C/F} \beta_{PS}^{C/F} k^B. \quad (9)$$

The scaling factors identified above have been calculated from independent information. Equation (9) thus represents a true prediction and does not rely on *a priori* information on weathering rates at field scale, i.e. no 'calibration' of weathering rates is necessary.

The factor $\alpha^{C/F}$ is hydrological and accounts for the potential existence of preferential flow paths and immobile water. For the Aitik case, it has been experimentally quantified (Eriksson *et al.* 1997) as the ratio between mobile and total water content in hydrological tests using unreactive tracers. The value for the field site is $\alpha^F = 0.65$ and the value for the column experiments is $\alpha^C = 1$ (Eriksson *et al.* 1997). The different β factors represent a correction of the batch weathering rate, k^B, for temperature (T), pH and particle size distribution (PS). This factor accounts for weathering that occurs in zones of immobile water where, although solutes are released to solution, they are not transported with the site discharge and so do not contribute to the observed weathering rate at field scale, which is based on solute mass flows in the discharge.

Weathering rates are known to depend in a predictable way on temperature based on activation energy effects for the breaking of chemical bonds in the crystal lattice that are associated with ion release to solution. The correction for temperature is quantified using the Arrhenius equation (equation 10), with E_a being the activation energy, R_g the universal gas constant and T the absolute temperature at each scale:

$$\beta_T^{C/F} = \exp\left[-\frac{E_a}{R_g}\left(\frac{1}{T^{C/F}} - \frac{1}{T^B}\right)\right]. \quad (10)$$

Weathering of both sulphide and silicate minerals is also known to depend on proton activity with rates being higher at lower values of pH. The pH effect is accounted for using equation (11), with n being the reaction order with respect to protons, within the pH range being considered:

Table 4. Characteristics and prevailing conditions for the three observation scales, in terms of total waste rock mass M, water flow Q, temperature T, pH and volumetric mineral content, γ_m, with the index $m = pyr$, $chalc$, $biot$, $plag$ denoting the minerals pyrite, chalcopyrite, biotite and plagioclase, respectively

	Rock mass M (kg)	Water flow Q (m³ s⁻¹)	Temperature T (°C)	pH	Mineral content γ_{pyr} (m³ m⁻³)	γ_{chal} (m³ m⁻³)	γ_{biot} (m³ m⁻³)	$\gamma_{plag}{}^a$ (m³ m⁻³)
Batch[b]	0.15	—	20–23	3.3	0.011	0.0025	0.15	0.20
Column[c]	1.82×10^{3d}	9.2×10^{-9}	4–10	3.5	0.011	0.0025	0.15	0.20
Field[e]	9.5×10^{10d}	0.2	1–4	3.8–4.2	0.0057	0.0009	0.080	0.19

[a] Plagioclase with a composition corresponding to 70% anorthite and 30% albite.
[b] Strömberg & Banwart (1999a).
[c] Strömberg & Banwart (1999b).
[d] Calculated as: $M = HA(1-\epsilon)\rho_s$ where H is height, A is total area, ϵ is porosity and ρ_s is density of the solid material; in the column experiments (10), $H = 2$ m, $A = 0.5$ m², $\epsilon = 0.35$ and $\rho_s = 2.8 \times 10^3$ kg m⁻³; in the field (8), the average $H = 20$ m, $A = 2.6 \times 10^6$ m², $\epsilon = 0.35$ and $\rho_s = 2.8 \times 10^3$ kg m⁻³.
[e] Strömberg & Banwart 1994.

Table 5. Mineral weathering rates as estimated from observed tracer release rates, in terms of the logarithms of the mass normalized weathering rate R and the associated rate coefficient $k = R/a_{sm}$, with a_{sm} being the specific surface area for mineral m. The labels B, C and F refer to the batch, column and field scale, respectively

Mineral	log $R^{B a,b}$ (mol kg⁻¹ s⁻¹)	log $R^{C a,c}$ (mol kg⁻¹ s⁻¹)	log $R^{F a,d}$ (mol kg⁻¹ s⁻¹)	log k^B (mol m⁻² s⁻¹)	log k^C (mol m⁻² s⁻¹)	log k^F (mol m⁻² s⁻¹)
Pyrite	−8.9[e]	−10.1[e]	−10.9[e]	−9.9	−11.1	−11.7
Chalcopyrite	−9.5	−11.0	−12.3	−10.0	−11.4	−12.2
Biotite	−9.7	−10.6	−11.5	−11.9	−12.8	−13.4
Plagioclase	−10.5	−10.9	−11.3	−12.8	−13.2	−13.5

[a] Estimated assuming stoichiometric mineral dissolution with Na⁺, Mg²⁺, Cu²⁺ and SO₄²⁻ originating mainly from plagioclase ((Na$_{0.7}$Ca$_{0.3}$)Al$_{1.3}$Si$_{2.7}$O$_8$), biotite (K(Mg$_{1.5}$Fe$_{1.5}$)AlSi$_3$O$_{10}$(OH)$_2$), chalcopyrite (CuFeS$_2$), and pyrite (FeS$_2$), Table 4, average of triplicates).
[b] Tracer release rate from Strömberg & Banwart (1999a, Table 4).
[c] Tracer release rate from Strömberg & Banwart (1999b, Table 4).
[d] Tracer release rate estimated as $\sum Q_i C_i / M$ where Q_i is water flow and C_i is tracer concentration in drainage ditch i of the two drainage ditches at the Aitik site reported by Strömberg & Banwart (1994; their Table 1) and M is the total waste rock mass (Table 4).
[e] Corrected for Chalcopyrite dissolution.

$$\beta_{pH}^{C/F} = 10^{n(pH^B - pH^{C/F})} \quad (11)$$

For the individual minerals, we use activation energies and reaction orders from the literature (Malmström et al. 2000, table 3 in the supporting information).

The 'particle-size cut-off' described above is presumed to have a significant effect on the weathering rates when compared across the three scales. The batch reactors largely comprise rock mass from the <4 mm size fraction, which is shown to be more reactive, while the column experiments contain the size fraction <10 cm and the field site consists of particle sizes up to boulders of 1 m or more in nominal diameter. The column and field should thus, show progressively slower weathering rates, when normalized to rock mass, as column and field contain progressively greater mass within the less reactive size fraction >4 mm.

The correction for the particle size effect is quantified by equation (12) with x_i being the particle size fraction that is associated with the corresponding weathering rate k_i:

$$\beta_{PS}^{C/F} = \frac{\sum_i^{C/F} x_i k_i}{k^B} \quad (12)$$

The product $x_i k_i$ is summed over the entire particle size distribution at the considered scale (Malmström et al. 2000, Fig. 1 in the supporting information). The rates, k_i, were obtained from the batch experiments, in which individual size fractions were studied separately (Strömberg & Banwart 1999a). Moreover, the quantification of the specific mineral surface area, $a_{sm}^{C/F} = \gamma_m^{C/F} a_s$, where γ_m is the mineral content and a_s is the total specific surface area, implies an additional scaling factor accounting for differences in mineral distribution between the observation scales. For the Aitik case, γ_m has been determined by point counting using light microscopy (Strömberg & Banwart 1994 and included references) and a_s has been determined using the N_2 adsorption method and is constant on all scales (Strömberg & Banwart 1999a; $a_s = 1000\,m^2\,kg^{-1}$), yielding the scaling factor $a_{sm}^{C/F}/a_{sm}^B = \gamma_m^{C/F}/\gamma_m^B$.

A comparison of predicted and measured weathering rates at mesoscale and field scale

Figure 7 compares weathering rates for the column experiments and the field investigations with those predicted from the laboratory rates using equation (9). The rates are predicted well considering the relatively small number of bulk physico-chemical parameters considered. These results suggest that these factors do, to a large degree, explain the discrepancy between laboratory and field observations of weathering rates. Because these factors are common for subsurface environments, it is likely that they will play a similar role for a much wider range of sites than the one considered here.

A further strength of the proposed scaling procedure is that it appears to capture essential features of the observed scale dependence for both sulphide and silicate minerals. This suggests that the procedure is likely to be robust with respect to the range of minerals that may be considered in other applications, as the detailed chemical mechanisms of oxidative weathering of sulphide minerals, and the hydrolysis of the crystal lattice network for silicate minerals, are fundamentally different.

The insets to Fig. 7 show the relative contribution of the various factors to the observed scale dependence. For both column and field rates, the largest contributions to the discrepancy arise from temperature and particle size effects. Additional contributions when scaling from column to field are the lower mass fraction of sulphide-bearing rock at the field site and the presence of immobile water within the heaps. Inspection of Fig. 7 also shows that chalcopyrite is not predicted particularly well, compared to the other minerals. This suggests that additional factors may have a significant impact on the scale dependence of chalcopyrite weathering rates.

A likely explanation is that dissolved Cu, used as a tracer for chalcopyrite weathering in the column and site discharges, is being removed by sorption processes at the site. Analysis of secondary hydrous ferric oxide precipitates collected from the columns and Aitik mine site do show a small fraction of Cu associated with this mineral phase (Strömberg & Banwart 1999b). This scavenged Cu represents a potentially significant sink within the site and would introduce a corresponding error to the estimated weathering rate. In general, processes that affect the solute mass flows, besides the weathering of primary minerals, may also prevail under other site-specific conditions and must then be conceptualized and modelled.

For the site-specific quantification of the temperature and pH factors, it is assumed, based on previously published results (Malmström et al. 2000 and included references), that the weathering processes are rate limited by abiotic surface chemical weathering reactions. It must be noted that sulphide weathering rates, in the pH range described in Table 2, will generally be microbially mediated. However, a particular characteristic of the Aitik site is the very low

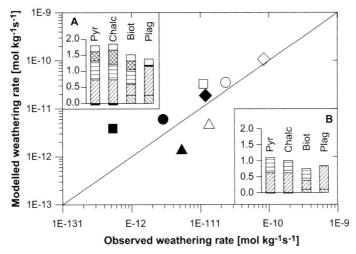

Fig. 7. Modelled field- and column-scale weathering rates plotted against observed weathering rates ($R^{C/F}$ in Table 2), i.e. against the rates that were estimated directly from observed tracer release rates. Modelled field and column rates were predicted according to the proposed equation (9), which scales up the batch rate coefficients, k^B, listed in Table 5. Open and filled symbols denote the column and the field scale, respectively ($\diamond$ pyrite, $\square$ chalcopyrite, ○ biotite, $\triangle$ plagioclase). The solid line represents the ideal case, "perfect prediction", where modelled rates equal observed rates. The inserts show the logarithmic values of the scaling factors for the field (insert A) and the columns (insert B) according to: $\triangle - \log \beta_{pH}^{C/F}$, $\triangle - \log \beta_{T}^{C/F}$, $\triangle - \log \beta_{PS}^{C/F}$, $\triangle - \log \gamma_{m}^{C/F}/\gamma_{m}^{B}$ and $\square - \log \alpha^{C/F}$; $\square$ denotes overlap of the $-\log beta_{pH}^{C/F}$ and the $-\log beta_{T}^{C/F}$ due to negative $-\log beta_{pH}^{C/F}$ values for the sulphides. Additional detail and numerical results not listed here are available as supporting information in Malmström et al. (2000).

annual average temperature that inhibits microbially mediated sulphide mineral weathering (Strömberg & Banwart 1994). Application of the proposed scaling procedure to other sites and conditions, where microbially mediated weathering may predominate, would still require correction for all factors included in equation (9). However, the actual quantification of the temperature and pH factors must then be based on parameters and mechanistic or empirical relations that are valid for the specific biologically mediated reactions at the site.

The quantification of the effect of preferential flow is independent of the nature of the weathering process and would be the same for biotic and abiotic weathering processes. Also, the quantification of the particle size effect is empirical and site specific. Experimental determination of this effect would thus apply also for biotic weathering processes, as well as implicitly account for potential diffusional resistance to weathering in large particles.

Closing comments

The scaling procedure described above is, to our knowledge, the first relatively successful attempt to account specifically for the physical and geochemical factors giving rise to the commonly observed discrepancy in weathering rates between laboratory and field observations. It represents a relatively general set of factors that must be accounted for. Some, such as the ratio of immobile to mobile water and the particle size distribution, are empirical and site specific. However, application of the procedure to further site data may indicate whether or not the procedure is sensitive to these factors, and if these factors are observed to vary significantly between sites. If variation between sites is small and the procedure is insensitive to these factors, the results observed here may, in fact, be quite general.

This scaling procedure accounts for retention of dissolved weathering products in immobile water when determining weathering rates from solute mass flows at the field site. Although this helps to remove the discrepancy when predicting rates of weathering at field sites from laboratory data, it is important to recognize that the accumulated, although immobile, solutes also contribute to the contamination legacy of the site. This contamination must be considered within the risk assessment framework, in addition to the mobile contamination that leaves the site with the discharge. This applies as well to any contamination that is sorbed or is accumulated as secondary mineral phases and thus immobilized within the site.

The hydrochemical approach advocated here is the simplest possible, and is appropriate for Tier 1 assessment. A major advantage of the approach is its transparency. Results can be obtained easily with spreadsheet calculations with input varied quickly in order to test a range of conceptual site models.

It is clear that more advanced modelling approaches can be applied. These include application of computational thermodynamic geochemical codes such as PHREEQC (see Appelo & Postma 1993), which include mineral solubility equilibria as additional constraints for interpretation of solute mass flows. However, such codes do require a higher level of geochemical understanding of the system. Their greatest advantage is the ability to explore a wider range of reactions, including sorption and secondary mineral formation, when testing the conceptual site model. In order to provide quantitative results, they generally require additional site data, such as characterization of secondary minerals, that we consider at present to be more appropriate for Tier 2 assessment. If neglecting reactions such as sorption and secondary mineral formation is found to generally create a significant error with respect to application to Tier 1 assessment, then such processes will need to be conceptualized and included in the scaling procedure.

As stated in the opening section of this paper, general conceptual models generally allow broader application to a wider range of site, and allow greater depth of site interpretation for the same allocation of resources. In addition, conceptual models that are based on fundamental descriptions of reaction and transport processes allow greater use of tabulated fundamental data when parameterizing mathematical models of reactive transport. For chemical kinetic processes, such as sulphide and silicate mineral weathering, fundamental data for specific minerals and weathering mechanisms are available including rate constants, reaction orders with respect to protons and activation energies.

The scaling procedure described above includes these factors. The greatest question at present is the generality of the approach. The procedure can be tested by prediction of weathering rates at other mine sites and for other systems, such as surface water catchments, where suitable datasets exist. Key uncertainties include whether the fraction of immobile water and thus the relative retention of dissolved weathering products is similar between sites, and whether neglecting retention of sorbed weathering products and secondary minerals will create sufficient errors at other sites to ensure that they must be included in a general scaling procedure. The relatively good agreement between laboratory and field results for minerals other than chalcopyrite suggests this is not the case for the Aitik site.

The authors wish to acknowledge the original work and publications cited here on resolving the scale dependence of mineral weathering rates between laboratory and field. The work originated from research collaboration on the Aitik mine site by S. Banwart during his appointment at the Royal Institute of Technology (KTH), Stockholm. The project was led by G. Destouni (KTH) with significant research input from M. Malmström (KTH), B. Strömberg (KTH, now at the Swedish Nuclear Energy Inspectorate) and N. Eriksson (KTH, now at Boliden Minerals Inc.). K. Evans is supported by the NERC Environmental Diagnostics Thematic Programme. S. Croxford is supported by a Hussein–Farmey Endowed PhD Studentship at the University of Sheffield.

References

APPELO, C.A.J. & POSTMA, D. 1993. *Geochemistry, Groundwater and Pollution*. A.A. Balkema, Rotterdam.

ASTM, 1995. *ASTM Designation: E1739–95 – Guide to Risk-based Corrective Action at Petroleum Release Site*, American Society for Testing and Materials, West Conshohocken, PA.

ASTM, 1996. *ASTM Designation: D5744–96–Standard Test Method for Accelerated Weathering of Solid Materials Using a Modified Humidity Cell*, American Society for Testing and Materials, West Conshohocken, PA.

AXELSSON, C.L., BYSTRÖM, J., HOLMÉN, J. & JANSSON, T. 1992. *Efterbehandling av sandmagasin och gråbergsupplag i Aitik, Hydrogeologiska förutsättningar för åtgärdsplan*. Golder Associates AB, Report 927-1801, (in Swedish).

BANWART, S.A. & MALMSTRÖM, M. 2001. Hydrochemical modelling for preliminary assessment of mine water pollution. Special Issue on Mine Water Geochemistry. *Journal of Geochemial Exploration*, **74**, 73–97.

BANWART, S., DESTOUNI, G. & MALMSTRÖM, M. 1998. Assessing mine water pollution: from laboratory to field scale. In: KOVAR, K. (ed) *Groundwater Quality: Remediation and Protection*, Publication, International Association of Hydrological Sciences, Wallingford, **250**, 307–311.

BLOWES, D.W., REARDON, E.J., JAMBOR, J.L. & CHERRY, J.A. 1991. The formation and potential importance of cemented layers in inactive sulfide mine tailings. *Geochimica Cosmochimica Acta*, **55**, 965–978.

BONE, B. 2001. *Environment Agency and Remediation of Mines and Quarries*. Environment Agency of England and Wales, National Groundwater & Contaminated Land Centre, Sulihull.

ERIKSSON, N., GUPTA, A. & DESTOUNI, G. 1997. Comparative analysis of laboratory and field tracer tests for investigating preferential flow and transport in mining waste rock. *Joural of Hydrology*, **194**, 143–163.

FINKELMAN, R.B. & GIFFIN, D.E. 1986. Hydrogen peroxide oxidation: an improved method for rapidly assessing acid-generation potential of sediments and sedimentary rocks. *Reclamation and Revegetation Research*, **5**, 521–534.

HARRIES, J.R. & RITCHIE, A.I.M. 1985. Pore gas composition in waste rock dumps undergoing pyritic oxidation. *Soil Science*, **140**, 143–152.

MALMSTRÖM, M., DESTOUNI, G., BANWART, S. & STRÖMBERG, B. 2000. Resolving the scaledependence of mineral weathering rates. *Environmental Science & Technology*, **34**, 1375–1377.

NORDSTROM, D.K. & SOUTHAM, G. 1997. Geomicrobiology of sulfide mineral oxidation. Geomicrobiology: interactions between microbes and minerals, *Reviews in Mineralogy*, **35**, 361–390.

O'SHAY, T.A., HOSSNER, L.R. & DIXON, J.B. 1990. A modified hydrogen peroxide oxidation method for determination of potential acidity in pyritic overburden. *Journal of Environmental Quality*, **19**, 778–782.

PRICE, W.A. 1997. *DRAFT Guidelines and Recommended Methods for the Prediction of Metal Leaching and Acid Rock Drainage at Minesites in British Columbia*. British Columbia Ministry of Employment and Investment, Energy and Minerals Division, Smithers, BC.

SCHNOOR, J.L. 1990. Kinetics of chemical weathering: A comparison of laboratory and field weathering rates. In: STUMM, W. (ed.) *Aquatic Chemical Kinetics*, John Wiley, New York.

SOBEK, A.A., SCHULLER, W.A., FREEMAN, J.R. & SMITH, R. M. 1978. *Field and Laboratory Methods Applicable to Overburdens and Minesoils*. US Environmental Protection Agency Publication, EPA-600/2-78-054.

STRÖMBERG, B. & BANWART, S. 1994. Kinetic modelling of geochemical processes at the Aitik mining waste rock site in northern Sweden. *Applied Geochemistry*, **9**, 583–595.

STRÖMBERG, B. & BANWART, S. 1999a. Experimental study of acidity consuming processes in mining waste rock: Some influences of mineralogy and particle size. *Applied Geochemistry*, **14**, 1–16.

STRÖMBERG, B. & BANWART, S. 1999b. Weathering kinetics of waste rock from the Aitik copper mine, Sweden: scale dependent rate factors and pH controls in large column experiments. *J. Cont Hydrol.*, **39**, 59–89.

STUMM, W. & MORGAN, J.J. 1990. *Aquatic Chemistry*, (3rd edn). John Wiley, New York.

WHITE, A.F. & BRANTLEY, S.L. (eds) 1995. *Chemical Weathering Rates of Silicate Minerals*, Reviews in Mineralogy, vol **31**.

WHITE, A.F. & PETERSON, M.L. 1990. In: MELCHOIR, D.L. & BASSETT, R.L. (eds) *Chemical Modeling of Aqueous Systems II*. Symposium Series, 416, American Chemical Society, Washington, DC.

YOUNGER, P.L. 2000. Predicting temporal changes in total iron concentrations in groundwaters flowing from abandoned deep mines: a first approximation. *Journal of Contaminant Hydrology*, **44**, 47–69.

YOUNGER, P.L. & BANWART, S.A. 2002. Time-scale issues in the remediation of pervasively-contaminated groundwaters at abandoned mine sites. In: *Groundwater Quality*, International Association of Hydrological Sciences, Wallingford, in press.

YOUNGER, P.L., BANWART, S.A. & HEDIN, R.S. 2002. Mine Water Hydrology, Pollution, Remediation. In: ALLOWAY, B.J. & TREVORS, J.T. (eds) *Environmental Pollution Series*; Kluwer, Dordrecht.

The hydrogeochemical dynamics of mine pit lakes

R. J. BOWELL

SRK Consulting, Windsor Court, 1–3 Windsor Place, Cardiff CF10 3BX, UK

Abstract: On cessation of mining open pits or opencast workings that extend below the water table are likely to fill with water and thus develop a mine pit lake (MPL). This body of water remains as a permanent feature on the mine site and as such becomes a closure issue with respect to water quality and potential to degrade groundwater. Further, it may present a risk to the environment through the development of poor quality water with elevated concentrations of metals, metalloids, sulphate and depressed pH.

The prediction of future pit lake water quality within a MPL is, therefore, essential in considering environmental impact on a closed or abandoned mine facility. The controls on a MPL will vary over time, and will involve chemical, biological and physical processes. Localized and regional-scale processes affect these in turn. Consequently, in order to predict pit lake water quality it is essential to understand the hydrogeological, geochemical and limnological processes that influence water quality.

Open-pit mining has become a common place method for extraction in the mining industry in recent years due to the advancement in mineral processing, which allows the economic utilization of near-surface low-grade high tonnage ores. This is particularly true for gold and copper operations in many parts of the world (Peters 1987).

If mining has occurred below the water table, then on closure and abandonment of these facilities, the pit fills with water producing a mine pit lake (MPL). The length of time needed for the lake to reach full depth will depend on several factors including pit dimensions, rainfall–evaporation budget, groundwater inflow and groundwater regime. By comparison, MPLs tend to show a greater depth to surface area ratio than most natural lakes, with implications for water circulation. Generally, pit lakes are smaller than natural closed lake basins and thus the depth to surface area ratio is usually greater than 1 for a MPL (Table 1).

MPLs can and do exhibit thermal and chemical stratification, and show changes related to seasonal turnover. However, the greater depth to surface ratio limits water circulation and thus a permanent stratified or anaerobic layer remains preserved through the development of a chemo-line (Fig. 1). Another major difference is that, for natural lakes, recharge is generally through surface water, whereas, for MPLs, groundwater flow is generally more important (Atkins *et al.* 1997).

In many cases the requirement for predictions of water quality is made during the design and permitting stage, prior to mining. Consequently, the predictions tend to be hypothetical, based on numerical predictive modelling and laboratory assessment of wall-rock behaviour (Pillard *et al.* 1995; Kirk *et al.* 1996; Davis & Eary 1996). It follows that, in some cases, the predictions can be in error due to limitations in the input data available for the model (Kempton *et al.* 1997). Methodologies for the prediction of pit lake chemistry involve construction of a series of linked empirical and physical models to represent the major controlling physio-chemical processes (Bird *et al.* 1994; Miller *et al.* 1996; Havis & Worthington 1997; Kempton *et al.* 1997; Murray 1997; Eary 1998). These processes include groundwater inflow, direct precipitation and pit wall–precipitation contact or run-off water, wall-rock–groundwater interaction, evaporation, and limnological and geochemical processes operating within the MPL itself. The latter includes oxidation–reduction, desorption or adsorption of metals and metalloids with mineral surfaces, precipitation and dissolution, as well as organic processes such as methylation or the activity of sulphate-reducing or sulphide-oxidizing bacteria (Fig. 2). The geochemical enrichment of potentially toxic elements in a MPL can lead to organisms having a harmful level of exposure. For example, within the Berkeley Pit, 342 migratory geese were exposed to a lethal concentration of metals when they used the pit lake on a seasonal migration.

The closure of open-pit mining operations requires careful planning to avoid environmental

Table 1. *Comparison of hydromorphic properties of pit lakes with those of Natural lakes (After Tempel et al. 1999 with additional data)*

	Surface area (km^2)	Depth (m)	Surface area/depth ratio
Natural lakes			
Lake Superior	82,350	395	208 481
Lake Victoria	69,485	455	152 714
Lake Baikal	30,510	1367	22 319
Pyramid Lake	438	102	4 294
Mine pit lakes			
Berkeley Pit	0.32	259	1.23
Summer Camp Pit	0.113	21.33	5.38
Ruth Pit	0.222	40	5.55
Crone Bane	0.084	17	4.94
Magcobar	0.158	70	2.26

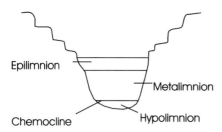

Fig. 1. Schematic representation showing nomenclature used in referring to the zones within a stratified pit lake (after Drever 1988).

impact. Thus, the formation of a MPL can affect both the availability and the quality of water resources in a mining area. Consequently, the assessment of water chemistry is a critical aspect of a open-pit closure programme. This paper presents an invited review of the current status of knowledge regarding the major hydrogeochemical processes that operate within a MPL.

It should be noted that hydrogeological and hydrological processes are also important but are outside the scope of this paper and will not be discussed in detail here. The reader is referred to Fetter (1994), Havis & Worthington (1997), Shevenell & Pasternak (2000) and papers elsewhere in this volume.

Conceptual model

In considering the interaction of a MPL with its environment, the fundamental consideration is what is the relationship of the MPL to the environment. If the MPL acts as a terminal sump,

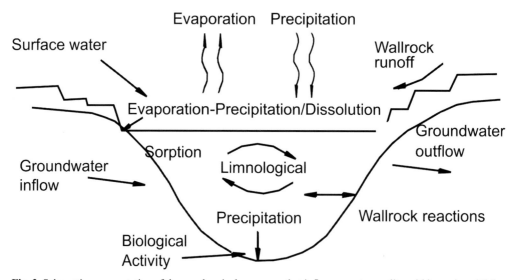

Fig. 2. Schematic representation of the geochemical processes that influence water quality within a mine pit lake.

i.e. the lowest hydrological point within a basin, then it acts as an evaporation pan with no net outflow other than evaporation (Fig. 3).

However, if the MPL is not a sump then 'flow-through' can occur within the MPL and it can interact with the surrounding groundwater environment (Robins *et al.* 1997; Davis *et al.* 1998). The Berkeley Pit at Butte, Montana has been one of the most studied MPLs (Davis & Ashenberg 1989; Miller *et al.* 1996; Robins *et al.* 1997). This pit lake is acidic and has high metal and sulphate concentrations (Table 2), and has impacted the immediate groundwater environment along with water draining old underground workings (Robins *et al.* 1997). In the Robinson Mining District, Ely, Nevada, groundwater has been sufficiently influenced by reaction with oxidized sulphide-bearing host rocks both in underground workings and open pits that distinct hydrogeochemical groundwater blocks can be mapped in the district, and groundwater shows a similar chemistry to MPLs (Davis *et al.* 1998).

Mine pit lake chemistry

The chemistry of a MPL is often, erroneously, assumed to be acidic, such as those pit lakes developed at Parys Mountain or Berkeley (Table 2). In many cases MPL chemistry can be circumneutral and have low metal chemistry, such as those as Copper Flat, Yerrington, and Getchell. In some cases, the pH of the pit lake can exceed pH 7, such as at Magcobar mine in Ireland (Table 2).

It should be noted that in this paper the metal and oxyanion concentrations are quoted from filtered samples (at 0.45 μm) rather than total, unless otherwise stated. This is because total metal concentrations will include particulate metals as well and thus the influence of chemical release processes could be overestimated. The 0.45-μm fraction includes all metals mobilized as free ions, chelated species and in a colloidal form.

In high pH environments, high metalloid concentrations can occur (Fig. 4), such as in the North Pit on the Getchell Mine in Nevada, where historically As levels can reach $10\,\mathrm{mg\,l^{-1}}$ (Miller *et al.* 1996). Certainly in these waters evapoconcentration is an important mechanism for enrichment of As and also for Se (Eary 1998). By contrast, low pH environments are characterized by high concentration of dissolved metals as hydrated cations and chelates (Fig. 5) or anions (Fig. 6). The major mineralogical controls on pit lake chemistry, based on predictive thermodynamic modelling for a range of Nevada pit lakes, has recently been published by Eary (1999). These are shown in Table 3 and are supplemented with phases that are important in reducing environments, such as the basal sediments of some pit lakes. Not all of these controls will be relevant to every site, due to local variations in hydrochemistry, but these represent the more common ones.

The chemistry of a MPL may change with time and reflects a dynamic rather than static equilibrium. For example, at the Summer Camp Pit site in Nevada, where water chemistry has changed from acidic to circumneutral (see the case study below).

High sulphide (particularly pyritic) wall-rocks, such as those exposed in the Parys Mountain and Berkeley pits, tend to produce poor water quality (Table 2). Oxidized wall-rocks that contain appreciable carbonate tend to produce better quality pit lakes due to the abundance of reactive buffering in the host rocks, such as at Magcobar in Ireland.

Consequently the chemistry of MPL can vary widely and this can be related to three major groups of controls:

- limnological processes – this relates to the lake's physical processes and how they will influence geochemical reactions;
- geochemical constraints – this is related to the minerals present, the chemistry of reactive water and the environment's physico-chemical status in controlling

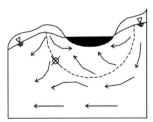
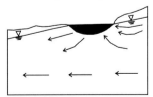

Fig. 3. Hydrogeological scenarios for a mine pit lake interaction with the the groundwater environment.

Table 2. *Representative analysis of mine pit lakes*

Parameter	Parys Mountain	Crone Bane, Ireland	Magcobar, Ireland	Corta Atalaya, Rio Tinto, Spain	Berkeley Pit, Montana	Ruth Pit, Nevada (1993)	North Pit, Getchell Mine, Nevada (1982)	Cortez Pit, Cortez, Nevada
Ref.	Bowell et al. (1996)	Unpublished data	Unpublished data	Unpublished data	Davis & Ashenberg (1989)	Miller et al. (1996)	Miller et al. (1996)	Miller et al. (1996)
pH	2.3	5.77	7	1.6	2.8	3.9	7.67	8.07
TDS, (mg l^{-1})	6000	3750	1210	15 000	–	4150	2420	432
Alkalinity (mg l^{-1})	0	0	512	0	0	0	–	–
SO$_4$ (mg l^{-1})	5200	2530	563	14 300	5740	2840	1570	90.2
Ca (mg l^{-1})	21.6	586	161	379	462	558	530	45.2
Ba (mg l^{-1})	<0.001	<0.004	0.071	<0.001	–	0.008	–	0.06
Al (mg l^{-1})	108	11	0.6	2400	–	23.7	–	–
As (mg l^{-1})	0.77	<0.005	<0.005	22.4	0.05	<0.005	0.38	0.038
Cu (mg l^{-1})	69.6	23.4	0.006	245	156	37.1	<0.005	–
Fe (mg l^{-1})	1300	0.08	5.1	4700	386	6.32	0.16	0.134
Pb (mg l^{-1})	<0.001	<0.005	0.032	4.11	–	0.007	<0.005	0.004
Mn, mg/L	4.8	65.2	0.73	344	95	68.6	0.13	0.002
Se (mg l^{-1})	0.02	<0.005	<0.001	8.9	–	<0.005	0.003	–
Zn (mg l^{-1})	54	29.5	6.3	435	280	32.2	0.02	0.002

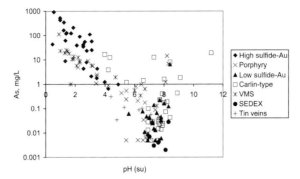

Fig. 4. Variation in arsenic within mine pit lakes as a function of ore deposit type (using the classification of Du Bray 1995) and pH. Data from published and unpublished sources.

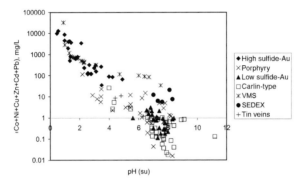

Fig. 5. Flicklin diagram assessing variation in divalent metal cations with pH and ore deposit type (Flicklin *et al.* 1992). Data from published and unpublished sources.

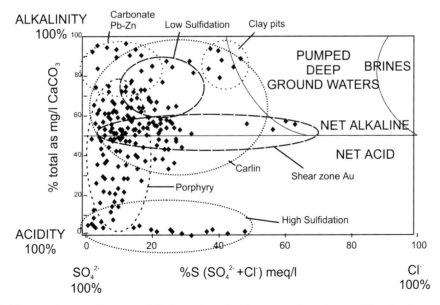

Fig. 6. Younger diagram (Younger 1995) for various pit lake waters. Data from published and unpublished sources.

Table 3. *Geochemical controls on major and important trace components in a mine pit lake system (based on a scheme formulated by Eary 1999). Prediction of mineral phase control based on thermodynamic calculations as well as mineralogy of pit lake sediments*

Component	Acidic (pH <4.5)	Circumneutral (pH 4.5–7.5)	Alkaline (pH >7.5)
Aluminium	Alunite, basaluminute	Gibbsite	Gibbsite
Alkalinity	Not applicable	Calcite	Calcite
Arsenic	Scorodite/adsorption onto HFO	Adsorption onto HFO/Fe-sulphides*	None identified/As- and Fe-sulphides*
Barium	Barite	Barite	Barite/witherite
Cadmium	None identified	Adsorption onto HFO/Zn-sulphides*[†]	Otavite/Zn-sulphides*
Calcium	Gypsum	Gypsum	Gypsum/calcite
Copper	Chalcanthite/cupromelanterite	Brochantite/covellite/Fe-sulphides*	Malachite/brochantite/covellite*
Fluoride	Fluorite	Fluorite	Fluorite
Iron	Melanterite/halotrichite/Fe-sulphides*	HFO/ferrihydrite/Fe-sulphides*	HFO/ferrihydrite/Fe-sulphides*
Lead	Anglesite/chloropyromorphite	Anglesite/chloropyromorphite/galena*	Cerrusite/chloropyromorphite
Manganese	Manganite/birnessite	$MnHPO_4$	Rhodochrosite
Nickel	Morenosite/Fe-minerals[†]	Morenosite/Fe-minerals[†]	Gaspéite/ni-oxides, nickeline
Selenium	Gypsum	Adsorption onto HFO	None identified
Sulphate	Gypsum	Gypsum	Gypsum
Zinc	Zinc melanterite	Hydrozincite/smithsonite/$ZnSiO_3$/Zn-sulphides[†]*	Zincite/Zn_2SiO_4/Zn-sulphides*

* In reduced environments.
[†] As a trace element.

the extent to which release and attenuation mechanisms can proceed;
- geological controls – this is related to the characteristics of individual mineral deposits and will influence the extent to which geochemical reactions will occur.

Limnological processes

Pit lakes in temperate climatic zones can develop vertical density stratification that may be seasonal or permanent. The density of water is a function of both its temperature and its salinity or total dissolved solids (TDS) content. Fresh water is densest at a temperature of about 4°C. At a given temperature, water density increases with increasing TDS. As its TDS increases, the temperature of maximum density of water also decreases.

Natural lake processes

Thermally induced seasonal density stratification in mid-latitude lakes of uniform dilute chemistry (uniform low TDS) is due to increases in ambient temperature in the spring and summer. As the air temperature and solar radiation increase, the surface water is heated (Fig. 7). This heating causes the density of the surface layer to decrease. The lake then has a surface layer ('epilimnion') of uniform lower density and higher temperature, and an underlying layer ('hypolimnion') of higher density and lower temperature. A zone called the 'thermocline' or metalimnion, in which the temperature decreases rapidly with depth, separates these layers (Fig. 1). Above the thermocline, the surface water is mixed by wind or surface inflow and is typically in equilibrium with atmospheric oxygen. The epilimnion is typically less than about 10 m deep, because wind-induced mixing in mid-latitude pit lakes seldom reaches beyond this depth. Below the thermocline, oxygen may gradually become depleted by oxidation of dead algae and lake fauna that fall into the hypolimnion from the surface.

At some point in the late autumn or early winter, the surface water temperature drops below that of the underlying water and the surface water density is now greater than that of the underlying water. The surface water sinks until it reaches the level of its new density. Eventually, the entire water column may overturn, depending on the temperature structure of the lower layer and the amount of cooling at the surface. Overturn through the water column could take less than 1 day or occur over several weeks. Overturn causes mixing that replenishes the oxygen throughout the depth of overturn, potentially to the bottom of the lake.

The water behaviour during winter depends on whether ice forms. If ice forms, the water below the ice may gradually lose its oxygen by the same processes that occur in the lower layer in summer. Below the ice, the water may become temperature stratified with 4°C water at the bottom or it could remain well mixed due to heating from the sides and bottom of the pit. Which process occurs depends on the relative temperatures of the air, water, ground and ground thermal conductivity. If no ice forms and the air temperature keeps the surface water in the vicinity of 4°C, the entire lake may remain well mixed and oxygenated throughout the winter.

In spring, the air temperature increases. If ice had formed in the winter it gradually melts and the water temperature increases from the surface downward, as well as from the bottom up. When the surface water temperature reaches 4°C, the lake may again overturn and reoxygenate due to the denser surface water. If no ice had formed and the lake had continuously overturned in the winter, warmer temperatures in the spring would repeat the cycle of thermally induced seasonal density stratification.

Mine pit lake

The seasonal cycle just discussed assumes that the pit lake has essentially no TDS and its density is unaffected by precipitation or evaporation. For example, if spring snowmelt and run-off have a significantly lower TDS than the lake water TDS, the addition could induce stratification near the lake surface, even when no temperature differences exist due to the lower density of the snowmelt. In dry temperate climates that experience positive net evaporation in the summer, evapoconcentration at the lake surface can increase the TDS and thus the density of the surface water. This density increase can potentially offset the density decrease of the surface water due to thermal heating in the summer. In such cases, the lake may overturn in late summer or fall even before the surface water temperature drops below the temperature of the underlying water, as long as the density of the surface water exceeds that of the underlying water.

Long-term (multi-year) or permanent density stratification can occur if a lake has a significant vertical variation in TDS due to large differences in the TDS of various source waters to the lake and/or to processes in the lake that increase the TDS. This in turn affects the density of the deeper water (Fig. 8). For example, if the lake contains enough organic matter to deplete oxygen

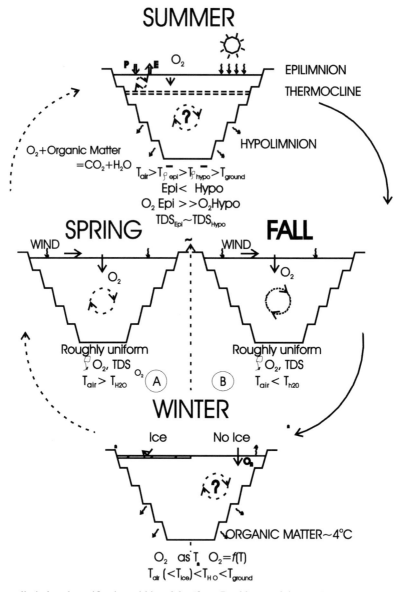

Fig. 7. Thermally induced stratification within a lake (from Parshley *et al.* in prep.).

in the hypolimnion, then during the summer ferric hydroxide that precipitates at the surface will sink, become reduced and dissolve in the basal anoxic water, raising the TDS content and the density of the bottom water.

Typically, a chemically stratified lake has an upper layer of lower TDS water overlying a layer of higher TDS water ('hypolimnion'). A zone called the 'chemocline', in which the TDS increases rapidly with depth, separates the two layers (Table 4). If the bottom layer is sufficiently deep (well below the depth of summer wind mixing), a seasonal middle layer ('metalimnion') may form below the summer thermocline and above the denser hypolimnion. Overturn in the autumn–winter usually occurs throughout the metalimnion, but not the hypolimnion. Thus, once the hypolimnion becomes anoxic, it remains so and will continuously dissolve any ferric hydroxide precipitates falling into it from above. This process further increases the TDS of the hypolimnion and strengthens the density

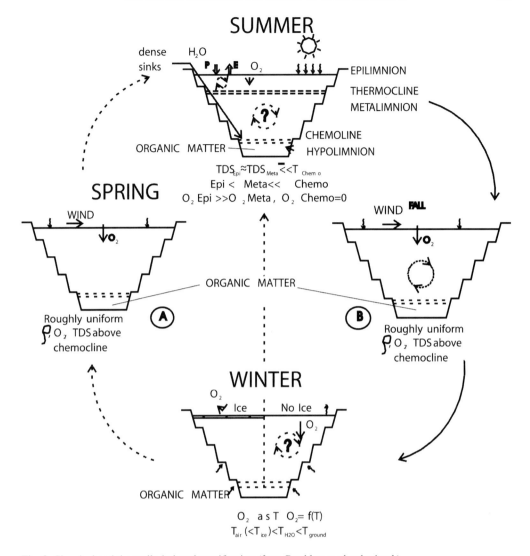

Fig. 8. Chemical and thermally induced stratification (from Parshley et al. submitted.).

gradient between it and the overlying layer, perpetuating the stratification. Sulphidization, in the hypolimnion, will lead to natural attenuation of metals and metalloids as well as sulphur (see below).

Few studies reporting MPL site-specific limnological data have been published to date (Gannon et al. 1996; Atkins et al. 1997; Parshley et al. submitted). These studies deal with pit lakes in Nevada in the western USA and display similar trends indicating that turnover occurs in MPLs similar to natural lakes.

Geochemical controls on MPL chemistry

In considering the geochemical processes that influence MPL chemistry (Fig. 9), processes can be viewed in terms of:

- Release processes – those that release elements into the pit lake;
- Attenuation processes – those that remove elements from the hydrosphere either as precipitation or through adsorption;
- Concentration processes – those that lead to accumulation of elements in the MPL.

Table 4. Examples of stratified pit lakes

Date	Magobar February 1998	Magobar February 1998	SCP 25 February 1998	SCP 25 February 1998	SCP 13 June 1998	SCP 13 June 1998
Depth, (m)	0	50	0	23	0	20
pH	6.7	6.1	8.01	3.67	8.54	7.5
Eh (mV)			278	−108	256	160
TDS	555	885	714	3490	245	922
Temperature °C	7.6	7.9	3.7	6.8	16.4	15.9
Dissolved O$_2$	1.5	0.15	4.8	1.2	3.9	2.4
Alkalinity	451	753	74	0	154	85
Chloride			46	51	13	59
Sulphate			370	2200	22	420
Na	26	30	33	36	25	34
Mg	52	82	16	46	7.5	19
Ca	142	242	140	400	30	160
Mn	6.2		0.06	1.1	0.21	0.19
Fe (II)	<0.2		0.14	414	0.1	158
Fe (total)	0.078	26	0.65	416	1.2	160
Ni		0.21	0.058	3.2	0.018	0.96
Cu			0.002	0.018	0.013	0.003
Zn	5.8	8	0.01	16	0.03	4
As (III)			0.01	6.6	0.01	15.3
As (total)	<0.005	<0.005	1	7.53	0.025	16
Cd	0.016	0.007	0.002	0.6	0.01	0.006
Pb	0.007	0.004	<0.003	<0.003	<0.001	0.001

* All concentrations in mg l^{-1} unless otherwise stated.

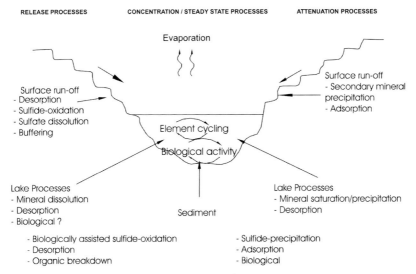

Fig. 9. Relationship between release–attenuation–concentration processes.

Principally these involve sorption onto colloids or evapoconcentration of salt components.

Release processes

This involves mineral–water reactions such as sulphide oxidation, sulphate dissolution or mineral buffering of water chemistry. Within many hard rock pits the most important reactions are those involving sulphide-bearing rocks.

The primary leaching processes for metals include sulphide oxidation and associated mineral buffering, both of which increase the total dissolved solid load in the resulting water. The rates at which such reactions occur are important as they can occur over a time scale of seconds to millions of years (Langmuir 1997). A generalized set of relative half-time reaction rates are shown in Table 5. As can be observed, hydrolysis reaction rates occur at relatively fast rates while mineral dissolution can take much longer. The wide range of precipitation rates reflect

Table 5. *General reaction rates important in mine pit lake systems (based on a compilation by Langmuir 1997)*

Reaction type	Reaction time frame
Solute reactions:	
$H_2CO_3 = H^+ + HCO_3^-$	$C.10^{-6}$ s
Solute-water reactions:	
$CO_2 + H_2O = H_2CO_3$ (hydration)	$C.0.1$ s
$Cu^{2+} + 2H_2O = Cu(OH)_2 + 2H^+$ (hydrolysis/complexation)	$C.10^{-10}$ s
$Fe^{3+} + 3H_2O = Fe(OH)_3 + 3H^+$ (hydrolysis/complexation)	$C.10^{-6}$ s
$Al_{n+m}(OH)_{3n+2m}^{+m} + m\,H_2O = (n+m)Al(OH)_3(s) + m\,H^1$ (multivalent ion hydrolysis)	h–year
Adsorption/desorption reactions:	
$X(AsO_4^{3-}) + PO_4^{3-} = X(PO_4^{3-}) + AsO_4^{3-}$, where X is the adsorption site	$C.$ s–h
Oxidation-reduction:	
$Fe^{2+} + 0.25O_2 + 2.5H_2O = Fe(OH)_3 + 2H^+$	min–h
Mineral–water equilibria:	
$Ca^{2+} + HCO_3^- = CaCO_3 + H^+$	week–year
Isotopic exchange:	
$^{34}SO_4^{2-} + H^{32}S^- = H^{34}S^- + {}^{32}SO_4^{2-}$	Year(s)
Mineral recrystallization:	
$Fe(OH)_3 \cdot n\,H_2O = \alpha\text{-FeO·OH (goethite)} + (n+1)H_2O$	Year(s)

the variation between precipitation of amorphous solids that can occur in a matter of minutes to formation of crystals that occurs on a geological time scale.

When sulphides are exposed to oxygen and water in the presence of a catalyst they undergo oxidation. The reaction leads to release of metals/metalloids, sulphate and possibly protons, or saturation to form secondary minerals (Thornber 1975, 1992; Nickel & Daniels 1985; Alpers *et al.* 1994; Miller *et al.* 1996; Nordstrom & Alpers 1999*a*; Bowell *et al.* 2000). Essentially, the determination of the resulting water chemistry in most hard rock scenarios will depend on:

- the concentration and presence of acid-generating phases;
- the concentration and presence of acid-consuming phases.

Sulphide–water reactions can produce acid pH water, most noticeably in the case of MS_2 (where M is a metal cation, eg Cu, Mo) and/or with iron-bearing sulphides (Thornber 1992). The mechanisms of sulphide oxidation involve the transfer of electrons because most sulphide minerals are electrical conductors in the semiconductor–metallic range (Thornber 1983, 1992). The rate of sulphide oxidation can be controlled by the rate at which oxygen is supplied and reduced at the cathode–solution interface. The separation of the cathodic oxygen-consuming, alkali-producing reaction from the anodic, oxidizing, acid-producing reaction will have a major control on the mineralogy of the resulting assemblage. The greater the distance between cathode and anode, then the more extensive the conducting area. Consequently, this leads to greater potential for sulphide oxidation. Anodic reactions can occur deep within cracks, fissures and along grain boundaries where solutions can penetrate without the necessity for dissolved oxygen (Thornber 1975, 1992). The oxidation of pyrite by ferric iron, can be given as:

$$FeS_2 + 14Fe^{3+} + 8H_2O = 15Fe^{2+} + 2SO_4^{2-} + 16H^+. \quad (1)$$

This is a first-order reaction with a rate given as $1 \times 10^{-4} s^{-1}$ (McKibben & Barnes 1986; Moses *et al.* 1987). With oxygen as the oxidizing agent the rate can be considerably faster by orders of magnitude (Karamenko 1969; Langmuir 1997).

However, it should be noted that not all sulphides on oxidation generate acidity (Thornber 1992; Jennings *et al.* 2000). Indeed, sulphides of the type M_2S such as chalcocite actually consume H^+ on oxidation. To reflect this the oxidation of chalcocite can be written as:

$$2Cu_2S + 6H_2O + 5O_2 = 4Cu^{2+} + 2SO_4^{2-} + 4H^+ + 8OH^- \quad (2)$$

Dissolution or oxidation of pyrite initially produces Fe^{2+} that is almost instantly oxidized to Fe^{3+}, which is either precipitated as an oxyhydroxide or reduced by pyrite generating more Fe^{2+} and increased acidity.

On weathering, sulphides can release all acid potential precipitating secondary minerals (Williams 1990; Nordstrom & Alpers 1999*a, b*). Alternatively, they can release only a portion of the total acidity and store some acidity in secondary salts, which are stable only in oxidizing acidic pH environments, for example the formation of melanterite:

$$FeS_2 + \frac{7}{2}O_2 + 8H_2O = Fe^{2+}SO_4 \cdot 7H_2O + SO_4^{2-} + 2H^+. \quad (3)$$

For each mole of pyrite oxidized, only a portion of the available hydrogen is released. The rest is stored as partly oxidized metal sulphate minerals. These minerals are highly soluble so can represent an instantaneous source of acidic, metal sulphate-rich water upon dissolution and hydrolysis, for example the dissolution of melanterite:

$$4Fe^{2+}SO_4 \cdot 7H_2O + O_2 = 4FeO \cdot OH + 4SO_4^{2-} + 8H^+ + 22H_2O. \quad (4)$$

Oxidation of ferrous iron and hydrolysis of ferric iron at pH > 2 provide the additional source of acidity through the reactions:

$$4Fe^{2+} + 10H_2O + O_2 = 4Fe(OH)_3 + 8H^+. \quad (5)$$

The ability of these minerals to react with water will depend on solubility. Hence, these minerals are important as both sinks and sources of acidity, sulphate and, possibly, metal ions on precipitation and rapid release on exposure to moisture (Nordstrom 1982; Fillipek *et al.* 1988; Cravotta 1994).

Acid-neutralization reactions result from mineral buffering of H^+ in drainage. This buffering is frequently accompanied by the precipitation of secondary minerals (Kwong & Ferguson 1997; Lawrence & Wang 1997; Nordstrom & Alpers

1999a). These reactions can reduce acid generation by forming an inhibitory surface coating on the reactive sulphides. Under acidic conditions, carbonate minerals (e.g. calcite, dolomite and magnesite) readily dissolve and provide bicarbonate alkalinity that results in neutralization of acid and precipitation of metal hydroxides. The order of carbonate-neutralizing capacity is calcite > dolomite > ankerite > siderite. In the case of siderite and, to a lesser extent, ankerite the reason for the limited neutralizing capacity is that ferrous iron in these minerals is an additional source of acidity due to the strong hydrolysis of ferrous iron in solution. This order of reactivity is partly controlled by equilibrium mass action constraints and partly by kinetic limitations (Morse 1983). Carbonate minerals (especially calcite) have often erroneously been thought of as the only geological source of neutralization potential (NP). However, carbonates dominate only limestone, dolomite and marble rock types, whilst the majority of geological materials are composed of silicates and hydroxide–oxide minerals.

Silicate weathering as a proton sink has been demonstrated in previous studies (Sverdrup 1990; Bhatti et al. 1994; Moss & Edmunds 1992; Kwong & Ferguson 1997). To assess the buffering capacity of mine wastes, silicate and hydroxide minerals therefore must also be considered. From soil acidification studies, Sverdrup (1990) divided the most common minerals into six groups according to pH dependency of their dissolution rate (Table 6). From the relative weathering rates of the mineral groups shown (Table 6), minerals in the poor–negligible neutralizing categories are unlikely to react due to their sluggish reaction rates. Even for minerals in the intermediate and fast mineral weathering groups, they will not be practical neutralizing materials unless they occur in excess of approximately 10% (Sverdrup 1990).

Attenuation processes

The accumulation of solutes in solution will lead to saturation with respect to some species. Consequently, in response to either saturation or destabilization as aqueous species, these compounds precipitate as secondary minerals such as arsenates, phosphates, carbonates, sulphates or hydroxides. An important control on the diversity of the precipitated mineral assemblage is pH. At low pH, oxyhydroxides and sulphates are commonly the main precipitates, while at higher pH other salts such as carbonates and hydroxides become more abundant; for example in the case of copper (Fig. 10).

Some solutes can be attenuated through adsorption onto mineral surfaces, noticeably iron hydroxides and clays. This is the process of element binding at the mineral–solution interface and, like solubility, is pH-dependent (Sigg & Stumm 1980; Stumm 1992; Deng & Stumm 1994; Dzombadt & Morel 1990). Many oxide surfaces change from being positive at low pH (thus attracting anions) to negative at high pH (attracting cations).

Pit lake chemistry, and particularly the level of As and heavy metals, has been shown to be influenced by adsorption onto precipitated hydrous ferric oxide, or HFO (this may also include minerals like schwertmannite, goethite and jarosite amongst others; Scott 1987; Fuge et al. 1994; Bigham 1994; Bowell et al. 1996). It should, however, be noted that in many acidic environments, flushing or dissolution of these HFO can lead to high As concentrations in solution, as well as competition from complexing ions mobilizing As-oxyanions.

As described above, as water pH increases above 3, hydrous ferric oxides precipitate that may ultimately crystallize to form goethite or a similar ferric hydroxide (Bigham 1994). As pH increases, ferric hydroxide solubility decreases with a minimum being around pH 6–7. At low pH, precipitated HFO tends to scavenge negatively charged oxyanions, as the surface of the HFO is positively charged in the Helmholtz layer (Deng & Stumm 1994). In low pH environments these HFO particles are usually colloidal sized and have a high reactivity proportional to surface area (Fig. 11). As the pH increases and colloid particles aggregate as Fe–OH bonds become longer and more rigid, due to the excess of hydroxyl molecules, the surface pH of the particles changes, and becomes, negative. In the case of goethite this occurs at a pH of between 6 and 9 (Parfait 1978; Hiemstra & van Riemsdijk 1996). The point at which this occurs is termed the point of zero charge (Stumm 1990). As pH increases the surface of the HFO particles attracts metallic cations and releases oxyanions (Fig. 11). In circumneutral–alkaline oxic environments, As and Se form species such as $H_nAsO_4^{-(3-n)}$ and $H_nSeO_4^{-(2-n)}$ respectively (Bowell 1994). These molecules do form sparingly soluble solids and, as they are not strongly adsorbed, their dissolved concentrations can increase in the pit lake from continued release and evaporation.

Attenuation of metals also occurs through sulphidization in the hypolimnion. Mineralogical analysis of pit lake sediments at Summer Camp Pit, Nevada (see case study below) has identified authigenic sulphides not observed in bedrock geology, including mackinawite ($[Fe, Ni,Co,Zn]_9S_8$)

Table 6. *Grouping of minerals according to their neutralization potential (Sverdrup 1990)*

Group name	Typical minerals	Buffering pH range*	Approx. NP range*	Relative reactivity[‡]
1. Dissolving	Calcite, aragonite, dolomite, magnesite, aragonite, portlandite and brucite	6–11.2	7.8–14.8	1.0
2. Fast weathering	Anorthite, nepheline, olivine, garnet, jadeite, leucite, spodumene, kutnahorite diopside, siderite and wollastonite	5.5–11	2.8–0.6.2	0.6
3. Intermediate weathering	Epidote, zoiste, enstatite, hyperthene, augite, hedenbergite, hornblende, glaucophane, tremolite, actinolite, anthophyllite, serpentine, chrysotile, talc, chlorite, biotite.	4.8–7.3	1.7–5.8	0.4
4. Slow weathering	Albite, oligoclase, labradorite, vermiculite, montmorillonite, manganite, goethite, gibbsite and kaolinite	2.4–5.1	0.5–2.9	0.02
5. Very slow weathering	K-feldspar, ferrihydrite and muscovite	2.2–4.1	0.2–0.6	0.01
6. Inert	Quartz, hematite, rutile and zircon	3.3–3.5	<0.01	0.004

* Buffering pH range evaluated by crushing 5 g of pure mineral and mixing with 5 ml of distilled water and left to react for 30 min. The pH of the distilled water was 3.4.
[†] NP range assessed as equivalent buffering potential of 10 g of pure mineral to calcite and titrated with hydrochloric acid. So, for example, 10 g of portlandite $(Ca(OH)_2)$ was found to have the equivalent capacity to neutralize HCl acid as 14.8 g of calcite. Whereas 10 g of hornblende was required to buffer HCl acid to a similar pH to only 3.1 g of calcite.
[‡] Calculated from Sverdrup's (1990) equation, see below and based on 100% mono-mineral sample.

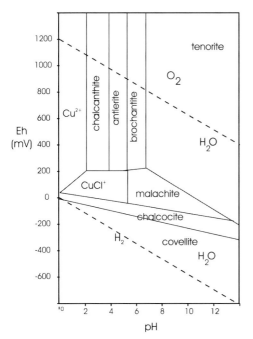

Fig. 10. Simplified Eh–pH diagram for the system Cu–S–Cl–O–H at 298K showing the dominance fields of copper minerals.

and greigite ($Fe^{2+}Fe_2^{3+}S_4$). Similar processes have been observed in deep-sea anoxic sediments (Morse 1994) and natural wetlands (Kwong & Van Stempvoort 1994; Rees 1998).

Concentration processes

In dry environments, where evaporation exceeds precipitation, the latter can exert an important control on water chemistry. The influence of evapoconcentration on brines is well understood (Garrels & Mackenzie 1967; Eugster & Hardie 1978; Drever 1988). As binary salts are precipitated the result is that further evaporation will result in an increase in the relative dissolved concentration of the ion or molecule present in greater concentration. This concept of 'chemical divides' has become the basis of understanding the influence of evaporation on natural lake chemistry in a closed basin (Eugster & Hardie 1978). If a MPL is a terminal sink then evapoconcentration can be predicted to occur over time. On the basis of this approach, Eary (1998) assessed the influence on MPL. As most pit lakes are circumneutral to alkaline, water chemistry is generally dominated by carbonate alkalinity and can evaporate until $CaCO_3$ becomes saturated and calcite is precipitated. Generally, this is the first chemical divide observed in the development of lake chemistry at equilibrium with atmospheric CO_2 (Drever 1988). This can be expressed as:

$$Ca^{2+} + 2HCO_3^- = CaCO_3 + CO_2 + H_2O. \quad (6)$$

Consequently, the concentrations of Ca and bicarbonate (and thus pH) will be controlled by calcite precipitation and evapoconcentration. Where the pit lake initially has $2m_{Ca} < m_{alkalinity}$ (here alkalinity is the sum of $m_{HCO_3} + m_{CO_3}$) then with continued evaporation, calcite will precipitate (Eary 1998). When Ca concentration in minimal then calcite precipitation will reduce to such a small amount that carbonate–bicarbonate molecules will concentrate in solution and exsolve CO_2 with further evaporation. This will increase water pH and the resulting highly saline alkaline fluid will have a chemistry of $Na–K–HCO_3–CO_3 \pm Cl \pm SO_4$. Such conditions are observed in several natural lakes such as Pyramid Lake, Nevada (Drever 1988). Where pit lake chemistry shows $2m_{Ca} > m_{alkalinity}$, then evapoconcentration increases until all available carbonate is removed by calcite precipitation and Ca concentrations increase with evaporation. The next 'chemical divide', which in the case of most MPL is that of gypsum, will be the next control on pit lake chemistry, by a reaction such as:

$$CaCO_3 + SO_4^{2-} + H^+ + 2H_2O$$
$$= CaSO_4 \cdot 2H_2O + HCO_3^-. \quad (7)$$

This is the probably the most important control for MPL water quality (Fig. 12). For most MPL in Nevada, Eary (1998) predicted that equilibrium chemistry would be of the type circumneutral $Ca–Na–SO_4–Cl$ chemistry. On the basis of this, Eary (1998) demonstrated that two differing evapoconcentration paths would be followed by oxyanions dependent on the major chemical controls in the pit lake. Where $2m_{Ca} < m_{alkalinity}$ then a $Na–HCO_3–CO_3 \pm Cl \pm SO_4$ type chemistry will result and pH can evolve to highly alkaline conditions (pH > 9). In this environment oxyanions are not adsorbed and can accumulate over time. By contrast, where $2m_{Ca} > m_{alkalinity}$ then pH is buffered to be circumneutral and oxyanions are strongly adsorbed and thus are not accumulated in the pit lake. In this scenario, pit lake chemistry evolves over time to become circumneutral $Ca–Na–SO_4–Cl$. On the basis of most case studies, the latter case will be the more common. In very acidic pit lakes the major control will relate to

Table 7. *Hydrogeochemistry of Summer Camp Pit (depth, 0–10ft)*

Parameter*	15/05/90	18/03/91	03/04/91	24/04/91	31/05/91	26/06/91	05/12/91	04/03/92	04/05/92	04/06/92	27/07/92	03/08/92	06/10/92	03/11/92	02/12/92	06/10/93	17/05/93	05/10/93	08/12/94	02/05/95	01/03/96	14/05/96	11/06/96	01/10/96	01/12/96	03/10/97
pH	7.7	7.7	7.3	7.2	3.2	7.2	7.86	7.1	3.67	3.48	3.22	3.18	3.96	3.08	3.29	6.92	7	5.97	8.06	7.58	7.23	7.82	7.53	9.36	8.02	6.93
TDS (mg l^{-1})	265.5	295	295	271	649	342	265	384	555	625	738	680	531	814	856	566	620	590	789	867	747	858	870	851	955	1717
Alk. (mg l^{-1})	149	155	187	111	0	109	160	93	0	0	0	0	0	0	0	79	14	6	64	68	51	58	64	30	81	80
Sulphate (mg l^{-1})	37	97	nd	Nd	nd	Nd	nd	nd	196	209	588	560	317	572	522	386	360	488	400	450	430	490	430	430	510	1000
Fe (mg l^{-1})	nd	nd	nd	Nd	nd	Nd	nd	nd	3.9	5.6	14	17.2	4.7	47	49.3	0.56	0.74	1.5	0.15	0.48	0.31	0.4	0.13	0.12	0.52	0.37
As (mg l^{-1})	0.021	0.053	0.074	0.27	1.2	0.15	0.023	0.086	0.044	0.11	0.32	0.41	0.12	0.29	0.81	0.04	0.07	0.019	0.53	0.7	2.5	11	10.1	2.4	10	1.4
Cu (mg l^{-1})	<0.02	<0.02	0.03	0.03	1.06	0.06	<0.02	0.03	0.35	0.47	0.49	0.51	0.29	0.71	0.74	0.05	0.05	0.22	<0.02	<0.02	<0.02	<0.02	<0.02	<0.02	<0.02	<0.02
Zn (mg l^{-1})	<0.01	0.32	0.69	0.76	6.4	1.06	0.01	5.2	11.6	10.8	13	13.7	18.2	26.8	21.2	7.35	5.1	5.5	<0.02	1.7	1.4	<0.02	<0.02	<0.02	<0.02	0.11
Ca (mg l^{-1})	52.4	nd	nd	Nd	nd	Nd	nd	nd	nd	nd	nd	nd	nd	nd	nd	nd	99	nd	Nd	164	360	190	180	nd	1.5	nd
Mn (mg l^{-1})	nd	nd	nd	Nd	nd	Nd	nd	nd	nd	nd	nd	nd	nd	nd	nd	nd	0.04	nd	Nd	nd	3.7	0.09	0.05	nd	nd	nd

* Alk., alkalinity; nd, not determined; TDS, total dissolved solids.

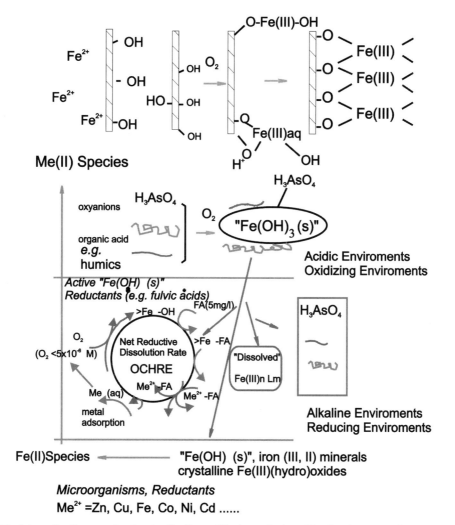

Fig. 11. Schematic diagram showing implications of hydrous ferric oxide chemistry on solute transport at different pH (from Deng & Stumm 1994).

sulphuric acid equilibrium. In the latter, the main controlling salts will be iron sulphates such as melanterite or gypsum. In either case, but particularly in the former, evapoconcentration can produce highly saline, highly acidic metal-rich brines (Nordstrom & Alpers 1999b).

Steady-state cycling

Within an established MPL, release and attenuation processes will reach a steady state or equilibrium at some stage. This steady state may be imbalanced by seasonal effects or by evapoconcentration processes. The speciation and distribution of elements within the pit lake may change greatly. Over time, if a lake deepens, then the effects of seasonal turnover will become less as turnover effects do not extend as deep into the lake and, consequently, a permanent hypolimnion becomes established.

Geological controls

Another factor influencing the formation and characteristics of secondary minerals is the deposit geology in which they form. As the major geochemical reactions in a 'sulphide rock' mine are related to acid generation and acid neutralization, so these in turn are related to the abundance of primary sources of these processes, namely sulphides, particularly pyrite, and buffering minerals, dominantly calcite. Over the last

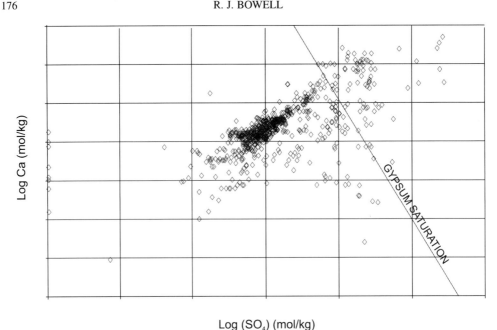

Fig. 12. Relationship between Ca and sulphate in mass-balanced mine pit lakes showing the majority have as an upper control saturation with respect to gypsum. Those above were all below pH > 2 and have an iron sulphate–sulphuric acid control.

few years several studies have examined the use of environmental geological models to provide some form of initial prediction mechanism for understanding potential impacts anticipated from mining sulphide-bearing mineral deposits (Ficklin et al. 1992; Plumlee 1994; Du Bray 1995; Plumlee & Logsdon 1999; Shevenell et al. 1999; Bowell et al. 2000).

Mineral deposits do show a variation in metal and anion hydrogeochemistry related to the distinct geological characteristics of these deposits (Figs 4–6). On the basis of the above discussions, it can be observed that the important aspects of metal deposits that influence MPL hydrogeochemistry will be the speciation and proportion of sulphides, and the presence and abundance of buffering materials, principally calcite. Owing to the nature of mineral deposits, only a few are exploited by open-pit mining and thus the data are biased towards copper porphyry deposits, Carlin-type gold deposits of the western United States and volcanogenic massive sulphide (VMS) deposits (Price et al. 1995).

Thus, it can be observed that the high sulphidation epithermal systems and VMS deposits that host tens of per cent sulphides, dominantly pyrite, and are usually devoid of calcite or other buffering agents result in the more acidic metal-rich pit lakes (Fig. 5). These deposits also tend to occur on a large scale and, as such, can cause significant impact, for example the Summitville high sulphidation system in Colorado (Gray et al. 1994).

Where carbonates form a large portion of the ore, even where sulphides are present in relatively high concentration pit lake, pH is considerably higher and metal concentration lower. A good example are the disseminated sediment-hosted micron gold deposits of the Carlin belt, Nevada that tend to show low base metal concentrations in the ore and a high proportion of calcite in the alteration zones related to carbonate host rocks (Hofstra et al. 1995). Although little data are available for pit lakes developed for sedimentary base metal sulphide deposits, what there is reflects as would be predicted, circumneutral low metal chemistry.

The majority of porphyry pit lake results reflect low metal; non to low-acid generating, but some zones within porphyrys can produce a high potential for acid generation and metal leaching (Bowell et al. 2000). Typically, these are from the silicified sulphide-bearing zones that have been partially oxidized and so have associated with the sulphides a series of Cu–Fe sulphate salts and other sources of secondary acidity, such as jarosite. Where they form a significant part of the exposed pit then an acid pit lake can prevail. Another sulphide zone that exists within porphyry is the chalcocite blanket or zone of

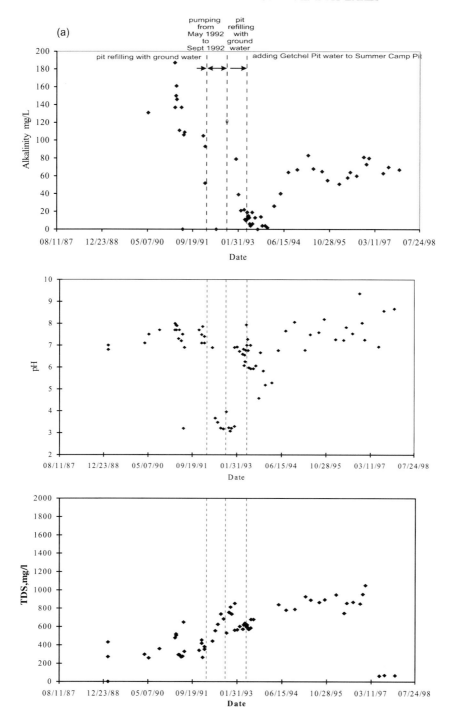

Fig. 13. Geochemical trends over time for Summer Camp Pit lake, 1990–1998.

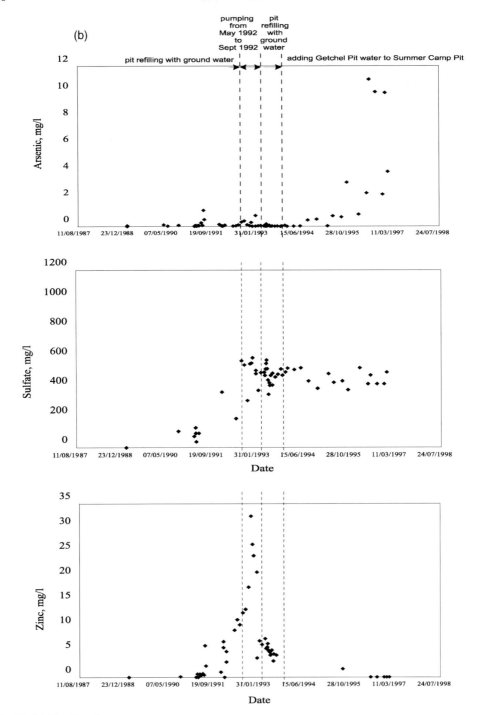

Fig. 13. (b) *(Continued).*

supergene enrichment, this typically comprises of goethite, cuprite, chalcocite and covellite, with minor jarosite and relict pyrite (Lowell & Guilbert 1970; Guilbert & Lowell 1974). If these zones are present then metal leaching can occur at relatively high pH (pH > 4–5). Owing to the configuration of chalcocite, it does not produce protons on oxidation but hydroxyl ions and consequently produces alkaline drainage.

Case study of pit lake chemistry: Summer Camp Pit, Nevada

Few detailed examples of pit lake evolution have been published, except for the pit lakes on the Getchell Mine, Nevada that have been subject to several studies from 1995 to 1999 (Barta et al. 1997; Tempel et al. 1999; Shevenell & Pasternak 2000; Parshley et al. submitted). For one of these pits, the period from pre-mining through to cessation of mining and formation of a MPL has been constantly monitored from 1990 to the present (Bowell et al. 1998).

Prior to mining in 1990 the Summer Camp Pit (SCP) area was undisturbed and as such was most likely in equilibrium with groundwater. Natural groundwater has an alkaline Ca–Mg–HCO$_3$ chemistry with variable amounts of arsenic and sulphate, dependent on contact with gypsum and sulphides (Table 4). Possibly this reflects the migration of oxygenated groundwater along the mineralized shear zone in which water reacted with primary sulphides, chiefly arsenian pyrite. The pH is circumneutral, suggesting that either pyrite oxidation was not appreciable or that groundwater alkalinity and host rock acid consumption was sufficient to buffer generated H$^+$.

During pit development in 1990 water quality remained fairly constant, with a small pool developing in the west of the pit. This had measurable levels of As (up to 0.082 mg l^{-1}), Se (up to 0.009 mg l^{-1}), Zn (up to 0.12 mg l^{-1}), low sulphate (110 mg l^{-1}) and TDS (360 mg l^{-1}). At this time pH of water around the SCP was circumneutral to alkaline (pH = 7.7–8.3). Water chemistry shows low salt load and trace element concentrations, with As below the primary drinking water standard (0.05 mg l^{-1}) in the pit sump. Pool alkalinity in March 1991 ranged from 146 to 187 mg l^{-1} (Fig. 13a) and pH from 7.3 to 8 (Fig. 13b). Slightly elevated metal/metalloid levels in the pool were recorded (Table 7). By May 1991 acid generation had exceeded available alkalinity in the sump and water pH was acidic (Table 7). The acidic pH was accompanied initially by a low TDS, due to precipitation of evaporative salts like gypsum, melanterite (Fe^{2+}SO$_4$·7H$_2$O), halotrichite (Fe^{2+}Al$_2$(SO$_4$)$_4$·22-H$_2$O) and copiapite (Fe^{2+}Fe$_4^{3+}$(SO$_4$)$_6$(OH)$_2$·20-H$_2$O) (Bowell et al. 1998). Adsorption of arsenate at mildly acidic pH (4–6) limits As hydromorphic dispersion but at lower pH (<3) arsenate can be mobilized, in response to hydrogen ion activity stabilizing arsenic acid (H$_3$AsO$_4$) and the increased solubility of HFO and stability of Fe^{3+} in solution. From June 1991 a gradual increase occurred towards the end of operations in December 1991 (Fig. 14).

After pumping was terminated at the end of operations in December 1991 alkalinity was

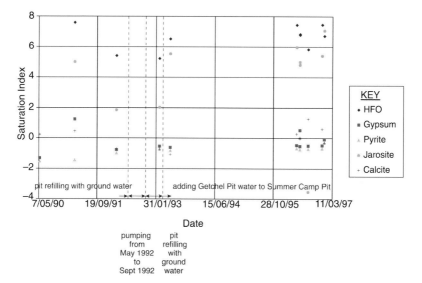

Fig. 14. Saturation Indices for major minerals in Summer Camp Pit lake, 1990–1998.

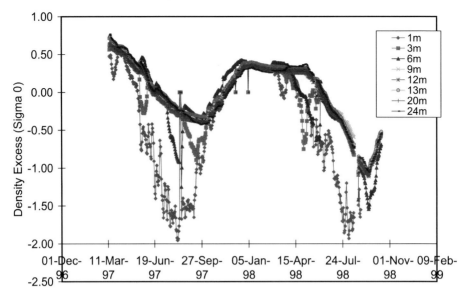

Fig. 15. Density (sigma excess) plot with time for the SCP. Depths are set at various intervals to record changes within the epilimnion (above 9 m), metalimnion and hypolimnion (below 20 m).

steadily consumed until April 1992, when all available alkalinity was consumed as acid generation exceeded alkalinity recharge (Table 7). The zero alkalinity observed is most probably not due to cessation of buffering but, rather, due to increased rates of acid generation. This increase in acid generation correlates to flushing of secondary acid salts as the water level recovers in the pit. During this period pH fell from 7–7.5 to 6.89 in April 1992 and by May 1992 decreased to 3.67 (Fig. 13b). In the absence of excess carbonate alkalinity, water pH was buffered from May 1992 by hydroxides and silicates, so remained in the range pH 3–4 (Sverdrup 1990; Kwong & Ferguson 1997).

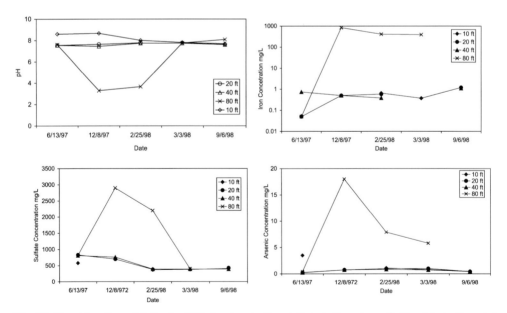

Fig. 16. Chemistry–time plots for the SCP showing depth variations in the values of pH, sulphate, Fe and As.

The slow accumulation of dissolved iron compared to sulphate is most probably due to a greater portion of iron being reprecipitated as secondary minerals such as hydrous ferric hydroxide. Arsenic and selenium during this period do not show significant increases, probably due to adsorption onto ferric hydroxide and clay minerals. Zinc levels show a steady increase from the end of operations, even at neutral pH. This is most probably due to secondary processes such as dissolution of soluble zinc minerals like smithsonite ($ZnCO_3$) or goslarite ($ZnSO_4 \cdot 7H_2O$) or ion exchange of Zn with other divalent cations during clay mineral–water interactions. Smithsonite dissolution might occur in response to increased acid production and be utilized in the buffering process along with calcite carbonate.

Water was abstracted from SCP for operational use from May 1992 to September 1992. The high rate of acid generation was maintained during this period. Most probably this is because pumping maintained the water level such that 'reactive' pit walls were exposed and these contributed metals, sulphate and acidity to the pit lake through sulphide oxidation and secondary mineral leaching due to water level recovery and rainwater surface run-off. The low pH environment was mitigated in late 1992 with the diversion of HCO_3-bearing water from other flooded open pits and the underground operations elsewhere on the Getchell property (Fig. 13a). Following the addition of alkaline water into the SCP, pH increased to circumneutral levels (Fig. 13a). Total dissolved solids have remained relatively constant, with the water chemistry being essentially a Na–Ca–SO_4 type with the

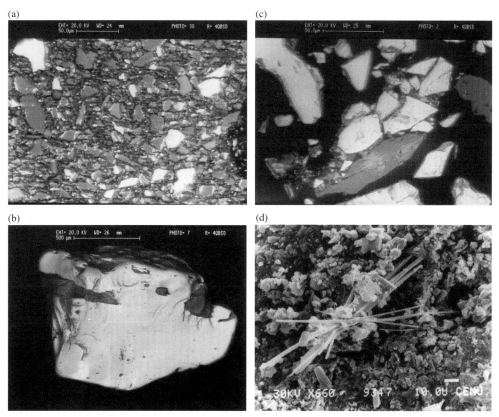

Fig. 17. Photomicrographs of sulphides from pit sediments and sulphates from pit walls. (**a**) Section showing euhedral to subhedral crystals of mackinawite ($(Fe,Ni,Co,Zn)_9S_8$), sphalerite (ZnS) and orpiment (As_2S_3) in kaolinite–quartz matrix, SCP hypolimnion sediment. Note sphalerite and orpiment are most probably detritus grains, whilst mackinawite is probably authigenic. (**b**) Backscatter image of mackinawite euhedral crystal showing negligible corrosion, SCP hypolimnion sediment. (**c**) Backscatter image of orpiment euhedral crystal showing negligible corrosion, SCP hypolimnion sediment. (**d**) Halotrichite, chalcanthite and gypsum on wall-rock. Sternberg lode, Copper Flat Pit lake, Hillsboro County, New Mexico.

main mineralogical control being gypsum saturation (Fig. 14).

Further variations in pit lake water quality reflect changes in the chemistry of pumped mine water rather than pit lake processes. Although not discussed here in detail, the nature of the stratified pit lake can be observed in the density plots for the pit lake set at various depths (Fig. 15). As can be observed for much of the year, the basal layer is considerably denser than the other two layers but in winter this changes as turnover occurs (Fig. 15). Over time the magnitude of this turnover would become less, as a permanent hypolimnion forms in the deep (20 m) pit lake. A crucial impact of turnover is that it acts as a 'self-cleansing' event and a post-turnover lake (such as the data shown for June 1998, Table 4) shows lower TDS water than pre-turnover (as shown for February 1998 for SCP, Table 4). Turnover events such as these could be both seasonal and, possibly, longer-term multi-year events. Similar processes will occur in other deep pit lakes, for example the Sleeper Mine in Nevada where the pit lake is over 60 m deep.

Chemical variations in the pit lake over the same period reflect its stratified nature, with low pH, high sulphate, As and metals in the basal layer for much of the year (Fig. 16). On turnover, the basal layer is briefly aerated resulting initially in higher pH and lower metals, but with the aeration comes oxygen and sulphide oxidation occurs in the basal sediments. This leads to low pH and high metals as stratification reforms. A more detailed study of the limnological–depth stratification of this pit lake is currently being prepared (Parshley et al. submitted).

Attenuation of metals occurs in the basal sediments of the SCP through sulphidization in the hypolimnion. In the hypolimnion of Summer Camp pit lake, sulphate-reducing bacteria have been identified (Gannon et al. 1996). Within this zone, sediments have been found through diagnostic leach studies to contain high metal concentrations associated with the reduced fraction (Bowell & Parshley 1999). The diagnostic leach scheme used in this study involved a modified Tessier selective extraction procedure (Bowell & Parshley 1999). From mineralogical analysis the phases in this fraction have been confirmed as authigenic sulphides not observed in bedrock geology, and include mackinawite ([Fe, Ni, Co, Zn]$_9$S$_8$) and greigite (Fe^{2+}Fe$_2^{3+}$S$_4$), as well as pararealgar, sphalerite and pyrite. The latter three all show authigenic morphologies (Fig. 17) but they do occur within the pit wall-rock and thus a detritus origin cannot be ruled out. Similar processes have been observed in deep-sea anoxic sediments (Morse 1994) and natural wetlands (Kwong & Van Stempvoort 1994; Rees 1998).

Utilizing all this information, a conceptual model was developed to assist understanding of the development and geochemical–limnological environment within the SCP lake (Fig. 18). This model reflects the internal and external processes that control the water quality observed within the mine pit lake. At the SCP an important external

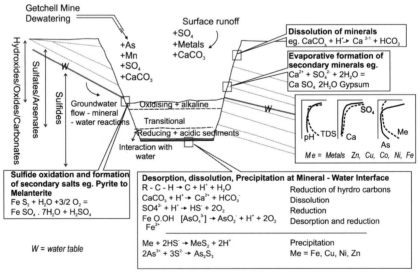

Fig. 18. Conceptual model developed for the Summer Camp Pit (SCP) based on the data collected in the field and by laboratory analysis to explain the complex interlinked limnological and geochemical mechanisms that control pit lake water quality.

control is the discharge of underground mine water into the pit, this practice has now ceased.

Conclusions

A review of the major processes influencing the hydrogeochemistry of MPLs has been presented. The major controls on MPL can be related to:

- limnology of the pit lake;
- geochemical processes within a pit lake;
- geology of the exposed wall-rocks and type of mineral deposit.

At the present time, methods for predicting the future water quality in mine pit lakes involve hypothetical mechanistic and empirical models to reflect the major chemical and physical processes operating within a MPL. These major processes include groundwater inflow, wall-rock leaching, pit wall run-off, precipitation, evaporation, lake hydrodynamics, and geochemical equilibrium and speciation. Most of these methods are still in the process of development. Refinement of such models and general understanding of the environmental assessment of pit lakes will be improved through characterizing the conditions in existing MPL.

This paper has made extensive use of field notes, reports, unpublished data and ideas held by SRK Consulting, and I acknowledge the debt owed numerous colleagues past and present for discussion, review and support particularly J. Parshley, B. Rees, R. Connelly, W. Harding, S. Day, D. Hockley, P. Sadler, K. Williams, M. Dey and L. Fillipek. I acknowledge the support of clients who provoked thought, improved the manuscript through review and graciously approved publication of the data within this paper, particularly Getchell Gold Corporation. Finally, I acknowledge the time and effort taken by L. Lovgren and C. Ayora in reviewing this manuscript, which resulted in a much improved final version. The views and opinions expressed within this paper are those of the author and not SRK or its clients.

References

ALPERS, C.A., NORDSTROM, D.K. & THOMPSON, J.M. 1994. Seasonal variations of Zn/Cu ratios in Acid Mine Water from Iron Mountain, California. In: ALPERS, C.A. & BLOWES, D.W. (eds) *Environmental Geochemistry of Sulphide Oxidation*, American Chemical Society Symposium Series, American Chemical Society, Washington DC, **550**, 324–345.

ATKINS, D., KEMPTON, J.H. & MARTIN, T. 1997. Limnologic conditions in three existing Nevada pit lakes: observations and modeling using CE-QUAL-W2. In: *Proceedings of the Fourth International Conference on Acid Rock Drainage, Vancouver, BC, 31 May–6 June*, Mine Environment Neutral Drainage Program, Ottawa, Canada, 699–713.

BARTA, J., BOWELL, R.J. & MASANARES, W. 1997. Geochemical controls on pit lake water quality. In: *Proceedings of Living with a Closed Basin. Nevada Water Association Annual Meeting, Elko, October 1997*, Nevada Water Association, Reno, Nevada.

BHATTI, T.M., BIGHAM, J.M., VUORINEN, A. & TUOVINEN, O.H. 1994. Alteration of mica and feldspars associated with the microbiological oxidation of pyrrhotite and pyrite. In: ALPERS, C.N. & BLOWES, D.W. (eds) *Environmental Geochemistry of Sulphide Oxidation*. American Chemical Society Symposium, Series, American Chemical Society, Washington DC, **550**, 90–107.

BIGHAM, J.M. 1994. Mineralogy of ochre deposits. In: JAMBOR, J.L. & BLOWES, D.W. (eds) *Environmental Geochemistry of Sulphide Mine Waste* Mineralogical Association of Canada, Ottawa, Canada, 103–131.

BIRD, D.A., LYONS, W.B. & MILLER, G.C. 1994. An assessment of hydrogeochemical computer codes applied to modeling post-mining pit water geochemistry. In: *Proceedings of Tailings and Mine Waste '94*. A.A. Balkema, Rotterdam, 31–40.

BOWELL, R.J. 1994. Arsenic sorption by Iron oxyhydroxides and oxides. *Applied Geochemistry*, **9**, 279–286.

BOWELL, R.J. & PARSHLEY, J.V. 1999. *Geochemical and Limnological Assessment of Summer Camp Pit, Nevada*. SRK Report to Getchell Gold Corporation.

BOWELL, R.J., FUGE, R., CONNELLY, R.J. & SADLER, P.K.J. 1996. Controls on ochre chemistry and precipitation in coal and metal mine drainage. In: *Minerals, Metals and the Environment II, Prague, September 1996*, I.M.M. London, 291–323.

BOWELL, R.J., BARTA, J., GINGRICH, M., MANSANARES, W. & PARSHLEY, J.V. 1998. Geologic controls on pit lake water chemistry: Implications for the assessment of water quality in Inactive open pits. In: *IMWA Symposium Proceedings, Johannesburg, RSA. Journal of IMWA*, **7**, 375–386.

BOWELL, R.J., REES, S.B. & PARSHLEY, J.V. 2000. Geochemical prediction of metal leaching and acid generation: geologic controls and baseline assessment. In: CLUER, J.K., PRICE, J.G., STRUHSACKER, E.M., HARDYMAN, R.F. & MORRIS, C.L. (eds) *Geology and Ore Deposits 2000: The Great Basin and Beyond: Geological Society of Nevada Symposium Proceedings, Reno/Sparks, May 2000*, Geological Society of Nevada, Reno, Nevada, 799–823.

CRAVOTTA, C.A. III 1994. Secondary iron-sulphate minerals as sources of sulphate and acidity. In: ALPERS, C.N. & BLOWES, D.W. (eds) *Environmental Geochemistry of Sulphide Oxidation*, American Chemical Society Symposium Series, American Chemical Society, Washington DC, **550**, 345–364.

DAVIS, A. & ASHENBERG, D. 1989. The aqueous geochemistry of the Berkeley Pit, Butte, Montana, USA. *Applied Geochemistry*, **4**, 23–36.

DAVIS, A. & EARY, L.E. 1996. Pit lake water quality in the western United States: An analysis of chemogenic trends. *Minerals Engineering*, **96**, 98–102.

DAVIS, A., NEWCOMB, B. & BYRNS, C. 1998. Discriminating between historical sources of solutes in the Robinson Mining District, Ely, Nevada. In: *Tailings and Mine Waste '98*, A.A. Balkema, Rotterdam, 631–638.

DENG, Y. & STUMM, W. 1994. Reactivity of aquatic iron (III) oxyhydroxides – implications for redox cycling of iron in natural waters. *Applied Geochemistry*, **9**, 23–36.

DREVER, J.I. 1988. *The Geochemistry of Natural Waters*, 2nd ed. Prentice-Hall, Englewood Cliffs, New Jersey.

DU BRAY, E.A. (ed.). 1995. *Preliminary Compilation of Descriptive Geoenvironmental Mineral Deposit Models*. US Geological Survey Open File Report, 95–831.

DZOMBAK, D.A. & MOREL, F.M.M. 1990. *Surface Complexation Modeling – Hydrous Ferric Oxide*, John Wiley, New York.

EARY, L.E. 1998. Predicting the effects of evapoconcentration on water quality in mine pit lakes. *Journal of Geochemical Exploration*, **64**, 223–236.

EARY, L.E. 1999. Geochemical and equilibrium trends in mine pit lakes. *Applied Geochemistry*, **14**, 963–988.

EUGSTER, H.P. & HARDIE, L.A. 1978. Saline lakes. In: LERMAN, A., (e.d). *Lakes, Chemistry, Geology, Physics*. Springer, New York, 237–293.

FETTER, C.W. 1994. *Applied Hydrogeology*, 3rd edn. Prentice-Hall, Englewood Cliffs, New Jersey.

FLICKLIN, W.H., PLUMBLEE, G.S., SMITH, K.S. & MCHUGH, J.B. 1992. Geochemical classification of mine drainage and natural drainage in mineralized areas. In: *Proceedings of the Seventh Water–Rock Conference, Utah*, A.A. Balkema, Rotterdam, 381–384.

FILLIPEK, L.H., NORDSTROM, D.K. & FICKLIN, W.H. 1988. Interaction of acid mine draianige with waters and sediments of West Squaw, West Shasta mining district, California. *Environmental Science & Technology*, American Chemical Society, Wasington DC, **21**, 388–396.

FUGE, R., PEARCE, F.M., PEARCE, N.J.G. & PERKINS, W.T. 1994. Acid mine drainage in Wales and influence of ochre precipitation on water chemistry. In: ALPERS, N. & BLOWES, D.W. (eds) *Environmental Geochemistry of Sulphide Oxidation*, American Chemical Society Symposium Series, American Chemical Society, Wasington DC, vol **550**, 261–274.

GANNON, J., WIELINGA, B., MOORE, J.M., POLICASTRO, P., MCADOO, D. & MEIKLE, T. 1996. *Field Investigation of the Sulphate Reduction Potential in the Summer Camp Pit Lake*. Report to Getchell Gold Corporation.

GARRELS, R.M. & MACKENZIE, F.T. 1967. Origin of the chemical composition of some springs and lakes. GOULD, R.F. (ed.) *Equilibrium Concepts in Natural Water Systems*, Advances in Chemistry Series, American Chemical Society, Washington, DC **67**, 223–242.

GRAY, J.E., COOLBAUGH, M.F., PLUMLEE, G.S. & ATKINSON, W.W. 1994. Environmental Geology of the Summitville, Colorado. *Economic Geology*, **89**, 2006–2014.

GUILBERT, J.M. & LOWELL, J.D. 1974. Variations in zoning patterns in porphyry ore deposits. *Candian Institute of Mining Bulletin*, **67**, 99–109.

HAVIS, R.N. & WORTHINGTON, S.J. 1997. A simple model for the management of mine pit-lake filling and dewatering. In: *Proceedings of Tailings and Mine Waste '97*, A. A. Balkema, Rotterdam, 478–549.

HIEMSTRA & VAN RIEMSDIJK 1996. Calculated values for point of zero charge for some common minerals. *Journal of Colloid and Interface Science*, **179**, 488–508.

HOFSTRA, A.H., LEVENTHAL, J.S., GRIMES, D.J. & HERAN, W.D. 1995. Sediment-hosted Au deposits. DU BRAY, E.A. (ed.) *Preliminary Compliation of Descriptive Geoenvironmental Mineral Deposit Models*, US Geological Survey Open File Report, United states Geological survey, Washington DC, **95–831**, 184–192.

JENNINGS, S.R., DOLLHOPF, D.J. & INSKEEP, W.P. 2000. Acid production from sulphide minerals using hydrogen peroxide weathering. *Applied Geochemistry*, **15**, 235–243.

KARAMENKO, L.Y. 1969. Biogenic factors in the formation of sedimentary ores. *International Geological Reviews*, **11**, 271–281.

KEMPTON, J.H., LOCKE, W.L., NICHOLSON, A.D., BENNETT, M., BLISS, L. & MALEY, P. 1997. Probabilistic prediction of water quality in the Twin Creeks Mine pit lake, Golconda, Nevada. In: *Fourth International Conference on Acid Rock Drainage Proceedings, Vancouver, BC*, Mine Environment Neutral Drainage Program, Ottawa, Canada 1729–1744.

KIRK, L.B., SCHAFER, W., VOLBERDING, J. & KRANZ, S. 1996. Mine lake geochemical prediction for the SPJV MacDonald Project. In: *Billings Symposium, Billings, Montana*, 393–403.

KWONG, Y.T.J. & FERGUSON, K.D. 1997. Mineralogical changes during NP determinations and their implications. In: *Proceedings ICARD '97, Vancouver*, Mine Environment Neutral Drainage Program, Ottawa, Canada, 435–447.

KWONG, Y.T.J. & VAN STEMPVOORT, D.R. 1994. Attenuation of acid rock drainage in a natural wetland system. ALPERS, C.A., & BLOWES, D.W. (eds) *Environmental Geochemistry of Sulphide Oxidation*, American Chemical Society Symposium Series, American Chemical Society, Wasington DC, **550**, 382–292.

LANGMUIR, D.I. 1997. *Aqueous Environmental Geochemistry*, Prentice-Hall, Englewood Cliffs, New Jersey.

LAWRENCE, R.W. & WANG, Y. 1997. Determination of neutralization in the prediction of acid rock drainage. *Proceedings ICARD '97, Vancouver*, Mine Environment Neutral Drainage Program, Ottawa, Canada, 451–464.

LOWELL, J.D. & GUILBERT, J.M. 1970. Lateral and vertical mineralization zoning in porphyry ore deposits. *Economic Geological*, **65**, 373–408.

MCKIBBEN, M.A. & BARNES, H.L. 1986. Oxidation of pyrite in low temperature acidic solutions: Rate laws and surface textures. *Geochimica et Cosmochimica Acta*, **50**, 1509–1520.

MILLER, G., LYONS, W.B. & DAVIS, A. 1996. Understanding the water quality of pit lakes. *Environmental Science & Technology*, **30**, 118–123A.

MORSE, J.W. 1983. The kinetics of calcium carbonate dissolution and precipitation. In: REEDER, R.J. (ed.) Reviews in Mineralogy, Mineralogical Society of America, **11**, 227–264.

MORSE, J.W. 1994. Release of toxic metals via oxidation of authigenic pyrite in resuspended sediments. ALPERS, C.A., & BLOWES, D.W. (eds) *Environmental Geochemistry of Sulphide Oxidation*, American Chemical Society Symposium Series, American Chemical Society, Wasington DC **550**, 289–297.

MOSES, C.O., NORDSTROM, K.K., HERMAN, J.S. & MILLS, A.L. 1987. Aqueous pyrite oxidation by disslved oxygen and by ferric iron. *Geochimica et Cosmochimica Acta*, **51**, 1561–1571.

MOSS, P.D. & EDMUNDS, W.M. 1992. Processes controlling acid attenuation in an unsaturated zone of a Triassic sandstone aquifer UK, in the absence of carbonate minerals. *Applied Geochemistry*, **7**, 573–583.

MURRAY, W.M. 1997. Are pit lakes susceptible to limnic eruptions? In: *Tailings and Mine Waste '97*, A. A. Balkema, Rotterdam, 543–548.

NICKEL, E.H. & DANIELS, J.L. 1985. Gossans. In: WOLF, K.H. (eds) *Handbook of Stratabound and Stratiform Ore Deposits*, Elsevier, Amsterdam Vol. 13, 261–390.

NORDSTROM, D.K. 1982. Aqueous pyrite oxidation and the consequent formation of secondary minerals. In: *Acid Sulphate Weathering*, Soil Science Society of America, New York.

NORDSTROM, D.K. & ALPERS, C.N. 1999a. Geochemistry of acid mine waters. In: PLUMLEE, G.S., & LOGSDON, M.J. (eds) *The Environmental Geochemistry of Mineral Deposits*, Reviews in Economic Geology, **6A**, 133–160.

NORDSTROM, D.K. & ALPERS, C.N. 1999b. Negative pH, efflorescent mineralogy and consequences for environmental restoration at the Iron Mountain Superfund site, California. *Proceedings of the National Academy of Sciences USA*, **96**, 33455–33462.

PARFAIT, R.L. 1978. Chemical properties of variable charge soils. In: THENG, B.K.G. (ed.) *Soils With Variable Charge*, New Zealand Soil Science Society, Special Publication, 167–194.

PARSHLEY, J.V., BOWELL, R.J. & FILLIPEK, L.L. Submitted. Geochemistry and limnology of Summer Camp Pit, Nevada. *Mine Water & The Environment*.

PETERS, W.C. 1987. *Exploration and Mining Geology*, 2nd edn. John Wiley, New York.

PILLARD, D.A., RUNNELLS, D.D., DOYLE, T.A. & YOUNG, J. 1995. Post-mining pit lakes: Predicting lake chemistry and assessing ecological risks. In: *Tailings and Mine Waste '96*, A. A. Balkema, Rotterdam, 469–478.

PLUMLEE, G. 1994. Environmental geology models of mineral deposits. *SEG Newsletter*, **16**, 5–6.

PLUMLEE, G.S., LOGSDON, & M.J. (eds) *The Environmental Geochemistry of Mineral Deposits*, Reviews in Economic Geology, **6A**.

PRICE, J.G., SHEVENELL, L., HENRY, C.D., RIGBY, J.G., CHRISTENSEN, L., LECHLER, P.J., DESILETS, M., FIELDS, R., DRIESNER, D., DURBIN, W. & LOMBARDO, W. 1995. Water Quality at Inactive and Abandoned Mines in Nevada. *Report of a Cooperative Project Amongst State Agencies*, Nevada Bureau of Mines and Geology Open File Report, Vol 95–4.

REES, S.B. 1998. Longevity of the Treatment Processes Operating in the Whitworth No.1 Constructed Wetland for Amelioration of Acid Mine Drainage MSc thesis University of Leeds.

ROBINS, R.G., BERG, R.B., DYSINGER, D.K., DUAIME, T.E., METESH, J.J., DIEBOLD, F.E., TWIDWELL, L.G., MITMAN, G.G., CHATHAM, W.H., HUANG, H.H. & YOUNG, C.A. 1997. Chemical, physical and biological interaction at the Berkeley Pit, Butte, Montana. *Tailings and Mine Waste '97*, A. A. Balkema, Rotterdam, 529–542.

SCOTT, K.M. 1987. Solid solution in, and classification of, gossan-derived members of the alunite–jarosite family, Queensland, Australia. *American Mineralogist*, **72**, 178–187.

SHEVENELL, L. & PASTERNAK, K.I. 2000. Modeled versus observed filling rates of the historical Getchell Pit Lakes, Nevada. CLUER, J.K., PRICE, J.G., STRUHSACKER, E.M., HARDYMAN, R.F., MORRIS, C.L. (eds) *Geology and Ore Deposits 2000: The Great Basin and Beyond: Geological Society of Nevada Symposium Proceedings, Reno/Sparks, May 2000*, Geological Society of Nevada, Reno, Nevada, 869–871.

SHEVENELL, L., CONNERS, K.A. & HENRY, C.D. 1999. Controls on pit lake water quality at sixteen open pit mines in Nevada. *Applied Geochemistry*, **14**, 669–687.

SIGG, L. & STUMM, W. 1980. The interaction of anions and weak acids with the hydrous goethite surface. *Colloidal Surface*, **2**, 101–117.

STUMM, W. 1992. *Chemistry of the Solid–Water Interface*, (Wiley-Interface, New York).

SVERDRUP, H.U. 1990. *The Kinetics of Base Cation Release due to Chemical Weathering*, Lund University Press, Lund.

TEMPEL, R.N., SHEVENELL, L.A., LECHLER, P. & PRICE, J. 1999. Geochemical modeling approach to predicting arsenic concentrations in a mine pit lake. *Applied Geochemistry*, **15**, 475–492.

THORNBER, M.R. 1975. Supergene alteration of sulphides, I: a chemical model based on massive nickel sulphide deposits at Kambalda, Western Australia. *Chemical Geology*, **15**, 1–14.

THORNBER, M.R. 1983. Mineralogical and electrochemical stability of the nickel–iron sulphides pentlandite and violarite. *Journal of Applied Electrochemistry*, **13**, 253–267.

THORNBER, M.R. 1992. Chemical processes during weathering. In: BUTT, C.R.M., ZEIGERS, H. (eds). *Handbook of Exploration Geochemistry, Regolith Exploration Geochemistry in Tropical Terrains*, Elsevier, Amsterdam Vol. 4, 65–99.

WILLIAMS, P.A. 1990. Oxide Zone Geochemistry. Ellis Horwood, Sussex.

YOUNGER, P.L. 1995. Hydrogeochemistry of minewaters flowing from abandoned coal workings in County Durham. *Quarterley Journal of Engineering Geology*, **28**, S101–S113.

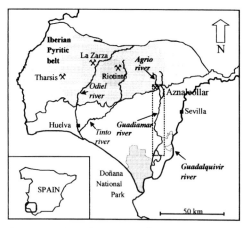

Fig. 1. Location of the Aznalcóllar mine and of the area affected by spill of sludge (rectangle).

In previous papers, the kinetics of sulphide dissolution and the behaviour of each contaminant present in the sludge has been investigated (Domènech et al. 2001a, b). It was observed that the diffusion of oxygen is the main parameter governing the dissolution of sulphides and release of metals, whereas the variation in pH is responsible for the precipitation of secondary minerals, which coprecipitate some of the contaminants.

In order to predict the evolution of the water in the pores of the soil, a reactive transport model has been constructed. A number of reactive transport modelling studies have been applied to pyrite oxidation and acid mine discharges. Among them, Walter et al. (1994a, b) described the code MINTRAN, which was applied to saturated aquifers. Wunderly et al. (1996) coupled the code PYROX, which described the oxygen diffusion and pyrite oxidation in an unsaturated zone, to MINTRAN. This coupled code was applied to study the effects of spatial heterogeneity on the water quality (Gerke et al. 1998) and to model the drainage of a mine impoundment (Bain et al. 2000). These codes assumed local equilibrium for all the chemical reactions. More recently, however, Mayer et al. (1999, 2000) developed the code MIN3P, which is able to model reactive transport in a saturated or an unsaturated zone without assuming local equilibrium between minerals and water.

Similarly to MIN3P, our model includes the basic processes required to describe the reactions taking place in an unsaturated zone: multiphase flow, heat transport, oxygen diffusion through the soil pores, sulphide oxidation, dissolution of soil minerals and precipitation of secondary phases. Mineral–water reactions are described by means of kinetics laws and no local equilibrium is assumed. This is especially important for silicates, which are reputed to be in disequilibrium with surface water. A realistic modelling of silicates is required because they may buffer the pH of the system and supply the cations required for secondary phases to precipitate. Equilibrium between aqueous redox species such as Fe^{2+} and Fe^{3+}, which is commonly assumed in conventional geochemical calculations, may not be true in nature (Nordstrom & Alpers 1999) and hence, is not assumed. A realistic estimate of Fe^{3+} concentration would be also of interest, as Fe^{3+} along with O_2 can oxidize pyrite as well as precipitate in secondary minerals.

First, we briefly describe the experiments performed under constrained conditions. Then, we use the model to simulate and discuss the results of the experiments. Once the main processes are modelled, we predict the behaviour of the pore water chemistry for different scenarios of sludge–soil ratios and mineral associations, such as those expected in the soil affected by the Aznalcóllar accident.

Column experiment

Two columns, 0.085 m in diameter, were filled with two mixtures of soil (90 wt%) and sludge (10 wt%). Prior to mixing, the soil samples were leached with distilled water. The mixture was made mechanically so that no structure was preserved. The two soil samples were regarded as being similar to the upper layer of the soils affected by the spill. During the cleaning works, the heavy machinery destroyed the soil structure and formed a layer of about 0.2 m of mixed soil and sludge. Another two columns with soil and without sludge were also prepared as a blank. A layer of quartz sand was deposited at the bottom of each column in order to facilitate the drainage (Fig. 2).

The chemical composition of the sludge used in the columns was that of sample PM1 of Alastuey et al. (1999). The chemical composition of the soil was obtained from the total digestion of the samples in a mixture of nitric and fluorhydric acids, following the method described in Querol et al. (1996). The mineralogy was determined by X-ray diffraction (XRD), using Cu radiation and graphite monochromator. The determination of the clay minerals required additional treatments of the sample with glycerol and with a temperature up to 550°C (Thorez 1976). The mass fraction of each mineral, ω_m, is found in Table 1. In the case of soils, the mass fraction was estimated from the net intensity of

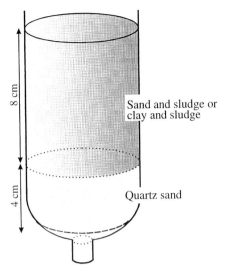

Fig. 2. Sketch of the columns used in the experiments.

and K were assumed to be entirely contained in the pyrite, sphalerite, galena, chalcopyrite, gypsum, chlorite and illite, respectively.

The columns were located outdoors on the roof of the laboratory in Barcelona for 15 months, and were leached 10 times with 100–250 cm^3 of Millipore MQ water. This is equivalent to 15–45 mm of rain. The leachates were recovered, filtered and analysed. Rainfall was generally not intensive enough to produce a leachate, with the exception of one particularly intensive rain shower. The concentration of Na, K, Ca, Mg, Al, Fe and Zn in solution was measured by means of ICP-AES (inductively coupled plasma-atomic emission spectrometry). The concentrations of As, Cu and Pb were determined by ICP-MS (inductively coupled plasma-mass spectrometry). Fe^{2+}, Fe^{3+} and total Fe concentrations were determined by colorimetry (To *et al.* 1999). The results of the analyses of the leaching solutions are given in Tables 2 and 3.

the more important peaks of the X-ray diffractograms (reference intensity method) and mixtures of pure minerals for calibration. In the case of the sludge, the mass fraction was calculated from the chemical composition (Alastuey *et al.* 1999), according to:

$$\omega_m = \frac{\chi_i M_m 10^{-6}}{M_i \nu_{m,i}} \quad (1)$$

where χ_i is the concentration of the element i in the analysis (ppm), M_m is the molecular mass of the mineral m, M_i is the atomic mass of the element i and $\nu_{m,i}$ is the stoichiometric coefficient of the element i in the mineral m. In the calculation, all the Fe, Zn, Pb, Cu, Ca, Mg

Modelling

O_2 is the ultimate driving force for the oxidation of pyrite (Nordstrom & Alpers 1999). The most important process that brings the O_2 from the atmosphere to the soil is the diffusion of oxygen in the gas phase. Therefore, the oxidation of sulphides and, as a result, the precipitation of sulphates and hydroxides will occur preferentially during dry weather when the soil contains more gas and less water. On the other hand, during wet weather, the lack of O_2 will inhibit the oxidation of sulphides and the precipitation of sulphates and hydroxides, while rainwater will dissolve the sulphates and hydroxides precipitated during

Table 1. *Mass fraction (ω_m) of each mineral in the sludge and in the soil used in the column experiments*

	Sludge $\omega_m \times 100$	Sandy soil $\omega_m \times 100$	Clayey soil $\omega_m \times 100$	Quartz sand $\omega_m \times 100$
Quartz	8	31	30	98
Albite		23	10	0.5
K-Feldspar		22		
Illite	5	7	35	1.5
Chlorite	5	7	35	
Hornblende		9		
Gypsum	6	1		
Dolomite			10	
Pyrite	73			
Sphalerite	1.4			
Galena	0.8			
Chalcopyrite	0.6			

Table 2. Chemical composition of the leachates of the column of sandy soil and sludge. Sample 9 is taken after an intensive rain-shower, the others after leaching events

	0	1	2	3	5	6	7	8	9	10	11	12
Date	2/03/99	2/03/99	15/03/99	22/03/99	6/04/99	26/04/99	21/06/99	10/09/99	15/09/99	9/12/99	7/03/00	5/06/00
t (days)	0	1	14	21	36	56	112	193	198	283	372	462
V_{added} (cm^3)*	100	200	200	200	200	200	250	200	–	200	200	200
$V_{collected}$ (cm^3)*	52	29	63	65	93	102	122	116	122	134	126	72
pH	7.6	6.8	5.4	4.4	4.3	3.0	2.5	2.0	2.3	1.1	2.4	2.1
Eh (mV)	–	–	–	–	605	630	660	–	679	–	–	–
Ca (mg l^{-1})	46	752	494	470	436	530	377	632	290	520	216	426
Mg (mg l^{-1})	6	255	247	134	148	271	194	797	283	640	175	160
Na (mg l^{-1})	25	16	11	7	7	12	1.5	<10	<10	<10	0.4	3.6
K (mg l^{-1})	7	11	10	10	8	6	<5	<10	–	<10	<10	<10
Al (mg l^{-1})	0.1	0.1	3	3	19	379	517	1581	578	1593	424	361
SO$_4^{2-}$ (mg l^{-1})	27	3 519	3 101	2 312	2 486	6 753	7 260	28 365	10 757	30 972	8 262	12 669
Fe$_{total}$ (mg l^{-1})	0.1	4	0.4	0.8	8	227	914	5 891	2 153	7 369	1 714	3 088
Fe^{2+} (mg l^{-1})	–	–	–	–	–	–	–	5 567	2 153	1 909	–	–
Zn (mg l^{-1})	1	358	427	285	361	913	376	714	225	491	312	723
Cu (mg l^{-1})	0.05	0.004	1	1	4	42	35	108	34	82	135	105
Pb (µg l^{-1})	0.8	<0.05	<0.05	<0.05	7	792	83	<0.05	<0.05	5	25	22
As (µg l^{-1})	5	9	46	26	29	126	2 677	36 521	19 599	68 678	17	17
											14 120	19 494

*V is a volume, see equation (3).

Table 3. Chemical composition of the leachates of the column of clayey soil and sludge. Sample 9 is taken after an intensive rain-shower, the others after leaching events

	0	1	2	3	5	6	7	8	9	10	11	12
Date	2/03/99	2/03/99	15/03/99	22/03/99	6/04/99	26/04/99	21/06/99	10/09/99	15/09/99	9/12/99	7/03/00	5/06/00
t (days)	0	1	14	21	36	56	112	193	198	283	372	462
V_{added} (cm^3)*	100	250	200	200	300	350	400	300	–	300	350	350
$V_{collected}$ (cm^3)*	48	29	91	21	71	96	79	75	111	114	100	90
pH	7.3	7.8	6.8	6.6	7.6	6.5	7.9	7.1	7.8	8.0	8.0	7.8
Eh (mV)	–	–	–	–	–	–	629	–	624	–	–	–
Ca (mg l^{-1})	36	660	530	610	645	626	650	890	630	600	610	120
Mg (mg l^{-1})	5	95	130	160	660	630	190	300	230	210	230	41
Na (mg l^{-1})	22	17	9	10	200	210	22	35	20	16	16	2.8
K (mg l^{-1})	15	14	8	10	16	22	13	27	17	<10	7	3
HCO$_3^-$ (mg l^{-1})	–	–	49	–	14	16	98	37	24	1	–	85
SO$_4^{2-}$ (mg l^{-1})	24	2 460	1 740	2 100	67	43	2 100	3 300	2 430	2 220	2 340	450
Al (μg l^{-1})	74	140	<30	<30	2 280	2 250	520	67	1 100	380	<30	<30
Fe (μg l^{-1})	130	37	57	31	130	300	100	61	4 200	420	12	26
Zn (μg l^{-1})	330	310	800	370	28	140	55	100	580	800	320	21
Cu (μg l^{-1})	23	30	20	15	1 100	900	34	46	87	60	71	7
Pb (μg l^{-1})	0.9	<0.1	<0.1	<0.1	35	50	<0.1	0.1	0.6	<0.1	1.5	0.7
As (μg l^{-1})	7	20	16	28	<0.1	<0.1	21	28	59	14	10	12

* V is volume, see equation (3).

the dry season and will transport the solutes to the run-off and groundwater.

Therefore, to describe the complete process of pyrite oxidation in the soil, the following processes have to be considered:

- flow of liquid due to a liquid pressure gradient;
- flow of gas due to gas pressure gradient (probably less important);
- transport of water in the liquid phase and in the gas phase (vapour). This also includes meteorological processes such as rain, evaporation and surface run-off;
- transport of gases both in the gas phase (diffusion) and in the liquid phase (less important);
- transport of heat both in the liquid and in the gas phase. This also includes meteorological processes (solar radiation and advective and convective heat transfer between the soil surface and the atmosphere);
- transport of solutes by advection, dispersion and diffusion in the liquid phase;
- chemical reactions such as aqueous speciation, oxidation–reduction, dissolution–precipitation and surface sorption.

Two types of models were developed: (1) transient flow and heat transport without solute transport and chemical reactions; and (2) steady-state flow and reactive transport. The objective of the transient flow and heat model was to study the evolution of the flow and heat parameters of the column experiments (flow, hydraulic saturation and temperature). These parameters were then used to establish the conditions for the steady-state reactive transport model. This procedure was a simplification of the real process and it was imposed by the present state of development of the codes. For all these calculations we built a new code that coupled a code for flow and heat, CODE-BRIGHT (Olivella *et al.* 1996), and a code for reactive transport, RETRASO (Saaltink *et al.* 1998).

Transient flow and heat transport model

CODE-BRIGHT models flow of water and gas and transport of heat. It calculates pressure of gas and liquid, hydraulic saturation and temperature from soil properties (porosity, retention curve, and relative and intrinsic permeability), and meteorological data (rainfall, atmospheric temperature, humidity, and hours of sun and wind velocity). Meteorological processes (rainfall, evaporation, solar radiation, heat exchange with the atmosphere, etc.) are simulated as boundary conditions. We adopted the same approach used by Carrera *et al.* (1992), which consisted of expressing the flow of liquid, gas and heat as a function of the state variables (temperature, and pressure of liquid and gas) and of the meteorological data. All equations are solved simultaneously by applying the Newton–Raphson iterative method. More information about the CODE-BRIGHT code can be obtained in Olivella *et al.* (1996).

In this study, the evolution of the columns with sand and sludge and clay and sludge during a period of outdoor exposure was modelled. Retention curves of the various materials (input for CODE-BRIGHT) were measured in the laboratory by means of a transistor psychrometer (Dimos 1991) and fitted to the Van Genuchten model (Van Genuchten 1978) (see equation 2; Fig. 3). Retention curves relate hydraulic saturation (i.e. the volume of water per volume of pores) and capillary pressure (i.e. pressure of gas minus pressure of liquid) for a given material:

$$S = \left(1 + \left(\frac{P_g - P_l}{P_0}\right)^{\frac{1}{1-\lambda}}\right) \quad (2)$$

where S is the hydraulic saturation, P_g and P_l are the pressure of gas and liquid, respectively, and P_0 and λ are retention curve parameters. The fine grains of the sludge (silt) make the saturation curve of the sandy soil close to that of the clayey soil. Another input parameter for CODE-BRIGHT is the intrinsic permeability. As no data on this parameter were available, different tentative values from literature were used. The calculated evolution of the hydraulic saturation was compared with the experimental moisture contents obtained by weighing the columns every 24 h after a leaching event. The best fits were

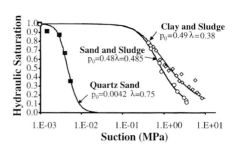

Fig. 3. Saturation curves for the mixture of clay and sludge, the mixture of sand and sludge and the quarsitic sand used in the modelling. The values of P_0 and λ are those that better fit the experimental values (dots, squares and rhombs) in equation (2) (see text).

obtained when intrinsic permeabilities of 10^{-17} and of 10^{-18} m² were used for the sandy and the clayey soil, respectively.

The model was run using small time steps (at most a few hours) in order to be able to see not only seasonal influence but also diurnal variations, leaching events and effects of individual rain showers. The models used a one-dimensional domain of 0.12 m. The first 0.08 m contained a mixture of sludge and sandy or clayey soil, and the remaining 0.04 m contained quartz sand only.

Figure 4 shows the results of the model for the sandy mixture. One can clearly observe the rise of hydraulic saturation (up to 1) after each leaching and rainfall event. With the exception of sample 9 (Tables 2 and 3), the infiltration of the rainfall was not enough to produce some leachate. As a consequence of the high saturation, evaporation rises. The high evaporation, in its turn, causes a drop in temperature. Furthermore, one can see the daily variation in saturation and, above all, temperature. Saturation is generally lower and shows more temporal variation at the surface than at a deeper level. Temperature, on the other hand, does not appear to vary with depth.

Figure 5 displays the results of the flux model of the column with the clayey mixture. The effects of rainfall and leaching events follow the same pattern as for the sandy column. However, the saturation is higher in the clayey column, both in depth and in surface. In this case, the column of clay achieves a saturation of nearly 0.4 only in the surface, while the saturation at the bottom of the column is nearly 1.

Reactive transport model

RETRASO simulates transport processes (advection, dispersion and diffusion) and chemical reactions (acid–base reactions, redox, complexation, adsorption, cation exchange, precipitation and dissolution of minerals). Chemical reactions can be assumed to follow either an equilibrium or a kinetic approach. The numerical solution is carried out using the global implicit or direct substitution approach. This means that all

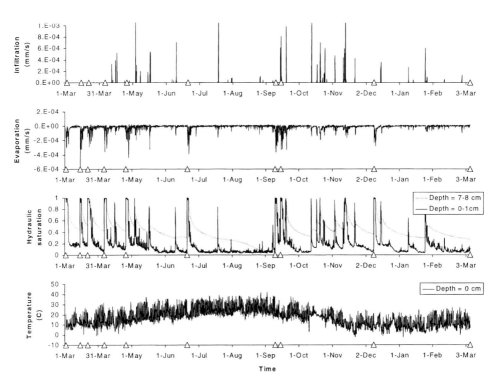

Fig. 4. Results of the flow and heat model for the sand–sludge mixture. Triangles indicate leaching events. Infiltration is defined as precipitation minus surface run-off. As flow entering the system is defined as positive, evaporation is commonly negative. Temperatures are given for one depth only because spatial variability is negligible, i.e. temperatures at different depth follow practically the same evolution.

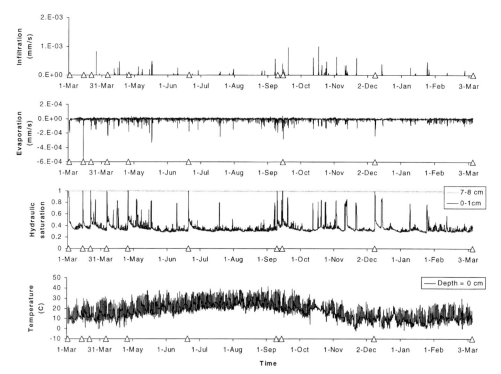

Fig. 5. Results of the flow and heat model for the clay–sludge mixture. Triangles indicate leaching events. Infiltration is defined as precipitation minus surface run-off. As flow entering the system is defined as positive, evaporation is commonly negative. Temperatures are given for one depth only because spatial variability is negligible, i.e. temperatures at different depth follow practically the same evolution.

equations are solved simultaneously by applying Newton–Raphson.

A simple time-lagged approach couples CODE-BRIGHT and RETRASO. This consists of first calculating pressures, temperatures and fluxes by means of CODE-BRIGHT and, second, of using the fluxes to calculate transport and chemical reactions by means of RETRASO. Examples of such parameters are porosity, which depends on precipitation or dissolution, and partial pressures in the gas phase depending on consumption of $O_2(g)$ or $CO_2(g)$. This simple method may be used when these parameters do not change significantly and/or small time steps are used, which is the case for our calculations.

The reactive transport model assumed that during the periods of outdoor exposure the flow of water could be neglected and that hydraulic saturation was constant in time, but variable in space. Hydraulic saturation was obtained by averaging the results from the flow and heat models (Fig. 6). This average is a simplification of the real process, which involves the evolution of the hydraulic saturation with time. The assumed average value tends to underestimate the saturation during the days after leaching. For the sandy mixture, except for the immediate hours after a leaching event, the hydraulic saturation was always below 0.6. For these values $O_2(g)$ is expected to diffuse freely in the gas phase. In the case of the clayey mixture, however, the hydraulic saturation was close to 1 in the deeper half of the column, suggesting that $O_2(g)$ did not easily diffuse.

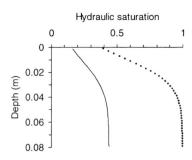

Fig. 6. Hydraulic saturation distribution averaged in time. The full line corresponds to the sandy soil and the dotted line to the clayey soil.

As stated above, the reactive transport model assumed no flux during the periods between leaching. Minerals reacted with the pore water and modified the concentration of solutes. The solution recovered after a rain event 5 days after leaching (sample 9 in Tables 2 and 3) showed concentrations higher than expected from the chemical reaction rates, calculated from kinetic rate laws (see the next section). This suggests than rather a pure advective displacement (piston model), mixing between the infiltrating and the pore water also takes place. By assuming total mixing, the total liquid concentration of a component in the leachate, c_L, can be calculated by:

$$C_L = \frac{c_W v_W + \sum_i v_i \phi S_i c_{P,i}}{v_W + \sum_i v_i \phi S_i} \quad (3)$$

where $\sum v_i \phi S_i c_{P,i}$ is the mass of solute and $\sum v_i \phi S_i$ the volume of water in the pores of the column just before a leaching event, where v_i the volume associated to node i, S_i its saturation and $c_{P,i}$ its total aqueous concentration as calculated by RETRASO. c_W is the concentration in the leaching solution (distilled water equilibrated with atmospheric CO_2) and v_W is the volume of the leaching water (v_{added} in Tables 2 and 3). The pH of the leachate can be calculated from c_L through a standard chemical speciation. For simplicity, we assumed that after each leaching event the chemical composition of the pore water returned to the same initial situation. This holds for the longer periods of outdoor exposure, because initial concentrations have a small influence but may give errors for the shorter periods (especially sample 9).

Geochemical model

Pyrite can be oxidized through two chemical reactions (Singer & Stumm 1970; Nicholson 1994; Nordstrom & Alpers 1999):

oxidation by oxygen:

$$FeS_{2(s)} + 3.5 O_{2(g)} + H_2O \rightarrow Fe^{2+}$$
$$+ 2 SO_4^{2-} + 2 H^+ \quad (R1)$$

and oxidation by Fe^{3+}:

$$FeS_{2(s)} + 14 Fe^{3+} + 8 H_2O \rightarrow 15 Fe^{2+}$$
$$+ 2 SO_4^{2-} + 16 H^+. \quad (R2)$$

Oxidation of Fe^{2+} by O_2 has to precede the last reaction:

$$Fe^{2+} + 0.25 O_{2(g)} + H^+ \rightarrow Fe^{3+} 0.5 H_2O. \quad (R3)$$

It is believed that the oxidation of pyrite by O_2 (reaction R1) is slower than the oxidation by Fe^{3+} (i.e., reactions R2 plus R3). Furthermore, the slowest reaction of the last two is (R3) and, therefore, is the rate-determining step. Microorganisms (*Thiobacillus ferrooxidans* amongst others) accelerate this reaction. These microorganisms tolerate high metal concentrations and are acidophiles, with an optimum growth at pH lower than 3. At higher pH bacteria do not significantly accelerate the reaction over and above the abiotic rate.

Besides pyrite, the sludge contained small amount of other sulphides. The dissolution rate of pyrite, sphalerite, galena and chalcopyrite was determined by Domènech et al. (2001b). Moreover, the dependence of the pyrite dissolution rate with pH and dissolved O_2 concentration was established. The same dependence has been assumed for the minor sulphides in the calculations (Table 4). Domènech et al. (2001b) interpreted that the slowest steps of the oxidation process of the sulphides correspond to the electron transfer from S^{2-} to SO_4^{2-}. This process is equal for all the sulphides.

The acid conditions generated by the oxidation of pyrite promoted the dissolution of the silicates in the soil. In the case of the clayey soil, they also promoted dolomite dissolution. Dissolution consumed protons and acted as a pH buffer of the soil pore water. Moreover, dissolution and evaporation caused the increase in ion concentrations and several new solid phases precipitated. The most abundant minerals identified in the sand column after more than 400 days of functioning are gypsum and jarosite. There were identified by XRD and scanning electron microscopy with microanalysis. From mass-balance calculations in the leachates and from the results of a sequential extraction and analysis of the solid, Domènech et al. (2001a) concluded that Cu coprecipitates with ferrihydrite in the initial stages of the experiment as $(Fe_{0.993}Cu_{0.007})(OH)_3$. Likewise, As and Pb coprecipitates as trace elements in the jarosite, which could have an average stoichiometry of $(Na,K,H,Pb_{0.018})Fe[(S_{0.984}As_{0.016})O_4]_2(OH)_6$. As no thermodynamic data are available for such a solid, the model included the Na, K and H pure phases, all of them with the same trace element proportion.

Table 4 lists the aqueous species, minerals and gases required for describing the geochemical

Table 4. *Aqueous species gases and minerals used in the calculations**

Aqueous primary species
Fe^{2+}, Fe^{3+}, SO_4^{2-}, Al^{3+}, H^+, $CO_2(aq)$, $SiO_2(aq)$, Na^+, K^+, Ca^{2+}, Mg^{2+}, $O_2(aq)$, $Zn^{2+\dagger}$, $Pb^{2+\dagger}$, Cu^{2+}, $H_2AsO_4^{-\dagger}$

Other aqueous species (equilibrium)
HSO_4^-, $AlSO_4^+$, $Al(SO_4)_2^-$, $Al(OH)_3(aq)$, $Al(OH)_2^+$, $FeSO_4^+$, $Fe(OH)_3(aq)$, $Fe(OH)_2^+$, HS^-, $H_2S(aq)$, HCO_3^-, CO_3^{2-}, OH^-

Gases (equilibrium)
$O_2(g)$, $CO_2(g)$

Precipitating minerals (equilibrium)

gypsum	Na-jarosite $(Na, Pb_{0.018})Fe_3[(S_{0.984}As_{0.016})O_4]_2(OH)_6$
$SiO_2(am)$	jarosite $(K, Pb_{0.018})Fe_3[(S_{0.984}As_{0.016})O_4]_2(OH)_6$
$Al(OH)_3(am)$	H-jarosite $(H, Pb_{0.018})Fe_3[(S_{0.984}As_{0.016})O_4]_2(OH)_6$
ferrihydrite	Dolomite†
$(Fe_{0.993}Cu_{0.007})(OH)_3$	

Dissolving minerals (kinetics)

Mineral	Kinetic law	Reference‡
Pyrite (R1)	$r = 7.9 \times 10^{-8} [O_2(aq)]^{0.49} a_H^{0.1} \sigma_{Py} (\Omega_{Py}-1)$	1
Pyrite (R2)	$r = 6.6 \times 10^{-9} a_{Fe(II)}^{-0.4} a_{Fe(III)}^{0.93} \sigma_{Py} (\Omega_{Py}-1)$	2
Fe^{2+}/Fe^{3+} (R3)	$r = 1.0 \times 10^{20} a_{Fe(II)} a_{O2(aq)} a_{OH}^2 + 1.3 \times 10^{-1} a_{Fe(II)} a_{O_{2(aq)}}$	3
Sphalerite	$r = 2.4 \times 10^{-7} [O_2(aq)]^{0.49} a_H^{0.1} \sigma_{Sph} (\Omega_{Sph} - 1)$	1
Galena	$r = 1.7 \times 10^{-7} [O_2(aq)]^{0.49} a_H^{0.1} \sigma_{Gn} (\Omega_{Gn} - 1)$	1
Chalcopyrite	$r = 4.9 \times 10^{-8} [O_2(aq)]^{0.49} a_H^{0.1} \sigma_{Cpy} (\Omega_{Cpy} - 1)$	1
Illite	$r = (4.2 \times 10^{-12} a_H^{0.38} + 1.5 \times 10^{-13} a_H^{0.09} + 1.1 \times 10^{-15} a_H^{-0.22}) \sigma_{ill} (\Omega_{ill} - 1)$	4
K-Feldspar	$r = (1.0 \times 10^{-10} a_H^{0.5} + 2.5 \times 10^{-17} a_H^{-0.45}) \sigma_{Fd} (\Omega_{Fd} - 1)$	5
Albite	$r = (7.5 \times 10^{-10} a_H^{0.5} + 1.2 \times 10^{-14} a_H^{-0.3}) \sigma_{Pl} (\Omega_{Pl} - 1)$	6
$Al(OH)_3(am)$	$r = 1.0 \times 10^{-11} a_H^{0.33} \sigma_{Gib} (\Omega_{Gib} - 1)$	7
Hornblende	$r = 1.8 \times 10^{-12} \sigma_{Horn} (\Omega_{Horn} - 1)$	8
Clinochlore	$r = (4.2 \times 10^{-12} a_H^{0.38} + 1.5 \times 10^{-13} a_H^{0.09} + 1.1 \times 10^{-15} a_H^{-0.22}) \sigma_{Cli} (\Omega_{Cli} - 1)$	4

* The equilibrium constants of the aqueous speciation, gas and mineral dissolution are those of the EQ3 database (Wolery 1992). Kinetic laws: r is the rate (mol m^{-2} s^{-1} for minerals and mol dm^{-3} s^{-1} for reaction R3), σ the reactive surface (m^2 m$_{soil}^{-3}$) and Ω the saturation index.
† Only for the model of clayey soil with sludge.
‡ References: (1) Domènech et al. 2001b; (2) Nicholson 1994; (3) Singer & Stumm 1970; (4) assumed equal to muscovite, Wieland & Stumm 1992; (5) Schweda 1989; (6) Chou & Wollast 1985; (7) assumed equal to gibbsite, Mogollón et al. 1996; (8) Swoboda-Colberg & Drever 1993; (9) Busenberg & Plummer 1982. The rate laws for reaction (R2) and (R3) have been changed as explained in the text.

model with a reasonable degree of completeness. Reactions between aqueous species were assumed to be in equilibrium except for the oxidation of Fe^{2+} to Fe^{3+} (R3 above mentioned), whose kinetic behaviour will be discussed below. The gas species and minerals precipitating were also assumed in equilibrium. The silicate minerals that dissolved, however, were assumed to behave according to kinetic laws extracted from experimental studies. All the laws (except that for R3) are multiplied by the reactive surface of the mineral and by a term dependent on chemical saturation, which accounts for a decrease in the rate as equilibrium is approached (Lasaga 1984). The reactive surface area of each mineral (σ_m m^2 m^{-3} soil) was roughly estimated on a geometrical basis. A spherical shape, homogeneous grain size and equal density for all the components (sulphides or silicates) were assumed:

$$\sigma_m = \frac{6(1-\phi)\omega_m}{d_m} \quad (4)$$

where ϕ is the porosity of the material and d_m is the grain size (m). An average grain size of 10^{-5} m was assumed for the sludge and the clayey soil, and 10^{-3} m for the sandy soil and quartz sand.

Calculations were performed at a constant temperature of 25°C. This temperature was slightly higher than the average of the temperatures predicted by the heat model (Figs 4 and 5). This led to an overestimation of about 10% of the value of the rate constants, according to the activation energies of the reactions involved, the average of which is between 60 and 80 kJ mol^{-1} (Lasaga 1984; Nicholson 1994; Nemati & Webb 1997).

Results and discussion

Sandy columns. Figure 7 shows the experimental and calculated values for the column with sandy soil and sludge. It can be seen that the modelled SO_4 and Fe concentrations rise and that pH drops due to the pyrite oxidation.

Although the model reproduces well the pH in the last periods of the experiment, the calculated pH values are much lower than the experimental data for the first 3 months. As the pH is underestimated, the modelled concentrations of Fe are higher than the experimental values because the precipitation of iron hydroxide is not predicted by the model. This disagreement is attributed to the fact that the model assumes a constant hydraulic saturation over time (Fig. 6). This saturation is not representative for the first stages of the experiment, where the most frequent watering made the soil more humid (Fig. 4). Wet conditions hindered the diffusion of O_2 into

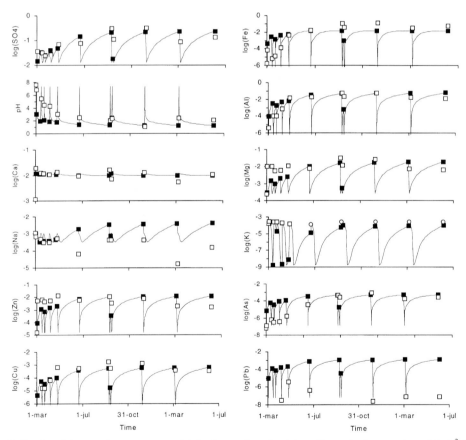

Fig. 7. Results of the reactive transport model for the sandy soil mixture. Concentrations are in mol dm^{-3}. Full squares represent the predicted concentrations in the leachate. Open squares represent the analytical values.

the column and, therefore, reduced the oxidation of pyrite. As a consequence, experimental pH decreased more slowly.

The low pH values of both model and experiment cause the dissolution of silicates and, consequently, the release of Al, Ca, Mg, Na and K. Similarly to Fe, the low pH values predicted by the model at the beginning of the experiment prevented the formation of the amorphous Al(OH)$_3$, and the calculated Al concentrations are overestimated. The modelled concentration of Ca is constant with time and agrees with the experimental values. This is attributed to the dissolution and later precipitation of gypsum, thus controlling the amount of Ca in solution. The modelled concentration of Mg matches the order of magnitude of the experimental values, except at the early stages. No explanation for the high Mg contents at the early stages is given.

Although the dissolution of silicates releases Na and K to the pore water, the amount of Na and K in solution did not increase due to the formation of jarosite. In the case of K, the predicted concentrations decreased, in accordance with the experimental values that were below the detection limit (c. 2×10^{-4} mol dm^{-3}). This indicates that Na, and especially K, control the formation of jarosite. The low K concentrations predicted for the early stages are due to the precipitation of jarosite favoured by the low pH of the calculations. However, the predicted concentration of Na did not decrease with time, and was higher than the experimental values. This disagreement is attributed to the use of thermodynamic values of pure-jarosite terms, whereas microscopic observation and microanalysis confirm the presence of Na- and K-bearing jarosites (Domènech et al. 2001a).

The concentration of Zn predicted by the model is similar to the experimental values, confirming that this metal is not retained in any solid phase. As for Mg, no explanation was found for the high Zn concentrations in the early samples. The concentrations of Cu and As predicted by the model are similar to the experimental values. This is consistent with the hypothesis that jarosite specifically retains As. Similarly to Fe, the overestimation of As concentration at the early stages can be explained by the absence of amorphous Fe(OH)$_3$ precipitation in the model, caused by the low modelled pH values. When formed, the Fe(OH)$_3$ is expected to retain As, either adsorbed and/or coprecipitated. The predicted concentrations of Pb are orders of magnitude higher than the analyses. This indicates that Pb is not coprecipitated as a trace element in the jarosite. Indeed, although its decrease seems to be related to jarosite (Domènech et al. 2001a), Pb was not detected in microanalysis of this phase. An alternative explanation could be the formation of an independent solid phase with a lower solubility product. Plumbojarosite has been described in the sulphide weathering environments (Scott 1987), but, unfortunately, no thermodynamic data are available.

It can be seen from Fig. 7 that the experimental data for Zn, SO$_4^{2-}$, Fe and Cu show a decrease in the last part of the experiment, which is not observed for the modelled values. This could be due to the exhaustion of the sulphides. This reduces the reactive surface (σ_m) and, consequently, reduces the dissolution rate according to the rate laws of Table 4, but is not represented well by the model.

Fe^{3+} was found both in the liquid and the precipitated form (jarosite). In order to reproduce this in our model we had to reduce the rate law constant of the oxidation of pyrite by Fe^{3+} (reaction R2) by a factor of 5000 and to increase that for oxidation of Fe^{2+} (reaction R3) by five orders of magnitude. Figure 8 shows the mismatch in the Fe^{2+}/Fe^{3+} ratio if rate laws from the literature are used. The difference between the calibrated rate law and the rate law from the literature for reaction (R3) can be explained by the acceleration of this reaction by microorganisms (Nordstrom & Alpers 1999). The difference of the rate law for reaction (R2) is more surprising. Moreover, with these calibrated rate laws, reaction (R2) is approximately as reactive as reaction (R3). This does not agree with the generally accepted theory that the oxidation of Fe^{2+} by O$_2$ (reaction R3) is the rate-determining step (Singer & Stumm 1970). According to this theory, reaction (R2) would consume all available Fe^{3+} leaving almost all Fe as Fe^{2+}, even with O$_2$ concentrations in equilibrium with the atmosphere. Our model predicts atmospheric O$_2$ concentrations (see below). As we did not carry out O$_2$ measurements, the model cannot verify whether atmospheric conditions prevail in reality. Nevertheless, if O$_2$ concentrations were lower, the rate law constants for reaction (R2) and (R3) would have to be, respectively, reduced and increased even more. An issue that we have not studied but that may affect the relation between Fe^{3+} and Fe^{2+} concentrations is the heterogeneity. Zones of Fe^{2+} near pyrite grains could exist along with zones of Fe^{3+} close to other grains. In that case, the rate-determining step would be the transport of solutes between these two zones.

Figure 9 shows profiles of various modelled concentrations and reaction rates for the sand–sludge column. The variation of the different

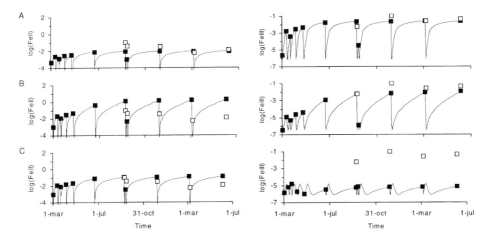

Fig. 8. Evolution of Fe^{2+} and Fe^{3+} concentrations for different rate laws. Full squares represent the predicted concentrations in the leachate. Open squares represent the analytical values.
(A) $r_2 = 6.6 \times 10^{-9} a_{Fe(II)}^{-0.4} a_{Fe(III)}^{0.93} \sigma_{Py} (\Omega_{Py} - 1)$ and $r_3 = 1.0 \times 10^{20} a_{Fe(II)} a_{O2(aq)} a_{OH}^2 + 1.3 \times 10^{-1} a_{Fe(II)} a_{O_2(aq)}$.
(B) $r_2 = 3.3 \times 10^{-5} a_{Fe(II)}^{-0.4} a_{Fe(III)}^{0.93} \sigma_{Py} (\Omega_{Py} - 1)$ and $r_3 = 1.0 \times 10^{20} a_{Fe(II)} a_{O2(aq)} a_{OH}^2 + 1.3 \times 10^{-1} a_{Fe(II)} a_{O_2(aq)}$.
(C) $r_2 = 3.3 \times 10^{-9} a_{Fe(II)}^{-0.4} a_{Fe(III)}^{0.93} \sigma_{Py} (\Omega_{Py} - 1)$ and $r_3 = 1.0 \times 10^{15} a_{Fe(II)} a_{O2(aq)} a_{OH}^2 + 1.3 \times 10^{-6} a_{Fe(II)} a_{O_2(aq)}$.

parameters in depth is, in general, small. Owing to the low hydraulic saturation of the model, O_2 and CO_2 can easily diffuse, so that atmospheric conditions prevail within the entire column and the concentration of $O_2(g)$ is high and that of $CO_2(g)$ is small. Observe that reaction (R1) is somewhat faster than reaction (R2). Furthermore, jarosite is more important as a secondary mineral than gypsum or ferrihydrite, the last not precipitating at all.

Clayey columns. Figure 10 shows the results of the model for the column with clayey soil and sludge. The only difference with the previous model is the hydraulic saturation used and the addition of dolomite to the set of minerals modelled. The high SO_4 concentrations indicate that oxidation of sulphides takes place, because no other sources (e.g. gypsum) were identified (Table 1), and blank experiments consisting of columns without sludge gave much lower sulphate concentrations in the leachates. As expected from the higher hydraulic saturation, the sulphate concentrations are lower than in the sandy soil. The dissolution of dolomite maintains a high pH, which, together with the elevated hydraulic saturation, causes a lower oxidation of pyrite compared to that observed in the column of sand and sludge. This high pH also promotes the oxidation of Fe^{2+} to Fe^{3+}, and the precipitation of amorphous $Fe(OH)_3$, which controls the amount of Fe^{3+} in solution. The precipitation of $Fe(OH)_3$ is also favoured by the rate of (R3) that is high enough to maintain the Fe^{2+} concentrations at very low values. At neutral pH, the dissolution of silicates is slow leading to low Al concentrations. As no jarosite precipitates, both K and Na remain constant throughout the experiment. Here again, the amount of Ca in solution is controlled by the precipitation of gypsum, regardless of an extra source of calcium (dolomite). In general, there is a good agreement between measured and calculated concentrations except for pH, as discussed below.

Contrary to the sandy column, the concentrations and reaction rates for the clay–sludge mixture do vary in depth (Fig. 9). The reason of this variation is the distribution of the hydraulic saturation, which reaches values near 1 at the lower half of the column. Under this wet condition, O_2 cannot easily diffuse downwards and CO_2, emanating from dolomite dissolution, cannot diffuse upwards. When the pore water is pulled out of the column during leaching, $CO_2(g)$ is released to the atmosphere (having a partial pressure of 0.0003 bar), producing an increase of the pH of water. Therefore, the measured pH was higher than the pH calculated for the pore water inside the column (Fig. 10). Observe that, within the dry half of the column, the reaction rate of (R1) is several orders of magnitude higher than the rate of (R2). Moreover, the reaction rates of (R1) and (R3) and the precipitation of ferrihydrite are almost identical. This means that all Fe^{2+} released from pyrite oxidizes to Fe^{3+} and

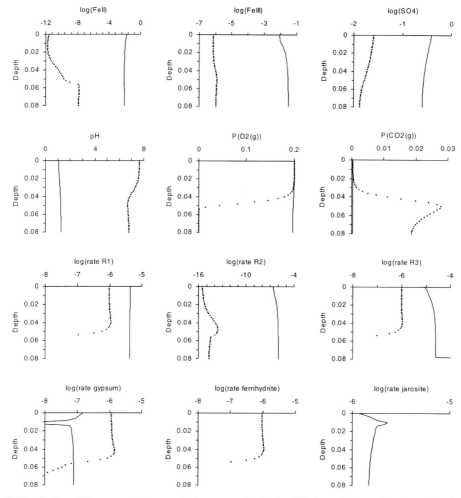

Fig. 9. Distribution of the concentrations and rates versus depth after 100 days of the sandy column (continuous lines) and the clayey column (dotted lines). No ferrihydrite precipitates in the sandy column and no jarosite precipitates in the clayey column. Rates are in mol m$_{column}^{-3}$ s^{-1}. Concentrations are in bar for gasses and in mol dm^{-3} for solutes.

precipitates as ferrihydrite. Likewise, SO$_4$ precipitates as gypsum.

Experimental results showed that the amounts of As, Cu, Zn and Pb released to the solution in the clayey column were low compared to those of the sandy column (Tables 2 and 3). This fact agrees with Voight et al. (1996), Roussel et al. (1999) and Savage et al. (2000), who suggested that those elements, especially As, can be retained by adsorption onto ferrihydrite or clay minerals. Tentative calculations using data of Dzombak & Morel (1990), Du et al. (1997) and Manning & Goldberg (1997) confirmed that As could be adsorbed onto Fe(OH)$_3$(a), and As and Cu onto illite, sufficiently to produce the observed leachate concentration. No explanation for Zn and Pb retention was found. Contrary to the sandy column, no experimental data are available on the solid phase after the experiment was completed. Therefore, we did not include these trace metals in the model for the clayey column.

Predictions. According to the experimental and modelling results described above, the immediate risk of water pollution could take place in sandy and gravel soils. Once the reactive transport model was able to reproduce the experimental results of the sandy soil column, it was used to predict the behaviour of the soil in two more realistic scenarios: a sandy soil containing a proportion of 1 wt% sludge and a gravel

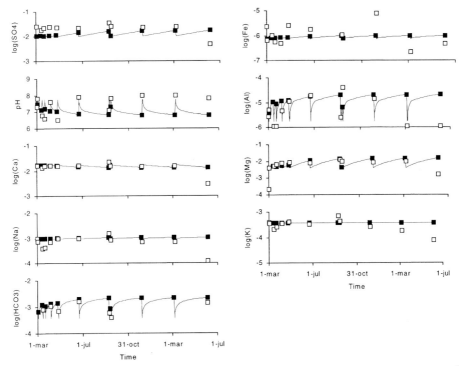

Fig. 10. Results of the reactive transport model for the clayey soil mixture. Concentrations are in mol dm^{-3}. Full squares represent the predicted concentrations in the leachate. Open squares represent the analytical values.

containing 10 wt% sludge. We will discuss the predictive results for Zn and As, the main pollutants of the system. The results for a period of 100 days without episodes of rain are represented in Fig. 11.

In the first case, as expected for a lower sulphide proportion in the soil, pH is higher, and is maintained close to 4 due to the buffering role of silicate dissolution. As less sulphide is dissolved, lower Zn and As concentrations are observed. Nevertheless, the concentrations of Zn and As reached in this scenario are also far above most statutory limits (10^{-5} and 10^{-6} mol l^{-1} for Zn and As, respectively).

In the second case, the gravel was calculated assuming a silicate reactive surface ten times lower than that for a sandy soil. Surprisingly, in the case of gravel, the pH does not decrease indefinitely but falls to 1.6, practically the same as for the sandy soil. This is due to the fact that at such low pH, jarosite precipitation controlled the pH of the solution, rather than silicate dissol-

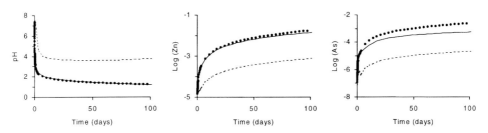

Fig. 11. Evolution of pH and the main contaminants for various scenarios of sludge–soil proportions. A period of 100 days with no rain was assumed. Full line: 10 wt% of sludge in a sandy soil, as described in the experiments. Spaced line: 1 wt% of sludge in a sandy soil. Dotted line: 10 wt% of sludge in a gravel (silicate reactive surface assumed 10-fold lower than the sandy soil). Concentrations are in mol dm^{-3}.

ution. In the case of gravel, however, less silicate dissolves, and less K and Na are released to the pore water. As these two solutes are limiting for jarosite to form, less mineral is formed. As less jarosite is formed, less As is coprecipitated and the concentrations in solution are higher than for the sandy soil (Fig. 11). On the other hand, Zn is not retained in jarosite and the predicted concentrations in gravel are similar to those for the sandy soil.

The results of the calculations are consistent with the field observations. Jarosite is the most abundant secondary mineral present in the weathered soils. The contaminated groundwater below these soils show pH values between 2 and 4, but no pH values lower than 1.9 have been measured. The Zn and As concentrations of contaminated waters range between 10^{-4} and 2×10^{-3}, and between 10^{-6} and 3×10^{-4} mol l^{-1}, respectively (Manzano et al. 1999), the same order of magnitude predicted by the model.

Conclusions

The flow model permitted a detailed description of the behaviour of the soil at a daily scale. The most important parameter extracted from the flow model and used in the reactive transport was the hydraulic saturation. This parameter controlled the amount of O_2 that could diffuse into the soil, which, in its turn, affected the rate of pyrite oxidation. The model assumed the hydraulic saturation to be constant in time. This has led to overestimation of the sulphide oxidation at the early stages of the experiment.

The results of the reactive transport model simulated the experimental data well, except during the first days of outdoor exposure, during which time hydraulic saturation was underestimated. Nevertheless, for predictive purposes, we feel that it is more important that the model simulation was good in the long term.

Although hydraulic saturation was high in the deep part of the clayey column, oxidation of pyrite took place in both columns, as shown by the high SO_4 concentrations of the leachates. We found, however, very different behaviour between the sandy and clayey columns. Rather than the saturation, it was the existence of dolomite or another carbonate minerals that caused the differences between the two columns. In the absence of carbonates, such as in the sandy column, pH dropped due to the oxidation of pyrite. As a result, silicate minerals dissolved, providing Na and/or K which precipitated together with Fe and SO_4 to jarosite. The high concentration of Zn in the leachates was consistent with that predicted from the sphalerite oxidation. The low As and Pb content of the leachates could be explained by the coprecipitation with jarosite. The presence of carbonates, such as in the clayey column, kept the pH high, impeding the dissolution of silicate minerals and precipitating amorphous $Fe(OH)_3$ instead of jarosite. The low concentration of trace metals in the leachate of the clayey column was not modelled because information on the solid phase subsequent to the experiment was not available.

The model could be used to predict the behaviour of other soil types and other sludge contents. According to the predictions, the precipitation of jarosite was very important to maintain the pH at a value of approximately 1.6 when silicate dissolution is not able to buffer a high input of acidity. It is interesting to note the non-lineal behaviour of the system. Thus, according to the calculations, the pH is not sensitive to the type of soil, sand or gravel, but the amount of pyrite. The amount of As in solution is predicted to vary with the extent of silicate dissolution, and the subsequent Na and K release. On the other hand, Zn is predicted not to depend on silicates, but only on the initial amount of sludge in the soil.

The model also reviewed rate laws proposed for sulphide oxidation and found that the oxidation of pyrite by Fe^{3+} was not faster than by O_2, contrary to earlier descriptions in the literature.

This work was supported by the contracts PIRAMID EVK1-1999-00097 of the European Union and REN2000-1003-C03-02/HID of the Spanish Government.

References

ALASTUEY, A., GARCÍA-SÁNCHEZ, A., LÓPEZ, F. & QUEROL, X. 1999. Evolution of pyrite mud weathering and mobility of heavy metals in the Guadiamar valley after the Aznalcóllar spill, southwest Spain. *Science of the Total Environment*, **242**, 41–55.

ANTÓN-PACHECO, C., JIMÉNEZ, M., GÓMEZ, J.A., GUMIEL, J.C., LÓPEZ-PAMO, E., DE MIGUEL, E., ORTIZ, G. & REJAS, J.G. 2001. Cartografía del lodo pirítico remanente en la cuenca del río Guadiamar mediante imágenes Daedalus-1268 ATM. *Boletín Geológico y Minero*, **112**, 107–114.

BAIN, J.G., BLOWES, D.W., ROBERTSON, W.D. & FRIND, E.O. 2000. Modelling of sulfide oxidation with reactive transport at a mine drainage site. *Journal of Contaminant Hydrology*, **41**, 23–47.

BUSENBERG, E. & PLUMMER, L.N. 1982. The kinetics of dissolution of dolomite in CO_2-H_2O systems at 1.5 to 65°C and 0 to 1 atm P_{CO_2}. *American Journal of Science*, **282**, 45–78.

CARRERA, J., ALFAGEME, H., GALARZA, G. & MEDINA, A. *Estudio de la infiltración a través de la cobertera de la F.U.A.* ENRESA (Publicación Técnica, núm. **02/92**.

CHOU, L. & WOLLAST, R. 1985. Steady state kinetics and dissolution mechanisms of albite. *American Journal of Science*, **285**, 963–993.

DIMOS, A. 1991. *Measurement of Soil Suction Using Transistor Psychrometer*.VIC ROADS, Internal Report IR/91–93.

DOMÈNECH, C., AYORA, C. & DE PABLO, J. 2001*a*. Sludge weathering and mobility of contaminants in soil affected by the Aznalcollar mining accident (SW Spain). *Chemical Geology*; accepted.

DOMÈNECH, C., DE PABLO, J. & AYORA, C. 2001*b*. Oxidative dissolution of the pyritic sludge from Aznalcóllar mine (SW Spain). *Chemical Geology*, accepted.

DU, Q., SUN, Z., FORSLING, W. & TANG, H. 1997. Adsorption of copper at aqueous illite surfaces. *Journal of Colloid and Interface Science*, **187**, 232–242.

DZOMBAK, D.A. & MOREL, F.M.M. 1990. *Surface Complexation Modeling: Hydrous Ferric Oxide*, John Wiley, New York.

GERKE, H.H., MOLSON, J.W. & FRIND, E.O. 1998. Modelling the effect of chemical heterogeneity on acidification and solute leaching in overburden mine spills. *Journal of Hydrology*, **209**, 166–185.

LASAGA, A.C. 1984. Chemical kinetics of water-rock interactions. *Journal of Geophysical Research*, **89**, 4009–4025.

LÓPEZ-PAMO, E., BARETTINO, D., ANTÓN-PACHECO, C., ORTIZ, G., ARRÁNZ, J.C., GUMIAL, J.C., MARTÍNEZ-PLEDEL, B., APARICIO, M. & MONTOUTO, O. 1999. The extent of the Aznalcóllar pyritic sludge spill and its effect on soils. *Science of the Total Environment*, **242**, 57–88.

MANNING, B.A. & GOLDBERG, S. 1997. Adsorption and stability of Arsenic (III) at the clay mineral–water interface. *Environmental Science & Technology*, **31**, 2005–2011.

MANZANO, M., AYORA, C., DOMÈNECH, C., NAVARRETE, P., GARRALÓN, A. & TURRERO, M.J. 1999. The impact of Aznalcóllar mine tailing spill on groundwater. *Science of Total Environment*, **242**, 189–209.

MAYER, K. U., BENNER, S. G. & BLOWES, D. W. 1999. The reactive transport model MIN3P: Application to acid mine drainage generation and treatment – Nickel Rim Mine site, Sudbury, Ontario. In: *Proceedings of Sudbury '99, Mining and the Environment* Aurentian Univesity Press, Sudbury, Ontario, Canada, 145–154.

MAYER, K. U., BLOWES, D. W. & FRIEND, E. O. 2000. Numerical modelling of acid mine drainage generation and subsequent reactive transport. In: *ICARD 2000, Fifth International Conference of Acid Rock Drainage, Denver, Colorado, USA*, 1–7.

MOGOLLÓN, J.L., GANOR, J., SOLER, J.M. & LASAGA, A.C. 1996. Column experiments and the full dissolution rate law of gibbsite. *American Journal of Science*, **296**, 729–765.

NEMATI, M. & WEBB, C. 1997. A kinetic model for biological oxidation of ferrous iron by *Thiobacillus ferrooxidans*. *Biotechnology and Bioengineering*, **53**, 478–486.

NICHOLSON, R.V. 1994. Iron-sulfide oxidation mechanisms: Laboratory studies. JAMBOR, J.L. & BLOWES, D.W. (eds) *Short Course on Environmental Geochemistry of Sulfide Mine-wastes*; Mineral Association of Canada, Water loo, Ontario, 163–183.

NORDSTROM, D.K. & ALPERS, C.N. 1999. Geochemistry of acid mine waters. In: PLUMLEE, G.S. & LOGSDON, M.S. (eds) *The Environmental Geochemistry of Mineral Deposits*; Reviews in Economic Geology, Society of Economic Geologist, Littletan, Colorado, vol **6A**, 133–160.

OLIVELLA, S., GENS, A., CARRERA, J. & ALONSO, E.E. 1996. Numerical formulation for a simulator (CODE-BRIGHT) for the coupled analysis of saline media. *Engineering Computations*, **13**, 87–112.

QUEROL, X., ALASTUEY, A., LÓPEZ-SOLER, A., MANTILLA, E. & PLANA, F. 1996. Min eralogy of atmospheric particulates around a large coal-fired power station. *Atmospheric Environment*, **30**, 3557–3572.

ROUSSEL, C., BRIL, H. & FERNANDEZ, A. 1999. Evolution of sulphides-rich mine tailings and immobilization of As and Fe. *Compete rendu de l'Académie de Sciences Paris, Earth and Planetary Sciences*, **329**, 787–794.

SAALTINK, M.W., AYORA, C. & CARRERA, J. 1998. A mathematical formulation for reactive transport that eliminates mineral concentrations. *Water Resources Research*, **34**, 1649–1656.

SAVAGE, K.S., TINGLE, T.N., O'DAY, P.A., WAYCHUNAS, G.A. & BIRD, D.K. 2000. Arsenic speciation in pyrite and secondary weathering phases, Mother Lode Gold District, Tuolumne County, California. *Applied Geochemistry*, **15**, 1219–1244.

SCHWEDA, P.S. 1989. Kinetics of alkali feldspar dissolution at low temperature. In: *Proceeding of the Sixth International of Water Rock Interaction*; Balkema, Rotterdam, 609–612.

SCOTT, K.M. 1987. Solid solution in, and classification of gossan-derived members of the alunite–jarosite family, northwest Queensland, Australia. *American Mineralogist*, **72**, 178–187.

SINGER, P.C. & STUMM, W. 1970. Acidic mine drainage: The determining step. *Science*, **167**, 1121–1123.

SWOBODA-COLBERG, N.G. & DREVER, J.I. 1993. Mineral dissolution rates in plot-scale field laboratory experiments. *Chemical Geology*, **105**, 51–69.

THOREZ, J. 1976. *Practical Identification of Clay Minerals*. G. Lelotte, Dison, Belgium.

TO, T.B., NORDSTROM, D.K., CUNNINGHAM, K.M., BALL, J.W. & MCCLESKEY, R.B. 1999. New method for the direct determination of dissolved Fe(III) concentration in acid mine waters. *Environmental Science & Technology*, **33**, 807–813.

VAN GENUCHTEN, R. 1978. *Calculating the Unsaturated Hydraulic Conductivity with a New Closed-form Analytical Model* Water Resource Program, Department of Civil Engineering Princeton University, Research Report, **78-WR-08**.

VOIGHT, D.E., BRANTLEY, S.L. & HENNET, R.J.C. 1996. Chemical fixation of arsenic in contaminated soils. *Applied Geochemistry*, **11**, 633–643.

WALTER, A.L., FRIND, E.O., BLOWES, D.W., PTACEK, C.J. & MOLSON, J.W. 1994. Modeling of multicomponent reactive transport in groundwater. 1. Model development and evaluation. *Water Resources Research*, **30**, 3137–3148.

WALTER, A.L., FRIND, E.O., BLOWES, D.W., PTACEK, C.J. & MOLSON, J.W. 1994b. Modeling of multicomponent reactive transport in groundwater. 2. Metal mobility in aquifers impacted by acidic mine tailings discharge. *Water Resources Research*, **30**, 3149–3158.

WIELAND, E. & STUMM, W. 1992. Dissolution kinetics of kaolinite in acidic solutions at 25°C. *Geochimica et Cosmochimica Acta*, **56**, 3339–3355.

WOLERY, T.J. 1992. *EQ3NR. A Computer Program for geochemical aqueous speciation-solubility calculations. Theoretical Manual, User's Guide and Related Documentation*. Lawrence Livermore National Laboratory, Livermore, California, UCRL-MA-110662 PT IV.

WUNDERLY, M.D., BLOWES, D.W., FRIND, E.O. & PTACEK, C.J. 1996. Sulfide mineral oxidation and subsequent reactive transport of oxidation products in mine tailing impoundments: a numerical model. *Water Resources Research*, **32**, 3173–3187.

Laboratory and numerical modelling studies of iron release from a spoil heap in County Durham

CATHERINE J. GANDY[1] & KATY A. EVANS[2]

[1]*Hydrogeochemical Engineering Research & Outreach (HERO), Department of Civil Engineering, University of Newcastle, Newscastle upon Tyne NEI 7RU, UK (e-mail: c.j.gandy@ncl.ac.uk)*

[2]*Groundwater Protection and Restoration Group, Department of Civil and Structural Engineering, University of Sheffield, Sheffield S1 3JD, UK (e-mail: k.evans@sheffield.ac.uk)*

Abstract: At present there is no suitable method to predict either the longevity of contaminant sources within spoil heaps, or the evolution of their strength over the contaminating life time of the sites. Existing acid–base accounting techniques provide little information relevant to the prediction of field contaminant concentrations and time scales. For robust prediction, the relative rates of contaminant generation and attenuation must be evaluated then extrapolated to the physical scale and environmental conditions of real field sites. Laboratory unsaturated column experiments on colliery spoil from a well-documented site in County Durham have been set up to assess its contamination potential, and the results compared with results from a mathematical model for contaminant release and transport. The random walk method, a form of particle tracking, is used to transport iron and sulphate 'particles', released by oxidative weathering of pyrite minerals. The model also includes the oxidation of ferrous iron to ferric iron in an attempt to account for contaminant 'sinks', for example where ferric iron spontaneously precipitates as ferric oxyhydroxide and is effectively removed from the transport process. In general, the modelled results compare favourably with the laboratory results and any discrepancies can be accounted for.

Spoil heaps often contain significant quantities of sulphide minerals, particularly pyrite, which are a potential long-term source of contamination for local water courses and groundwater. Oxidative weathering and dissolution of these sulphides releases metals (e.g. iron, zinc, copper), sulphate and, in the case of pyrite, acidity into percolating waters, which are then released into the wider environment. Effective remediation requires an understanding of long-term discharge evolution (Younger 1997). At present, there is no suitable method to predict either the longevity of contaminant sources, or the evolution of their strength over the contaminating life time of a spoil heap. Existing techniques, such as acid–base accounting, which use primitive laboratory assays to estimate the balance between acidity-producing minerals, such as pyrite, and alkalinity-producing minerals, such as calcite, are inadequate for a number of reasons, as described by Jambor (2001). Essentially, they cannot provide the basis required for predictive calculations as additional information on temporal changes arising from differential reaction rates and transient hydrological fluxes are also required. In reality, they are more concerned with the possibility of a problem occurring rather than how severe the problem will be over the long term. Comparison of time scales for transport and reactive processes provides a first step in determining both the timing and the duration of contamination.

Iron concentrations from abandoned mines have been shown to follow an exponential decline with time. Early workers (Frost 1979; Glover 1983) suggested that chronic pollution should last no longer than a few years. However, later studies by Younger (1997) proposed that the initial exponential decay was followed by the establishment of asymptotic levels. This relates to the initial 'flushing' of existing oxidation products, which may last up to 40 years, followed by the ongoing seasonal oxidation of source minerals, such as pyrite, within the oxic zone as the water table fluctuates. Long-term contamination will continue for many decades, if not millennia, until these source minerals are depleted (Younger & Harbourne 1995). Younger (1997) used the terms 'vestigial acidity' and 'juvenile acidity' to describe these two components, with vestigial acidity referring to the initial acidity resulting from the flushing process, and juvenile acidity referring to that generated during seasonal fluctuation of the water table.

Acidity generated by pyrite dissolution may be neutralized by carbonate dissolution, which can buffer the pH and increase alkalinity. Dissolution may also be physically inhibited by the armouring of pyrite mineral surfaces with iron hydroxides, substantially reducing oxygen supply to the reaction site and hence oxidation rate (Wood et al. 1999). Estimates of the lifetime of the contamination arising from spoil are necessary for an optimum choice of remediation options. Wood et al. (1999) suggest that planning should be based on intensive treatment of discharges using active processes, such as lime dosing, while the vestigial acidity is being depleted, followed by long-term passive treatment using constructed wetlands technology for the juvenile acidity.

As a prelude to a full-site investigation, laboratory unsaturated column experiments have been carried out at the University of Sheffield on colliery spoil from the Morrison Busty spoil heap in County Durham, UK, to assess the contamination potential of the spoil material. The results were then compared to the results from a one-dimensional mathematical model for contaminant release and transport developed at the University of Newcastle. The model uses the latest techniques in object-oriented programming to produce a computationally fast model capable of simulating the large time scales required. The random walk method, a form of particle tracking used successfully by Prickett et al. (1981) to model solute transport, is used to transport iron and sulphate 'particles', released by the oxidative weathering of pyrite minerals, through the spoil. The oxidation of ferrous iron to ferric iron is incorporated to provide an iron 'sink', where ferric iron spontaneously precipitates as ferric oxyhydroxide and is effectively removed from the transport process. Effluent iron and sulphate concentrations are subsequently calculated and are used to estimate the contamination potential of the spoil material.

Historical and geological overview

The Morrison Busty spoil heap, near Quaking Houses in County Durham, UK (Fig. 1), has received material from surrounding coal mines since 1922 (Kemp & Griffiths 1999). Mining ceased in the 1970s and top soil has since been applied to selected areas of the heap in an attempt to encourage revegetation, with limited success. The pollution emitting from the heap, which

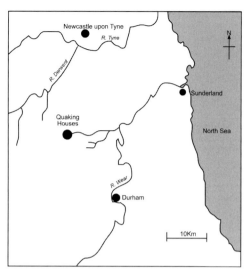

Fig. 1. Map showing location of Quaking Houses, County Durham, UK.

covers an area of approximately 35 hectares and varies in height from 4 to 10 m, is thought to have increased after a deep road cutting was made through the heap in 1986–1987 (Jarvis & Younger 1999). The spoil comprises colliery waste consisting of grey and black weathered shale, ash, coal and coal dust. The heap overlies glacial clay and drift deposits, which themselves overlie Carboniferous Coal Measures beds.

Acidic drainage water emanating from the Morrison Busty spoil heap enters the Stanley Burn, which flows into the River Wear and ultimately the North Sea (Fig. 1). Following the success of a pilot-scale wetland, a full-scale compost wetland was constructed in 1997 at the head of the Stanley Burn to treat the pollution and prevent migration into other water courses. This proved to be a singularly efficient constructed wetland (Jarvis & Younger 1999).

Laboratory methods

- Drill material from 6–7 m depth in the spoil heap was retrieved and stored without drying at $<4°C$.
- Mineralogical composition was characterized by X-ray diffraction (XRD).
- Surface mineralogy was investigated by scanning electron microscope (SEM) with an energy dispersive element analysis facility.
- The material was dried at 60°C and then packed into 10 cm diameter, 15 cm tall uPVC tube with sintered glass discs at the

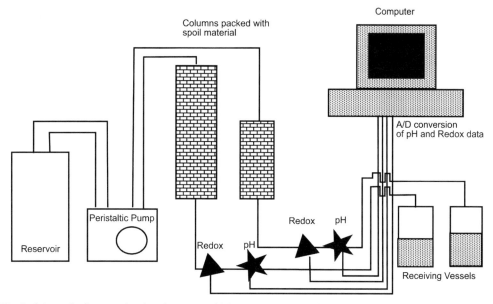

Fig. 2. Schematic diagram showing the set-up of laboratory unsaturated column experiments on colliery spoil.

top and bottom to distribute influent and effluent flow (Fig. 2).
- Deionized water influent was pumped onto the top sintered disc at a flow rate ranging between 0.05 and 0.3 ml min^{-1}.
- Flow rates were low enough to ensure unsaturated hydrological conditions in the column throughout the experiment.
- Effluent pH and redox conditions were recorded by probes housed in custom-made Teflon flow cells.
- The effluent was sampled daily into polypropylene tubes and acidified using concentrated nitric acid.
- Samples analysed for major cations by inductively coupled plasma-atomic emission spectroscopy (ICP-AES).
- The laboratory temperature was maintained at $20 \pm 1°C$.

Laboratory results

X-Ray diffraction and subsequent semi-quantitative analysis of traces with Siroquant (Taylor 1991) revealed that the spoil was composed of, in order of abundance, illite, quartz, kaolinite and pyrite. Scanning electron microscopy enabled observation of additional small quantities of jarosite and iron oxyhydroxides located on the surface of illitic particles. These secondary weathering products of pyrite and clay minerals were present in small quantities ($<2\%$) and were not detected by XRD. Calcite, the most common mineral capable of buffering significant amounts of acidity, was not observed and is, therefore, not included in the following model.

Initially, effluent concentrations of iron, aluminium and sulphate were high, although values rapidly declined over a period of 3 weeks (results not shown). These characteristics are typical of the vestigial flushing phase of Younger (1997). After this period the spoil was washed and dried, and the experiment restarted. Iron and sulphate concentrations were again initially high, but rapidly approached asymptotic levels (see Figs 5 and 6) as the proportion of contaminants released from juvenile sources began to dominate. Intermittent blocking of the column by iron oxyhydroxides forming in the effluent tubes occurred towards the middle of the run (results not shown), and partial blocking may account for the rise in iron concentrations observed between days 20 and 40. Otherwise, concentrations of iron and sulphate declined slowly over the course of the run. This pattern probably results from the combination of juvenile pyrite weathering with dissolution and gradual exhaustion of small quantities of relatively insoluble secondary weathering products, such as jarosite. Jarosite saturation indices calculated using PhreeqC (Parkhurst & Appelo 1999) are between -10

and 0, consistent with the dissolution of this mineral. Measured pH was relatively stable between 2 and 3 during the whole run. This value is significantly lower than that observed at the field site, which fluctuates between 4 and 5. This is probably because the experimental material was much fresher than the bulk of the spoil exposed to weathering *in situ*. The implications of this sampling artefact for tip modelling are discussed below.

Modelling methods

Contaminant transport

A one-dimensional particle tracking model has been developed to simulate contaminant transport through mine spoil, based on the random walk method (Prickett *et al.* 1981). Solute components are represented by sets of moving particles, with each particle assigned a mass that represents the total mass of chemical constituent involved. Iron, sulphate and oxygen particles are present within the model. The advective component of the advection–dispersion equation (equation 1) is solved by moving the particles at the same rate as the average linear velocity of the groundwater. Dispersion is considered to be a random process and is represented by an additional displacement to each particle without modifying the mass of the particle:

$$\frac{\partial}{\partial x}\left(\frac{D_L}{R_d}\frac{\partial C}{\partial x}\right) - \frac{V}{R_d}\frac{\partial C}{\partial x} - C_s Q = \frac{\partial C}{\partial t} \quad (1)$$

where V is the average linear groundwater velocity, D_L is the coefficient of longitudinal hydrodynamic dispersion, x is the space dimension, t is time, R_d is the retardation factor, C is the concentration of solute and $C_s Q$ is a sink function, with a concentration C_s and a flux rate Q.

Contaminant transport using the random walk method can be equated to a normal distribution, where advective movement represents the mean of a normally distributed random variable and dispersive movement the deviation from the mean (Zheng & Bennett 1995), as shown in Fig. 3. Each particle is therefore moved over a distance x to its position at time t according to equation (2) (Kinzelbach 1986):

$$x = Vt + Z\sqrt{2D_L t} \quad (2)$$

where Z is a normally distributed random variable with zero mean and unit variance, V is the average groundwater velocity and D_L the coefficient of longitudinal dispersion. The resulting path lengths are normally distributed with mean Vt and standard deviation $\sqrt{2D_L t}$.

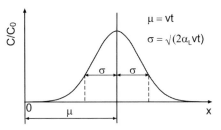

Fig. 3. Contaminant transport using the random walk method is characterized by a normal probability distribution.

The direct result of the random walk model is a particle (mass) distribution that can be easily converted to a concentration distribution. The major advantage of this method above others is that it can be implanted directly on top of any flow model and it does not involve numerical dispersion.

Oxygen diffusion

The longitudinal dispersion coefficient (D_L) in equation (1) is a term that includes both mechanical mixing and molecular diffusion. It is represented by:

$$D_L = \alpha_l V + D_D \quad (3)$$

where α_l is the longitudinal dispersivity, which is taken to be a 'characteristic length' of the spoil material, V is the average linear velocity and D_D is the coefficient of molecular diffusion in the spoil material.

The coefficient of molecular diffusion (D_D) is represented by:

$$D_D = \frac{D_O \tau}{n} \quad (4)$$

where D_O is the diffusion coefficient in air, τ is the tortuosity of the spoil material, which is an approximate expression of pore geometry, and n is the air-filled porosity of spoil material.

The molecular diffusion of oxygen in the gas phase is incorporated into the model by moving each oxygen gas particle according to equation (3), with the mechanical mixing component considered to be zero. The spoil is assumed to be homogeneous so that the diffusion coefficient in the air-filled pores is assumed to be constant with time and depth.

Weathering reactions

Pyrite is oxidized by dissolved oxygen according to:

$$FeS_{2(s)} + \frac{7}{2}O_{2(aq)} + H_2O_{(aq)} \rightarrow Fe^{2+}_{(aq)}$$
$$+ 2SO_4^{2-}{}_{(aq)} + 2H^+_{(aq)}. \quad (5)$$

The literature on pyrite oxidation by dissolved oxygen has been extensively reviewed and numerous studies on reaction rates have been published (e.g. Singer & Stumm 1970; Nordstrom 1982; McKibben & Barnes 1986; Moses et al. 1987; Moses & Herman 1991; Williamson & Rimstidt 1994). The rate equation used in this study uses the square root rate dependence on concentration of dissolved oxygen proposed by McKibben & Barnes (1986) which was also used by Strömberg & Banwart (1994). The rate law used is shown below:

$$-\frac{d[O_{2(aq)}]}{dt} = k[O_{2(aq)}]^{0.5} \quad (6)$$

where the square brackets indicate species activity and k, the rate constant, has a value of $3 \pm 2 \times 10^{-8}\,s^{-1}$ and incorporates scaling for pyrite surface area.

Dissolved ferrous iron is oxidized by dissolved oxygen according to:

$$Fe^{2+}_{(aq)} + \frac{1}{4}O_{2(aq)} + H^+_{(aq)} \rightarrow Fe^{3+}_{(aq)}$$
$$+ \frac{1}{2}H_2O_{(aq)}. \quad (7)$$

The conversion of ferrous iron to ferric iron is likely to be bacterially mediated. Bacterial catalysis by, for example, *Ferrobacillus ferrooxidans* or *Thiobacillus ferrooxidans*, can increase reaction rates by a factor of up to 10^6 (Singer & Stumm 1970). Upon hydrolysis, the ferric iron produced spontaneously precipitates as ferric oxyhydroxide according to:

$$Fe^{3+}_{(aq)} + 3H_2O_{(aq)} \leftrightarrow Fe(OH)_{3(s)} + 3H^+_{(aq)}. \quad (8)$$

The rate constant, and hence the rate expression, for equation (7) depends on pH and oxygen concentration. Several authors have studied the effect of pH on the oxidation of ferrous iron by dissolved oxygen (e.g. Sung & Morgan 1980; Millero 1985; Wehrli 1990). They all agree that in the pH range < 2 the rate equation is given by:

$$-\frac{d[Fe^{2+}_{(aq)}]}{dt} = k_0[Fe^{2+}_{(aq)}][O_{2(aq)}]. \quad (9)$$

For a pH in the range 2–5 the rate equation is given by:

$$-\frac{d[Fe^{2+}_{(aq)}]}{dt} = k_1[Fe^{2+}_{(aq)}][O_{2(aq)}][H^+_{(aq)}]^{-1}. \quad (10)$$

Between a pH of 5 and 8 the rate equation is given by:

$$-\frac{d[Fe^{2+}_{(aq)}]}{dt} = k_2[Fe^{2+}_{(aq)}][O_{2(aq)}][H^+_{(aq)}]^{-2}. \quad (11)$$

The rate constants k_0, k_1, and k_2 depend on oxygen concentration and are calculated for the corresponding oxygen concentration from published data for 1 atm partial pressure oxygen in Wehrli (1990). As the experiments from which these data were derived were performed in abiotic systems, it was necessary to account for bacterial catalysis occurring in the columns. This was achieved by arbitrarily multiplying the rate of ferrous iron oxidation by a factor of 10^3.

Conceptual model

In order to model the column of spoil material, several assumptions were made. Figure 4 summarizes the conceptual model.

Hydrological conditions

Flow is steady state and vertical. Water enters only at the top of the column, where it is evenly distributed across the spoil surface; it is assumed there is no influx of water through the side boundaries. The column is assumed to be unsaturated at all times and the flow rate to be constant. The material is also assumed to be homogeneous.

Contaminant sources

It is assumed that the spoil was not previously exposed to weathering processes, and the only source of iron and sulphate contamination is pyrite, which is evenly distributed throughout the column. Oxidation of organic matter is not considered. Although the assumptions above describe only a small fraction of spoil heap–mine environments, it was felt that the benefits of testing the basic model with as few parameters as possible far outweighed the disadvantages associated with 'calibrating' a multiparameter, multiprocess model. In this particular case, the applicability of the assumptions was ensured by the collection of experimental material from a deep, relatively unoxidized, position in the heap, and by pre-experimental treatment of material to largely preclude vestigial contamination sources produced during sample storage.

Pyrite is assumed to be oxidized by dissolved oxygen only, which is not involved in any other

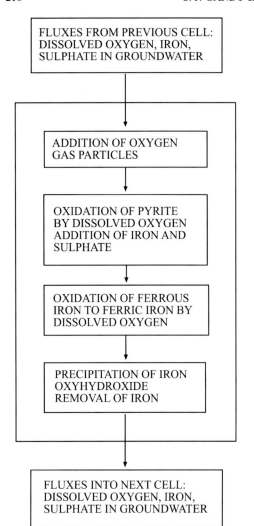

Fig. 4. Generalized conceptual model of a single finite-difference cell showing fluxes into and out of the cell, and reactions taking place within the cell, within a single time step.

difference grid at each time step. The gas then diffuses through the air-filled pores.

Contaminant sinks

Ferric oxyhydroxide ($Fe(OH)_3$) is assumed to be the only sink for iron in the model. No sink occurs for sulphate.

Results and discussion

Table 1 gives the model input parameters used in the simulation. These were either measured in the laboratory or taken from published data. Figure 5 shows the variation in observed and calculated iron concentrations for a 220 day simulation. Gaps in the laboratory data represent occasions when the column was saturated or blocked so that hydrological conditions were not steady state. The slightly rugged appearance of the modelled results is due to the random element in the model design, and is one of the disadvantages of using the random walk method for contaminant transport. Although it is possible to smooth the results, it was not felt to be necessary in this study as the general trends were evident. Figure 6 compares observed and predicted results for effluent sulphate concentrations from the column. The results show a similar pattern to that for iron release, with laboratory results exhibiting a pseudo-exponential decrease in sulphate concentration over the simulation period, whereas modelled results reach steady-state conditions almost immediately.

The order of magnitude agreement between predicted and observed iron and sulphate concentrations is consistent with the interpretation of reactions other than the oxidation of pyrite and the oxidation of ferrous iron. All ferrous iron produced by pyrite oxidation is available for oxidation by any remaining dissolved oxygen, while all ferric iron subsequently produced precipitates as ferric oxyhydroxide. Oxygen enters the column both as a gas and dissolved in the influent. Oxygen gas is assumed to diffuse between the spoil and the tube wall, and therefore enters the column through the side boundaries as well as at the top. This is incorporated into the model by adding oxygen gas particles at atmospheric pressure to each cell of the finite-

Table 1. *Model input parameters*

Parameter	Value
Number of nodes	5
Length of column (m)	0.125
Width of column (m)	0.089
Surface width of column (m)	0.089
Average Inflow (l day^{-1})	0.1
Time step (days)	1
Effective porosity	0.2
Average moisture content	0.1
Longitudinal dispersivity (m)	0.002
Partial pressure of atmospheric oxygen (atm)	0.2
Mass of pyrite (%)	3.74
Density of pyrite (kg m^{-3})	5000*
Average diameter of pyrite crystals (μm)	50*
Bacterial catalysis factor	10^3

*Parameters taken from Bronswijk *et al.* (1993).

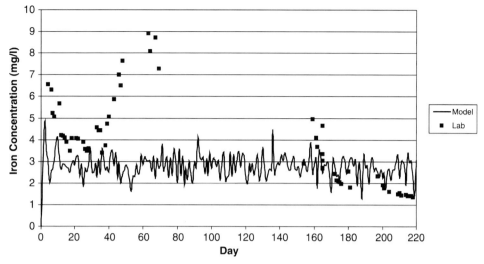

Fig. 5. Laboratory and modelled iron concentrations for a typical 220 day simulation.

laboratory results, i.e. that the bulk of the dissolved contaminants are derived from pyrite weathering (modelled), with an additional component derived from vestigial weathering of secondary minerals (not modelled). This interpretation is also supported by the agreement between iron and sulphate ratios, as large contributions from an unmodelled process would be unlikely to affect iron and sulphate equally, whereas a small contribution from jarosite, which contains both iron and sulphate, could produce the observed pattern.

Alternative explanations for the differences in behaviour between the predicted and observed results must also be considered. It is improbable that the downward trend in laboratory results is caused by the exhaustion of pyrite, as mass-balance calculations show that only a small proportion of the total is accounted for in the effluent. A reduction in reactive surface area could produce the same effect; however, extended preliminary testing and lack of sample preparation suggest that this is unlikely. It is also unlikely that dissolution rates decreased as a

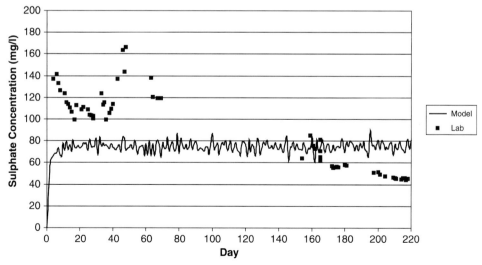

Fig. 6. Laboratory and modelled sulphate concentrations for a typical 220 day simulation.

result of pyrite armouring by iron oxide, as this has only been observed to any significant degree in the presence of limestone or at neutral pH (Nicholson et al. 1990; Puura et al. 1999). Iron overprediction by the model in the later stages of the experiment could indicate that the ochreous precipitation observed in the output tubes represented a significant unmodelled contaminant sink. However, if this was the case, the overprediction by the laboratory results in the early part of the run would not be expected. It is likely that the ochreous material was deposited during periods when the solution was essentially static, and oxidation of ferrous iron occurred to a significant extent within the effluent tube.

The model does not incorporate pyrite oxidation by ferric iron, it assumes oxidation by dissolved oxygen only. In very low pH conditions ferric iron does not precipitate as ferric oxyhydroxide and is therefore available to react with pyrite. This represents the cycle suggested by Singer & Stumm (1970), whereby initial oxidation of pyrite is undertaken by dissolved oxygen to produce ferrous iron (equation 12) that is oxidized by oxygen to produce ferric iron (equation 13). The ferric iron subsequently oxidizes pyrite, thereby producing more ferrous iron to re-enter the cycle (equation 14):

$$FeS_{2(s)} + \frac{7}{2}O_{2(aq)} + H_2O_{(aq)} \rightarrow Fe^{2+}_{(aq)}$$
$$+ 2SO_4^{2-}{}_{(aq)} + 2H^+_{(aq)} \quad (12)$$

$$Fe^{2+}_{(aq)} + \frac{1}{4}O_{2(aq)} + H^+_{(aq)} \rightarrow Fe^{3+}_{(aq)}$$
$$+ \frac{1}{2}H_2O_{(aq)} \quad (13)$$

$$FeS_{2(s)} + 14Fe^{3+}_{(aq)} + 8H_2O_{(aq)} \rightarrow 15Fe^{2+}_{(aq)}$$
$$+ 2SO_4^{2-}{}_{(aq)} + 16H^+_{(aq)}. \quad (14)$$

Oxidation by ferric iron was omitted from the model because the overall objective of the research is to predict the long-term contamination from the spoil heap, and as the pH at a field scale is on the order of 4–5, oxidation of pyrite by dissolved oxygen is considered to be the dominant process. In the laboratory experiments, however, the pH within the column is significantly lower and the occurrence of pyrite oxidation by ferric iron is a possible reason for the model underprediction of effluent iron and sulphate concentrations.

Figure 7 shows a model prediction for the long-term depletion of pyrite within the column in terms of effluent iron and sulphate concentrations. The model shows an exponential decline in concentration with time, with iron remaining in the effluent for up to 134 years, but concentrations are predicted to fall below $1\,mg\,l^{-1}$ after approximately 55 years. The calculated half-life for iron is 38 years. Significant sulphate concentrations persist for

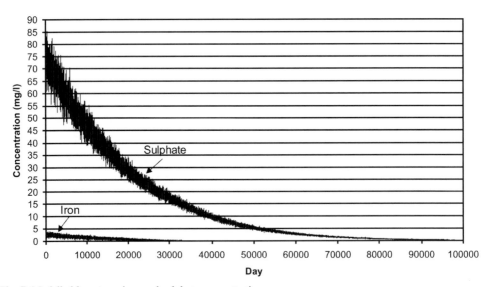

Fig. 7. Modelled long-term iron and sulphate concentrations.

longer and are predicted to remain in the effluent for as long as 273 years, although they become fairly insignificant ($<1\,\mathrm{mg\,l}^{-1}$) after 210 years. The calculated half-life for sulphate is 40 years. Half-lives for both iron and sulphate effluent concentrations are of the order of decades, suggesting that a small experimental column containing only 1 kg of spoil could, in fact, remain polluting for several centuries.

Conclusions

A method for predicting the contamination potential of mine spoil is proposed that is tested by modelling weathering rates from laboratory unsaturated column experiments. Results for simulations of iron and sulphate release from material from the Morrison Busty spoil heap in County Durham, UK, agree with experimental results on an order of magnitude scale. Differences between the dynamic response of the laboratory experiment and the steady-state output of the model indicate an oversimplification of the source–sink terms during model conceptualization. However, the order of magnitude agreement between observed and predicted results represents a significant potential in predictive capability for this type of system.

A model prediction is given for the long-term depletion of pyrite within the column of spoil material, showing that the material could remain polluting for several centuries. This emphasizes the long-term severity of mine water pollution and the contamination potential of mine spoil material.

The research project is funded by NERC (Grant reference GST/02/2060), under their Environmental Diagnostics Programme, and Northumbrian Water Group Research Centre Ltd.

The authors would like to acknowledge the contribution of P. L. Younger and other members of the Water Resource Systems Research Laboratory at the University of Newcastle, and S. A. Banwart and other members of the Groundwater Protection and Restoration Group at the University of Sheffield.

References

BRONSWIJK, J.J.B., NUGROHO, K., ARIBAWA, I.B., GROENENBERG, J.E. & RITSEMA, C.J. 1993. Modelling of oxygen transport and pyrite oxidation in acid sulphate soils. *Journal of Environmental Quality*, **22**, 544–554.

FROST, R.C. 1979. Evaluation of the rate of decrease in the iron content of water pumped from a flooded mine shaft in County Durham, England. *Journal of Hydrology*, **40**, 101–111.

GLOVER, H.G. 1983. Mine water pollution – an overview of problems and control strategies in the United Kingdom. *Water Science & Technology*, **15**, 59–70.

JAMBOR, J.L. 2001. The relationship of mineralogy to acid- and neutralisation-potential values in acid rock drainage. In: COTTER-HOWELLS, J.D., CAMPBELL, L.S., VALSAMI-JONES, E. & BATCHELDER, M. (eds) *Environmental Mineralogy: Microbial Interactions, Anthropogenic Influences, Contaminated Land and Waste Management*. Mineralogical Society of Great Britain & Ireland, London, 141–161.

JARVIS, A.P. & YOUNGER, P.L. 1999. Design, construction and performance of a full-scale compost wetland for mine-spoil drainage treatment at Quaking Houses. *Journal of the Chartered Institute of Water and Environmental Mangagement*, **13**, 313–318.

KEMP, P. & GRIFFITHS, J. *Quaking Houses. Art, Science and the Community: A Collaborative Approach to Water Pollution*; Jon Carpenter, Charlbury.

KINZELBACH, W. *Groundwater Modelling. An Introduction with Sample Programs in Basic*; Elsevier, New York.

MCKIBBEN, M.A. & BARNES, H.L. 1986. Oxidation of pyrite in low temperature acidic solutions: rate laws and surface textures. *Geochimica et Cosmochimica Acta*, **50**, 1509–1520.

MILLERO, F.J. 1985. The effect of ionic interactions on the oxidation of metals in natural waters. *Geochimica et Cosmochimica Acta*, **49**, 547–553.

MOSES, C.O. & HERMAN, J.S. 1991. Pyrite oxidation at circumneutral pH. *Geochimica et Cosmochimica Acta*, **55**, 471–482.

MOSES, C.O., NORDSTROM, D.K., HERMAN, J.S. & MILLS, A.L. 1987. Aqueous pyrite oxidation by dissolved oxygen and by ferric iron. *Geochimica et Cosmochimica Acta*, **51**, 1561–1571.

NICHOLSON, R.V., GILLHAM, R.W. & REARDON, E.J. 1990. Pyrite oxidation in carbonate-buffered solution: 2. Rate control by oxide coatings. *Geochimica et Cosmochimica Acta*, **54**, 395–402.

NORDSTROM, D.K. 1982. Aqueous pyrite oxidation and the consequent formation of secondary iron minerals. In: KITTRICK, J.S., FANNING, D.S. & HOSSER, L.R. (eds) *Acid Sulphate Weathering*. Soil Science of America, Madison, Special Publication, Vol 10, 37–56.

PARKHURST, D.L. & APPELO, C.A.J. *Users guide to PhreeqC (version 2): A Computer Program for Speciation, Batch Reaction, One Dimensional Transport and Inverse Geochemical Calculations. US Geological Services*, Water-Resources Investigations Report, 99–4259.

PRICKETT, T.A., NAYMIK, T.G. & LONNQUIST, C.G. 1981. *A 'Random Walk' Solute Transport Model for Selected Groundwater Quality Evaluations*. Illinois State Water Survey, Champaign, Bulletin, **65**, 1–103.

PUURA, E., NERETNIEKS, I. & KIRSIMAE, K. 1999. Atmospheric oxidation of the pyritic waste rock in

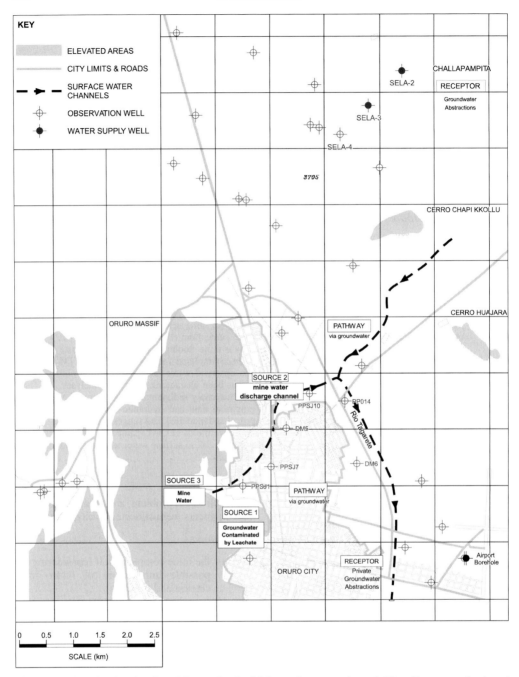

Fig. 1. Site plan showing the city of Oruro, San José Mine, mine water channel, River Tagarete and selected monitoring wells.

Challapampa wellfields), supplying Oruro with its drinking water (see Fig. 1);
- other public and private groundwater abstractions;
- the aquifer complex as a potential future groundwater resource.

The postulated risk scenario is defined as follows:

- pumping ceases at San José Mine. Mine water rises until it overflows at the lowest available mine entrance;
- the Khala Kaja–Challapampita wellfields create an area of depressed piezometric heads in the Quaternary aquifer complex, which may reach as far as Oruro;
- there may thus be a head differential between the water level in the flooded mine and the head in the Quaternary aquifer system;
- mine water may thus enter the Quaternary aquifer system via natural fracture systems or via unmapped mine openings and migrate towards the Khala Kaja–Challapampita wellfields;
- in addition, groundwater contaminated by mine waste run-off, beneath the city of Oruro, may migrate towards the wellfields through groundwater in the Quaternary aquifers.

This paper focuses on characterization (in terms of flows and chemical quality) of source terms for the risk assessment. Other aspects of the risk assessment are briefly touched on here, and will be dealt with in more detail in future papers. The objective of this paper is demonstrate the construction of a conceptual hydrogeological and hydrochemical model of the mine. The paper will also illustrate the range of parameters that need to be quantified to generate the source term for a risk assessment.

The San José Mine – an introduction

Oruro lies on the Bolivian Altiplano, a high-level plain (*c.* 3700 m asl) situated between the Cordilleras Occidental and Oriental of the Andes. The San José Mine is located adjacent to Oruro city, in a series of low hills (referred to hereafter as the 'Oruro massif'), rising approximately 300 m above the Altiplano around Oruro. The mine's 'level 0' is defined at 3800 m asl. Climatic data for the area are given in Table 1.

The San José Mine has a history of over 400 years, primarily as a tin and silver mine. The first mine was that of San Miguel, in the area now occupied by the Plaza of the Church of the Virgin of Socavón, the focus of the world-renowned Oruro carnival. Modern tin and zinc–silver–lead exploitation commenced in 1948, when control of the mine passed to the 'Banco Minero de Bolivia'. The mine was nationalized in 1952 (along with other major Bolivian mines) and passed into the control of the Corporación Minera de Bolivia (COMIBOL). For economic reasons, production was halted in July 1992.

Most of the deep mining has taken place in the latter half of the twentieth century and hence is relatively well mapped. Shallow-level mining, both above and below level 0, has a history of several centuries, however, and its exact extent is not well documented.

Three shafts allow access to the deep mine interior from the surface (see Fig. 2). The Itos shaft extends from the surface down to below the -460 m level on the western side of the Oruro massif. The Santa Rita and San José shafts allow access to the -340 m level from the eastern side. Galleries at the -340 and -460 m levels (Flores 1998) connect the Itos, Santa Rita and San José workings. An interior shaft, the Auxiliar Shaft, provides additional access to the -460 m level from the -340 m level.

Since closure in 1992, the workings have been kept dry by pumped dewatering to the -460 m level (3340 m asl). Water entering the mine below level -340 m is lifted from level -460 m through the Auxiliar Shaft to level -340 m. Water entering the mine above -340 m is allowed to drain down to -340 m. The total water make of the mine is thus collected at the -340 m level, before being pumped up the Santa Rita Shaft to the Santa Rita Mine entrance (level -30 m).

From January 1996 to August 1999, the average monthly pumping rates in the Santa Rita Shaft ranged from 5.7 to 13.7 l s^{-1} (average 8.1 l s^{-1}), although for the vast majority of months the rate was around 8 l s^{-1} (Fig. 3). Of this total, an average of 3.9 l s^{-1} was pumped up from levels deeper than -340 m (via the Auxiliar Shaft). There is remarkably little seasonal fluctuation.

Table 1. *Climatic data derived from Oruro Airport and Vinto Meteorological Stations (after SGAB-COMIBOL 1995a), with evaporation from SGAB (1996b). The bulk of the precipitation falls from October to March*

Average temperature	+10.5°C
Max. daily temp. 1986–1994	+24.5°C
Min. nightly temp. 1986–1994	-17.3°C
Average annual precipitation	345 mm
Max. 24 h precipitation (airport)	34 mm
Mean actual annual lake evaporation (Poopo)	1875 mm year^{-1}
Annual average relative humidity	43%

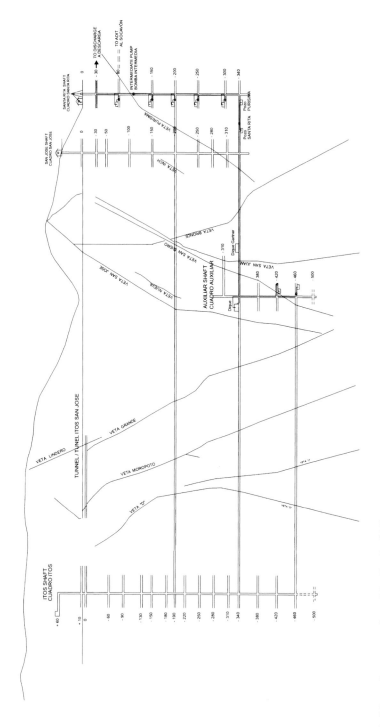

Fig. 2. Schematic cross-section of the San José Mine.

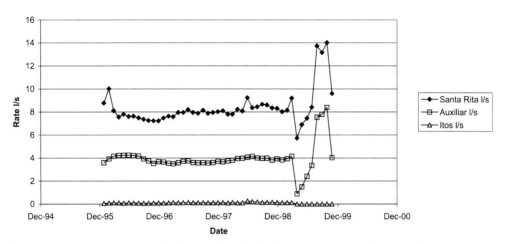

Fig. 3. Pumping rates from the San José Mine; 1996–1999. The pumping increase in August 1999 was related to the need to fully dewater the mine system to allow access for investigators, i.e. non-equilibrium pumping. ('Santa Rita' represents the total amount of water pumped from the mine, 'Auxiliar' represents the water ingress derived from below level -340 m.)

The (untreated) mine water discharges from the Santa Rita adit, via a mine water channel, into the Rio Tagarete system (Fig. 1). The Rio Tagarete is heavily polluted and discharges, south of Oruro, into the shallow Lake Uru-Uru, where considerable damage has been wrought on lake ecosystems and sediment quality (SGAB-COMIBOL 1995a; Pescod & Younger 1999).

Geological and hydrogeological setting

The San José Mine is located in an 'island' of Tertiary igneous rocks (the 'Oruro massif') intruding Palaeozoic metasedimentary bedrock, surrounded by the dominantly Quaternary sedimentary complex of the Altiplano.

The bedrock

The Tertiary igneous complex hosting the San José Mine is situated in black slates of the Silurian Uncía Formation. The San José deposit corresponds to a Tertiary (c. 16 Ma old; Avila Salinas 1990) igneous volcanic–intrusive caldera complex that has been eroded and partially removed, exposing the central conduit, composed of intrusive rocks (Long 1992; Flores 1998).

The San José ore

The San José deposit's genesis, geology and mineralogy are described in great detail by Turneaure (1960a, b, 1971) and Kelly & Turneaure (1970). It is classified as a Bolivian polymetallic vein (BPV) type deposit. BPV deposits typically exhibit a mineral assemblage dominated by pyrite, marcasite (both FeS_2) and pyrrhotite (FeS), and also including sphalerite (ZnS), galena (PbS), cassiterite (SnO_2), native gold (Au), arsenopyrite (FeAsS), argentite (Ag_2S), chalcopyrite ($CuFeS_2$), tetrahedrite ($(Cu,Fe)_{12}(Sb,As)_4S_{13}$), and many other sulphide and complex sulphosalt minerals.

At San José, two mineralization stages occurred, an early tin stage and a late silver sulphosalt stage (SGAB-COMIBOL 1995a). Tin mineralization predominates near the surface. With increasing depth, tin content diminishes and silver content increases. Ludington et al. (1992) regard Pb, Zn, Ag, Sn and Sb as the principal metals at San José, of which Pb, Ag and Sn were considered commercially important.

The Quaternary sedimentary complex of the Altiplano

The flat plains of the Altiplano are underlain by Quaternary sediments of varying thickness (due to buried bedrock ridges, alluvial troughs and shallow bedrock platforms). The sediments are typically fluvio-lacustrine (derived from erosion of adjacent mountains and hill ridges), with some glacial (Pleistocene) or possibly interbedded volcanic deposits (Dames & Moore 1967). The Quaternary complex contains valuable groundwater resources, which are believed to

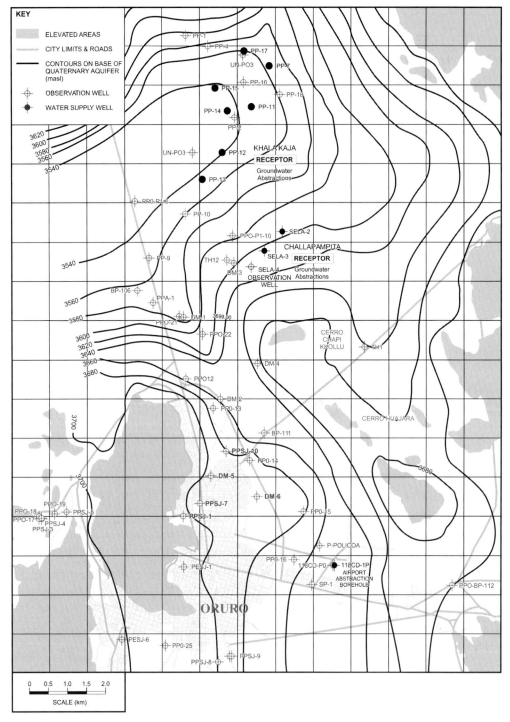

Fig. 4. Contour map (m asl) of the base of the Quaternary Altiplano aquifer complex. The map is based on interpretation of geophysical profiles and drilling logs.

be mostly recharged by run-off through coarse alluvial fan-like material at the foot of the mountain ranges of the Cordillera Oriental. Direct recharge through the surface of the Altiplano is believed to be very limited due to high evaporation and the low permeability of the surficial sediments. Abstractions of importance for the city of Oruro include the following, which are defined as risk receptors in the risk assessment approach:

- the wellfield at Khala Kaja, pumping water from a complex of semi-confined and confined aquifers, from around seven boreholes, at a rate of some $150 \, l \, s^{-1}$ (see Cortez & Torrez 1996);
- the well field at Challapampita, pumping water from two boreholes at a rate of some $40 \, l \, s^{-1}$;
- smaller production wells, such as borehole PP-117 at the airport (up to $8 \, l \, s^{-1}$ on average) and those operated by private concerns and industries.

A contour map of the base of the Quaternary aquifer (Fig. 4) complex was prepared on the basis of geophysical and drilling investigations. It appears that there is a buried channel more than 150 m deep in the Khala Kaja area. In this case, the combination of great depth and considerable thickness of permeable strata produces a structure that is a very effective 'sink' for abstracting groundwater.

Groundwater salinity (Fig. 5) in the aquifer complex varies considerably, both laterally and vertically. There is a tendency for fresher waters in the Khala Kaja area, possibly indicating that the Khala Kaja buried channel is a direct flow pathway for recharge waters from the foot of the Cordillera Oriental.

A map (Fig. 6) has also been prepared of the piezometric surface in the aquifer complex (it has not been possible to correct measured water levels for salinity due to the extreme lateral and vertical variability). The map indicates head gradients to be low. Test pumping of boreholes suggests that the hydraulic conductivity of the Quaternary aquifer sediments typically varies between 1 and c. $50 \, m \, day^{-1}$ (URS Dames & Moore Norge 2000b–d). The groundwater level map indicates that:

- there is an area of depressed groundwater head around the Challapampita–Khala Kaja wellfields;
- water abstracted at these wellfields is largely derived from the NW, north and NE;
- a small amount of groundwater is attracted to the wellfields from the south, creating

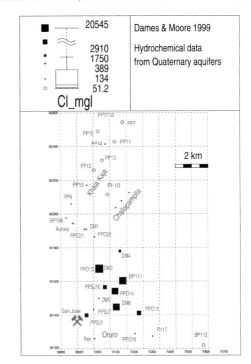

Fig. 5. Map showing concentrations of chloride in groundwaters from the Quaternary Altiplano aquifer complex in the vicinity of Oruro, based on sampling and analysis of 1999–2000. Symbol sizes are based on the boxplot (inset). Concentrations are in $mg \, l^{-1}$.

an ENE–WSW-oriented groundwater divide to the north of San José;
- the groundwater divide is *not* caused by any physical feature. It is a dynamic feature, caused by human abstraction. Should abstraction increase in the future, the divide would probably migrate south. Should abstraction decrease, the divide would move north;
- to the south of the groundwater divide, the main direction of groundwater flow is southwards towards Lake Uru-Uru;
- currently, the contaminated groundwater from mine wastes at San José, Santa Rita, San Miguel, etc., lies to the south of the divide. Groundwater flow lines constructed on the basis of the contours suggest that contamination would migrate east from Oruro and then south;
- currently, on the basis of flow dynamics, there is thus *no risk* from the mine wastes to the wellfields at Challapampita or Khala Kaja;
- the mine water discharge channel enters the Tagarete very close to the groundwater divide. If contaminated water from these

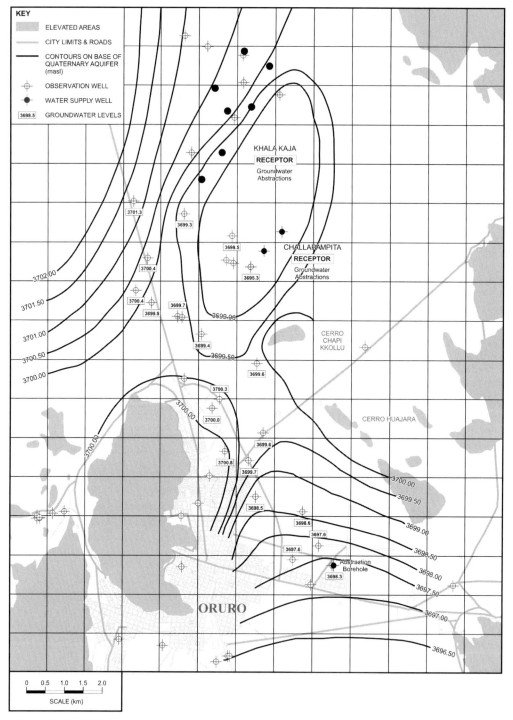

Fig. 6. Groundwater level contour map for the Quaternary Altiplano aquifer complex around Oruro. The map is based on empirical water levels measured in wells and boreholes in December 1999. Contours are in m asl. Note the very shallow gradients and the groundwater divide north of Oruro.

watercourses infiltrates the aquifer (and there is evidence that it does: URS Dames & Moore Norge 2000c), there is a possibility that it may flow slowly northwards toward the wellfields. This does *not* necessarily imply a risk, as there is evidence (URS Dames & Moore Norge 2000d) that the aquifer system attenuates mining-related contaminants extremely effectively;
- if abstraction is increased in the future from the Challapampa wellfields, the groundwater divide may move south, implying that contamination from the mine may then migrate northwards. Again, this does *not* necessarily imply a risk.

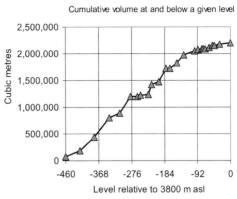

Fig. 7. Cumulative distribution diagram of mine void space (as void below a given level) in the San José Mine, based on the geometric approach (see text).

Mine water hydraulics

At present, the mine is being dewatered, and inward hydraulic gradients thus prevent direct leakage of mine water to regional groundwater. It is, however, necessary to consider a future scenario where the expensive pumping operation is closed down, and to estimate the rate at which the mine would subsequently overflow (if at all) at the surface and the rate at which it might leak to regional groundwater systems. To answer these questions, the MIFIM (Banks 2001) and MODFLOW (McDonald & Harbaugh 1988) models were employed. These models require the following input data:

- the distribution and total volume of void spaces in the mine;
- the locations and flows of major influxes of mine water to the mine (MIFIM);
- an estimate of bulk hydraulic conductivity of the mine host rock and the recharge rate (MODFLOW);
- the level of the most likely overflow horizon.

Void distribution in the mine

The volume of open space in the mine was estimated by three approaches.

The geometric approach. The volume of galleries and workings in veins was estimated on the basis of their physical dimensions, as measured from mine plans. This approach yielded a volume of $2\,196\,000\,m^3$ void below level 0, distributed according to Fig. 7.

The production approach. Mine production records for 1960–1992 were then examined to calculate the total mineral production of the mine ($4\,389\,825\,t$). Based on an assumed specific gravity of $3\,t\,m^{-3}$ for the extracted rock material, the total volume of material removed from the mine was $1\,463\,275\,m^3$. Given that intensive production below the $c.\,-80$-m level (potential overflow level – see below) is believed to have commenced in the 1940s or 1950s, one can extrapolate production backwards in time to around $2\,000\,000\,m^3$ from the deeper levels.

Waste production approach. SGAB (1996c) have estimated the total volumes of waste rock and tailings around the San José massif as $2\,662\,000\,m^3$. Assuming a porosity for the mining waste of 30%, and an original rock porosity of 0%, a volume of $2\,662\,000\,m^3$ mining waste corresponds to an original rock volume of $1\,863\,000\,m^3$.

All three methods of mine void calculation suggest that a floodable void space of some $2\,000\,000$–$2\,500\,000\,m^3$ is realistic.

Water influxes to the mine

There are several possible sources for the water being pumped from the mine:

- direct infiltration of precipitation from above, via fractures, but most probably through historic mine openings in the surface;
- inflow of groundwater from regional aquifer storage in saturated fracture systems in hard rock or (ultimately) from the surrounding Quaternary aquifers;
- inflow of saline formation water from deep levels in the Tertiary crystalline massif.

Undoubtedly, all three components exist, although salinity and isotopic evidence suggest a major component of deep-seated saline water – see below. Anecdotal evidence exists, however, for ingress of surficial waters, e.g. temporary changes in mine water flows related to rainfall events or works on surface water drainage systems (D. Salinas, COMIBOL pers. comm.).

Observations suggest that the majority of significant water inflows to the San José Mine are from the periphery of the mine system at the deeper levels (-340 m and below). This may reflect the fact that the upper levels of the mine are dewatered and any potentially transmissive fractures at higher levels are now dry.

Ingress of surficial waters (head – independent inflows). Conventional models for calculating recharge to the San José Mine from precipitation are of no value for several reasons:

- there is little soil cover and vegetation;
- there is a very rapid surface run-off response to rainfall;
- old workings of exposed veins have left wide openings on the hillside that provide the potential for direct inflows of rain water into the mine.

A field reconnaissance of exposed mine openings allowed calculation of open areas and the topographic 'catchments' of these openings on the mountainside. The total area draining towards mine openings and worked vein outcrops is estimated as some $207\,250\,\mathrm{m}^2$. If we assume that all the rainfall ($364\,\mathrm{mm\,year}^{-1}$: SGAB, 1996$b$) falling on the openings and their catchments runs into the mine, this totals an amount of $75\,440\,\mathrm{m}^3\,\mathrm{year}^{-1}$ or $2.4\,\mathrm{l\,s}^{-1}$. If, however, we assume that only a proportion of the water falling on the catchments runs into the openings, the rest being lost to evapotranspiration, the amount of run-in is less. The soil thickness and vegetation cover are low on the Oruro massif, and the topographic slope is high and the run-off coefficient is thus likely to be relatively high. Hydrological estimates suggest that the total recharge to the mine via run-in to mine openings and a small amount of direct ground infiltration is likely to be of the order of $1-2\,\mathrm{l\,s}^{-1}$. This corresponds well to estimates made from hydrochemical evidence (see below).

Water inflows to the mine. Flow gaugings have been carried out of the major water inflows in the mine. The results indicate that the inflowing water is derived from relatively few features:

- $c.\ 1.2\,\mathrm{l\,s}^{-1}$ from an exploration borehole in the Gartner Vein at the -340 level;
- $c.\ 1.8\,\mathrm{l\,s}^{-1}$ inflow to the -340 level near the Santa Rita Shaft;
- $c.\ 2.5\,\mathrm{l\,s}^{-1}$ inflow from Vein D at the -420 level;
- $c.\ 1.5\,\mathrm{l\,s}^{-1}$ unaccounted water inflow below from *below* level -340 (given that the total pumped amount from below -340 m in the Auxiliar shaft is $c.\ 4\,\mathrm{l\,s}^{-1}$);
- $c.\ 1\,\mathrm{l\,s}^{-1}$ unaccounted for water from level -340 or above (given that the total mine inflow is $8\,\mathrm{l\,s}^{-1}$ from pumping data).

Possible water overflows from the mine

Field surveys and examination of mine plans indicate the most probable pathway for water overflow from the mine in the case of pump switch-off to be via the Santa Rita Shaft through the so-called Socavón adit (Fig. 2) at the -80-m level. This currently contains a mining museum and has a sulphurous-smelling entrance from the Church of the Virgin del Socavón. The final and lowest adit exit is (ironically) within the grounds of the Geological Survey of Bolivia's (SERGE-OMIN) offices.

Given the unreliable or non-existent mapping of many of the older mine workings, the possibility of mine water overflowing elsewhere than at the Socavón adit cannot be excluded with certainty.

Modelling of mine flooding

In order to simulate possible future mine flooding, two different models have been employed. The MIFIM model of Banks (2001) has been codified for the Oruro mine in an EXCEL® spreadsheet, and the US Geological Survey MODFLOW model (McDonald & Harbaugh 1988) has also been employed. The two models use differing basic philosophies. The MIFIM model focuses on the geometry of the mine void and uses simple hydraulic algorithms, and assumes discrete water inflows (fractures, exploration boreholes, etc.). MODFLOW uses the porous medium assumption, and Darcy's equations for porous media, but does not realistically attempt to simulate the mine geometry.

Given the uncertainty surrounding the origin of mine inflow, modelling has employed a range of scenarios. A *best guess* assumes that $1-2\,\mathrm{l\,s}^{-1}$ is derived from surficial *head-independent* sources (F_i), while $6-7\,\mathrm{l\,s}^{-1}$ is derived from *head-dependent* groundwater inflows ($F_1\ldots F_n$), which decrease as water head in the mine rises above them. A *worst case* assumes $4 \, l \, s^{-1}$ to be head-independent.

The MIFIM model. The MIFIM concept, which is fully explained in a paper by Banks (2001), builds strongly on the GRAM (Groundwater Rebound in Abandoned Mine-workings) model of Sherwood (1993) and Sherwood & Younger (1997). In many ways, it is less sophisticated than GRAM (no Bernoulli-flow along roadways and no interchange of water between mine ponds). However: (i) MIFIM is able to consider non-constant inflow to the mine, i.e. head-dependent inflows; and (ii) it is able to use a reference table to account for variations in mine volume with depth. A volume input to MIFIM was based on the cumulative volume distribution shown in Fig. 7.

As regards water inflows, the following were assumed:

- of the $1.5 \, l \, s^{-1}$ unaccounted water from levels *below* -340, $0.5 \, l \, s^{-1}$ is arbitrarily assigned to level -460, $0.5 \, l \, s^{-1}$ to level -420 and $0.5 \, l \, s^{-1}$ to level -380;
- the 'missing' $1 \, l \, s^{-1}$ from the -340 m level or above was assigned to the -340 level.

This results in the following distribution of the $8 \, l \, s^{-1}$ inflow, as a baseline case (Run 1 in Table 2):

- -340 m level: $4.0 \, l \, s^{-1}$;
- -380 m level: $0.5 \, l \, s^{-1}$;
- -420 m level: $3.0 \, l \, s^{-1}$;
- -460 m level: $0.5 \, l \, s^{-1}$.

It was also assumed that a certain proportion ($x\%$) of all inflow is head-independent (derived from surface-water run-in) and that this is evenly distributed amongst all the inflows.

The model has thus been run with the flow distributions shown in Table 2. The external driving heads $H_1 - H_n$ for the head-dependent inflows $F_1 - F_n$ are all assumed to be equal to the head in the regional Altiplano groundwater reservoir, namely 3700 m asl. The time step employed is between 20 and 50 days. The mine overflow level is set to 3720 m asl.

Six runs of the model were performed. It will be noted that the head-independent inflow F_i is the crucial factor in determining rate of filling and final discharge flux.

- Runs 1 and 5 are extreme (unrealistic) cases where all the inflow is derived from deep, head-dependent flow and surficial head-independent inflow, respectively.
- Runs 2 and 3 represent low and high *best-guess* estimates of the situation, where between 1 and $2 \, l \, s^{-1}$ of mine water are derived from head-independent surficial water sources. In these cases (Fig. 8), flooding of the mine to level -80 occurs within between 5490 and 7520 days (15.0 and 20.6 years, respectively). Overflow from the Socavón adit is estimated at between 0.5 and $1.6 \, l \, s^{-1}$, while losses to regional groundwater are estimated as less than $0.5 \, l \, s^{-1}$. Of these figures, the losses

Table 2. *Input data and key modelling results for six runs of the MIFIM mine water filling model at San José*

	Level	Elevation (m asl)	Run number					
			1	2	3	4	5	6
Head-independent flow								
x (% of total)			0	12.5	25	50	100	25
F_i ($l \, s^{-1}$)			0	1	2	4	8	2
Head-dependent flows ($l \, s^{-1}$)								
F_1	-200	3600	0	0	0	0	0	0.75
F_2	-340	3460	4.00	3.50	3.00	2.00	0	2.25
F_3	-380	3420	0.50	0.44	0.38	0.25	0	0.38
F_4	-420	3380	3.00	2.63	2.25	1.50	0	2.25
F_5	-460	3340	0.50	0.44	0.38	0.25	0	0.38
Results								
Flooding time (days)			∞	7520	5490	4050	2980	5310
Final overflow ($l \, s^{-1}$)			0	0.49	1.56	3.71	8	1.47
Discharge to groundwater ($l \, s^{-1}$)			0	0.51	0.44	0.29	0	0.53

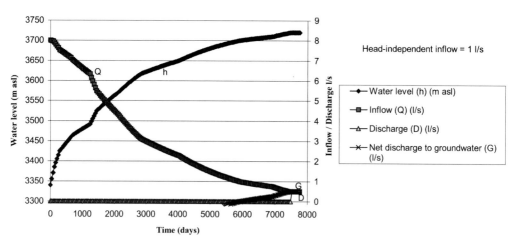

Fig. 8. Output for the MIFIM model for Run 2, where $F_i = 1\,1\,s^{-1}$, illustrating changes in mine water level, inflow to the mine, leakage to regional groundwater and overflow, with time.

to groundwater are most uncertain, due to the simplified hydraulic algorithms employed and to the fact that there may exist dewatered conductive fractures at high levels in the mine that MIFIM cannot simulate.

- Run 4 represents a *worst-case* scenario where substantially more mine water is derived from surficial sources than the available evidence would suggest. Mine overflow occurs here in 4050 days (11.1 years), at a rate of $3.7\,1\,s^{-1}$.
- Run 6 is an alternative to Run 3, where, instead of assigning the 'missing' (undocumented) $1\,1\,s^{-1}$ mine water flux to level −340, it is assigned to a higher-level fracture system at −200 m. It can be seen that this has relatively little effect on the results.

Note that: (i) if the head-independent inflow F_i is low enough (less than around $0.5\,1\,s^{-1}$; see Run 1), or (ii) if dewatered highly transmissive connections with regional groundwater systems exist in the upper levels of the mine, the mine water may not overflow at all at the −80 level Socavón adit. It may, instead, attain an equilibrium with regional groundwater heads.

The MODFLOW model. For characterization of contaminant transport pathways in the Quaternary Caracollo–Challapampa–Oruro aquifer complex, a model was developed for the assessment of regional groundwater hydraulics using repetition MODFLOW (McDonald & Harbaugh 1988). The development, calibration and application of this regional model are documented by URS Dames & Moore Norge (2000d). The Oruro massif was incorporated as a subsection in the model, and it was possible to investigate the effects of cessation of pumping of the San José Mine. Figure 9 presents a schematic representation of the hydrogeological system showing the main inflows to the mine areas. The dewatering of the mine to approximately the −460 m level was simulated as a fixed head boundary (of limited extent) at 3340 m asl. Regional infiltration was estimated to be 0.01% of direct precipitation. However, in the mined area of the Oruro massif, the presence of mine openings and fractures has led to an area of the model with increased recharge such that a total of some $1\,1\,s^{-1}$ is recharged in the vicinity of the mine.

The hydraulic conductivity of the bedrock is one of the unknowns in the model and the main objective of a steady-state calibration was to provide an estimate of this value. Literature reviews and packer testing of cored boreholes (URS Dames & Moore Norge 2000c) indicate that hydraulic conductivities of the Palaeozoic–Tertiary (i.e. 'bedrock') lithologies present near the mine workings typically range from 10^{-9} to $10^{-5}\,m\,s^{-1}$, depending on degree of weathering, fracture density and aperture.

Based on the known volume of the floodable mine workings ($c.\ 2 \times 10^6\,m^3$) and an estimate of the rock volume containing the floodable workings ($532 \times 10^6\,m^3$), the mine voids were

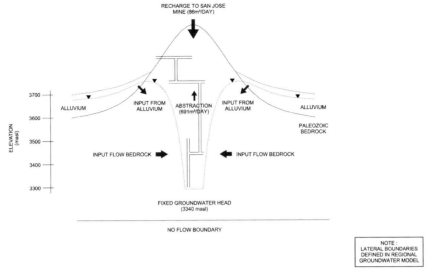

Fig. 9. Schematic section through the Oruro massif illustrating the conceptual model behind the MODFLOW simulation.

simulated in the model by applying a 'bulk porosity' of 0.4% to the cells occupied by the mine. The base of the model was designed as a 'no-flow' boundary. A steady state calibration was undertaken of the San José area, by varying the hydraulic conductivity of the bedrock until calculated outputs from the model approximated to the following calibration targets:

- area of mine dewatering restricted to the immediate area of the mine workings;
- abstraction from the mine for a fixed (pumped) head at level -460 should be $8 \, \text{l s}^{-1}$.

To achieve calibration, a bulk hydraulic conductivity of $3 \times 10^{-8} \, \text{m s}^{-1}$ was required. The results of the steady-state calibration are presented in Fig. 10 as a cross-section through the dewatered area of the mine. The model suggests that approximately 70% of the $8 \, \text{l s}^{-1}$ pumped mine water is derived from inflow from the Quaternary aquifer complex entering the Palaeozoic bedrock in the area around San José. This is a negligible flux compared to total groundwater flows in the Quaternary aquifer complex and there are no significant effects of the mine dewatering on the simulated groundwater heads in the Quaternary aquifer. The remaining 30% is provided directly from the surrounding Palaeozoic bedrock (20%), and direct recharge in the area of dewatering (10%). It should be noted, however, that the model does not permit any inflow of groundwater from deep in the bedrock underlying the mine. Chemical and temperature evidence suggests that a substantial part of the San José mine water is derived from deep saline formation water sources, which are *not* adequately simulated in the model.

In the transient version of the model, the initial level of dewatering of the mine was simulated as a general head boundary at 3340 m asl (-460 level). To simulate cessation of pumping, the hydraulic conductivity of the general head boundary was decreased to an infinitely small value, to prevent any further removal of water from the system. The model was then run for 100 years to assess the rebound in water levels, and the results are shown in Fig. 11. The model predicts that the water level in the mine will reach a steady-state level of 3710 m asl after 95 years. Although this is below the overflow level of 3720 m asl, the final level reached is dependent on the estimated infiltration of surficial water, currently set at $1 \, \text{l s}^{-1}$. The MIFIM model is probably better suited than the MODFLOW model at simulating the detailed rates of filling of mine void. It is, however, reassuring that both models predict rebound to c. 3700 m asl within a few tens of years.

Simulation of groundwater leakage from a flooded mine. In the MODFLOW model, an overflowing mine was simulated by a fixed head at 3720 m asl (approximate level of the Socavón adit). In this case, the model estimates

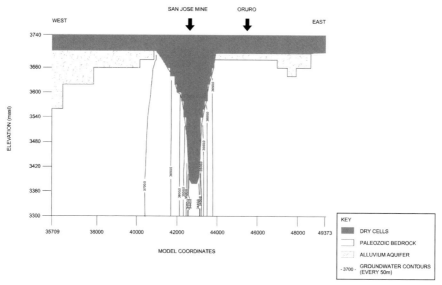

Fig. 10. Schematic section through the Oruro massif illustrating the results of steady-state calibration of the MODFLOW simulation.

the magnitude of groundwater flow from the mine workings to the surrounding Altiplano Quaternary aquifer system to be some $0.5 \, l\,s^{-1}$. This corroborates the results of the MIFIM model (see above).

Mine water quality

Hydrochemical characteristics

Analyses of water from San José are published by SGAB-COMIBOL (1995a, b), Flores (1998), Pescod & Younger (1999). The mine water being pumped from San José Mine is of extremely poor quality, with pH around 1.5, up to $3 \, g\,l^{-1}$ iron, around $8 \, g\,l^{-1}$ sulphate and high concentrations of a range of heavy metals and other potentially toxic elements. According to analyses cited by Flores (1998), some 50–70% of the total iron is in ferrous (+II oxidation state) form.

Approximate loadings ($kg\,day^{-1}$) of metals discharged in the San José mine water have previously been calculated by SGAB-COMIBOL (1995a) and are reproduced in Table 3.

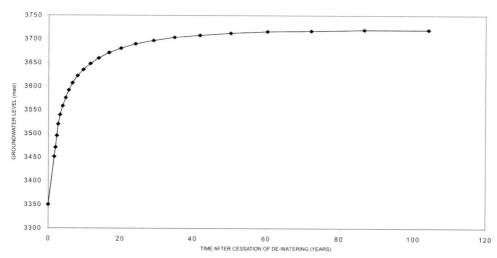

Fig. 11. Rise in mine water levels in the San José Mine, in years after pump shut-off, with the original mine water level at level −460 m, according to the MODFLOW simulation.

Table 3. *Load of dissolved contaminants discharged from San José Mine in kg each day, based on 9 l s^{-1} pumping rates (after SGAB-COMIBOL 1995a)*

Parameter	Discharge (kg day^{-1})
Ag	1
As	23
Cd	4
Cr	3
Cu	33
Fe	3850
Ni	2
Pb	23
Sb	8
Zn	117

During the course of the current study, the mine water emerging from the Santa Rita access adit was repeatedly sampled every 1–2 weeks. Analytical results and a summary of methods are shown in Table 4. Full sampling and analytical protocols are provided by URS Dames & Moore (2000a).

For the first sample (DM002, 5 October 1999), an additional aliquot was collected. This parallel sample (DM002-N) was sent to Norges geologiske undersøkelse (NGU), Trondheim for control purposes. A comparison of the NGU results with those provided via the contracted laboratory of DM002 were broadly comparable (Table 4), except for the case of chloride (where the ion chromatography methods employed by NGU are probably to be preferred) and the following elements: Ba, Be, Ce, La, Mo, Sn, V and Zr. In addition, discrete mine water leakages within the mine were also sampled and analysed. The results are shown in Table 5.

The new results confirm that the pumped mine water is very aggressive (low pH, high concentrations of dissolved sulphate, iron and metals) compared with waters from mines in many parts of the world. This is indicative of:

- large exposed areas of sulphide (especially pyrite) mineralization, leading to rapid oxidation kinetics;
- low water through-flow relative to quantities of oxidized sulphides;
- lack of buffering species in the ore, host rock and ambient groundwater.

The low pH of the water leads to dissolution of silicate and carbonate (if present) mineral phases such as feldspars, which release significant quantities of silicon (Si) and the major cations (Ca, Mg, Na, K) to the water. High concentrations of Th (up to 41 μg l^{-1}) and Tl (up to 124 μg l^{-1}) are recorded. The analyses also confirm the presence of around 1 mg l^{-1} Ag in the mine water. Gold (Au) was also detected in three samples at 0.2, 0.3 and 0.7 μg l^{-1}.

Possibly the most interesting feature of the mine water is the extremely high concentration of sodium chloride salinity detected in the water (over 30 000 g l^{-1} chloride, i.e. almost twice the concentration in seawater). In fact, this is a higher salinity than most of the Quaternary aquifer groundwaters in the Oruro area (see Fig. 5) and the salinity is unlikely to be derived from infiltration from these aquifers. There are no obvious anthropogenic or mineralogical sources for sodium chloride in the mine. Analyses (Flores 1998) suggest that the main inflow of saline water is from around level −420. A substantial proportion of the water infiltrating into the mine is deep bedrock groundwater (saline and possibly slightly thermal) from the Tertiary volcanogenic massif.

The most saline water inflows recorded in the mine have chloride concentrations of some 49 000 mg l^{-1}, temperatures of over 25°C and conductivities of up to 128 000 μS cm^{-1} (the very small inflow DM-Int-6 has even higher salinity and temperature, see Table 5). The pumped mine water has chloride concentrations of around 30 000 mg l^{-1}, temperatures of some 21°C and conductivities of around 90 000 μS cm^{-1}. Thus, the proportion of deep saline water in the final pumped water is over 50%, and probably nearer 60–70%, relative to 'fresh' water derived from surface infiltration.

Speciation modelling of mine water quality

The speciation of elements present in the mine water has been investigated using the Davies activity algorithm in the US EPA (1991) code MINTEQA2, for the sample DM002-N (Table 4). Input parameters were temperature, pH, Na$^+$, K$^+$, Ca^{2+}, Mg^{2+}, Cl$^-$, SO$_4^{2-}$, Br$^-$, Al^{3+}, Fe (60% as Fe^{2+}, 40% as Fe^{3+}), Mn^{2+}, Cu^{2+}, Zn^{2+}, Ni^{2+}, Ba^{2+}, Sr^{2+}, Ag$^+$, Li$^+$, Pb^{2+}, Cd^{2+}, Si (as H$_4$SiO$_4$). Results (Table 6) indicate that the mine water is highly undersaturated relative to primary silicates and clay minerals, and that only a relatively small number of minerals would be expected to precipitate from the water at its initial pH level. These include silica, gypsum–anhydrite, strontium as celestite, lead as anglesite, iron as jarosite and goethite (probably causing the bright yellow colour observed in the mine water channel) and barium as barite. The other heavy metals (apart from lead) are not expected to precipitate.

X-Ray diffraction analysis of precipitates in the mine water channel proved the presence of

Table 4. Analytical results for composition of pumped mine water at Santa Rita entrance

Sample ID:	Unit	DM002-N 05/10/99	DM002 05/10/99	DM006 19/10/99	DM007 25/10/99	DM009 05/11/99	DM011 25/11/99	DM012 06/12/99
Field measurements								
pH		1.72	1.72	1.47	1.42	1.40	1.0*	1.52
Temp.	°C	21.0	21.0	20.8	22.7	21.8	20.5	21.4
EC	$\mu S\,cm^{-1}$	95 000	95 000	84 000	106 000	96 000	87 500	84 100
TDS, TSS and anions (laboratory determinations)								
TDS	$mg\,l^{-1}$	na	74 636	75 282	88 156	69 045	77 674	71 888
TSS		na	16.75	227.3	14.4	47.2	1350	15.2
Cl^-	$mg\,l^{-1}$	27 000	41 950	32 670	44 230	31 010	40 600	36 313
SO_4^{2-}		8200	7927	8477	8470	13 410	623	7586
NO_3^-		<0.4	na	na	na	na	na	na
Major cations and Si (laboratory determinations)								
Ca	$mg\,l^{-1}$	1720	1740	1780	1750	1460	2280	1660
Mg		769	873	838	975	686	807	654
Na		18 800	23 970	17 256	23 440	18 570	18 433	20 325
K		274	274	293	517	264	349	264
Fe		2220	1820	2460	2930	3080	2350	1880
Mn		18.0	19.1	27.4	20.4	19.3	17.7	20.3
Al		489	473	559	1330	875	542	458
Si		18.2	24.3	23.3	25.4	25.5	10.8	17.3
Minor elements (laboratory. determinations)								
Ag	$\mu g\,l^{-1}$	1100	921	1610	1060	3840	860	2180
As		24 500	24 000	34 600	38 900	40 900	33 800	28 000
Au		na	0.3	<0.2	<0.2	0.2	<0.2	0.7
B		9270	na	na	na	na	na	na
Ba		103	896	123	288	102	150	139

Be	<1	44	54	102	56	34	37
Cd	2000	1890	2750	3540	2730	1900	2040
Co	732	611	811	1050	971	647	527
Cr	233	245	320	526	441	412	309
Cu	17000	15500	21900	21600	19200	17200	16000
Hg (ICPMS)	na	<20	na	na	na	na	na
Hg (CVAA)	0.56	<1	1.4	1.2	1.6	<0.001	<0.001
La	428	10.5	12.2	17.6	14.5	9.3	7.6
Li	72000	50300	49600	66600	34300	36400	41200
Mo	226	<10	15	<10	24	<10	20
Ni	797	1580	1620	2720	2330	1160	1010
Pb	26400	32700	52200	34000	31000	36800	47500
Pt	na	<2	<2	<2	<2	<2	<2
Sb	1930	2020	3160	2680	3720	2600	3090
Se	1.7	<20	<20	<20	62	<20	37
Sn	19.8	153	293	692	467	280	226
Sr	26500	30900	26400	36400	21200	31200	29100
Th	na	16.5	22.0	40.7	27.7	16.3	14.6
Ti	62.1	37	107	221	126	75	95
Tl	na	50.2	83.0	124	84.6	51.5	60.0
U	na	7.2	11.8	26.3	8.8	7.9	22.3
V	881	271	357	893	700	488	355
Zn	78200	67800	79400	124000	98000	62600	52900

Samples for metals and trace elements filtered at 0.45 μm. NB: *DM002-N*=duplicate determination by Geological Survey of Norway, with following methods: Hg by cold vapour atomic absorption (CVAA); Pb, As, Cd, Se, Sb, Sn by AA; Cl, NO_3^-, SO_4^{2-} by ion chromatography; Si, Al, Fe, Ti, Mg, Ca, Na, K, Mn, P, Cu, Zn, Ni, Co, V, Mo, Cr, Ba, Sr, Zr, Ag, B, Be, Li, Sc, Ce, La, Y by inductively coupled plasma-atomic emission spectroscopy. Other samples by SpectroLab/ActLabs, Canada, with following methods: All trace elements by inductively coupled plasma-mass spectrometry (ICP-MS) (except where stated for Hg, cold vapour AA); major cations and Si by ICP-MS (except where concentrations too large, when determined by AA, e.g. Na); Cl^- by Moor titration with silver nitrate; SO_4^{2-} by gravimetry, checked by photometric methods.
*pH measured using pH paper.
na, not analysed.
TDS, total dissolved solids; TSS, total suspended solids.

Table 5. Results of analyses of waters (DM-Int-1 – DM-Int-6) sampled in the interior of San José Mine between 18 November 1999 and 1 December 1999

No.	Level	Field			Laboratory determinations								
		T (°C)	EC ($\mu S\,cm^{-1}$)	pH	EC ($\mu S\,cm^{-1}$)	pH	Na ($mg\,l^{-1}$)	K ($mg\,l^{-1}$)	Ca ($mg\,l^{-1}$)	Mg ($mg\,l^{-1}$)	Fe ($mg\,l^{-1}$)	Cl ($mg\,l^{-1}$)	SO_4^{2-} ($mg\,l^{-1}$)
Int-1	at −420 m	21.6	100 700		101 000	3.0	27 400	408	2797	740	128	49 070	1692
Int-2	pumped from −420 and −460 m	20.6	97 300		104 700	2.0	24 600	318	2256	810	876	41 755	5977
Int-3	at −460 m	14.5	109 900		111 000	3.0	24 900	398	2707	825	366	45 815	1660
Int-4	at −340 m	25.2	127 700		63 200	3.0	13 060	246	963	410	226	22 820	1675
Int-5	at −340 m	18.6	59 800		74 470	1.5	17 100	189	849	625	1904	18 305	19 479
Int-6	at −460 m	26.5	125 000	6.75	132 000	7.0	34 703	562	4951	1150	1.6	65 870	625

gypsum, quartz and halite. As the mine water mixes with other surface waters (such as the rather alkaline headwaters of the Rio Tagarete), the pH might be expected to rise, and other metals precipitate more efficiently.

Isotopic characteristics

Several isotopic studies have been carried out of the waters of the Oruro region (Salazar Delgado undated; SGAB 1996a). Mine waters are found to have δ^2H and $\delta^{18}O$ signatures below −120 and below −16‰, respectively, whereas groundwaters of the Quaternary system have signatures in the range −80 to −120 and −10 to −16‰, respectively (Salazar Delgado undated). This indicates that mine water is dominantly not immediately derived from rainfall or from rapid, direct infiltration of groundwaters. Intriguingly, the data of SGAB (1996a) does not support the observations of Salazar Delgado, reporting $d^{18}O$ values of only around −13 to −14‰ for the mine water.

Mine waters have negative $d^{34}S$ signatures, as do observation boreholes impacted by mine water or mine waste tip pollution (SGAB 1996a). Other groundwaters in the Quaternary aquifer system possess signatures ranging from 0 to +23‰. These signatures are, however, related to the origin of dissolved sulphate and not to the origin of the water per se.

Prognosis for overflowing mine water

The current quality of pumped mine water cannot be taken as representative of water overflowing from a flooded mine, for several reasons:

- Rising mine water will dissolve intermediate sulphide oxidation products and efflorescences (mineral growths on rock surfaces formed by the evaporation of mineral-saturated water) within dewatered rocks, fractures and mine voids, resulting in a 'first flush' of highly contaminated water. Large quantities of such secondary minerals can be observed in the mine as efflorescences and stalactites. X-Ray diffraction analyses of selected samples are presented in Table 7.
- The flooding of the mine prevents oxygen from reaching flooded sulphide bodies. Thus, following flooding, the rate of new pyrite oxidation will be considerably reduced (Younger 1997, 2000). Thus, the 'first flush' of highly contaminated mine water will tend to experience exponential decay of contaminant concentrations as clean

Table 6. *Common minerals that are oversaturated ((Saturation index) SI > 0.5), saturated ($-0.5 <$ SI < 0.5) or undersaturated (SI < -0.5) with respect to the mine water sample DM002-N, calculated using the Davies algorithm in MINTEQA2 (US EPA 1991)*

Undersaturated minerals	Saturated minerals	Oversaturated minerals
Amorphous $Al(OH)_3$	$AlOHSO_4$	Barite ($BaSO_4$)
Alunite ($KAl_3(SO_4)_2(OH)_6$)	Anhydrite ($CaSO_4$)	Chalcedony (SiO_2)
Epsomite ($MgSO_4 \cdot 7H_2O$)	Celestite ($SrSO_4$)	Gypsum ($CaSO_4 \cdot 2H_2O$)
Ferrihydrite ($Fe_{4-5}(O,OH)_{12}$)	Goethite ($FeO(OH)$)	Na-Jarosite ($NaFe_3^{III}(OH)_6(SO_4)_2$)
$Fe_2(SO_4)_3$	Amorphous silica (SiO_2)	K-Jarosite ($KFe_3^{III}(OH)_6(SO_4)_2$)
Gibbsite ($Al(OH)_3 \cdot 3H_2O$)	Anglesite ($PbSO_4$)	H-Jarosite ($HFe_3^{III}(OH)_6(SO_4)_2$)
Halite (NaCl)		Quartz (SiO_2)
Melanterite ($FeSO_4 \cdot 7H_2O$)		
Mirabilite ($Na_2SO_4 \cdot 10H_2O$)		
(All common minerals of Mn, Cu, Zn, Cd, Ni, Ag, Pb, except anglesite)		
Cerargyrite (AgCl)		
Primary silicates		
Kaolinite		
Montmorillonite		

recharge water flushes out the intermediate pyrite-weathering phases. In a mine such as San José, located in a semi-arid area with low recharge and a high volume of workings, the rate of concentration decay would probably be slow.

- In the long term, concentrations will approach a stable value, representing the steady-state weathering of sulphides in the flooded mine. In the case of San José, this concentration may still be significant given the large volume of old workings above the probable level of flooding.
- As the mine floods, deep inflows of groundwater will be suppressed. Thus, parameters related to these deep inflows (e.g. salinity) should decline.

Mine wastes as a source of risk

Four hundred years of mining have resulted in large volumes of mining waste (ranging from coarse-grained rock waste to fine-grained tailings from ore milling and processing) being disposed of on and around the Oruro massif (SGAB-COMIBOL 1995a). During heavy rainfall, sulphide oxidation products are washed out of these wastes to form torrents of contaminated water which, in the worst storms, rush through the streets of the town. SGAB-COMIBOL (1995a) allege that the acidic mine waters and contaminated groundwaters have a corrosive effect on underground infrastructure in Oruro, including potable water and sewage pipeline networks.

Samples of leachate from the Frankeita tailings, the Itos tailings and the San José waste rock were collected on 7 October 1999, on a morning after heavy rain. The data from these samples are presented in Table 8. Note in particular, the substantial quantities of uranium (U), thorium (Th) and thallium (Tl) present in the leachates. Of the three leachates sampled, that from Itos is by far the most aggressive, with low pH (0.72) and extremely high loadings of iron, sulphate and heavy metals.

The leachate from the mine wastes also appears to drain into the shallow Quaternary aquifers below the city of Oruro. Data from observation wells (Fig. 12a) suggest that groundwater contamination by a range of mining-related elements (Ag, Al, Be, As, Bi, Ca, Cd, Co, Cr, Cs, Cu, rare earth elements, Fe, Mg, Mn, Ni, Pb, Rb, Sb, Se, Si, Sn, Sr, SO_4^{2-}, Th, Tl, U, V, Zn) occurs immediately down-gradient of mine waste tips. This appears to be rapidly attenuated in the geological environment. Table 9 shows data from a series of observation boreholes in the Quaternary aquifer complex, located on a postulated flow-path down-gradient of the San José and Frankeita mine wastes (see Figs 1 and 12). It will be seen that concentrations generally decrease rapidly with distance through PPSJ1-PPSJ7-DM5. Elevated concentrations of a range of contaminants in the area of DM6 and PPSJ10 may be due to direct infiltration of mine water from the mine water channel and River Tagarete in this area.

URS Dames & Moore Norge (2000b) also note the likelihood of several other possible

Table 7. *Mineralogy (XRD) of samples of secondary mineral efflorescences collected at the −340-m level of the San José Mine complex on 30 September 1999 pH obtained on 1:20 mixture/dissolution of sample in distilled water is also cited*

Sample No. and description	Dominant mineralogy	pH on dissolution
E1: Pale blue (-green), fibrously macrocrystalline material	Melanterite. $FeSO_4 \cdot 7H_2O$	3.7
E2: As E1 and also microcrystalline white, pink and yellow material	Römerite. $Fe^{II}Fe_2^{III}(SO_4)_4 \cdot 14H_2O$	1.3
E3: Bright orange–yellow efflorescence. Damp	Metasideronatrite. $Na_4Fe_2^{III}(SO_4)_4(OH)_2 \cdot 3H_2O$	2.5
E4: Mushy, cream–white efflorescence (and soft brownish white stalactite). Damp	Halotrichite. $FeAl_2(SO_4)_4 \cdot 22H_2O$ Pickeringite. $MgAl_2(SO_4)_2 \cdot 22H_2O$	2.8
E5: Powdery yellow efflorescence. Dry	Aluminocopiapite. $(Mg, Al)(Fe^{III}, Al)_4(SO_4)_6(O,OH)_2 \cdot 20H_2O$	2.0

sources of contamination (sewage and solid waste) contributing towards poor groundwater quality under certain areas of Oruro, especially in the areas around DM5, DM6 and PPSJ-10, where elevated concentrations of, for example, nitrogen species are noted (Fig. 12b).

Summary – risk source characterization

For the purposes of risk assessment, the sources of contamination have thus been characterized as follows:

- in the case of pumping cessation and future mine flooding, a *best-guess* scenario envisages the overflow of $1-2 \, l \, s^{-1}$ of mine water from the Socavón adit, which should be collected and treated. An uncontrolled leakage of $0.5 \, l \, s^{-1}$ mine water to regional groundwater was estimated. A *worst-case* scenario envisaged the break-out of $4 \, l \, s^{-1}$ mine water from an undiscovered low-level adit or fracture, discharging to the Quaternary aquifer complex;
- mine water quality following flooding would be a 'first-flush', assumed to contain double the maximum concentrations of contaminants recorded in analyses of pumped mine water;
- in the case of leachate from mine wastes, the semi-arid climate and episodic rainfall events made it impossible to adequately characterize leachate quality and quantity at source. Instead, groundwater quality in borehole PPSJ1 was regarded an effective source term (albeit displaced 'downstream' from the true source term), as it represents the integration of pollutant inputs over a long period.

Risk assessment – an overview

Space does not permit a full description of the risk assessment methodology. However, in order to simulate transport of contaminants in groundwater, two modelling approaches were utilized:

- the ConSim model (Golder Associates & Environment Agency 1999), which simulates attenuation by sorption and dispersion through a porous aquifer;
- the MPATH (Lichtner 1992, 1996) reactive transport model, which calculates rates of precipitation and dissolution of master species such as K-feldspar, alunite, sericite, kaolinite, gibbsite and chalcedony along a one-dimensional (1-D) flow path

Table 8. *Analyses of samples of leachate taken from Frankeita tailings, San José Mine wastes and Itos tailings following heavy rain*

Sample ID:	Unit	DM-003 Frankeita 07/10/99	DM-004 San José 07/10/99	DM-005 Itos 07/10/99
Field measurements				
pH		2.03	1.00	0.72
Temp	°C	9.4	17.5	26.6
EC	$\mu S\,cm^{-1}$	2200	50 000	75 000
TDS, TSS and Anions (laboratory determinations)				
TDS	$mg\,l^{-1}$	2155	115 593	193 782
TSS		<5	<0.5	<0.5
Cl^-		125	310	246
SO_4		988	67 650	109 410
Major cations and Si (laboratory determinations)				
Ca	$mg\,l^{-1}$	140	296	202
Mg		20.6	140	133
Na		131	20 100	27 960
K		6.5	24.5	120
Fe		132	11 440	20 890
Mn		4.7	19.4	11.3
Al		50	1,300	1,560
Si		11.0	44.5	37.2
Minor elements (laboratory determinations)				
Ag	$\mu g\,l^{-1}$	0.4	35	62
As		68.9	277 000	711 000
Cd		291	1230	9490
Cu		1280	35000	77600
Hg (CVAA)		<1	<1	<1
Pb		327	2510	4390
Sn		<0.1	1490	18 000
Th		6.0	136	603
Tl		0.16	63.0	314
U		4.3	47.5	82.0
Zn		7220	24 400	317 000

All trace elements by inductively coupled plasma-mass spectroscopy (ICP-MS) (Hg by cold vapour atomic absorption (CVAA)); major cations and Si by ICP-MS, except where concentrations too large, when determined by atomic absorption (AA) (e.g. Na). Cl^- by Moor titration with silver nitrate, SO_4^{2-} by gravimetry. Samples for metals and trace elements filtered at 0.45 μm.
TDS, total dissolved solids; TSS, total suspended solids.

and concomitant changes in pH. (Calcite was not included as no conclusive evidence of its presence in Quaternary sediments was obtained. Its presence would, however, render attenuation of most species even more effective.) pH, in turn, controls the solubility of the majority of potentially toxic elements in the mine water.

Input data included real hydraulic head gradients measured from monitoring wells (Fig. 6), hydraulic conductivity values based on pumping tests and contaminant concentrations derived from the source characterization (see above). Recipients were defined as private and public groundwater abstractions with Bolivian drinking water norms as the primary failure criterion.

Both models predicted the extremely rapid attenuation of most contaminant parameters in the Quaternary aquifer complex, with its high pH. The performance of the models could be assessed against empirical data, namely contaminant concentrations along the flow path (mine wastes–PPSJ1–PPSJ7–DM5). The ConSim model simulated adequately the attenuation of most contaminant concentrations along this path. It predicted only a slow migration of the contamination front from the mine wastes, at a rate of up to 100 m every 50 years. The MPATH model predicted that buffering reactions within the aquifer cause a pH-related contaminant attenuation 'front' which will remain within 500 m of the source for a time scale of 100 years. Within the following 200 years, the 'front' would

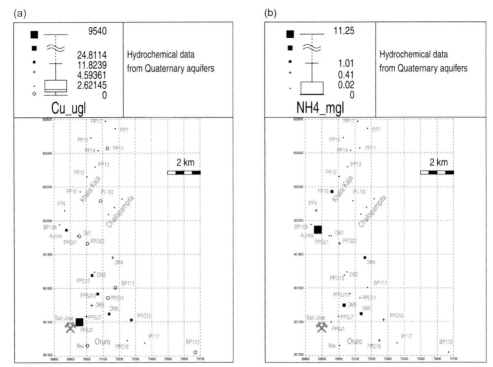

Fig. 12. (**a**) Map showing concentrations of copper (μg l^{-1}) in groundwaters from the Quaternary Altiplano aquifer complex in the vicinity of Oruro, based on sampling and analysis of 1999–2000. Symbol sizes are based on the boxplot (inset). In the case of copper, elevated concentrations around PPSJ-10, DM-6, PPO-15 and PPO-13 may be the result of direct infiltration of mine water from the mine water channel and/or Tagarete. (**b**). Map showing concentrations of ammonium (mg l^{-1}) in groundwaters from the Quaternary Altiplano aquifer complex in the vicinity of Oruro, based on sampling and analysis of 1999–2000. Symbol sizes are based on the boxplot (inset). Elevated ammonium around DM-5 and DM-6 is probably from waste water. That at 'Aurora' is probably natural and related to reducing aquifer conditions.

migrate only a matter of a few tens of metres further (URS Dames & Moore Norge 2000*d*).

Conclusions

The investigation has demonstrated that it is possible, with conventional techniques of investigation and sampling, to construct a conceptual model of a mine system for the purposes of risk assessment. The key information required to quantify this conceptual model includes:

- detailed mine plans (supplemented by data on ore and waste production);
- knowledge of the quantities and locations of water pumped;
- survey to locate the sites of the main underground water leakages;
- estimates of surficial recharge to the mine (mine openings, meteorological data);
- sampling and analysis of pumped mine water and individual underground leakages (not neglecting major ion, isotope and salinity parameters);
- anecdotal evidence from pump operators and miners concerning the behaviour of underground leakages.

Accurate prediction of mine water quality following flooding could not be predicted quantitatively on the basis of existing pumped water quality. Existing water quality did, however, place a lower limit on the expected concentrations of contaminants. A definitive test of the validity of the conceptual mine model will, of course, only take place once the mine ceases pumping and the rise of water within the mine void is monitored for the purpose of refining and calibrating the model.

Assuming that $1-2$ l s^{-1} of the mine water are derived from infiltration of superficial waters (e.g. rainfall), with $6-7$ l s^{-1} being derived from infiltration of deep groundwater, modelling suggests that:

Table 9. *Groundwater chemistry in the wells PPSJ1, PPSJ7, DM5, DM6 and PPSJ10 (data from URS Dames & Moore Norge 2000c, d). na, not analysed. Samples for metals and trace elements filtered at 0.45 μm*

Well	Unit	PPSJ-1	PPSJ-7	DM5	DM6	PPSJ-10
Temp.	°C	16.3	20.4	16.3	14.1	14.4
pH		3.23	6.25	6.34	7.21	6.37
EC	$\mu S\,cm^{-1}$	16 900	5570	6640	40 700	19 600
Alkalinity	$meq\,l^{-1}$	0	4.13	7.07	9.03	18.53
Cl^-	$mg\,l^{-1}$	6550	1350	1690	16 310	7040
SO_4^-	$mg\,l^{-1}$	4045	855	1380	3350	1380
$NO_3^- + NO_2^-$	$mg\,N\,l^{-1}$	<0.5	<0.5	9.8	3.42	<0.5
Hg by CVAA	$mg\,l^{-1}$	<0.001	<0.001	<0.001	<0.001	<0.001
Na by AA	$mg\,l^{-1}$	2835.8	730.5	1163.8	11 288.4	3791.2
NH_4^+	$mg\,l^{-1}$	<0.5	<0.05	2.29	1.7	0.18
Ag	$\mu g\,l^{-1}$	455	<0.8	<0.2	<8	<4
Al	$\mu g\,l^{-1}$	375 000	36	<2	<80	53
As	$\mu g\,l^{-1}$	38.6	1.43	3.75	23.1	1.4
Be	$\mu g\,l^{-1}$	42.9	0.5	<0.1	<4	<2
Ca	$\mu g\,l^{-1}$	1 220 000	414 000	484 000	346 000	677 000
Cd	$\mu g\,l^{-1}$	1890	130	63.1	45.9	1.6
Co	$\mu g\,l^{-1}$	582	4.79	10.8	8.2	0.9
Cr	$\mu g\,l^{-1}$	339	<2	<0.5	31	<10
Cu	$\mu g\,l^{-1}$	9540	15.2	11.8	152	43
Fe	$\mu g\,l^{-1}$	232 000	23 100	971	904	2780
K	$\mu g\,l^{-1}$	7480	21 900	30 700	151 000	51 600
La	$\mu g\,l^{-1}$	59.2	0.07	0.059	6.6	<0.1
Li	$\mu g\,l^{-1}$	7930	1050	1180	38 200	10 100
Mg	$\mu g\,l^{-1}$	397 000	120 000	107 000	283 000	438 000
Mn	$\mu g\,l^{-1}$	112 000	6310	24 600	6410	171
Mo	$\mu g\,l^{-1}$	0.9	12.5	1.4	6	<2
Ni	$\mu g\,l^{-1}$	973	96.7	201	733	16
Pb	$\mu g\,l^{-1}$	82.7	3.9	<0.1	34	4
Sb	$\mu g\,l^{-1}$	3.95	0.8	0.41	4.9	1.1
Sc	$\mu g\,l^{-1}$	51	10	8	<40	<20
Se	$\mu g\,l^{-1}$	23	4.6	10.1	11	<4
Si	$\mu g\,l^{-1}$	67 200	17 800	24 400	20 500	18 500
Sn	$\mu g\,l^{-1}$	0.9	<0.4	<0.1	<4	<2
Sr	$\mu g\,l^{-1}$	3730	3310	3440	14 400	15 700
Th	$\mu g\,l^{-1}$	0.6	0.024	<0.003	<0.12	<0.06
Ti	$\mu g\,l^{-1}$	27.7	8.5	8.2	31	15
Tl	$\mu g\,l^{-1}$	4.58	<0.02	0.066	1.2	0.5
U	$\mu g\,l^{-1}$	13.6	0.322	1.07	111	25.2
V	$\mu g\,l^{-1}$	85.8	<0.2	0.51	21	<1
Y	$\mu g\,l^{-1}$	281	0.48	0.08	8.49	0.13
Yb	$\mu g\,l^{-1}$	14.92	0.005	0.002	0.24	<0.02
Zn	$\mu g\,l^{-1}$	31 700	1950	1.5	329	138

na, not analysed.

- the bulk hydraulic conductivity of the Palaeozoic–Tertiary crystalline bedrock is some $3 \times 10^{-8}\,m\,s^{-1}$;
- heads in the Quaternary aquifer system of the Altiplano are not significantly affected by dewatering of the mine;
- mine water rebound following pump switch-off occurs on a time scale of a few tens of years;
- according to the MIFIM model, overflow of the mine via the Socavón adit occurs within 15–21 years of pump cessation, at a

rate of $0.5-1.6 \, l \, s^{-1}$. There is, however, a possibility that the mine may not overflow at all, but reach an equilibrium situation at a level slightly below the overflow point;

- when fully flooded, leakage of mine water through the bedrock to the surrounding aquifer system, will be of the order of only some $0.5 \, l \, s^{-1}$.

The risk assessment concluded on the basis of:

- effective geochemical contaminant attenuation within the Quaternary aquifer complex

and

- the fact that the bulk of the mine area lies to the south of a groundwater divide between the city of Oruro and the Challapampa wellfields

that neither the mine wastes nor a future flooded mine pose a risk to the public supply wellfields. However, it was concluded that continued discharge of mine water to the Rio Tagarete posed a potential risk to private groundwater abstractions in the vicinity of the Tagarete channel (as well as obvious impacts on the surface water environment). Continued infiltration of mine waste leachate also presents potential risks to structures and services, as well as to potential future groundwater resources, in the immediate vicinity of the sources.

The authors wish to acknowledge the support of a large team of staff who worked on the San José project, in particular: J. P. Aabel, C. Arispe, E. Bakke, E. Baldellón, J. Baldolomar, C. Borrovich, M. Condori, L. Cortez, R. Diaz, A. Eltervåg, G. Encinas, M. Fuentes, C. Gonzales, G. Gutierrez, D.Harpley, D. Herrington, A. Huerta, A. Izquierdo, M. Jakobsen, F. Ledezma, P. López, V. Lopez, J. Mendez, M. Palacios, R. Renjel, Jr, A. Salas, R. Sanhueza, J. Santivanez, P. Schuster, F. Sempertegui, H. Silvennoinen, P. Sinton, G. Soliz, A. Soliz, C. Ulrich, N. Vallejos, W. Veizan, I. Walder and G. Zamora. O. Eppers of SpectroLab and B. Davidsen of Norges Geologiske Undersøkelse were of great assistance in providing timely analyses, while P. Younger kindly reviewed much of the work. Throughout the project, the client, COMIBOL, has been extremely supportive in terms of provision of information and access to the mine. In this connection we would particularly thank D. Condori and D. Salinas.

References

AVILA SALINAS, W.A. 1990. Petrologia del domo resurgente del Cerro de San Pedro (Oruro, Bolivia). *Revista Tecnica de YPFB*, **11**, 139–149.

BANKS, D. 2001. A volume-variable, head-dependent mine water filling model. *Ground Water*, **39**, 362–365.

CORTEZ, L. & TORREZ, J. 1996. *Programa de abastecimiento de agua y saneamiento para Oruro convenios BOL/88/004. Abastecimiento de agua potable para la ciudad de Oruro paquete abapo 1.1. Informe tecnico de los pozos NU-II, NU-III, NU-V, NU-VI, NU-VII, NU-VIII, NU-P02, NU-P03.* Report of the Servicio Geologico de Bolivia 'GeoBol', Departamento de Hidrogeologia, Cochabamba, April 1996.

DAMES & MOORE. 1967. *Phase II, Comprehensive Evaluation of the Ground-Water Conditions, Challapampita Area, Oruro, Bolivia.* Servicio Local de Acueductos y Alcantarillado, Oruro.

FLORES, J. 1998. *Caracteristicas Hidrogeologicas del Yacimiento de San José – Oruro.* COMIBOL Report.

GOLDER ASSOCIATES & ENVIRONMENT AGENCY. 1999. *ConSim v. 1.04 Users' Manual.* Golder Associates (Nottingham, UK) and Environment Agency, UK.

KELLY, W.C. & TURNEAURE, F.S. 1970. Mineralogy, paragenesis and geothermometry of the tin and tungsten deposits of the Eastern Andes, Bolivia. *Economic Geology*, **65**, 609–680.

LICHTNER, P.C. 1992. *Multiple Reaction Path Model: Version 2.0.* Gruppe Gestein-Wasser-Wechselwirkung des Geologischen und des Mineralogisch-petrographischen Instituts, Universität Bern, Bern, Switzerland.

LICHTNER, P.C. 1996. Continuum formulation of multicomponent–multiphase reactive transport. In: LICHTNER, P.C., STEEFEL, C.I. & OELKERS, E.H. (eds) *Reactive Transport in Porous Media* Reviews in Mineralogy, Mineralogical Society of America, Washington DC, **34**, 1–81.

LONG, K. R. 1992. Mines, prospects, and mineral occurrences, Altiplano and Cordillera Occidental, Bolivia. Appendix A. In: *Geology and Mineral Resources of the Altiplano and Cordillera Occidental, Bolivia.* US Geological Survey and Servicio Geologico de Bolivia. US Geological Survey Bulletin, **1975**.

LUDINGTON, S., ORRIS, G.J., COX, D.P., LONG, K.R. & ASHER-BOLINDER, S. 1992. Mineral deposit models. In: *Geology and Mineral Resources of the Altiplano and Cordillera Occidental, Bolivia.* Geological Survey and Servicio Geologico de Bolvia. US, Geological Survey Bulletin, **1975**.

MCDONALD, M.G. & HARBAUGH, A.W. 1988. A modular three-dimensional finite-difference ground-water flow model. In: *US Geological Survey Techniques of Water-Resources Investigations*, Book 6, Chap. A1, US Geological Survey, Denver, USA.

PESCOD, M.B. & YOUNGER, P.L. 1999. Sustainable water resources. In: NATH, B., HENS, L., COMPTON, P. & DEVUYST, D. (eds) *Environmental Management in Practice: Volume 2* UNESCO/Routledge, London, Chap. 3.

SALAZAR DELGADO, J. Undated. *Estudio hidrologico con isotopos en la subcuenca Oruro-Caracollo*; Editorial Universitaria, Oruro, Bolivia.

SGAB. 1996a. *Proyecto Piloto Oruro: Impact of Mining and Industrial Pollutants on Groundwater*. Swedish Geological AB Report, R-Bo-E-9.45-9702-PPO9616.

SGAB. 1996b. *Proyecto Piloto Oruro: Hydrology of the PPO area*. Swedish Geological AB Report, R-Bo-E-9.45-9605-PPO9606.

SGAB. 1996c. *Proyecto Piloto Oruro: Mineral Waste Deposits in the PPO area*. Swedish Geological AB Report, R-Bo-E-9.45-9612-PPO9611.

SGAB-COMIBOL. 1995a. *Environmental Audit of the San José Mine*. Swedish Geological AB and Corporación Minera de Bolivia, Report.

SGAB-COMIBOL. 1995b. *Anexos*. Swedish Geological AB and Corporación Minera de Bolivia, Report Annexes.

SHERWOOD, J.M. 1993. *A Lumped Parameter Model of the Groundwater Rebound Associated With the Imminent Closure of Mines in the Durham Coalfield*. Unpublished MSc thesis Dept. of Civil Engineering, University of Newcastle on Tyne.

SHERWOOD, J. M. & YOUNGER, P. L. 1997. Modelling groundwater rebound after coalfield closure. In: Chilton, P.J. *et al.* (eds) *Groundwater in the Urban Environment Volume 1: Problems, Processes and Management. Proceedings of the XXVIIth Congress of the International Association of Hydrogeologists, Nottingham, UK*, 21–27 September 1997. A.A Balkema, Rotterdam, 165–170.

TURNEAURE, F.S. 1960a. A comparative study of major ore deposits of central Bolivia, Part 1. *Economic Geology*, **55**, 217–254.

TURNEAURE, F.S. 1960b. A comparative study of major ore deposits of central Bolivia, Part 2. *Economic Geology*, **55**, 574–606.

TURNEAURE, F.S. 1971. The Bolivian tin-silver province. *Economic Geology*, **66**, 215–225.

URS DAMES & MOORE NORGE. 2000a. *Sub-Project 7. Hydrogeological Study of the San José Mine and Adjacent Aquifers Supplying Water to the City of Oruro: Report 1 – Risk Source Characterisation: San José Mine*. Report by URS Dames & Moore Norge to COMIBOL.

URS DAMES & MOORE NORGE. 2000b. *Sub-Project 7. Hydrogeological Study of the San José Mine and Adjacent Aquifers Supplying Water to the City of Oruro: Report 2 – Risk Transport Pathway: Regional Hydrogeology*. Report by URS Dames & Moore Norge to COMIBOL.

URS DAMES & MOORE NORGE. 2000c. *Sub-Project 7. Hydrogeological Study of the San José Mine and Adjacent Aquifers Supplying Water to the City of Oruro: Report 3 – Results of Field Investigations/Hydrogeochemical Interpretation*. Report by URS Dames & Moore Norge to COMIBOL.

URS DAMES & MOORE NORGE. 2000d. *Sub-Project 7. Hydrogeological Study of the San José Mine and Adjacent Aquifers Supplying Water to the City of Oruro: Report 4 – Numerical Modelling and Risk Assessment*. Report by URS Dames & Moore Norge to COMIBOL.

US EPA. 1991. MINTEQA2/PRODEFA2, *A Geochemical Assessment Model for Environmental Systems: Version 3.0 user's manual*. US Environmental Protection Agency Report, **EPA/600/3-91/021**. NTIS accession No., **PB91 182 469**.

YOUNGER, P.L. 1997. The longevity of mine-water pollution: a basis for decision-making. *Science of the Total Environment*, **194/195**, 457–466.

YOUNGER, P.L. 2000. Predicting temporal changes in total iron concentrations in groundwaters flowing from abandoned deep mines: a first approximation. *J. Contaminant Hydrology*, **44**, 47–69.

Secondary minerals in the abandoned mines of Nenthead, Cumbria as sinks for pollutant metals

C. A. NUTTALL & P. L. YOUNGER

Hydrogeochemical Engineering Research & Outreach (HERO), Department of Civil Engineering, University of Newcastle, Newcastle upon Tyne NE1 7RU, UK

Abstract: Direct observations made during underground hydrogeochemical surveys of abandoned lead–zinc mines has highlighted the precipitation of secondary zinc minerals within abandoned lead–zinc mine workings in the north Pennines. Chemical analysis of mine waters has shown that molar concentrations of sulphate exceed those of zinc by two or three orders of magnitude, although they are released in equimolar proportions following the weathering of sphalerite. The excess of sulphate over zinc indicates that there must be significant sinks for zinc within the mine workings. Secondary zinc mineral sinks (principally hydrozincite and smithsonite) are the most likely explanation for the deficit in molar zinc concentrations and these minerals have been identified underground. In addition to the secondary zinc minerals, secondary calcite and aragonite from the workings have also been shown to provide sinks for zinc (by coprecipitation and solid-solution incorporation of zinc in these minerals). Calculation of the molar quantities of zinc and sulphate involved showed that as little as 5% of the zinc, weathered daily from the mineral deposits within the workings, is found to leave the mine dissolved in the mine water. However, this is sufficient to adversely impact the ecology of the receiving waters of the River Nent, which currently receives five circumneutral zinc-rich mine water discharges and drainage from a disused aqueduct.

The River Nent lies in NE Cumbria on the Alston Block, the most northern block of the North Pennine Orefield as described by Dunham (1990). The area has experienced over two centuries of intensive galena (PbS) and sphalerite (ZnS) mining, as lead and zinc and mining continued in the area until the early twentieth century. During this time over 90 adits were created throughout the valley, some of these adits currently discharge mine water into the River Nent, which has its headwaters above the village of Nenthead (NY 7810 4370) and flows approximately NW for 8 km to join the South Tyne at the town of Alston (NY 7170 4560) (see Fig. 1).

The River Nent has a poor abundance and diversity of fish and invertebrates because of the aquatic concentrations of zinc. Figure 2 shows how zinc concentrations along the River Nent (measured monthly at Nenthead, Nenthall and Nent Force over a 1-year period) typically exceed the environmental quality standard (EQS) for salmonids of 0.5 mg l^{-1} for a river of this hardness, i.e. 200 mg l^{-1} as $CaCO_3$ (Fig. 2) (Mance & Yates 1984). Other ecotoxic metals are also present, for example lead and cadmium are usually found at concentrations of less than 0.1 mg l^{-1}. Water samples were also analysed for arsenic, but this was below detection (or at very low concentrations). However, zinc concentrations exceed those of the other ecotoxic metals by several orders of magnitude (the mine water discharges typically have zinc concentrations in the range $2-10 \text{ mg l}^{-1}$). This is due to the different solubilities of the mineral oxidation products involved. Sphalerite (ZnS) becomes oxidized to zinc sulphate, which is soluble, and galena (PbS) is oxidized to lead sulphate, which is insoluble and coats the ore mineral sealing it from further oxidation.

The River Nent receives five mine water discharges and a metal-rich drainage from a disused aqueduct at Rampgill. Figure 1 shows where these discharges enter the main river. Characterization of the mine waters over a period of 2 years has been carried out at monthly intervals by Nuttall & Younger (1999). In addition to these point sources of metal contamination, contaminated river sediments, bank deposits and tailings material means that there are diffuse inputs of metal contamination occurring along the entire length of the river.

Zinc is an essential micronutrient for mammals and birds (Förstner & Wittman 1981) and the limit for zinc in potable water supply is set at

From: YOUNGER, P.L. & ROBINS, N.S. (eds) 2002. *Mine Water Hydrogeology and Geochemistry.* Geological Society, London, Special Publications, **198**, 241–250. 0305-8719/02/$15.00
© The Geological Society of London 2002.

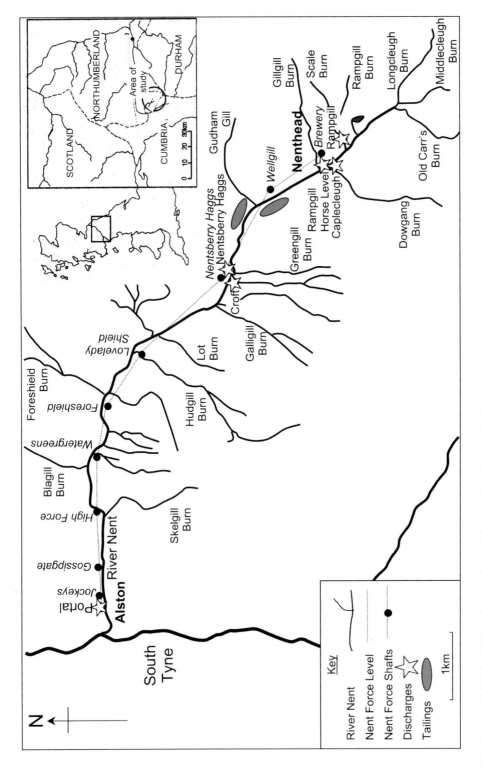

Fig. 1. Map of the Nent Valley showing the main inputs of metal contamination.

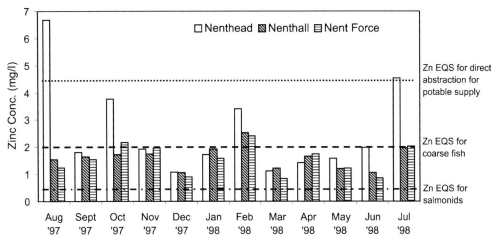

Fig. 2. Graph showing zinc concentrations at various points along the River Nent with respect to various regulatory limits.

4.5 mg l^{-1} (this is shown on Fig. 2). Above this concentration it is not harmful to humans but would affect the taste of the water. However, zinc can be very toxic to fish and invertebrates, with the free zinc ion (Zn^{2+}) being the most toxic species (Alabaster & Lloyd 1980). This species predominates in low pH waters, but geochemical modelling (carried out by the authors using WATEQ4F, Ball & Nordstrom 1991) has shown that around half of the zinc present in the Nent Valley mine waters is present in the Zn^{2+} form. In fish, zinc binds to the layer of mucus that normally coats the body surface. This causes excessive amounts of mucus to be produced and eventually the fish will suffocate (Handy & Eddy 1990). Zinc concentrations in the main river are typically 1–2 mg l^{-1} and zinc-tolerant filamentous green algae are the only thriving species. The algae become especially prolific during the summer months because there are few invertebrates present to graze upon it. It is postulated that their fungus-based digestive system cannot cope with zinc (which has fungicidal properties), therefore they are not present in the quantities and diversity expected for an unpolluted river of this type (A. Lewis, Environment Agency pers. comm.).

Geological setting

The rock types present and the structures throughout the North Pennine Orefield control the nature and spatial distribution of the mineralization (refer to Fig. 3 for a stratigraphic column). This area was so mineralogically productive due

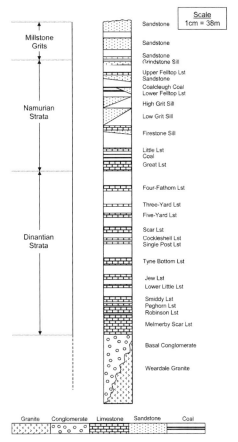

Fig. 3. Generalized vertical section showing the Carboniferous sequence within the Alston Block.

to the presence of specific rock types (notably limestone in this area), which are capable of transporting mineral bearing solutions due to their physical and chemical properties (e.g. the existence of joints and faults and the chemical reactivity of limestone are both important). The Carboniferous rocks contained within the Alston Block dip at up to 10° E and rest unconformably on basement rocks of Lower Palaeozoic folded slates (formed from Ordovician and Silurian mudstones and volcanic ash deposits) and greywackes (Dunham 1981).

Volcanism during the Lower Devonian caused the basement rocks of the Alston Block to be intruded by granitic stocks and batholiths, one of which is the Weardale Granite (Johnson 1981). These stocks were not directly responsible for orefield mineralization, as they pre-date the mineralized Carboniferous rocks and were exposed to surface weathering during the early Carboniferous. Evidence for former surface exposure of these granites comes from borehole data obtained from Rookhope during the 1960s (Dunham et al. 1965). Consequently, the only contributions these igneous rocks may have made towards the mineralization of the area is by forming an effective heat conductor during orefield mineralization and by releasing metals to hydrothermal fluids circulating within them.

Transgression and regression of the Lower Carboniferous Sea created coal, sandstone and limestone cyclothems. The rocks deposited during the Lower Carboniferous up to the base of the Great Limestone are termed 'the Dinantian strata' (Fig. 3). Rocks deposited from the Great Limestone through to the Millstone Grits are loosely referred to as 'the Namurian strata' (Fig. 3). Above the Namurian strata, numerous coal seams of the Coal Measures were formed during more than 30 cyclic depositional events (Johnson 1981).

The mineral deposits at Nenthead can be classified as Pennine-type, strata-bound deposits (Evans 1993). They provide important resources worldwide for the metals lead and zinc, and are also economic sources of barite and fluorite. In the Alston Block, faults became active due to crustal movements from the end of the Carboniferous creating a conduit network for mineralization. Hot fluids emanating from a deep source (in this case, a postulated mantle hotspot), circulated upwards through fissures, scavenging metals and fluorine from the pre-existing sedimentary and igneous rocks (Dunham 1981). The mineralization resulted from low-to medium-temperature (50–219°C) hydrothermal solutions emplaced around approximately 270 Ma (Dunham 1990). A series of productive veins and flats were produced throughout the orefield with the Great Limestone providing an especially mineralogically productive host.

Sampling methods

Mine waters were sampled at the discharge point every month over a two year period. For every water sample taken, field measurements were taken for temperature, pH, conductivity and alkalinity. Three samples were collected from each discharge in polythene bottles: one sample for anions, one unfiltered sample for cations and one sample filtered in the field (at 0.45 μm) for cations. Analysis for the anions fluoride, chloride, sulphate and nitrate was carried out using a Dionex DX-100 ion chromatograph. Cations, dissolved silica and hardness data were provided following analysis by the Environment Agency (EA). For the Caplecleugh and Nent Force Level discharges, flow was measured using a velocity area method, chosen as the only easy approach. All the other discharges were measured using a bucket and stopwatch. Water sampled underground was taken according to the method used for the surface samples. Underground minerals were collected in bags, some of each sample was ground for X-ray diffraction (XRD) analysis and the remainder of the sample was mounted onto slides in preparation for analysis by the scanning electron microscope (SEM).

Rampgill Mine

The underground surveys were carried out in Rampgill Mine due to relative ease of access. Water and mineral samples were collected over the course of several different visits. Rampgill Level was driven in 1736 and was widened to a horse level in 1800. The mine was active until 1886, by which time over 140 000 t of lead ore had been raised (Fairbairn 1993). The high productivity of the mine is due to the fact that it exploited 'flats' within the Great Limestone. Rampgill Horse Level (Fig. 1) drains the top northeastern part of the Nent Valley. It receives water from the Rampgill Mine complex and also from the adjacent workings of Smallcleugh Mine. The Rampgill Horse Level also contains the most upstream shafts to reach the Nent Force Level (marked on Fig. 1): Brewery Shaft (NY 4355 7833) and Engine Shaft (NY 7896 4361). Both of these shafts drain water from the Rampgill workings into the Nent Force Level below.

Groundwater percolating through the roof and dripping from an ore hopper in the vicinity of the Engine Shaft were sampled and analysed for zinc. Both of these seepages contain relatively high zinc concentrations (6.5 and 10 mg l^{-1} Zn, respectively), which perhaps indicates the presence of mineralized strata above this part of the mine. Indeed, these workings have the Rampgill Firestone Level workings directly above them (Critchley 1998). The concentration of zinc in these groundwater roof seepages shows that water is percolating through the workings and dissolving zinc.

Primary and secondary zinc minerals found in the Nent Valley

Sphalerite (ZnS) is a primary zinc mineral that can be found at Nenthead. It formed following precipitation of hydrothermal brines in veins and flats within the Carboniferous host rocks.

Some secondary zinc minerals are also found at Nenthead, and it is thought that these minerals may be partly responsible for the attenuation of zinc concentrations within mine workings. Secondary minerals have been directly observed by the authors within mine workings (e.g. in Rampgill Horse Level) and forming crusts and cements on spoil heaps and tailings dams (e.g. Brownley Hill Tailings Dam; XRD has shown that these precipitates are composed of hydrozincite). The secondary zinc minerals hemimorphite ($Zn_4[Si_2O_7](OH)_2.H_2O$), smithsonite ($ZnCO_3$) and hydrozincite ($5ZnO.2CO_2.3H_2O$) have all been found in mines in the Nenthead area (Dunham 1990). Hydrozincite and smithsonite were directly observed by the authors, their presence being confirmed following analysis by XRD. Hydrozincite is believed to form closes to mine entrances where ventilation is better, and smithsonite precipitation is usually favoured deeper within the mine workings where concentrations of carbon dioxide are greater (T. E. Bridges pers. comm.).

Other mineral sinks for zinc

Geochemical modelling using the WATEQ4F code of Ball & Nordstrom (1991) was performed on mine water data from Rampgill Horse Level. The following minerals were found to be at saturation and could theoretically precipitate from solution as secondary minerals: aragonite, calcite, dolomite, ferrihydrite, fluorite, goethite, quartz and amorphous zinc oxide (hydrozincite was not predicted because it is not included in the WATEQ4f database).

Calcite, aragonite, hydrozincite, smithsonite and goethite were identified by XRD in secondary precipitates from this mine. Most of the stalactites analysed were composed of calcite, aragonite and hydrozincite. The presence of aragonite is not surprising because it may precipitate chemically in preference to calcite from a solution that is charged with strontium, lead or zinc ions, presumably due to the metal ions causing distortions within the crystal lattice (Mondadori 1988). Mondadori (1988) also suggests Alston Moor as a good locality for aragonite. Coetzee et al. (1998) discovered that aragonite was forming in preference to calcite as a scale deposit during physical water treatment when zinc was present in the water. Zinc in particular, was found to decrease the rate of calcite nucleation and promote the crystallization of calcium carbonate in the aragonite form (Coetzee et al. 1998). XRD analysis also showed that the tailings dams (see Fig. 1) had secondary precipitates of hydrozincite. During war time, spoil reworking zinc oxide was also found in some of the older spoil heaps. This was noted as it reduced the efficiency of the reprocessing by lowering the flotability of the zinc ore (Dawson 1947).

The secondary deposits of calcareous minerals from Rampgill Mine were examined qualitatively using SEM. Zinc was found in all of these minerals (see Fig. 4 for an example). This suggests that the precipitation of all of these underground minerals (and especially aragonite) form interstitial sinks for zinc (i.e. the zinc is incorporated within the crystal lattice of the mineral). This is in addition to the inferred 'main phase sink' of hydrozincite and smithsonite within the workings. Attenuation of zinc concentrations may also occur by sorption onto mineral surfaces underground.

Zinc deficits in the Nent Valley mine waters

The concept of zinc deficits has been addressed by Younger (1999), who studied the mine water chemistry of a fluorspar (previously lead–zinc) mine, Frazer's Grove in Weardale County Durham (see also Johnson & Younger 2002), and found that molar concentrations of sulphate are in excess of zinc concentrations by two or three orders of magnitude. The excess of sulphate over zinc indicates that there must be significant sinks for zinc within the mine workings, as both components are sourced in equimolar proportions by sphalerite weathering. Weathering of other sulphide minerals such as galena (PbS), chalcopyrite ($CuFeS_2$) and pyrite (FeS_2) account for only a small amount of the total sulphate produced daily (less than 1%). The system studied by Younger (1999) showed that

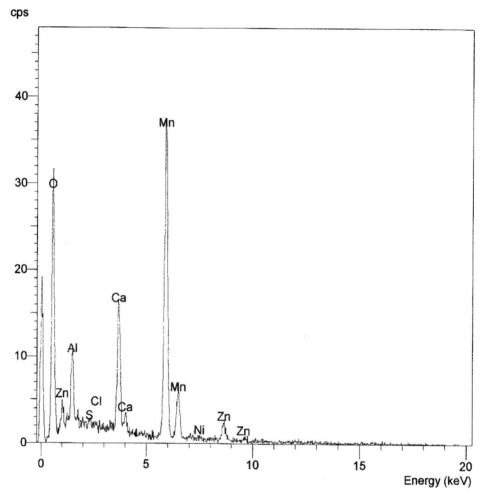

Fig. 4. Output from the qualitative SEM analysis of one of the calcareous secondary minerals found underground showing incorporated zinc.

attenuation of zinc was happening over relatively short distances within the mine. This suggests the presence of a carbonate sink for zinc that has rapid precipitation kinetics.

The main assumptions made whilst calculating zinc deficits are as follows:

- Sulphate is transported conservatively from the sites of sphalerite oxidation, evidence for this comes from geochemical modelling (using WATEQ4F, Ball & Nordstrom 1991) which shows that the feasible sinks for sulphate (gypsum and goslarite) are undersaturated and therefore not precipitating in large quantities underground.
- Significant pyrite oxidation and subsequent precipitation as ochre is not taking place (this would lead to an overestimation of the sulphate attributed to sphalerite oxidation). Evidence for this comes from the lack of ochre noticed on underground observations (it occurs in localized patches). In addition, there is no evidence for significant iron attenuation because iron concentrations found in water samples taken deep within the workings (usually less than $1 \, \text{mg} \, l^{-1}$) are comparable with those found at the mine entrance.

The zinc deficit data for the Nent Valley mines are calculated using mine water flow rates and zinc and sulphate concentrations to calculate the amount of zinc and sulphate produced per day in $\text{mg} \, l^{-1}$. The number of moles of zinc and sulphate produced per day was calculated

Table 1. *Results of zinc-deficit calculations for mines in the Nent Valley (24 February 1998)*

Mine	Moles of SO_4^{2-} day^{-1}	Moles of Zn day^{-1}	Moles of SO_4^{2-} day^{-1} other sources	Moles of SO_4^{2-} as Zn deficit	Molar ratio S:Zn	Zn dissolved (kg/day^{-1})	Kg Zn leaving mine per day	% Zn lost from mine per day
Caplecleugh	2946	172	2.65	2772	17	191	11.18	5.8
Rampgill Horse Level	1404	53	21	1331	26	90	3.44	3.8
Haggs	647	21	0.8	626	31	42	1.43	3.2
Croft	351	9	1.7	340	39	23	0.59	2.6
Nent Force	4478	103	7	4268	43	290	6.71	2.3

by dividing the amount produced per day (in mg) by the gram formula weight (i.e. by 65 and 96 for zinc and sulphate, respectively, for this example data from samples taken on 24 February 1998 were used). The amount of sulphate sourced from other sulphide minerals (e.g. galena and pyrite) was also calculated using the concentrations of other metals (e.g. iron, copper and cadmium) bearing in mind that pyrite (FeS_2) is a bisulphide (i.e. it has the molar ratio 1:2). This amount of sulphate attributable to weathering of the other sulphide minerals was then subtracted from the total amount of sulphate present. The remaining value then represents the amount sulphate attributed to the zinc deficit. As zinc and sulphate are sourced in equimolar amounts, the number of moles of sulphate created must equal the number of moles of zinc produced. When the molar data are converted back into a mass for zinc (by converting the number of moles produced into kg values) the deficits of zinc become apparent. The results obtained from calculating molar ratios of S:Zn are presented in Table 1.

The results show that very little zinc (up to 6% calculated as a percentage of the total amount of zinc dissolved per day) actually leaves the mine in comparison to the amount released per day from sphalerite oxidation. The missing zinc must be forming minerals within the workings. Smithsonite has been recorded by Dunham (1990) as being present in the mines, and the mineral hydrozincite has also been directly observed by the author underground in Scraithole Mine (NY 8032 4694) in the adjacent West Allen Valley, where large quantities of this mineral can be directly observed precipitating from mine water.

Zinc deficits can also be described by Fig. 5. The poor correlation between zinc and sulphate on the x–y plots for each site shows that there is little relationship between the amount of zinc and sulphate in the discharges (i.e. the maximum R^2 value is around 0.6 and, for Rampgill, it is as low as 0.04). This is perhaps due to water mixing from different branches in the mine in addition to mineral precipitation or sorption of zinc. When all the data are combined on an x–y plot the sites lie in distinct clusters (reflecting the individual chemistry of each mine) but there is no discernible trend in the relationship between zinc and sulphate in the mine waters (Fig. 6). If zinc and sulphate were released from the workings in a 1:1 ratio (i.e. attenuation of zinc was not taking place) we would expect to find a much more significant correlation between the zinc and sulphate concentrations.

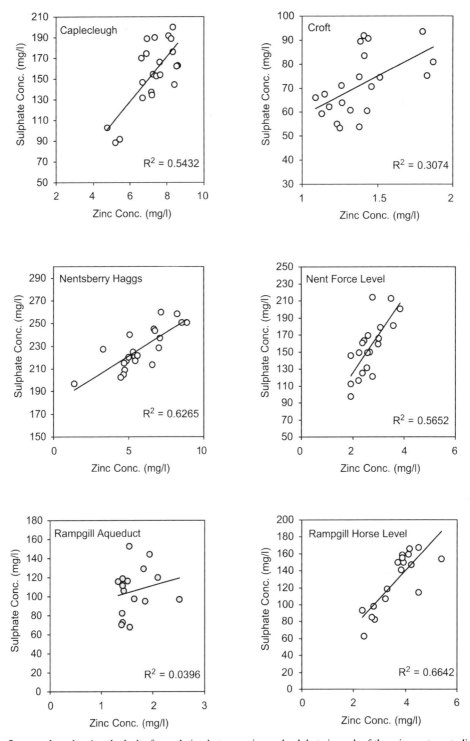

Fig. 5. x–y plots showing the lack of correlation between zinc and sulphate in each of the mine waters studied.

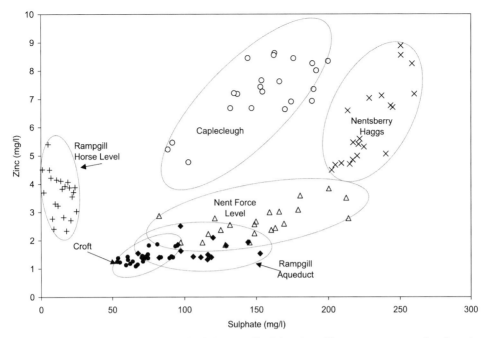

Fig. 6. Combined x–y plot for zinc and sulphate at all of the sites. Clusters are apparent but there is no discernible trend.

Table 2. Modelled mine water zinc concentrations in the absence of mineral sinks for zinc

Site	Average mine water Zn concentration (mg l^{-1})	Mine water Zn concentration if no 'sinks' existed (mg l^{-1})
Caplecleugh	7.35	126
Rampgill Horse Level	4.15	108
Haggs	5.06	159
Croft	1.15	45
Nent Force	2.59	112

Consequences for mine water chemistry in the absence of mineral sinks for zinc

When considering the data presented in Table 1, if the total amount of zinc weathered from sphalerite each day did not form secondary minerals, become incorporated into carbonate minerals or be adsorbed onto surfaces within the workings then mine water zinc concentrations would be almost two orders of magnitude greater than normal, by this theory.

Table 2, shows the likely zinc concentration of the five Nent Valley mine waters if mineral sinks were not in existence. These zinc-rich mine waters would then have a substantial impact on the Nent and are also likely to have a negative impact upon the South Tyne further downstream (Fig. 1), which is important for its fisheries and as a water resource. These calculations highlight the importance of metal sinks and show that, under the right conditions, large quantities of zinc can be removed from solution. Nuttall & Younger (2000) successfully emulated the precipitation of smithsonite in a passive treatment reactor as a means of treating mine waters to remove zinc by inducing a rise in pH (and, hence, smithsonite precipitation) within a closed limestone reactor.

Conclusions

Secondary zinc minerals and other carbonate minerals that sequester zinc interstitially precipitate within mine workings and on spoil heaps. These minerals provide effective 'sinks' for zinc because when comparing molar ratios of zinc and sulphate only around 5% of the zinc calculated to weather daily from the workings actually leaves the mine. Smithsonite precipitation has already formed a basis for a treatment system that passively removed zinc from these circum-neutral mine waters. The main phase 'sink' minerals appear to be precipitates of hydrozincite and smithsonite, with the other sinks being interstitial zinc within the carbonates calcite and aragonite and sorption of zinc within the workings. If these natural mechanisms of zinc removal were not occurring then the water quality emanating from the Nent Valley mines would contain increased concentrations of zinc and would have an even greater impact on the Nent River system.

The authors would like to express their gratitude to the Engineering and Physical Sciences Research Council and the Environment Agency who funded this work as part of a PhD project. We are also indebted to A. Lewis for her support and assistance in the field and staff at the Nenthead Mines Heritage Centre. We would also like to thank T. Bridges for his useful discussions on secondary mineralization. This paper reflects the views of the authors and does not necessarily represent the views of the Environment Agency.

References

ALABASTER, J.S. & LLOYD, R. 1980. *Water Quality Criteria for Freshwater Fish*, Butterworths, London.

BALL, J. W. & NORDSTROM, D. K. 1991. *User's Manual for WATEQ4F, With Revised Thermodynamic Data Base and Test Cases for Calculating Speciation of Major Trace and Redox Elements in Natural Waters*. US Geological Survey Open File Report, 91–183.

COETZEE, P.P., YACOBY, M., HOWELL, S. & MUBENGA, S. 1998. Scale reduction and modification effects induced by Zn and other metal species in physical water treatment. *Water South Africa*, **24**, 77–84.

CRITCHLEY, M.F. 1998. The history and working of the Nenthead Mines, Cumbria. *Bulletin of the Peak District Mines Historical Society*, **9**, 1–50.

DAWSON, E.W.O. 1947. War-time treatment of the lead–zinc dumps situated at Nenthead, Cumberland. *Transactions of the Institution of Mining and Metallurgy*, **56**, 587–605.

DUNHAM, K.C. 1981. Mineralisation and mining in the Dinantian and Namurian rocks of the North Pennines. In: SAY, P.J. & WHITTON, B.A. (eds) *Heavy Metals in Northern England: Environmental and Biological Aspects*. University of Durham, 7–18.

DUNHAM, K.C. *Geology of the North Pennine Orefield: Volume 1, Tyne to Stainmore. Economic Memior of the British Geological Survey*. HMSO, London.

DUNHAM, K.C., DUNHAM, A.C., HODGE, B.L. & JOHNSON, G.A.L. 1965. Granite beneath Viséan sediments with mineralisation at Rookhope, Northern Pennines 1965. *Quaterly Journal of the Geological Society of London*, **121**, 383–417.

EVANS, A.M. 1993. *Ore Geology and Industrial Minerals: An Introduction*, 3rd edn. Blackwell Science, Oxford.

FAIRBAIRN, R.A. 1993. *The Mines of Alston Moor. British Mining No. 47*, Northern Mine Research Society, UK.

FÖRSTNER, U. & WITTMAN, G.T.W. 1981. *Metal Pollution in the Aquatic Environment*, 2nd edn. Springer, Berlin.

HANDY, R.D. & EDDY, F.B. 1990. The interactions between the surface of Rainbow Trout, *Oncorhunchus mykiss*, and waterborne metal toxicants. *Functional Ecology*, **4**, 385–392.

JOHNSON, G.A.L. 1981. An outline of the geology of North East England. In: SAY, P.J. & WHITTON, B.A. (eds) *Heavy Metals in Northern England: Environmental and Biological Aspects*. University of Durham, 1–6.

JOHNSON, K. & YOUNGER, P.L. 2002. Hydrogeological and geochemical consequences of the abandonment of Frazer's Grove carbonate-hosted Pb–Zn fluorspar mine, north Pennines, UK. In: YOUNGER, P.L.K. & ROBINS, M.S. (eds) *Mine Water Hydrogeology and Geochemistry*, Special Publications, 198. Geological Society, London, 347–364.

MANCE, G. & YATES, J. 1984. Proposed Environmental Quality Standards for List II Substances in Water: Zinc. MAFF Technical Report, **209**.

MONDADORI, M. (ed.) *The Macdonald Encyclopedia of Rocks and Minerals*, Macdonald, London.

NUTTALL, C.A. & YOUNGER, P.L. 1999. Reconnaissance hydrochemical evaluation of an abandoned Pb – Zn orefield, Nent Valley, Cumbria, UK. *Proceedings of the Yorkshire Geological Society*, **52**, 395–405.

NUTTALL, C.A. & YOUNGER, P.L. 2000. Zinc removal from hard, circum-neutral mine waters using a novel closed-bed limestone reactor. *Water Research*, **34**, 1262–1268.

YOUNGER, P.L. 1999. Nature and practical implications of heterogeneities in the geochemistry of zinc-rich, alkaline mine waters in an underground F–Pb mine in the UK. *Applied Geochemistry*, **15**, 1383–1397.

The importance of pyritic roof strata in aquatic pollutant release from abandoned mines in a major, oolitic, berthierine–chamosite–siderite iron ore field, Cleveland, UK

PAUL L. YOUNGER

Hydrogeochemical Engineering Research and Outreach (HERO), Department of Civil Engineering, University of Newcastle Newcastle upon Tyne NE1 7RU, UK
(e-mail: p. l.younger@ncl.ac.uk)

Abstract: The Cleveland Ironstone Field (NE England) is a major sedimentary iron orefield in which the principal ore minerals are iron silicates (berthierine, chamosite) and carbonates (siderite). The siderite in this area is known to be rich in Mg and Mn in solid solution with Fe. Although this ore assemblage would not normally be expected to give rise to acid mine drainage phenomena, a number of discrete ferruginous mine water discharges (totalling some 6.5 million litres (Ml) day^{-1}) have been identified flowing from abandoned underground mine workings and old spoil heaps in the ore field. Some of these discharges are extremely acidic (pH ≥ 3.3), with total Fe reaching 1220 mg l^{-1}. At the point of first emergence to the surface, most of the discharges are so highly charged with dissolved CO_2 that they effervesce. Upon degassing, one of the discharges precipitates a ferroan calcite deposit, which is most unusual as a mine water discharge precipitate. All the other discharges precipitate more usual ferrihydrite and goethite 'ochres'. The geochemistry of these waters supports the view that oxidation of pyrite in the roof strata initiates dissolution of siderite in the old workings, releasing CO_2, Fe and Mg to solution. This chain of reactions results in these waters having SO_4^{2-} as their major anion (from pyrite weathering) with Mg as the major cation (except where Fe exceeds Mg in concentration). Two of the near-coastal discharges have Na as the major cation, and elevated Cl concentrations, suggesting a seawater component. However, SO_4/Cl ratios suggest that sea water can account for no more than 20% of these waters. Most of the Cleveland mine waters have significant environmental impacts, ranging from ecological damage to receiving water courses to flooding problems caused by the clogging of surface sewers with mine water precipitates. A range of remedial measures are proposed.

Mine water pollution is a widespread phenomenon in present and former mining districts of the world (e.g. Singer & Stumm 1970; Banks *et al.* 1997 Nordstrom *et al.* 2000; Younger *et al.* 2002), with numerous instances of severe water pollution having been reported from:

- base metal mines – such as those of southern Spain (Manzano *et al.* 1999), northern California (Nordstrom *et al.* 2000), Norway (Banks *et al.* 1997) and Sweden (Malmström *et al.* 2000);
- gold mines – for instance in Ghana (Smedley *et al.* 1996), South Africa (Wood & Reddy 1998) and Nevada (Shevenell 2000);
- coal mines – with particularly extensive records from the eastern United States (e.g. Ahmad 1974; Aljoe 1994; Demchak *et al.* 2000), as well as Poland (e.g. Rózkowski & Rózkowski 1994), South Africa (Hattingh *et al.* 2002), and many other countries.

However, only one example of mine water pollution associated with abandoned mines in sedimentary iron ore bodies has yet appeared in the international literature (Razowska 2001), and in that instance the polluted water is migrating into an adjoining aquifer rather than giving rise to ecologically damaging surface discharges.

The literature concerning mine water pollution in the UK reflects this international pattern, with many publications in the last decade documenting aquatic pollution from:

- the major coalfields (e.g. Robins 1990; NRA 1994; Younger 1995*a*, 1997, 1998, 2000*a*; Wood *et al.* 1999);
- most of the base metal ore fields, such as those of Cornwall (Banks *et al.* 1997; Bowen *et al.* 1998), upland Wales

(McGinness & Johnson 1993; Fuge et al. 1994) and northern England (Nuttall & Younger 1999; Younger 2000b).

However, there has so far been only a single, brief mention of pollution from a flooded sedimentary ironstone mine in Cleveland (Younger 1995b), and one published analysis (without geochemical interpretation) of a Scottish ironstone mine spoil leachate (Heal & Salt 1999).

Why should the literature on mine water pollution contain so few references to sedimentary iron ore fields? The obvious answer is that sedimentary iron ores are generally hydrogeochemically 'benign', being predominantly composed of carbonates, oxides and/or silicates of iron which are slow to dissolve and release little acidity. (This is in marked contrast to the more polluting base-metal ore bodies and coal measures, which are typically rich in iron sulphide minerals.) Nevertheless, non-sulphidic iron ore bodies do occasionally give rise to subtle water quality problems, such as the desorption of arsenic from haematitic ore bodies, which has been detected in settings as diverse as the banded-iron formations of the Quadrilátero Ferrífero (Minas Gerais, Brasil) and the limestone-hosted haematite ore bodies of west Cumbria (UK) (author's own unpublished data). Nevertheless, it is a reasonable generalization to state that dramatic, ferruginous discharges of the type associated with polluted mine waters in other geological settings do not generally emerge from abandoned haematite mines.

However, in many mining settings it is not so much the economic minerals that give rise to polluted leachates, but the enclosing rock mass. This is true, for instance, of certain fluorite mines of the North Pennine Orefield, in which the main source of contamination is accessory sphalerite present in unworked veinlets peripheral to the principal ore bodies (Younger 2000b). In this paper, cases of serious pollution and flooding problems related to ferruginous discharges from long-abandoned ironstone mines in Cleveland, NE England, are documented, illustrating the importance of unworked roof measures in driving pollutant release. As such, it is hoped that this paper may be useful as a cautionary tale, providing an antidote to possible complacency about the pollution risks that may be associated with similar sedimentary iron ore bodies elsewhere in the world.

Methods

The data analysed in this paper were collected as follows. Flow was measured by the best means possible at each site, given limitations of access, etc. In practice, this meant using a purpose-installed thin-plate weir at Eston, by means of a bin-and-stopwatch at Skinningrove and New Marske, and by the velocity–area method (with an OTT™ impeller flow meter) at Saltburn. The following parameters were invariably measured on site:

- alkalinity (by titration with 1.6 N H_2SO_4 using a HACH hand-held digital titrator);
- pH, temperature, Eh (the oxidation–reduction potential) and conductivity (all using a Myron L Company (Carlsbad, CA) 'Ultrameter 6P' multiparameter meter, which was laboratory-calibrated before each site visit).

Samples collected for subsequent analysis in the environmental engineering laboratories of the University of Newcastle were all filtered at the time of collection using 0.45 μm Whatman filters and syringes, and expressed into plastic 500 ml bottles. Each sample comprised two bottles, in one of which a few drops of concentrated nitric acid were present to preserve metals in the dissolved state.

Total acidity (A_T) values (expressed according to the 'mg l^{-1} as $CaCO_3$' equivalent) were calculated using the following formula (see Hedin et al. 1994; Younger 1995a; Younger et al. 2002):

$$A_T = 50[2(Fe^{2+}/56) + 3(Fe^{3+}/56) + 2(Mn^{2+}/55) + 3(Al^{3+}/27) + 2(Zn/65) + 1000(10^{-pH})] \quad (1)$$

where all cation symbols (Fe^{2+}, etc.) represent the total dissolved concentrations of those cations in mg l^{-1}. The above expression correctly recognizes that the total acidity in mine waters is dominated by the activities of 'hydrolysable metals' such as Fe, Al, Mn and Zn that can consume hydroxyl ions to form (hydr)oxide solids, consuming alkalinity and releasing further protons in the process (Younger et al. 2002). For example, for the case of ferric iron, the appropriate acid-generating hydrolysis reaction can be written:

$$Fe^{3+}_{(aq)} + 3H_2O \rightarrow Fe(OH)_{3(s)} + 3H^+_{(aq)}. \quad (2)$$

By contrast, the alkalinity of mine waters is primarily provided by bicarbonate (Younger et al. 2002), as in most natural groundwaters. The bicarbonate present in a mine water can counteract the lowering of pH by reactions such as reaction (1) by reacting with the liberated protons to form carbonic acid, which is a relatively weak acid compared to that represented by free protons:

$$HCO_3^- + H^+ \rightarrow H_2CO_3. \qquad (3)$$

Given that acidity and alkalinity are governed by different chemical species in mine waters, these two characteristics can co-exist in a single water. (Readers seeking further elaboration of this point are referred to the works of Hedin *et al.* 1994; Younger 1995*a*; Younger *et al.* 2002.)

Speciation and mineral equilibrium calculations were obtained using the well-known geochemical equilibrium modelling code WATEQ4F (Ball & Nordstom 1991). As detailed descriptions of the application of WATEQ4F to mine waters have recently appeared in the open literature (e.g. Chen *et al.* 1999; Razowska 2001), no further detail of its use or interpretation in this context is needed here.

The Cleveland Ironstone Field

Mining history

The industrial development of the Teesside conurbation has been inextricably linked to the discovery and exploitation of iron ore reserves beneath the picturesque plateaux and ridges of the Cleveland Hills (Whitehead *et al.* 1952; Hemingway 1974; Goldring & Juckes 2001). Although ironstone had been worked sporadically since ancient times, and more systematic exploitation of coastal outcrops was underway by the early 1800s (Mead 1882), large-scale deep mining did not commence until the middle of the nineteenth century (around 1848 at Skinningrove and 1850 at Eston; for localities see Fig. 1). By 1886 regional mapping had revealed that the ironstone beds underlie as much as 900 km^2 of the North York Moors, although economically workable reserves were restricted to around 20% of this total area (Kendall 1886). With such vast dimensions, the Cleveland Ironstone Field is certainly one of the largest continuous iron ore bodies in the world. Between 1850 and 1912, Cleveland effectively dominated UK iron ore production (Whitehead *et al.* 1952), with peak output reaching 6.5 million tonnes (Mt) in 1883 (Tuffs 1997). Although terminal decline set in after the First World War, ironstone mining persisted in Cleveland until the closure of North Skelton Mine (Fig. 1) on 17 January 1964 (Tuffs 1997).

The Cleveland ironstones were deep-mined by the 'bord-and-pillar' method, in which a

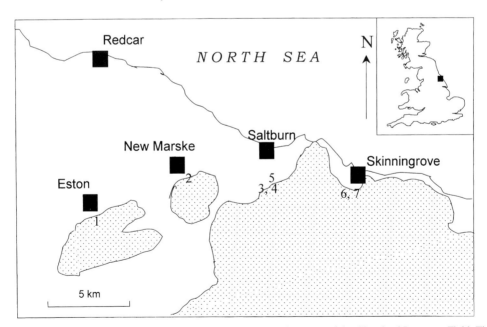

Fig. 1. Sketch map of the northern (and most historically productive) area of the Cleveland Ironstone Field. The stippled areas are those underlain by the Main Seam (mapping after Whitehead *et al.* 1952). Main towns and villages mentioned in the text are shown as black squares. The numbers correspond to the following major mine water discharges (see Tables 1 and 2 for more information): 1, Trustee Drift, Eston; 2, Upleatham Main Winning, New Marske; 3, Tributary discharge, Saltburn; 4, Borehole discharge, Saltburn; 5, Old Shaft discharge; Saltburn. 6, Carlin How Staple Shaft; 7, Loftus Horse Drift.

rectilinear network of passages is driven in the payable rock, while pillars of intact rock are left in place to support the roof (for discussion of the hydrogeological implications of bord-and-pillar mining, see Younger & Adams 1999). While extraction of pillars during the retreat from an area of workings was certainly undertaken in some places, it seems that the majority of pillars were left in place, as economic circumstances frequently led to the closure of mines well before they would ordinarily have been worked out. This has the consequence that vast networks of open voids still remain in the now-flooded workings beneath the Cleveland Hills.

Geological framework

The iron ores worked in Cleveland are sedimentary in origin (Hemingway 1974). They are characteristically oolitic in texture. Ooids are considered to be marine in origin, and to have formed in shallow inshore waters (Hemingway 1974; Scrutton 1994). There are three principal iron minerals forming the ooids of the Cleveland ores. Two of these are silicates:

- chamosite – a ferruginous chlorite, with the approximate chemical formula: $(Fe_5^{2+}Al)(Si_3Al)O_{10}(OH)_8$ (Young 1989);
- berthierine – an iron silicate of the serpentine family of minerals, which displays a range of chemical compositions (Velde 1989) approximating to the following formula: $(Fe^{2+}, Fe^{3+}, Mg)_{2-3}(Si,Al)_2O_5(OH)_4$.

It is likely that these minerals were formed by replacement of more common species (such as calcium carbonate or ferric hydroxide) as the sediment pile became buried (Hemingway 1974; Velde 1989; Young 1989). In particular, it is now thought that berthierine is the silicate mineral formed where temperatures during burial remained below 100°C, whereas chamosite is favoured where temperatures exceeded this figure (Velde 1989).

The third common iron mineral in the Cleveland ironstones is a carbonate:

- siderite ($FeCO_3$) – this typically occurs as an inter-ooid cement, but is also found as a replacement of pre-existing minerals within ooids (Whitehead et al. 1952; Kearsley 1989).

The oolitic ironstones of Cleveland are of Lower Jurassic age (Pliensbachian Stage, Middle Lias; Kent 1980). Seven seams are known altogether (Whitehead et al. 1952), although only three have been deep-mined, and only one of these (the Main Seam) has been widely worked (Mead 1882). The Main Seam varies in thickness from a maximum of about 5 m at Eston (location on Fig. 1) to around 1.7 m in the southernmost area of working. The roof strata of the Main Seam are also ferruginous, but are mineralogically distinct from the seam strata in that they are very rich in the iron sulphide, pyrite (FeS_2) (Mead 1882; Whitehead et al. 1952; Spears 1989). This has important environmental consequences, as will be explained in the following section.

Likely pollutant generation and attenuation reactions

The severity of the surface water pollution arising from abandoned mines in the Cleveland Ironstone Field has surprised many people with intimate knowledge of the mine workings. Former miners recall that the waters pumped from the Loftus Mine, for instance, was of good quality and supported a healthy trout fishery in the adjoining Kilton Beck. After flooding of the mine, however, the former miners were astonished to observe that the water that began to decant from the old workings was highly polluting, staining the receiving water course bright orange (T. Evans, Skinningrove Link pers. comm. 2000).

Such surprise is entirely understandable in geochemical terms, for the flow of water through the ore body itself during mining cannot have produced very severe pollution. This is because:

- the silicates berthierine and chamosite are relatively insoluble;
- whilst siderite is more soluble, above the water table it tends to oxidize *in situ* to form ochre (ferric hydroxide, $Fe(OH)_3$):

$$FeCO_{3(s)} + 1.5H_2O + 0.25O_2$$
$$\rightarrow Fe(OH)_{3(s)} + CO_{2(g)}. \qquad (4)$$

(Note that in reaction 3 the iron is in the ferrous (Fe^{2+}) state within the siderite, but has been oxidized to the ferric (Fe^{3+}) state in the ochre.) This *in situ* alteration of siderite accounts for the characteristic orange–brown colour of the Cleveland Ironstones at outcrop.

Although water quality during working appears not to have displayed much evidence of sulphide oxidation, geotechnical problems with the roof strata were widely attributed to destabilization due to weathering of pyrite (Mead 1882). Thus, although pyrite was being weathered, the water-sparse conditions obtaining in the ceiling of dewatered workings resulted in most of the oxidation products remaining *in situ*, in the form of ferrous and ferric hydroxy-sulphate salts, such as römerite:

$$3FeS_{2(s)} + 11.5O_2 + 15H_2O$$
$$\rightarrow Fe^{(2+)}Fe_2^{(3+)}(SO_4)_4 \cdot 14H_2O_{(s)}$$
$$+ 2SO_4^{2-} + 2H^+ \quad (5)$$

where FeS_2 is pyrite and $Fe^{(2+)}Fe_2^{(3+)}(SO_4)_4 \cdot 14H_2O$ is römerite. Römerite is only one of several dozen minerals that form by oxidation of pyrite under these conditions (see Bayless & Olyphant 1993; Younger 2000a). These minerals form conspicuous white and yellow surficial encrustations on pyritic beds within mines. They are highly soluble when immersed in water, and release much acidity upon dissolution, principally in the form of ferric iron which hydrolyses to release protons (i.e. H^+ ions, as shown by reaction 1). Where the water contains insufficient alkalinity to counteract this release of protons (i.e. reaction 2 cannot consume enough of the liberated protons), this reaction will cause the pH of the water to drop. Because of this property, these minerals have been collectively termed 'acid-generating salts' (Bayless & Olyphant 1993; Younger 2000a, b; Younger et al. 2002). When the Cleveland ironstone mines flooded, large quantities of acid-generating salts in the roof strata must have dissolved. The protons subsequently released by reaction (1) (and analogous reactions for Al^{3+} and Mn^{4+}) would have been available to attack siderite present in the Main Seam workings, releasing ferrous iron and dissolved carbon dioxide into the mine water:

$$FeCO_{3(s)} + 2H^+ \rightarrow Fe^{2+} + H_2O + CO_{2(aq)}. \quad (6)$$

It should be noted that the siderite of Cleveland is not pure iron carbonate, but contains significant quantities of magnesium, manganese (and possibly calcium) in solid solution (Whitehead *et al.* 1952). Hence, reaction (5) will yield dissolved Mg, Mn and Ca in addition to Fe.

While the iron silicate minerals will also be susceptible to attack by protons, they are not likely to dramatically affect water quality because:

- the rates of dissolution of these minerals are likely to be several orders of magnitude lower than those of pyrite and siderite;
- unlike siderite, which dissolves congruently (releasing all of its component elements to solution), the iron silicate minerals will dissolve incongruently, leaving solid aluminosilicate clay mineral phases in their place.

Finally, non-ferrous minerals will also react to neutralize acidity, most notably calcite ($CaCO_3$):

$$CaCO_{3(s)} + 2H^+ \rightarrow Ca^{2+} + H_2O$$
$$+ CO_{2(aq)}. \quad (7)$$

The other common carbonate, dolomite ($CaMg(CO_3)_2$), is not found in the Cleveland Ironstone sequence. The net effect of reactions (1)–(6) would be expected to be a water with relatively high concentrations of dissolved ferrous iron and dissolved CO_2, but with little proton acidity, and therefore with circum-neutral pH at the point of emergence at the ground surface. The degree to which this expectation is realized will become apparent in the following sections.

Where the roof strata were removed from the mine as waste material and deposited in spoil heaps, the pyrite will oxidize freely in the presence of rain water to yield acidic ferruginous waters, according to the well-known summary equation for pyrite oxidation by atmospheric oxygen:

$$FeS_{2(s)} + \frac{15}{4}O_{2(g)} + \frac{7}{2}H_2O \rightarrow Fe(OH)_{3(s)}$$
$$+ 2SO_4^{2-} + 4H^+. \quad (8)$$

As will be seen, the occurrence of this type of reaction is documented from one of the old spoil heaps in the ore field, and is doubtless occurring at many more.

Current mine water discharges

Table 1 summarizes all mine water discharges in the Cleveland Ironstone Field currently known to the author. No doubt others exist, but these are likely to be in very remote locations where their impacts on human activities are presumably minimal.

To judge from the known discharges and the likely frequency of discharges in unsurveyed areas, it is estimated that the total flow of mine water from the abandoned ore field is on the order of $6500 \, m^3 \, day^{-1}$. Given the long periods since mine abandonment, it is reasonable to assume that this rate of discharge reflects the mean annual recharge rate to the abandoned workings, such that the groundwater system is in a quasi-steady state. (Possible exceptions to this steady-state assumption are highlighted

Table 1. *Known discharges of polluted mine drainage in the Cleveland Ironstone Field*

Mine associated with discharge	Location	Approx. grid reference	Observations	pH*	Estimated flow rate
Eston Mine	Near Eston Equitation Centre	NZ 561181	Large ferruginous discharge from collapsed portal of old mine adit	6.7	Flow measured by rectangular thin-plate weir; in the range 1050–1200 m^3 day^{-1}
Upleatham Mine, Main Winning	New Marske	NZ 624625	Acidic, ferruginous spoil drainage at the edge of New Marske village	3.4	Flashy flow of 5–30 m^3 day^{-1}
North Skelton and Longacres mines	Saltburn Gill	NZ 676199	Three highly polluted discharges into Millholme Beck, an unnamed tributary and the Saltburn Gill	6.7† 5.9 5.6	Old shaft: approx. 50 m^3 day^{-1} Borehole: approx. 100 m^3 day^{-1} Tributary discharge: approx. 260 m^3 day^{-1}
Liverton Mine	Confluence of Whitecliff and Kilton Becks	NZ 710185	Adit intersecting main shaft of Liverton Mine; ferruginous but has little impact on Beck	7	Estimated flow rate approx. 150 m^3 day^{-1}
Carlin How Mine	Near entrance to Skinningrove village	NZ 712192	Overflowing staple shaft at roadside, seriously polluting Kilton Beck	6.8	2073 m^3 day^{-1}
Loftus Mine	Discharge from adit downstream from Mining Museum	NZ 712193	Adit discharge into Kilton Beck, exacerbating impacts of Carlin How discharge upstream	6.7	1900 m^3 day^{-1}
Sandsend, Goldsborough and Kettleness mines	Sandsend Cliff	NZ 861135	Many small discharges from cliff-line into the sea. Natural wetland treats discharge entering the local beck	6.8	Flow rates of the order of 50–100 m^3 day^{-1}
Warren Moor Mine	New Row	NZ 625088	Orange staining of the bed of the River Leven, access difficult	–	No estimate available
Chaloner Mine	Dunsdale	NZ 600191	Ochre staining of Dunsdale Beck	–	No estimate available
Coate Moor Mine and Lonsdale Vale Mine	Lonsdale Farm	NZ 613017	Orange staining of river bed	6.9	No estimate available

* At point of first emergence.
† Values for old shaft, borehole and tributary discharges, respectively.

below.) If recharge equals discharge, then for a total undermined area of around 80 km^2 the observed mine drainage rate equates to around 30 mm of recharge. Given that the mean annual rainfall for this area is around 800 mm, and the mean annual run-off is estimated to be around 500 mm, the Cleveland mine waters might appear to be a negligible fraction of the local water budget (accounting for less than 4% of the total rainfall, and no more than 6% of the total run-off in the district). In reality, the mine waters are non-negligible, for two reasons:

- The mine water discharges are not distributed evenly among the surface water catchments of the mined districts, and they locally account for a much higher proportion of mean annual flow in their receiving water courses than would be anticipated from the rough water budget figures given above. For instance, at Eston, mine waters account for about 95% of the total flow entering the receiving water course (a largely culverted, unnamed tributary of the Tees). Spot gauging in the Kilton Beck at Skinningrove in September and October 1996 showed that the Carlin How and Loftus mine waters (which both flow into the Beck as it flows through the village) accounted for 42% of the total flow in the Beck at that time. At times of dry weather flow, these two mine waters are thought to account for as much as 70% of the total flow in the Kilton Beck at Skinningrove.
- The waters are of far poorer quality than most other run-off from the Cleveland Hills and therefore give rise to environmental impacts out of all proportion to their volume. Of the 10 discharges listed in Table 1, four sites are particularly problematic and are therefore discussed in greater detail in the following sections.

Eston: A major discharge from Cleveland's premier mine

Eston was the largest and most productive mine in the North Pennine Orefield. Located on the northern scarp of the Cleveland Hills, working the Main Seam under the hill known as Eston Nab, this mine was very close to the major iron and steel works which lined the southern bank of the Tees estuary in the vicinity of Redcar (Fig. 1). Eston Mine worked from 1850 until 16 September 1949, by which time it had produced more than 63 Mt of ore (Pepper 1996). Taking advantage of the outcrop pattern of the Main Seam, Eston worked primarily by means of three near-horizontal adits driven into the northern flank of Eston Nab. It is one of these adits, the Trustee Drift (Site 1, Fig. 1), which today yields some 1100 m^3 day^{-1} of ferruginous mine water to an otherwise diminutive stream draining from the hillside. There is no known record of when the mine water began to discharge from the Trustee Drift, although presumably some time elapsed between the cessation of pumping in the southernmost (and deepest) portions of the workings and the commencement of gravity-driven flow from the Drift portal. The volume of flooded workings is large, however, and gauging of flows in the summer of 1998 revealed a 2-week lag between rainfall events and peaks in mine water discharge rate.

The chemistry of the Eston mine water is shown in Table 2 and is plotted relative to the other mine waters on a Piper diagram (Fig. 2). The water is of Ca–SO$_4$ facies, which is probably the most common facies in mined systems generally (e.g Younger 1995a; Nuttall & Younger 1999), but is actually anomalous when compared with the other waters in this ore field (Table 1 and Fig. 2). The fact that SO$_4$ is the principal anion is consistent with pyrite oxidation in the roof measures driving siderite dissolution (by a sequence of reactions 7 → 5).

The concentrations of Na and Cl in the Eston discharge are far greater than in most inland mine waters in NE England (e.g. Younger 1995a). There are three possible explanations for this:

- intrusion and mixing of modern sea water;
- upconing of brines from the deeply-buried Permo-Triassic evaporites that underlie Teesside; or
- slow release of ancient brackish pore waters from shales within the sequence.

Sea-water intrusion seems an unsatisfactory explanation, for the following reasons:

- the outcrop of the discrete outlier of ironstone worked by Eston Mine never crosses the coast, and indeed lies more than 7 km distant from it at all points;
- the deepest workings at Eston were still above sea level and therefore a hydraulic gradient from the sea to the workings cannot have existed even when dewatering was at a maximum;

Table 2. Hydrochemistry of major mine water discharges in the Cleveland Ironstone Field sampled by the author. Alkalinity and acidity are in mg l^{-1} as $CaCO_3$. All ion concentrations in mg l^{-1}. 'n.d.' = not determined. Where the number of analyses available (n) is greater than 1, the range of measured values on other dates is given in the second line of each record (for the Saltburn sites, these other measurements were made by the Environment Agency
(a) Physico-chemical parameters, anions, alkalinity, acidity (calculated) and deduced facies

Name	Grid Ref	Date of sample	pH	Eh (mV)	Conductivity (µS cm^{-1})	Temp. °C	SO_4	Cl	Alkalinity	Acidity	Facies
Eston	NZ 561181	14/7/98	7.7	n.d.	3270	14.5	1456	148	459	14	Ca–SO_4
		ranges (n = 10):	7.5–7.8		2940–3600	13–15	1452–2124	89–390	456–468		
New Marske	NZ 624625	14/7/98	3.6	n.d.	5800	13	3228	759	0	1657	Mg–SO_4
		ranges (n = 10):	3.3–3.7		4200–5830	10–13	2316–6143	157–1354	0		
Saltburn Borehole	NZ 676198	05/09/2000	5.9	+87	7965	12.8	4339	228	340	1702	Mg–SO_4
		ranges (n = 4):					4330–5940	220–395	100–340		
Saltburn Old Shaft	NZ 676199	05/09/2000	6.7	+67	n.d.	12.1	3161	853	428	565	Mg–SO_4
		(n = 1)									
Saltburn Tributary	NZ 675198	05/09/2000	6.5	+180	7816	13.5	5341	273	320	1838	Mg–SO_4
		ranges (n = 4):	5.6–6.5				5300–6400	270–400	145–320		
Carlin How Mine	NZ 712192	30/11/99	6.7	–80	11250	14.1	2548	2048	934	23	Na–Cl–SO_4
		ranges (n = 5):					2252–2600	1817–3125	928–1051		
Loftus Mine	NZ 712193	27/6/2000	7.1	–168	10890	16.7	1214	1689	980	14	Na–Cl–SO_4
		(n = 1)									

(b) Metals

Name	Total Fe	Mn	Al	Zn	Ca	Mg	Na	K
Eston ranges ($n = 10$)	7.1 0.9–7.1	0.8 0.003–1.2	0.01 <0.001–2	0.03 <0.001–0.06	455 340–454	184 161–200	303 280–503	16 13–72
New Marske ranges ($n = 10$)	317 74–317	92 41–92	164 85–205	2.3 2.0–4.7	424 334–1179	765 175–2155	151 50–427	90 2–632
Saltburn Borehole	934	17.4	0.5	0.3	503	857	583	89
Saltburn ranges ($n = 4$)	930–1080	16–18	0.02–0.5		425–505	780–860	400–590	
Saltburn Old Shaft ($n = 1$)	308	5.4	1	0.3	627	652	354	18
Saltburn Tributary	1000	28	0.3	0.7	544	853	630	25
ranges ($n = 4$)	1000–1220	5–31	0.1–1.7		506–627	650–850	350–650	
Carlin How Mine	12.6	0.01	0.001	0.01	396	132	1349	39
ranges ($n = 5$)	1.2–17.5	0.01–0.3			382–396	129–142	1342–1350	24–40
Loftus Mine	15.3	0.6	n.d.	0.007	356	153	2587	18.8

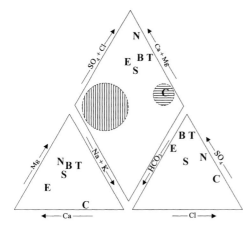

Fig. 2. Piper diagram showing the hydrochemical affinities of the Cleveland mine waters listed in Table 2. E, Eston; B, Borehole discharge at Saltburn; C, Carlin How discharge, Skinningrove; N, New Marske spoil heap; S, Old Shaft discharge at Saltburn; T, Tributary discharge, Saltburn. The field with vertical hatching is that occupied by shallow fresh groundwaters in major aquifers of this region (Younger 1995a, 1998). The field with horizontal hatching is that of sea water (after Younger 1998).

- the strata that underlie the ironstone beds are of very low permeability, so that, despite their stratigraphic continuity with submarine outcrops, they would have been unlikely to have transmitted large quantities of water so far inland, even if hydraulic conditions had been favourable.

Upconing of brines from Permo-Triassic beds at depth also appears unlikely, for at least two reasons:

- hundreds of metres of low-permeability shales lie between the Main Seam and the uppermost halite beds;
- the Na/Cl molar ratio of the Eston waters is on the order of 3.1, whereas equivalent ratios for the brines encountered in the Permo-Triassic strata at the nearby Boulby Mine are in the range 0.4–1.0 (Bottrell et al. 1996).

For this reason, the third explanation (slow release of ancient brackish pore waters) appears the most credible.

WATEQ4F modelling (Table 3) showed the water to be supersaturated with respect to most carbonate minerals at ambient Eh (+70 mV), and supersaturated with respect to ochre phases (goethite, ferrihydrite, jarosite) once Eh is raised to values consistent with aeration after discharge at the surface. These results fully explain the precipitation of ochreous calcite in the culverts downstream, and further suggest that reactions, such as reactions (1), (5) and (6), dominate water quality development within the mine system.

The mean loading of iron leaving the mine is around 8 kg day^{-1}, and most of this load is deposited as precipitates within a culverted water course that runs beneath residential areas of Eston and Grangetown. Constriction of the culvert by the mine water precipitates has led to flooding of properties on a number of occasions in the past. With the transfer of responsibilities for the sewers from the local authority to Northumbrian Water Ltd in 1997, a search for a more sustainable solution than endless pipe clearing is now being sought. The preferred option (subject to detailed feasibility studies) would be to install a simple gravity-driven, wetland-type passive treatment system upstream of the culverted section of the water course, to precipitate all of the minerals likely to form from the water before it enters the sewer system.

The vast majority of ferruginous mine waters precipitate ferric hydroxide 'ochre' (see, for instance, McGinness & Johnson 1993; NRA 1994; Younger 1998). It was therefore presumed that the problematic precipitates forming from the Eston mine water would also be ferric hydroxide. However, most ferric hydroxide is very soft and readily amenable to jetting where it accumulates in sewers. In the Eston case, the precipitates in the sewer are typically so hard that they have to be drilled out of the pipes. X-Ray diffraction analysis of recently removed material revealed that the bulk of the precipitates are, in fact, ferroan calcite. While this is consistent with the WATEQ4F model output (Table 3), it is extremely unusual in a ferruginous mine water, with no other case being known in the UK.

New Marske: acidic spoil drainage

Clogging of surface water sewers by mine water precipitates has also led to intermittent flooding problems in the village of New Marske (Site 2, Fig. 1). When the sewer clogging problems at New Marske were investigated, it was found that the precipitates responsible are soft (in distinction to the consolidated deposits which clog the pipes at Eston). X-Ray diffraction and chemical analyses showed these precipitates to be ferrihydrite, goethite and white aluminium sulphate deposits. While ferrihydrite and goethite can precipitate from alkaline mine waters, aluminium sulphate precipitates are particularly indicative of *acidic* mine drainage

Table 3. Selected results of WATEQ4F modelling of mineral equilibria for major Cleveland mine waters, as log (IAP/K_T) values for the minerals named. Interpretation of the values is thus: if $\log(IAP/K_T) = 0$, the water is at equilibrium ('saturated') with respect to the mineral in question. If $\log(IAP/K_T) < 0$, the water is undersaturated with respect to that mineral, which will tend to dissolve in the water if it is available. If $\log(IAP/K_T) > 0$, the water is supersaturated with respect to that mineral, such that it will tend to precipitate from these waters

Mine water Mineral	Eston Trustee Drift	New Marske spoil heap	Saltburn Borehole	Saltburn Old Shaft	Saltburn Tributary	Carlin How Staple Shaft, Skinningrove
Aragonite ($CaCO_3$)	0.9	<-20	-1.3	-0.1	-0.6	-0.6
Calcite ($CaCO_3$)	1.1	<-20	-1.1	0.0	-0.4	-0.4
Dolomite ($CaMg(CO_3)_2$)	1.4	<-20	-18.6	-0.4	-1.1	-1.1
Siderite* ($FeCO_3$)	1.0	<-20	1.0	1.5	1.6	0.4
Goethite[†] ($FeOOH$)	10.2	7.2	11.1	11.5	11.7	11.7
Ferrihydrite[†] ($Fe(OH)_3$)	4.3	1.4	5.2	5.6	5.8	5.8
K-Jarosite[†] ($KFe_3(SO_4)_2(OH)_6$)	4.7	8.9	13.9	11.7	13.7	13.7
Gypsum ($CaSO_4.2H_2O$)	-0.2	-0.2	0.0	0.0	0.1	0.1

*Values quoted are for the water pre-aeration (Eh at field values where available (Table 2), or else set to the median value +70 mV).
[†] Values for the water post-aeration (Eh = +150 mV).

(Younger et al. 1997), as aluminium is not significantly mobilized in solution until pH drops below about 4.5.

Investigation of the source of the water reveals it to be seeping from the toe of a large mine spoil heap, and it is therefore presumed to originate as perched groundwater within the spoil. This spoil heap is one of the few visible remains of the former Upleatham Mine, the principle entrance to which was an adit known as the 'Main Winning', which was driven into the hillside to the south of New Marske (Tuffs 1997). The spoil heap from which leachate is emerging was formed by tipping of waste roof strata brought out of the Main Winning during development work (Tuffs 1997). As such, it offers us some insights into the nature of roof-weathering reactions operating in the now inaccessible underground workings.

The spoil leachate is of $Mg-SO_4$ facies (Table 2 and Fig. 2), which is unusual nationally, but apparently the norm in this ore field (Table 2). The origin of the magnesium is interesting, as the most common source of magnesium in groundwaters, dolomite, has never been found in these strata (Whitehead et al. 1952). However, magnesium occurs within the siderite of this ore field in solid solution, and also occurs within the berthierine. A Mg/Ca molar ratio of 2.97 for this water is very similar to the value of 2.99 calculated from an analysis of a magnesium-rich siderite ore from this area presented by Whitehead et al. (1952, p. 40). The high SO_4 is explicable in terms of oxidative dissolution of pyrite (reaction 7), which also accounts for the high iron and acidity concentrations. The high zinc probably arises from oxidation of sphalerite (ZnS) and the aluminium from the action of acid water on clay minerals (Younger et al. 1997).

The Na and Cl in the New Marske spoil drainage are elevated well above typical surface water concentrations. As cross-flow from deep evaporites is utterly impossible in this setting, the inference must be that these two elements are slowly leaching from the shale clasts in the spoil heap. This corroborates the interpretation of the Na and Cl values at Eston, outlined above.

Fortunately, the water has a very low flow rate, although it varies significantly in response to rainfall. There is a lag of about 3 days between rainfall and peak flows, which is consistent with the discharge arising from a shallow water table perched within the spoil. While regrading and capping of the spoil heap could be implemented to reduce infiltration, and thus minimize the contaminant release, the mature vegetation cover on the heap and its proximity to housing would make this an unpopular option. A discharge with very similar hydrochemical characteristics has been successfully passively treated in recent years at Quaking Houses, County Durham (Younger et al. 1997; Jarvis & Younger 1999), by means of an anaerobic (compost) wetland. If suitable land already identified at New Marske could be acquired, a similar solution could be implemented here.

Saltburn: an enigmatic new discharge

The Saltburn Gill is an attractive stream draining a narrow, forested valley (which is a designated Site of Special Scientific Interest on account of its trees and understorey vegetation), which flows down to meet the Skelton Beck in the principal cove of the town of Saltburn, whence its waters enter the North Sea via the main recreational beach of the district. The catchment of the Saltburn Gill was subjected to significant ironstone mining activity between 1865 (when Longacres Mine opened) and 17 January 1964 (when North Skelton Mine closed). Minor ferruginous discharges have long flowed into the Millholme Beck (an upstream tributary of the Gill) at the site of an old exploration borehole (Site 4 on Fig. 1), and into the Saltburn Gill itself at the site of a shallow flooded shaft (Site 5 on Fig. 1), apparently associated with the main drainage level of Longacres Mine (Tuffs 1997). (In accordance with local convention, these two discharges are hereafter (and in Tables 1 and 2) referred to as the 'Borehole' and 'Shaft' discharges, respectively.) Ferruginous drainage from the borehole and the shaft have long caused ochre staining of the Saltburn Gill, although in all but the periods of lowest receiving-water flow, discoloration of the streambed rarely extended as far as the confluence with the Skelton Beck.

In May 1999 the situation changed dramatically when a subsidence hollow suddenly appeared alongside a minor tributary of the Saltburn Gill, only a few tens of metres from the pre-existing discharges. Water immediately began to flow from this hollow at a rate of about $180 l\,min^{-1}$, constituting the so-called 'Tributary' discharge (Site 3, Fig. 1). As shown in Table 2, the tributary discharge is little different in chemistry from the pre-existing borehole discharge (which continues to flow unabated). However, the flow rate of the tributary discharge is considerably greater, and results in severe discoloration not only of the entire length of Saltburn Gill, but also of the Skelton Beck, as far as the sea. Under certain combinations of tide and river flow, a plume of ochreous water extends out into the marine waters (D. Mason pers. comm. 2000). This is causing serious concerns for the agencies responsible for the bathing beach and the regeneration of Saltburn as a tourist destination. Meanwhile, both the flow rate and the iron concentrations (around $1000\,\text{mg}\,l^{-1}$) of the tributary discharge have remained more-or-less steady since May 1999, so that the iron loadings entering the Saltburn Gill 3 years after the commencement of the discharge remain as high as ever.

There are a number of as-yet-unanswered questions relating to the tributary discharge, including:

- Why did the discharge only commence in 1999 when the last working mine in the area closed 35 years previously? It seems inconceivable that the workings took that long to flood, for several reasons:
  - there is a long-established discharge of similar chemistry in the vicinity (the borehole discharge);
  - the even more extensive workings of Eston, Carlin How and Loftus all flooded up to surface within a few years of mine closure.
- Why is the discharge not displaying the usual exponential decrease in iron concentrations which has been observed in virtually every other case? (cf. Younger 1997.)

The most likely explanation is that the tributary discharge represents drainage from the same workings as the borehole discharge, and that outflow from the latter has become so throttled over the decades due to clogging with ochre that a build-up of head has finally led to a discharge at a higher location. An ongoing investigation by D. Mason (Environment Agency) will, hopefully, shed further light on these questions and others, and thus facilitate planning for the long-term remediation of this discharge.

As at New Marske, all three Saltburn discharges are of $Mg-SO_4$ facies (Table 2 and Fig. 2), suggesting a powerful influence of roof strata weathering processes on the overall hydrochemistry of the water. The Mg/Ca molar ratios of the Borehole discharge (2.8) and the Tributary discharge (2.6) suggest that the Mg is sourced from dissolution of an Mg-rich siderite, as was deduced at New Marske. However, a lower value for the Old Shaft (1.7) is closer to the mean Mg/Ca ratio for the ore field as a whole (1.2, calculated from the mean analysis given on p. 62 of Whitehead *et al.* 1952). This reflects the fact that the strata penetrated by the Old Shaft lie stratigraphically below those accessed by the Borehole and the subsidence hollow at the Tributary discharge, and are presumably mineralogically distinct from them. The sulphate concentrations are so high that the waters are shown by WATEQ4F modelling to be saturated with respect to gypsum (Table 3). It is likely that gypsum precipitation is exerting an upper limit on dissolved SO_4 and Ca. The Tributary and Borehole discharges contain such high concentrations of dissolved CO_2 that they effervesce at their point of first emergence. The elevated alkalinity that this represents results also in reasonably high pH values at their points of first emergence, but this is deceptive as all three discharges are strongly net-acidic. This is because the extreme iron concentrations (up to $1100\,\text{mg}\,l^{-1}$ Fe; Table 2) drive reaction (1), which ensures that pH drops dramatically (reaching 2.6 in the case of the tributary discharge; D. Mason pers. comm.) once the waters are aerated and allowed to settle.

Skinningrove: two overflowing mines

The village of Skinningrove (Fig. 1) lies in the steep-sided valley of the Kilton Beck. Beneath the northern flank of this valley, the Carlin How Mine (Site 6, Fig. 1) worked from 1873 to 1954 (latterly as part of Lumpsey Mine), while beneath the southern flank, the Loftus Mine (Site 7, Fig. 1) worked until 1958 (Tuffs 1997). Both mines have long-since flooded up to the surface, and they each give rise to a major discharge. The two discharges are very similar in both flow rate and chemistry (Tables 1 and 2), both containing between 15 and $18\,\text{mg}\,l^{-1}$ Fe, both being brackish and both being of $Na-Cl-SO_4$ facies. This results in these two waters plotting in the sea-water field on the Piper diagram (Fig. 2), which prompts speculation as to whether there is a component of marine water in these two mine systems (Younger 1995*b*). Certainly, the Na and Cl concentrations are much higher than those at Eston and New Marske, which were explained by drainage of ancient shale pore waters. Furthermore, the Na/Cl molar ratios in these two waters (2.4 in the Loftus waters, 1.0 in the Carlin How waters) are much closer to the mean sea-water ratio of 0.85 than are the Eston waters, with a ratio of 3.1. As the Main Seam does not extend offshore in this area, a direct connection with the sea is not possible. However, as the deepest workings were 60 m below sea level, it is possible that sea water could have flowed upwards into the ironstone workings via the underlying Staithes Sandstone (an aquifer which does extend offshore).

The sulphate concentrations in the two Skinningrove mine waters are very high for net-alkaline mine waters (cf. Younger 1995*a*), suggesting a contribution from sea water. Molar ratios of SO_4/Cl provide a simple but robust means of identifying marine influences in coastal ground waters (Richter & Kreitler 1993), which have previously been applied to coal mine waters by Younger (1998). If we compare molar ratios of SO_4/Cl for sea water (0.05) with those for the Loftus and Carlin How discharges (0.26 and 0.45, respectively), it is clear that mixing with sea water cannot account for more than 10 (Loftus) to 20% (Carlin How) of the SO_4 in these mine waters. This means that 80–90% of the sulphate is from another source, which is almost certainly pyrite oxidation. Thus, even in long-flooded workings, sufficient oxidation of roof-strata pyrite is still driving pollutant release from the sideritic orebody. As subaqueous oxidation of pyrite by dissolved oxygen initially present at saturated concentrations (around $12\,mg\,l^{-1}$) can only account for a few tens of $mg\,l^{-1}$ of SO_4, it is likely that the oxidation of pyrite is most active within and above the zone of water table fluctuation, generating acid recharge to the flooded workings below.

As such acid waters enter the flooded workings, they will be aggressive towards siderite, which can be expected to dissolve briskly in accordance with reaction (3). The CO_2 liberated from siderite presumably accounts for the extremely high concentrations of this gas observed in both Skinningrove mine waters, which are often seen to effervesce at first emergence. The elevated dissolved CO_2 ensures that, prior to aeration, the water is saturated with respect to aragonite, calcite, dolomite and siderite (Table 3). When the waters are aerated, CO_2 degasses and siderite saturation is lost; iron then becomes insoluble as the 'ochre' phases (ferrihydrite, goethite, etc.), which are all supersaturated in aerated Skinningrove waters (Table 3). Ochre precipitation causes vivid staining of the Kilton Beck throughout its passage through the village. Apart from this visual impact, the ferruginous waters also act as a significant barrier to migratory salmonid fish returning to the Beck to spawn.

The law in force in the 1950s allowed the former mine owners to evade legal liability for the mine water pollution. In the 1970s the district council undertook a project named 'Operation Eyesore', in which the discharges were collected in a pipeline and taken out to sea. In the event, the Beck was spared the discharges for only 2 days before multiple leaks occurred. No attempt was ever made to fix the pipeline, and the rusting pipes themselves contribute to the current eyesore in the Beck. In recent years pressure has grown to address the pollution and a voluntary grouping including local residents, the Loftus Development Trust and Newcastle University has obtained funding for a novel treatment facility. While these waters would be amenable to wetland treatment, there is no flat land available. Hence, an alternative solution has been developed in which surface-catalysed oxidation of ferrous iron will be promoted within a chamber filled with trickle filter media. A pilot plant that operated from October 1999 for 6 months achieved 50% iron removal (Younger 2000c), and provided design criteria for the full-scale system, which is under construction at the time of writing (i.e. October 2001).

Conclusions

The abandoned Cleveland Ironstone Field emits waters with a wider variety of hydrochemical facies than is encountered in nearby coalfields (Younger 1995a) and Pb–Zn orefields (Nuttall & Younger 1999). In detail, the hydrochemistry is seen to reflect weathering of the ore body, which is a rich source of Mg, prompted by oxidation of pyrite in the roof measures (the source of SO_4 and Fe). These ferruginous waters give rise to a similarly wide range of problems, including visual, ecological, amenity and flooding impacts. A range of measures to address these problems is proposed, ranging from passive treatment at Eston and New Marske, through to semi-passive treatment at Skinningrove, to pumping and active treatment at Saltburn.

This paper presents a hydrogeological interpretation of data previously gathered for other purposes (i.e. pollution impact assessment and design of remedial works), and I am therefore indebted to the following friends and colleagues for help with sampling and analysis: S. Ross and H. Ward (former Masters students at Newcastle University), D. Elliott (Senior Lecturer at Newcastle), A. Snape (Northumbrian Water Ltd), and T. Evans and T. Richardson (residents of Skinningrove). The data collection at Eston and New Marske was funded by Northumbrian Water Ltd, through the MISS project (Minewater Impacts on Surface-water Sewers). D. Mason of the Environment Agency kindly showed me the Saltburn discharges and provided me with some archival water quality records. The interpretations presented are, however, entirely my own and should not be construed as representing those of any organizations or individuals named above.

References

AHMAD, M.U. 1974. Coal mining and its effect on water quality. Water resources problems related to mining. *Proceedings No 18,* American Water Resources Association, pp 138–148.

ALJOE, W.W. 1994. Hydrologic and water quality characteristics of a partially-flooded, abandoned underground coal mine. In: *Proceedings of the International Land Reclamation and Mine Drainage Conference and the Third International Conference on the Abatement of Acidic Drainage, Pittsburgh, PA, April 1994, Volume 2*, US Bureau of Mines, Pittsburgh, Special Publication, **SP 06B-94**, 178–187.

BALL, J.W. & NORDSTROM, D.K. 1991. *User's Manual for WATEQ4F, With Revised Thermodynamic Database and Test Cases for Calculating the Speciation of Major, Trace and Redox Elements in Natural Waters*. US Geological Survey Open File Report, **91–183**.

BANKS, D., YOUNGER, P.L., ARNESEN, R.-T., IVERSEN, E.R. & BANKS, S.D. 1997. Mine-water chemistry: the good, the bad and the ugly. *Environmental Geology*, **32**, 157–174.

BAYLESS, E.R. & OLYPHANT, G.A. 1993. Acid-generating salts and their relationship to the chemistry of groundwater and storm runoff at an abandoned mine site in southwestern Indiana, USA. *Journal of Contaminant Hydrology*, **12**, 313–328.

BOTTRELL, S.H., LEOSSON, M.A. & NEWTON, R.J. 1996. Origin of brine inflows at Boulby potash mine, Cleveland, England. *Transactions of the Institution of Mining and Metallurgy (Section B: Applied Earth Sciences)*, **105**, B159–B164.

BOWEN, G.G., DUSSEK, C. & HAMILTON, R.M. 1998. Pollution resulting from abandonment and subsequent flooding of Wheal Jane Mine in Cornwall, UK. In: MATHER, J., BANKS, D., DUMPLETON, S. & FERMOR, M. (eds) *Groundwater contaminants and their migration*. Geological Society, London. Special Publications, **128**, 93–99.

CHEN, M., SOULSBY, C. & YOUNGER, P.L. 1999. Modelling the evolution of minewater pollution at Polkemmet Colliery. Almond catchment, Scotland. *Quarterly Journal of Engineering Geology*, **32**, 351–362.

DEMCHAK, J., SKOUSEN, J., BRYANT, G. & ZIEMKIEWICZ, P. 2000. Comparison of water quality in fifteen underground coal mines in 1968 and 1999. In: *Proceedings of the Fifth International Conference on Acid Rock Drainage, (ICARD 2000), Denver, Colorado, 21–24 May, 2000, Volume II*, 1045–1052.

FUGE, R., PEARCE, F.M., PEARCE, N.J.G. & PERKINS, W.T. 1994. Acid-mine drainage in Wales and the influence of ochre precipitation on water chemistry. *American Chemical Society Symposium Series*, **550**, 261–274.

GOLDRING, D.C. & JUCKES, L.M. 2001. Iron ore supplies to the United Kingdom iron and steel industry. *Transactions of the Institution of Mining and Metallurgy (Section. A: Mining Technology)*, **110**, A75–A85.

HATTINGH, R.P., PULLES, W., KRANTZ, R., PRETORIUS, C. & SWART, S. 2002. Assessment, prediction and management of long-term post-closuer water equality: a case study – Hbbane Collery. South Africa. In: YOUNGER, P.L. & ROBINS, N.S. (eds) *Mine Water Hydrogeology and Geochemistry*, Geological Society, London, Special Publications, Vol 198, 297–314.

HEAL, K.V. & SALT, C.A. 1999. Treatment of acidic metal-rich drainage from reclaimed ironstone mine spoil. *Water Science & Technology*, **39**, 141–148.

HEDIN, R.S., NAIRN, R.W. & KLEINMANN, R.L.P. 1994. *Passive treatment of polluted coal mine drainage*. U.S. Bureau of Mines Information Circular, **9389**.

HEMINGWAY, J.E. 1974. Ironstone. In: RAYNER, D.H. & HEMINGWAY, J.E. (eds) *The Geology and Mineral Resources of Yorkshire*, Yorkshire Geological Society, Hull, 329–335.

JARVIS, A.P. & YOUNGER, P.L. 1999. Design, construction and performance of a full-scale wetland for mine spoil drainage treatment, Quaking Houses, UK. *Journal of the Chartered Institution of Water and Environmental Management*, **13**, 313–318.

KEARSLEY, A.T. 1989. Iron-rich ooids, their mineralogy and microfabric: clues to their origin and evolution. In: YOUNG, T.P. & TAYLOR, W.E.G. (eds) *Phanerozoic Ironstones*, Geological Society, London Special Publications, vol **46**, 141–163.

KENDALL, J.D. 1886. On the iron ores of the English Secondary Rocks. *Transactions of the North of England Institute of Mining and Mechanical Engineers*, **35**, 113 ff.

KENT, P. 1980. *Eastern England from the Tees to the Wash*, 2nd ed. British Regional Geology. Institute of Geological Sciences, London.

MALMSTRÖM, M., DESTOUNI, G., BANWART, S. & STRÖMBERG, B. 2000. Resolving the scale-dependence of mineral weathering rates. *Environmental Science & Technology*, **34**, 1375–1377.

MANZANO, M., AYORA, C., DOMENECH, C., NAVARETTE, P., GARRALON, A. & TURRERO, M.-J. 1999. The impact of the Aznalcóllar mine tailing spill on groundwater. *Science of the Total Environment*, **242**, 189–209.

MEAD, R. 1882. The coal and iron industries of the United Kingdom. In: *Yorkshire – North Riding (Cleveland District) Iron Industries*, Chap. III. Republished 1997 as: TUFFS, P. (ed.) *Cleveland Ironstone Mines and Iron Industry* Peter Tuffs, Guisborough.

MCGINNESS, S. & JOHNSON, D.B. 1993. Seasonal variations in the microbiology and chemistry of an acid mine drainage stream. *Science of the Total Environment*, **132**, 27–41.

NORDSTROM, D.K., ALPERS, C.N., PTACEK, C.J. & BLOWES, D.W. 2000. Negative pH and extremely acidic mine waters from Iron Mountain, California. *Environmental Science & Technology*, **34**, 254–258.

NRA. 1994. *Abandoned Mines and the Water Environment*. Report of the National Rivers Authority of England and Wales. Water Quality Series, **14**.

NUTTALL, C.A. & YOUNGER, P.L. 1999. Reconnaissance hydrogeochemical evaluation of an abandoned Pb–Zn orefield, Nent Valley, Cumbria, UK. *Proceedings of the Yorkshire Geological Society*, **52**, 395–405.

Pepper, R. 1996. *Eston and Normanby Ironstone Mines*; Industrial Archaeology of Cleveland, Cleveland Ironstone Series, Peter Tuffs, Guisborough, IMWA, Johannesburg.

Razowska, L. 2001. Changes of groundwater chemistry caused by the flooding of iron ore mines (Czestochowa Region, southern Poland). *Journal of Hydrology*, **244**, 17–32.

Richter, B.C. & Kreitler, C.W. 1993. Geochemical Techniques for Identifying Sources of Ground-Water Salinization. C.K. Smoley, Boca Raton, FL.

Robins, N.S. *Hydrogeology of Scotland*. HMSO for British Geological Survey, London.

Rózkowski, A. & Rózkowski, J. 1994. Impact of mine waters on river water quality in the Upper Silesian Coal Basin. *Proceedings of the 5th International Minewater Congress, Nottingham, UK*. Volume 2, IMWA, Nottingham, 811–821.

Scrutton, C.T. (ed.) 1994. *Yorkshire Rocks and Landscape. A Field Guide*, Yorkshire Geological Society, Ellenbank Press, Maryport.

Shevenell, L.A. 2000. Water quality in pit lakes in disseminated gold deposits compared to two natural, terminal lakes in Nevada. *Environmental Geology*, **39**, 807–815.

Singer, P.C. & Stumm, W. 1970. Acid mine drainage: the rate limiting step. *Science*, **167**, 1121–1123.

Smedley, P.L., Edmunds, W.M. & Pelig-Ba, K.B. 1996. Mobility of arsenic in groundwater in the Obuasi gold-mining area of Ghana: some implications for human health. Appleton, J.D., Fuge, R. & McCall, G.J.H. (eds) *Environmental Geochemistry and Health*, Geological Society, London,Special Publication, **113**, 163–181.

Spears, D.A. 1989. Aspects of iron incorporation into sediments with special reference to the Yorkshire ironstones. In: Young, T.P. & Taylor, W.E.G. (eds) *Phanerozoic Ironstones*, Geological Society, London, Special Publications, **46**, 19–30.

Tuffs, P. 1997. *Catalogue of Cleveland Ironstone Mines*, Industrial Archaeology of Cleveland, Cleveland Ironstone Series, Peter Tuffs, Guisborough.

Velde, B. 1989. Phyllosilicate formation in berthierine peloids and iron oolites. In: Young, T.P. & Taylor, W.E.G. (eds) *Phanerozoic Ironstones*, Geological Society, London, Special Publications, vol **46**, 3–8.

Whitehead, T.H., Anderson, W., Wilson, V., Wray, D.A. & Dunham, K.C. 1952. *The Liassic Ironstones*, Memoirs of the Geological Survey of Great Britain, HMSO, London.

Wood, A. & Reddy, V. 1998. Acid mine drainage as a factor in the impacts of underground minewater discharges from Grootvlei Gold Mine. In: *Proceedings of the International Mine Water Association Symposium on 'Mine Water and Environmental Impacts', Johannesburg, South Africa, 7–13 September 1998, Volume II*; 387–398.

Wood, S.C., Younger, P.L. & Robins, N.S. 1999. Long-term changes in the quality of polluted minewater discharges from abandoned underground coal workings in Scotland. *Quarterly Journal of Engineering Geology*, **32**, 69–79.

Young, T.P. 1989. Phanerozoic ironstones: an introduction and review. In: Young, T.P. & Taylor, W.E.G., (eds) *Phanerozoic Ironstones*, Geological Society, London, Special Publications, vol **46**, ix–xxv.

Younger, P.L. 1995*a*. Hydrogeochemistry of minewaters flowing from abandoned coal workings in the Durham coalfield. *Quarterly Journal of Engineering Geology*, **28**, S101–S113.

Younger, P.L. 1995*b*. Minewater pollution in Britain: past, present and future. *Mineral planning*, **65**, 38–41.

Younger, P.L. 1997. The longevity of minewater pollution: a basis for decision-making. *Science of the Total Environment*, **194/195**, 457–466.

Younger, P.L. 1998. Coalfield abandonment: geochemical processes and hydrochemical products. In: Nicholson, K. (ed) *Energy and the Environment. Geochemistry of Fossil, Nuclear and Renewable Resources*, Society for Environmental Geochemistry and Health, McGregor Science, Aberdeenshire, 1–29.

Younger, P.L. 2000*a*. Predicting temporal changes in total iron concentrations in groundwaters flowing from abandoned deep mines: a first approximation. *Journal of Contaminant Hydrology*, **44**, 47–69.

Younger, P.L. 2000*b*. Nature and practical implications of heterogeneities in the geochemistry of zinc-rich, alkaline mine waters in an underground F–Pb mine in the UK. *Applied Geochemistry*, **15**, 1383–1397.

Younger, P.L. 2000*c*. Holistic remedial strategies for short- and long-term water pollution from abandoned mines. *Transactions of the Institution of Mining and Metallurgy Section A: Mining Technology*, **109**, A210–A218.

Younger, P. L. & Adams, R. 1999. *Predicting mine water rebound*. Environment Agency R&D Technical Report, **W179**.

Younger, P.L., Curtis, T.P., Jarvis, A.P. & Pennell, R. 1997. Effective passive treatment of aluminium-rich, acidic colliery spoil drainage using a compost wetland at Quaking Houses, County Durham. *Journal of the Chartered Institution of Water and Environmental Management*, **11**, 200–208.

Younger, P.L., Banwart, S.A., Hedin, R.S. 2002. *Mine Water: Hydrology, Pollution, Remediation*, Kluwer, Dordrecht.

Arsenic removal by oxidizing bacteria in a heavily arsenic-contaminated acid mine drainage system (Carnoulès, France)

MARC LEBLANC, CORINNE CASIOT, FRANÇOISE ELBAZ-POULICHET & CHRISTIAN PERSONNÉ

UMR Hydrosciences, University Montpellier 2, 34095 Montpellier, France

Abstract: In the Carnoulès Pb–Zn mining site (Gard, France), abandoned 40 years ago, acidic waters (pH 3) with an extremely high As content (80–350 mg l^{-1}) emerge from the base of a tailings stock containing As-rich pyrite (2–4% As). From the acidic spring, the oxidation – reduction potential (Eh) and O_2 parameters strongly increase within a few metres and a Fe–As-rich (up to 22% As) material precipitates and covers the bed of the acidic creek. Consequently, there is a sharp decrease in arsenic concentration of the acidic waters downflow (<10 mg l^{-1} As) and of the Fe-rich precipitates (down to 2% As). Seasonal variations in dissolved arsenic concentrations of the spring waters are important. Furthermore, the Fe–As-rich stream sediments that were stored, during drought periods, are reworked and transported downflow during rainy periods. The annual fluxes of total arsenic comprise between 2 and 6 t; the lifetime of the As-releasing system is that of several centuries.

Fe and As speciation measurements have been carried out. Fe(II) and As(III) dominate all along the acidic stream. Fe(III) is rapidly precipitated. The removal of total dissolved arsenic mostly corresponds to a decrease in dissolved As(III) and results in the formation of As-rich ferric precipitates. A further decrease in arsenic concentration in water can be attributed to adsorption mechanisms on ferric hydroxides.

In the Fe–As-rich products, the most common bacteria are long rod-shaped phenotypes of the genus *Thiobacillus ferrooxidans*. Laboratory experiments have been carried out to investigate the potential catalytic role of these acidophilic oxidizing bacteria on the removal of arsenic: in the biotic systems, 60–80% As were removed in a few days against 5% for abiotic systems.

Arsenic, a naturally occurring toxic element, is mostly present in sulphide minerals. Since the Industrial Revolution, human activity has largely contributed, mainly through mining and metallurgy, to the dispersal of arsenic in the environment (North *et al.* 1997). The mining of sulphide ores releases acid mine drainage with a high metal and arsenic content.

Arsenic exists in surface waters primarily as arsenite (As(III)) and arsenate (As(V)). These inorganic species are 100 times more toxic than organic species (Ferguson & Gavis 1972; Squibb & Fowler 1983). Furthermore, As(III) is relatively more mobile and toxic than As(V), which tends to precipitate with ferric iron. Depending on the physico-chemical conditions (Eh, pH, Fe and As concentrations), As(V) may either form specific minerals, such as ferric arsenates, or adsorb onto ferric hydroxides. These oxidation reactions result in a marked decrease in arsenic concentrations of the dissolved phase. Nevertheless, oxidation reactions generally have slow kinetics, unless they are catalysed. In this respect, chemolithotrophic micro-organisms that use inorganic chemicals as energy sources play an important role. The most common genus of oxidizing bacteria in acid mine waters is *Thiobacillus ferrooxidans*. These acidophilic chemolithotrophic bacteria oxidize both iron and sulphur, enhancing sulphide oxidation in mining wastes; they also oxidize Fe(II) in acid mine drainage. Over the past three decades, evidence has been growing that such micro-organisms are involved in the geochemical cycle of arsenic, despite its toxicity (Osborne & Ehrlich 1976).

The acidic waters (pH 3) surging from the Carnoulès mining site (Gard, France) are characterized by very high concentrations of arsenic (80–350 mg l^{-1} As) and by the presence of As-rich stream sediments (up to 22% As), which often display bacterial structures of stromatolitic type (Leblanc *et al.* 1996). This affords a unique opportunity to investigate the downstream evolution of arsenic concentrations and the role of bio-oxidizing bacteria, and to consider possible microbial remediation strategies.

Sampling and analytical methods

The main physico-chemical parameters of the acid mine drainage system (Eh, pH, conductivity, temperature) were measured in the field at least twice a month with an Ultrameter™ Model 6P (Myron L Company, Camlab, Cambridge, UK). Dissolved oxygen was measured on the field with CHEMets tests (CHEMetrics, Calverton, VA, USA) based on colorimetric detection. Water samples were collected in poyethylene bottles and filtered immediately through 0.45-μm Millipore membranes. Fe(II) was determined immediately after filtration. Samples for total Fe and As determination were acidified with HNO_3 and stored at 4°C. Samples for Fe and As speciation and sulphate determination were analysed within 24 h or immediately deep-frozen.

Total As determination was performed by inductively coupled plasma-mass spectrometry (ICP-MS); high As concentrations need a dilution factor of 1000 and cancels interference due to formation of ArCl in the plasma. For dissolved As(III) determinations, a hydride generation system was coupled to the ICP-MS. The detection limit was $75\,ng\,l^{-1}$ and the precision better than 5%. Dissolved As(V) was calculated as the difference between total dissolved As and As(III).

Total dissolved iron was determined by flame atomic absorption spectrometry. For dissolved Fe(II) determinations, filtered samples were buffered to pH 4.5 in the field and Fe(II) complexed by adding 1 ml of a 0.5% (w/w) phenantroline solution to 10 ml of sample. Analyses were undertaken by colorimetry at 510 nm. The precision was better than 5%.

Sulphate was determined by precipitation of $BaSO_4$ with $BaCl_2$ and spectrophotometric measurement at 650 nm.

Samples of suspended matter (corresponding to 500 ml of water filtered through 0.45 μm) and samples of sediments were dried then mineralized using 20 ml of concentrated HNO_3 before being analysed for total As, Fe and SO_4 concentrations. Suspended matter and bacterial sediments and mats were examined by scanning electronic microscopy (SEM) coupled with energy-dispersive X-ray spectrometry analysis (EDS).

Site description (Fig. 1)

Carnoulès was a relatively important stratabound Pb–(Zn) ore body from the Pb–Zn belt of the southern border of the Massif Central, France (2.5 Mt containing 3.5% Pb and 0.8% Zn). Lower Triassic conglomerates, unconformably covering the Palaeozoic basement, host the Carnoulès mineralization; it comprises a pyrite–galena(sphalerite)–barite mineral association (Alkaaby et al. 1985). The Carnoulès ore body was mainly worked as an opencast operation and was abandoned 40 years ago, leaving $1\,km^2$ of quarries and a tailings stock (1.5 Mt, 10–20 m in thickness) stored behind a dam built on the uppermost course of the small Reigous Creek. Acidic water (pH 2.2–4) surges from the base of the tailings stock, constituting the present spring of the Reigous Creek. The flow rate at the source of the Reigous Creek is between 0.2 and $1.3\,l\,s^{-1}$, the variation being mainly dependent on rainfall. This acid stream collects other acidic seepage waters (pH 1.8) downstream from the surrounding mine quarries (Fig. 1). The waters of the Reigous Creek are cloudy, yellow-coloured, and precipitate light yellow–orange sediments. The Reigous Creek joins, 1.5 km away, the Amous River which is a tributary of the Gardon River, itself a tributary of the Rhône River. From their junction with the Carnoulès Creek, the fresh and clear waters of the Amous River (pH 8.4) are contaminated over 5 km downstream: they become cloudy, beige-coloured and host no fish.

The Carnoulès acid mine drainage system (Fig. 1)

The tailings are a very fine-grained material (40 μm on average) resulting from ore milling and flotation treatments. Quartz is the main mineral component, but the tailings also contain 5–10% wt of arsenic-rich pyrite (2–4% As). Thus, they correspond to a stock of about 3000 t As. The tailings stock constitutes an efficient hydrobiogeochemical reactor, where pyrite oxidation reactions allow the formation of acidic waters (pH 3.3 on average) with high concentrations (Table 1) of sulphate, iron and arsenic (2000–7500, 750–2700 and $80-350\,mg\,l^{-1}$ respectively). Nevertheless, the structure of this reactive system is uncommon: the oxidation zone is not located along the uppermost part of the tailings stock but along its floor. The very low permeability of the tailings material ($10^{-6}\,m\,s^{-1}$) and the presence of a thin clay cover prevent the penetration of rain water. Fresh water probably comes from a spring buried under the tailings stock. The generated acidic waters circulate horizontally along the basal part of the tailings stock through old draining pipes. The temperature of the acidic spring is relatively stable all

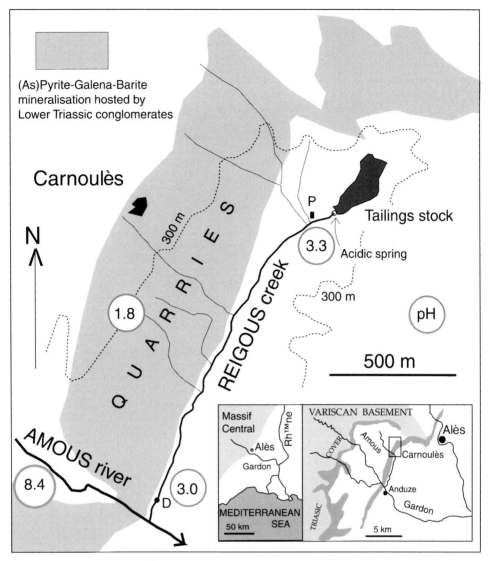

Fig. 1. Sketch map of the Carnoulès acid mine drainage system. A few pH mean values are given (see Table 1). P, pilot-scale ponds (experimental station); D, downflow monitoring point.

over the year (14–16°C). This is in agreement with an aquifer-like system.

Seasonal variations of the physico-chemical parameters of the acidic spring water are observed (Fig. 2): pH increases from 2.3 in winter to 3.6 in summer, while O_2 concentration decreases from 1.8 to 0.08 mg l^{-1}. In the winter rainy season, the input of subsurface water bringing O_2 to the tailings influences pyrite oxidation rate and therefore pH. Fe concentration at the spring is below 1200 mg l^{-1} in winter and 1.8 times higher in summer. In the same way, As increases up to 350 mg l^{-1} in summer, then decreases down to 80 mg l^{-1} during the rainy season. The molar ratio of Fe/As ranges between 7 and 13; these variations are ascribed to the heterogeneity in tailings composition and the different routes of water inside the tailings. Fe(II) is the only dissolved iron species detected at the spring, while As(III) is the only, or the main, dissolved arsenic species. As(V) can reach up to 40% of total As in winter. There are no organic forms of arsenic.

The arsenic content of the Reigous Creek decreases quickly downstream (Fig. 3). A sharp concentration decrease is observed along the first 40 m from the spring (Table 1). This decrease can

Table 1. *Mean values and standard deviation (SD) of the main characteristics of the acidic waters of Carnoulès (cf. Figs 1 and 4), calculated from 30–80 measurements made from June 1998 to June 2001.*

	Spring water of the Reigous Creek	Reigous water 40 m downstream	Reigous water 1500 m downstream
pH	3.36 (SD = 0.56)	3.3 (SD = 0.59)	3.0 (SD = 0.16)
Eh (mV)	266 (SD = 85)	299 (SD = 54)	447 (SD = 26)
Oxygen (mg l^{-1})	0.33 (SD = 0.3)	4.46 (SD = 1.2)	6.10 (SD = 1.0)
Conductivity (µS cm^{-1})	3637 (SD = 1703)	4055 (SD = 759)	1463 (SD = 343)
Total As in solution (mg l^{-1})	189 (SD = 82)	136 (SD = 78)	4.6 (SD = 3.1)
% As(III)	86 (SD = 14)	84 (SD = 17)	67 (SD = 20)
Total Fe in solution (mg l^{-1})	1390 (SD = 479)	1205 (SD = 745)	78 (SD = 22)
% Fe(II)	100 (SD = 0)	100 (SD = 0)	66 (SD = 22)
SO$_4$ (mg l^{-1})	4286 (SD = 1546)	3426 (SD = 1960)	1100 (SD = 170)

only be ascribed to precipitation reactions, as no dilution occurs along this uppermost part of the acidic creek. Up to 50% As and 40% Fe can be removed along this uppermost part of the acidic creek. The highest removal efficiency for both Fe and As are obtained in winter. Efficiencies were below 25% for both Fe and As in May and June, despite the fact that dissolved Fe and As concentrations in spring water are higher during this period. Along the uppermost part of the acidic creek, As(III) remains the dominant dissolved arsenic form and Fe(II) the only dissolved iron form (Table 1). Consequently, the total dissolved arsenic decrease is due only to the As(III) decrease. The Fe decrease is ascribed to Fe(II) oxidation resulting in the formation of insoluble Fe(III) precipitates. Along this uppermost part of the creek, the main changes in the physico-chemical characteristics of the acidic water (Table 1) result from exposition of the effluent to the air. There is a major increase in O$_2$ concentration (from 0.3 to 4.5 mg l^{-1}) and about a 40 mV increase in Eh. These oxidizing conditions allow the formation and deposition of As-rich ferric sediments.

Just before the confluence with the Amous River, 1.5 km away from the spring, dissolved arsenic concentrations in the Reigous Creek are lowered down to 2–9 mg l^{-1}, less than 5% of the initial concentration from the spring (Fig. 3C).

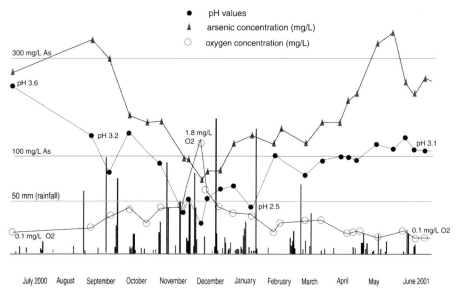

Fig. 2. Monitoring of the acidic spring (Fig. 1) showing the variations in pH values (filled circles), dissolved arsenic concentrations (filled triangle) and dissolved oxygen concentrations (open circles) in relation to the rainfall events (vertical bars). The decrease in pH and As in winter, which is the main rainy period, strongly suggests that both oxidizing reactions and dilution occurs during the input of oxygen-rich waters in the tailings stock.

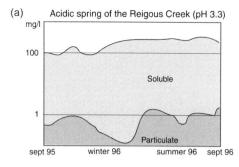
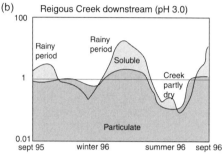
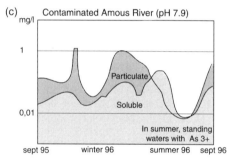

Fig. 3. Seasonal variations in soluble arsenic (light grey) and particulate arsenic (grey) concentrations (mg l^{-1}; log scale): (**a**) of the acidic spring; (**b**) of the downstream part of the Reigous acidic Creek; and (**c**) of the Amous River contaminated with acidic Reigous waters, just after the confluence (Fig. 1). Note, from (**a**) to (**b**), the sharp decrease in total arsenic concentration (two orders of magnitude), and the relative increase of particulate arsenic. From (**b**) to (**c**), at the confluence of the Reigous Creek and the Amous River, there is a dilution of the acid Reigous waters resulting in a further decrease of arsenic concentrations. Seasonal variations become important (**b**, **c**) as the rainy periods of the Mediterranean climate (usually spring and autumn) allow a reworking and a transport downflow of the As-rich stream sediments.

This decrease may be partly ascribed to dilution of the Reigous waters with the waters of small effluents and quarries seepages; nevertheless, the dilution ratios cannot be accurately calculated due to the non-conservative behaviour of most dissolved elements. There is also a change in Fe and As speciation, as the oxidized species are present in a significant amount; there is 20–70% Fe(III) and 10–60% As(V). Eh is still higher than the Eh of the uppermost zone (of about 150 mV), whereas pH remains stable or slightly decreased (Table 1).

In the acidic spring water (Fig. 3A), arsenic is essentially in solution (99% dissolved As). Downstream (Fig. 3B), an increasing proportion of total arsenic, up to 40%, is transported by the suspended load (filtered on a 0.4 μm filter). The suspended load comprises flocks resulting from the precipitation of amorphous As-rich ferric phases (10–18% As). The settling of these flocks allows the formation of As-rich sediments. The solid phases have the stochiometry of ferric arsenates (Fe/As: 1.5–2). The richest ochres (18–22% As) are located along the uppermost part of the Reigous Creek; they are yellow-coloured. Downstream, the arsenic concentration in the suspended load decreases to less than 1% As (Fe/As > 100). This suspended load mainly consists of amorphous ferric hydroxides, which are known to coprecipitate with and adsorb arsenic (Raven *et al.* 1998). The corresponding sediments are orange coloured because of their high iron oxide content (Fe/As: 12–15). They contain more As (2–5% As) than the suspended load, probably by *in situ* As adsorption processes. During drought periods, the As-rich sediments of the acidic Reigous Creek accumulate in place and only soluble arsenic is transported, whereas during rainy periods the As-rich sediments are reworked and transported downflow (Fig. 3B).

At the confluence of the acidic waters of the Reigous Creek (pH 3) and the slightly alkaline waters of the Amous River (pH 8.4), Fe concentrations in the Amous water decrease further (<0.5 mg l^{-1}) and dissolved Fe is mostly in the oxidized form (30–90% Fe(III)). This decrease is partly due to dilution; the flow rate ratio between the Amous River and the Reigous Creek is between 8 and 12. In addition, there is a strong precipitation of amorphous ferric hydroxides from the dissolved iron still present in the Reigous water (50–100 mg l^{-1}). These two processes would normally tend to decrease the As concentrations, but they are partly counterbalanced by the desorption of As from the ferrihydrite material transported by the Reigous due to pH increase (Raven *et al.* 1998). Consequently, the suspended load of the Amous River may have higher a As content (2–8%) than that of the Reigous. During rainy periods (Fig. 3C), when the suspended load is high, 80% of As is in the particulate phase and dissolved arsenic concentration remains over 0.5 mg l^{-1}, the present legal limit for As concentration in natural water. During dry periods, total arsenic concentrations are 10 times lower and most of the As is in solution; dissolved arsenic still comprises 30–50% As(III).

The total arsenic flux flowing from the Reigous Creek has been calculated for two different hydrological years (a dry and a rainy one), and varies from 2 to 6 t year^{-1}. Considering the available arsenic mass from the tailings stock (3000 t) and assuming a constant arsenic release, the lifetime of the Carnoulès arsenic-producing system is at least 500 years.

Arsenic bio-oxidation

In a previous study (Leblanc et al. 1996), a possible As bio-oxidation in the Carnoulès drainage waters was pointed out on the basis of Eh–pH field measurements and of the abundance of As-rich bacterial mats.

Oxygen concentration and redox potential (Eh) of the acidic waters both increase along the stream of the Reigous Creek (Fig. 4). Oxygen concentrations increase, on average, from 0.3 to 6.1 mg l^{-1} and Eh from 265 to 450 mV (Table 1). This clearly shows that oxidation reactions take place from the acidic spring of the Reigous Creek to the Amous confluence. Nevertheless, measured Eh has no significance in fresh water because it is influenced by several redox couples; furthermore, thermodynamic equilibrium is rarely met in mine drainage waters because energetically favoured redox reactions are kinetically slow. Speciation analysis is therefore a useful tool to study As and Fe distribution and behaviour in this kind of system. Fe is present totally as Fe(II) at the acidic spring. As soon as the effluent from the tailings stock is exposed to the air, the redox conditions rapidly change from reducing to oxidizing and Fe(II) is partly oxidized and precipitated as hydrous ferric hydroxide. Measured values of Eh coincide fairly well with the redox potential calculated according to the Fe(II)/Fe(OH)$_3$ couple, suggesting that the systems tend towards equilibrium regarding Fe(OH)$_3$. Fe removal is clearly due to an oxidation reaction. The interpretation of As decrease along the Reigous Creek may be also ascribed to oxidation. As is mostly present as As(III) in the acidic spring water (60–100% As(III)). The decrease of total dissolved As concentration observed downflow corresponds mainly to a decrease in As(III) concentration (Table 1). Consequently, the As(III)/total As ratio does not change in the uppermost part of the Reigous Creek (85% As(III) on average), then decreases slowly downflow (67% AS(III)) (Fig. 4). The strong positive correlation between Fe and As concentrations along the Reigous Creek suggests either adsorption of the As(V) formed onto iron hydroxides or coprecipitation with Fe(III). The Fe/As ratio of suspended particles (Fe/As = 1.5) in the upper part of the Reigous Creek should correspond to the formation of ferric arsenates. The corresponding sediments have a very high As content (15–21% As). In the lower part of the Reigous Creek, the Fe/As ratio increases leading to the formation of As-rich iron hydroxides. In fact, the interactions between iron and arsenic oxidation reactions are probably complex. For example, ferric iron may contribute, as an electron acceptor, to the oxidation of As(III).

The As-rich sediments of the Reigous Creek have been described as bacterial mats made up of an association of coated bacteria cells and colloidal flocks (Leblanc et al. 1996). After a flood event, bacterial colonies, in the form of light yellow dome-shaped dots (1–3 mm), settle in a few days and grow rapidly forming small ridges that then cover the creek bed. The most abundant bacteria are long rod-shaped (3–8 μm in length and 1–2 μm in diameter) and are

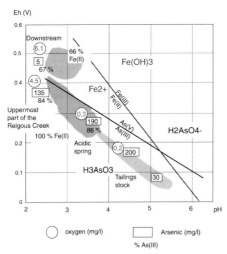

Fig. 4. Eh–pH diagram for the Carnoulès acidic waters (calculated for 10^{-2} M Fe and 10^{-3} M As). The fields of the waters from different parts of the system are in grey; the given numbers are mean values (Table 1). In the tailings stock, there are different types of water depending on their location and on the water–tailings interactions. Along the stream of the Reigous Creek, from the acidic spring to the downstream part of the Reigous Creek, the surface waters show a sharp increase in oxygen concentration (mg l^{-1}, circle), and a strong decrease in total dissolved arsenic concentration (mg l^{-1}, rectangle). The main form of dissolved arsenic is arsenite, as shown by the high ratio As(III)/total As (%). The only dissolved form of iron is Fe(II), except downstream.

strongly coated with amorphous As–Fe-rich material (up to 22% As). These bacteria are acidophilic chemolithotroph Gram-negative bacteria; they may be a phenotype of *Thiobacillus ferrooxidans* (Lundgren et al. 1972). This species is very common in acid mine drainage, where it oxidizes ferrous iron as its primary energy-generating process and which, consequently, strongly catalyses the oxidation reactions (Brock et al. 1994). The long rod-shaped forms present in the Reigous Creek might correspond to a genetic adaptation of *Thiobacillus ferrooxidans* to acidic waters with extremely high arsenic concentrations. Preliminary experiments have shown that these long rod-shaped bacteria can grow on culture medium without iron but with thiosulphate. Investigations are in progress: (i) to check if these bacteria really are *Thiobacillus ferrooxidans* phenotypes; and (ii) to isolate pure strains in order to evaluate their respective efficiency in As removal.

Laboratory experiments were carried out to investigate the potential catalytic role of bacteria on the precipitation of As, using the bacteria that are naturally present in the acidic spring. The water used for the experiment was collected at the spring in February 2001; it contained 160 mg l^{-1} As, 74% of which was As(III). It was placed in a glass container (5 l) with an injection of air bubbles (80 l^{-1} h); laboratory conditions (temperature, light and oxygenation) reflected roughly the field conditions. For a period of 15 days, 10 ml of the water was collected each day for As(III), total As, Fe(II), total Fe and SO_4 determinations. Eh, pH, conductivity, temperature and dissolved O_2 concentration were measured before each sampling. At the end of the experiment, the sediments were dried, weighed then mineralized (using concentrated HNO_3) before being analysed. A blank experiment was carried out to follow the evolution of an abiotic system using the same spring water, but was filtered through a 0.2-μm membrane filter to avoid the presence of bacteria cells ($0.7-1.2 \times 2.5-8$ μm). After a few hours the biotic system became cloudy; then from day 1 to day 6 the water progressively became a yellowish colour and an abundant light-orange precipitate covered the bottom. However, the water of the abiotic system remained limpid and, at the end of the experiment (day 15), there was only a slight and whitish precipitate. pH decreased and Eh increased in the biotic system, from 3.2 to 2.6 and from 270 to 300 mV, respectively, whereas pH and Eh remained stable in the abiotic system. At the end of the experiment, 75% of the total dissolved arsenic was precipitated in the biotic system against 5% in the abiotic one. In the biotic system, As(III) concentration decreased, mainly after day 6. Thus, the decrease in total arsenic can be ascribed to the decrease in As(III). In the abiotic system, As(III) concentration remained roughly unchanged. The orange precipitate of the biotic system corresponds to an As-rich bacterial mat (25% As). From this preliminary experiment, the efficiency of the bio-oxidation processes to reduce arsenic concentration seems obvious. This kind of experiment allows not only the comparison of biotic/abiotic reactions but also the evaluation of the kinetics of bio-oxidizing reactions that start in a few hours and become very efficient within a few days. Two replicates were carried out using acidic spring waters with different As concentrations (95–200 mg l^{-1}), and a direct inoculation of the biotic system with a sample of bacterial mat from the spring was tested. The results were roughly similar with 60–80% of the total arsenic As removed from the biotic experiments, against 3–10% for the abiotic ones.

Pilot-scale ponds have been installed in the Carnoulès mining site (Fig. 1) to check, *in situ*, the bio-oxidizing processes and to define the limiting parameters. These small experimental structures (8 m^2) are supplied with water from the acidic spring. The water flow is low (2–10 l min^{-1}); the water's pathway is lengthened by PVC ridges. Oxygen diffusion is promoted by a thin water layer (<5 cm) and/or small cascade structures. Within a few days, the water surface was covered with a bacterial film and the bottom of the pond with yellow bacterial deposits containing 17–20% As. Nevertheless, from preliminary results, the efficiency of arsenic removal is still relatively low (10–20%). Additional work is necessary to constrain oxygenation, time residence and the effectiveness of the removal of arsenic.

Conclusions

- The mining site of Carnoulès is unique because of the very high As concentrations (80–350 mg l^{-1}) of its main acidic effluent (pH 3). This site, abandoned 40 years ago, can be considered as a model for studying the processes that control As transfer in acid mine drainage.
- Dissolved arsenic is mostly present as As(III) in the acidic spring water of Carnoulès. The downflow decrease of total dissolved arsenic mainly results from a decrease of As(III). The corresponding precipitates are As-rich ferruginous bacterial mats (18–22% As).

- From field observations, speciation analysis and laboratory experiments, the removal of arsenic can be ascribed to bio-oxidation processes. Acidophilic chemolithotroph bacteria (probably belonging to the species *Thiobacillus ferrooxidans*) catalyse the oxidation of arsenic. The efficiency of arsenic removal in such a bio-oxidizing system may be up to 80%. This natural depollution process may be considered in remediation strategies.

This work is supported by the European Programme PIRAMID (Passive In situ Remediation of Acidic Mine/Industrial Drainage), a research project of the European Commission Fifth Framework Programme (contract EVK1-CT-1999-000021).

References

ALKAABY, A., LEBLANC, M. & PERISSOL, M. 1985. Minéralisation diagénétique précoce (Pb–Zn–Ba) dans un environnement détritique continental: cas du Trias de Carnoulès (Gard France). *Compte rendu de l'Académie des Sciences de Paris*, **300**, 919–922.

BROCK, T.D., MADIGAN, M.T., MARTINKO, J.M. & PARKER, J. 1994. *Biology of Microorganisms*; 7th edn. Prentice-Hall, Englewood Cliffs, New Jersey.

FERGUSON, J.F. & GAVIS, J. 1972. A review of the arsenic cycle in natural waters. *Water Research*, **6**, 1259–1274.

LEBLANC, M., ACHARD, B., BENOTHMAN, D., BERTRAND-SARFATI, J., LUCK, J.M. & PERSONNÉ, CH. 1996. Accumulation of arsenic from acidic mine waters by ferruginous bacterial accretions (stromatolites). *Applied Geochemistry*, **11**, 541–554.

LUNDGREN, D.G., VESTAL, J.R. & TOBITA, F.R. 1972. The microbiology of mine drainage pollution. *In*: MITCHELL, R., (ed.) *Water Pollution Microbiology*. John Wiley, New York, 69–88.

NORTH, W., GIBB, H.J. & ABERNATHY, C.O. 1997. Arsenic: past, present and future considerations. *In*: ABERNATHY, CALDERON, & CHAPPELL, (eds) *Arsenic, Exposure & Health Effects*. Chapman & Hall, London, 406–423.

OSBORN, F.H. & ERRICH, H.L. 1976. Oxidation of assemite by a soil isolate of Alcaligenes. *Journal of Applied Bacteriology*, **41**, 295–305.

RAVEN, K.P., JAIN, A. & LOEPPERT, R.H. 1998. Arsenite and arsenate adsorption on ferrihydrite: kinetics, equilibrium and adsorption envelopes. *Environmental Science & Technology*, **32**, 344–349.

SQUIBB, K.S. & FOWLER, B.A. 1983. The toxicity of arsenic and its compounds. In: FOWLER, B.A. (ed.) *Biological and Environmental Effects of Arsenic*. Elsevier, New York, 233–269.

Fingerprinting mine water in the eastern sector of the South Wales Coalfield

M. M. E. BROWN[1], A. L. JONES[2], K. G. LEIGHFIELD[2] & S. J. COX[2]

[1]*CSMA Consultants Ltd, Trevenson Road, Pool, Redruth, Cornwall TR15 3SE, UK*
(e-mail: M.M.E.Brown@csm.ex.ac.uk)
[2]*Wardell Armstrong, 22 Windsor Place, Cardiff CF10 3BY, UK*
(e-mail: ajones@wardell-armstrong.com)

Abstract: Coal has been mined in the South Wales Coalfield for centuries. Substantial dewatering operations have attended these activities, particularly with the advent of steam and electric pumps that facilitated deeper mining operations. However, underground coal mining in South Wales has been in decline since 1913. The closure of the last deep mine in the eastern part of the coalfield in 1992 entailed a cessation of pumping over a large geographical area, with a resultant recovery of the groundwater. A consequence of this recovery is that emissions of mine water are occurring at the surface.

Surface water emissions of known provenance from the Upper Coal Measures were sampled in 2000 and characterized in terms of their major ion composition. Also, known emissions of mine water within the catchments of the Rivers Ebbw, Ebbw Fach, Rhymney and Taff Bargoed were sampled, and the major ion data plotted on trilinear Piper diagrams. The results indicate that two distinctive hydrochemical facies could be identified from these plots. By combining knowledge of the underground mine workings and lithology, and by comparison with the Upper Coal Measures 'fingerprint', it is possible to divide the mine water emissions into those sourced from the Upper and Middle and Lower Coal Measures, and at times to infer a degree of mixing between the two.

The conclusion is that in the eastern valleys of the South Wales Coalfield, the Upper Coal Measures water has a calcium–sulphate facies and the Middle and Lower Coal Measures water has a sodium–sulphate to sodium–bicarbonate facies. This information can be added to the available knowledge to aid in assessing the nature and the progress of groundwater recovery within the coalfield.

This paper refers to preliminary work that has been carried out in the South Wales Coalfield to attempt to 'fingerprint' mine water emissions to identify their underground source. This, in turn, provides valuable information on the hydrogeological regime and, coupled with the geological and mining setting, may further assist in identifying underground flow paths. This information helps to fill in some of the 'knowledge gaps' and provide a tool for use in the prediction of future surface mine water emissions and in the formulation of treatment–mitigation strategies. The current work uses observations and interpretations of the hydrochemistry of the coal Measures to develop a robust and simple to use method of characterizing and classifying mine water emissions within the South Wales Coalfield. It is anticipated that a similar interpretive process could be applied to mine water emissions in other coalfields, although it should be stressed that this cannot be considered as a 'stand alone' process and that the initial characterization of the mine water needs to be carried out with reference to the geological and mining setting for any particular area under consideration.

The South Wales Coalfield is an exposed coalfield with the structure of a synclinal basin covering an area of approximately $2200\,km^2$, approximately 87-km long by 31-km wide. The Coal Measures are Carboniferous in age and can broadly be subdivided into the Upper, Middle and Lower Coal Measures (Figs 1 and 2).

Older rocks of the Millstone Grit, Carboniferous Limestone and Old Red Sandstone underlie the coalfield. With the exception of drift and river deposits, younger rocks are not deposited in the coalfield, except for a limited region at the southern outcrop near Bridgend.

The coalfield is traversed by faults that are broadly related to three main geological structure systems. Traversing the coalfield from NNW to SSE are normal faults that often occur in pairs. These faults have often formed boundaries to

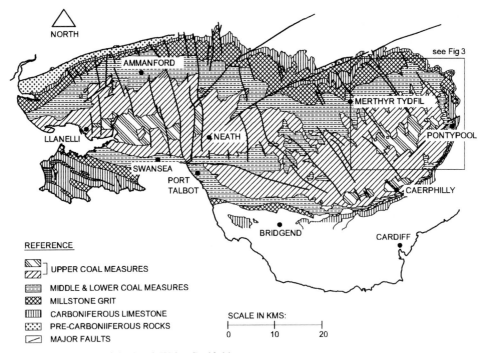

Fig. 1. Outline geology of the South Wales Coalfield.

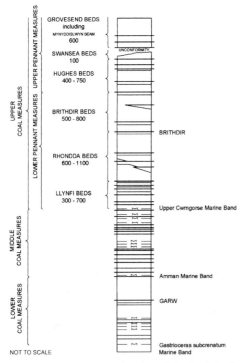

Fig. 2. Generalized vertical section of the South Wales Coal Measures.

coal mine workings and are more common in the central part of the coalfield. The southern sector of the coalfield is affected by faults traversing generally E–W. These are often accompanied with folding and overthrusting. The third system of faulting lies in the NW of the coalfield. These faults traverse in a NE–SW direction and are sometimes accompanied by overthrowing and lag faulting.

Upper Coal Measures

The Upper Coal Measures include the Grovesend Beds, Swansea Beds and Hughes Beds of the Upper Pennant Measures, and the Brithdir, Rhondda and Llynfi Beds of the Lower Pennant Measures. The Upper Cwmgorse Marine Band forms the base of the Upper Coal Measures. These Pennant Measures mainly include thickly bedded and jointed sandstones. The sandstones provide water flow pathways and localized areas of water retention, often because of the natural near-vertical jointing. The ability of the joints and fractures to allow water movement has been increased by mining activity, particularly in zones of tension. Sealing of these joints can also occur where clays and mudstones are present or within zones of compression induced by mining. Coal mine workings in the Upper Coal Measures are generally subjected to lesser strata pressure than in Middle and Lower Coal Measures because they are at shallower depths beneath the surface. Coupled with the massive nature and strength of the surrounding sandstones in the Upper Coal Measures, remnant voids from coal mining activities are more likely to remain partly open for many years after mining has ceased.

Middle and Lower Coal Measures

The Middle Coal Measures comprise predominantly shales, with insubordinate sandstones. The Lower Coal Measures are predominantly argillaceous measures of mudstones, siltstones and shales, with only occasional sandstone beds. The permeability of the Middle and Lower Coal Measures is expected to be reduced by the presence of impermeable shales and mudstones, and also reduced void spaces because of greater roof and floor strata pressures. Within the mining environment in the eastern sector of the South Wales Coalfield, water outbreaks into the Garw seam mine workings in the Lower Coal Measures were recorded. The outbreaks were rarely large enough to require pumping but were a problem to mining activities. The underlying Millstone Grit was considered to be an important source of the ingressing water, where the seat-earth below the Garw seam had been penetrated.

Mining position

Coal has been mined in the South Wales Coalfield for centuries, and in some areas coal has been mined in conjunction with ironstone and seat-earth. The dewatering requirements have necessitated pumping of substantial quantities of water, particularly from the Upper Coal Measures sandstones. The lower permeability of the predominantly argillaceous Middle and Lower Coal Measures meant that these coal seams were generally drier to work. Analysis of National Coal Board pumping data for deep mines in South Wales working in the early 1980s indicates that over 60% of the total water pumped was derived from workings in the Upper Coal Measures (Strong 1984).

In the eastern sector of the South Wales Coalfield the coal seams of the Upper, Middle and Lower Coal Measures have been extensively mined. In addition, ironstones and seat-earth (fireclay) have been mined separately or in conjunction with coal in localized areas.

The Mynyddislwyn and Brithdir coal seams crop out on the valley sides and have been extracted from inclined drifts from the coal seam outcrops. In some cases, shafts sunk from the valley floors to extract the deeper coal seams passed through or mined the Brithdir coal seam, but none of these collieries intersected or worked the Mynyddislwyn coal seam.

Much of the mining in the Brithdir coal seam took place many years ago. The *Abandoned Mines Catalogue* held by the Coal Authority indicate abandonment dates in the vicinity of Vivian and Six Bells shafts between 1896 and 1926. Because the Brithdir seam mine workings were in the sandstones of the Upper Coal Measures, large quantities of groundwater drained through the sandstones into the mine workings and was either pumped to the surface via the mine shafts or allowed to gravitate through drainage levels (such as the Plas-y-Coed drainage level) to discharge into rivers or streams. As mining progressed in the deeper coal seams it became evident that the groundwater regime associated with the Upper Coal Measures was significantly different from that associated with the Middle and Lower Coal Measures.

In the Six Bells locality, in the eastern sector of the coalfield for example, nine workable coal seams occur in the Middle and Lower Coal Measures within some 117 m of strata that is known to be predominantly mudstones

Subsequently to abandonment of mining and cessation of pumping in the eastern sector of the coalfield, groundwater recovery is occurring with mine water emissions at the surface. Despite previous historical efforts made to predict groundwater recovery, surface mine water emissions often occur in an unpredictable manner and in unexpected locations. Failure to be 100% accurate in predicting recovery and probable surface emission points is partly due to gaps in knowledge and available information. The problem has been exemplified by a lack of records regarding treatment of mine openings and is also partly due to geological complexity, particularly the different permeabilities of the varying lithologies of the Coal Measure strata and superimposed faulting.

Methodology

The work discussed in this paper was carried out on behalf of the Coal Authority by Wardell Armstrong as part of a regional study on mine water recovery in the eastern sector of the South Wales Coalfield. The principal aim was to obtain, from the chemistry of the mine water emissions sampled, a means of 'fingerprinting' mine water from different underground sources. In particular, the ability to differentiate between mine water sourced from the Upper Coal Measures and the Middle and Lower Coal Measures provides valuable information regarding the underground hydrogeological regime.

The field work component of this task comprised visiting and sampling 15 known mine water emissions within the catchments of the Rivers Ebbw, Ebbw Fach, Rhymney and Taff Bargoed in the summer of 2000 (Fig. 3). On-site measurements of pH, alkalinity, dissolved oxygen, redox potential, conductivity and temperature were taken. Two samples were taken at each site, one filtered through a 0.45 μm cellulose nitrate filter membrane and one unfiltered. The unfiltered sample was used in the laboratory, after splitting into subsamples and appropriate preparation, for measurement of all the major determinands (metals, major ions and suspended solids). The filtered sample was acidified in the laboratory and analysed for iron, to give a figure for dissolved iron at the point of emission.

Presentation of data

In order to assess water data, presentation in graphical form enables easy and quick comparison by eye, and also allows classification of water samples into various 'types'. One tool available for such graphical presentation is the trilinear, or Piper, diagram on which are plotted the ionic strengths of major cations and anions, in terms of the per cent composition of each ion in the water sample. When water data are plotted on a Piper diagram it allows identification of the hydrochemical facies of the water samples. The hydrochemical facies is a function of lithology, solution kinetics and flow patterns in the strata, and as such is a useful means of classifying water types and differentiating between water from different points of origin.

In the current study of mine water emissions within the eastern part of the South Wales Coalfield, use of Piper diagrams allows differentiation of the various water samples into two

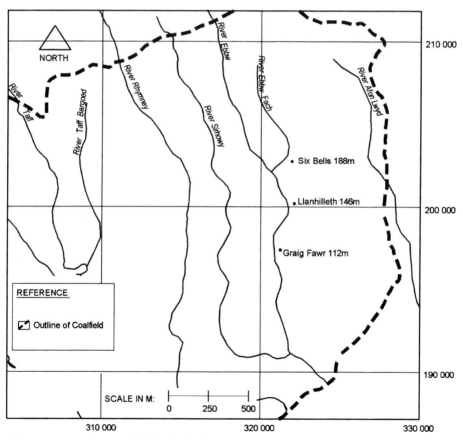

Fig. 3. Eastern sector of the South Wales Coalfield.

Table 1. *Water emissions considered to be associated with Upper Coal Measures strata*

Source of water sample	Major ions from laboratory analysis								On-site data						Metals from laboratory analysis			
	Na (mg l^{-1})	K (mg l^{-1})	Ca (mg l^{-1})	Mg (mg l^{-1})	Cl (mg l^{-1})	HCO$_3$ (mg l^{-1})	SO$_4$ (mg l^{-1})	Hardness (CaCO$_3$)	pH	Conductivity (μS cm^{-1})	DO$_2$ (mg l^{-1})	T (°C)	Redox (mV)	Alkalinity (CaCO$_3$)	Fe total	Fe dissolved	Mn	Al
Bensons Level	8.39	1.39	18.7	11.6	12	29.9	55.9	95	8.1	271	9.7	na	na	49	<0.01	na	<0.01	<0.01
Fothergills	4.39	4.27	52.5	25.7	9	91	106	237	6.55	484	3.07	14.5	138	150	0.03	na	<0.01	0.04
Plas y Coed	8.33	5.27	48.4	25.3	10	76	115	225	6.92	469	4.82	11.1	106	125	0.15	0.11	0.1	0.02
Hafodyrynys	6.82	3.51	69.3	27.5	7	107	115	287	7.7	532	2.03	9.8	62	175	1.71	0.33	0.49	0.01
Trelewis	8.67	9.84	82.8	29.9	12	153	136	330	7.08	685	3.12	13.5	−40	250	3.42	1.55	1.08	0.02
Aberbeeg North	12.3	10.3	100	47.9	11	152	232	447	7.2	874	5.08	11	34	250	1.81	0.52	0.99	0.02
Tillery adit	19.1	8.1	64.8	36.4	20	42.1	240	na	6.6	711	na	na	na	na	na	na	na	na
Llanhilleth Red Ash	15.6	7.62	117	49.3	12	152	268	496	7.19	1026	2.28	11.6	39	250	3.64	0.9	0.97	0.01
Taff Merthyr North	13.5	13.9	125	51	12	213	309	523	6.96	1051	0.87	14	−77	350	14.6	0.18	1.32	0.06
Taff Merthyr South	14.6	14.1	128	51.8	11	154	311	533	6.95	1059	2.71	14.7	−19	250	2.17	0.33	1.35	0.11
Graig Fawr	69.1	12.3	131	59.4	20	305	335	572	7.24	1398	1.2	12.1	−51	500	8.21	6.99	1.24	<0.01
Tir-y-Berth	18.8	11.2	130	63.7	14	136	395	587	6.84	1102	2.99	12.4	19	200	2.31	0.18	0.85	0.04

DO$_2$, dissolved oxygen.
na, not available.

Table 2. *Water emissions considered to be associated with upper Coal Measures strata*

Source of water sample	Major ions from laboratory analysis								On-site data						Metals from laboratory analysis		
	Na (mg l^{-1})	K (mg l^{-1})	Ca (mg l^{-1})	Mg (mg l^{-1})	Cl (mg l^{-1})	HCO$_3$ (mg l^{-1})	SO4 (mg l^{-1})	Hardness (CaCo$_3$)	pH	Conductivity (μS cm^{-1})	DO$_2$ (mg l^{-1})	T (°C)	Redox (mV)	Alkalinity (CaCo$_3$)	Fe total	Fe dissolved	Mn	Al
Crumlin	185	39.1	96.5	53.7	17	450	157	462	7.46	1544	1.71	19.3	−166	740	2.04	na	0.22	19.7
Six Bells	100	22.9	51.8	35	16	180	210	274	7.8	1016	6.28	12.4	−22	300	1.91	0.76	0.14	0.01
Vivian	440	87.8	200	160	33	470	1470	1160	6.93	3680	0.47	18	−85	771	48.8	47	1.5	0.05

DO$_2$, dissolved oxygen.

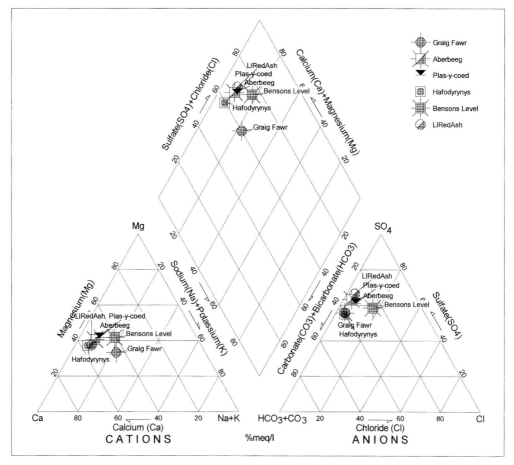

Fig. 4. Piper diagram showing CaSO$_4$ type water emissions considered to be associated with Upper Coal Measures strata.

distinctive types, or hydrochemical facies. When combined with knowledge of the geological, mining and hydrogeological setting of the area, this in turn allows interpretation of the data in terms of whether the water is sourced from the Upper or Middle and Lower Coal Measures. Other water chemistry data obtained in the course of the study can be used to supplement and corroborate interpretations made on the basis of major ion data.

In addition to the mine water emissions sampled in the summer of 2000, selected sites were also subsequently sampled over a period of 5 months from October 2000 to January 2001. This was undertaken to establish whether any changes may be occurring in the mine water conditions in the eastern sector of the South Wales Coalfield. The additional samples also give a measure of precision of sampling on a temporal basis. Mine water emissions located at Six Bells, Llanhilleth and Graig Fawr were sampled on a monthly basis for a five month period.

Results

The results from the 15 water emissions sampled in the initial baseline investigation are given in Table 1 and 2 and displayed in three Piper diagrams (Figs. 4, 5 and 7).

On each Piper diagram, each water sample is depicted as a unique point in the cation plot on the left, the anion plot on the right, and projected up to a single point on the diamond-shaped field between the two triangles. The circles surrounding the points on the diamond-shaped field represent the total dissolved solids of the sample, which can be measured off on the scale bar on the left.

Figure 4 indicates the results of emissions from Graig Fawr Shaft at Celynen North;

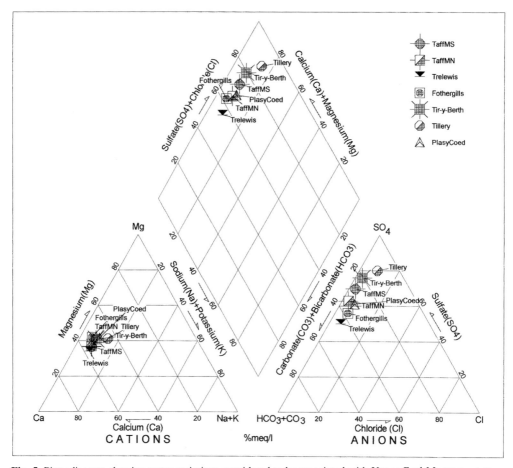

Fig. 5. Piper diagram showing water emissions considered to be associated with Upper Coal Measures strata.

Llanhilleth Red Ash; Aberbeeg North, Plas y Coed drainage level, Hafodyrynys and Bensons Level. The latter four plots are very similar, and can be classified as calcium-sulphate-type water. The Graig Fawr emission also falls within the calcium–sulphate field, but has slightly higher sodium and bicarbonate content that the other samples (refer to Fig. 6, which shows the data plotted as a three-dimensional (3-D) column chart, and from which the slightly different composition of Graig Fawr can clearly be seen). A possible explanation for this is explored below.

Using knowledge of the mining history of this area it can be confidently stated that the water emissions from Plas y Coed and Aberbeeg North originate in the Upper Coal Measures (note that they plot virtually on top of each other in the Piper diagram, thus confirming their similarity in terms of hydrochemical facies). Thus, these samples can be used as a 'fingerprint' of Upper Coal Measures water, and it can be proposed that water samples that plot near to them on a Piper diagram are also Upper Coal Measures water.

Figure 5 indicates the results of sampling of water emissions from Taff Merthyr Colliery North and South shafts, Trelewis Colliery site, Fothergills Shaft, Tir-y-Berth (River Rhymney) and Tillery adit near Vivian Shaft (this latter sample was taken by International Mining Consultants Ltd in December 1999). The results of the sampling at Plas-y-Coed are also shown on this plot to depict the typical Upper Coal Measures 'fingerprint'. These samples all plot within the calcium–sulphate field, with very similar cation content, and increasing bicarbonate content from the Tillery adit in the east to Trelewis further west (with Plas y Coed falling midway, plotting very close to the Taff Merthyr North Shaft). It can thus be proposed that all these water emissions are issuing from the Upper Coal Measures with subtle differences in

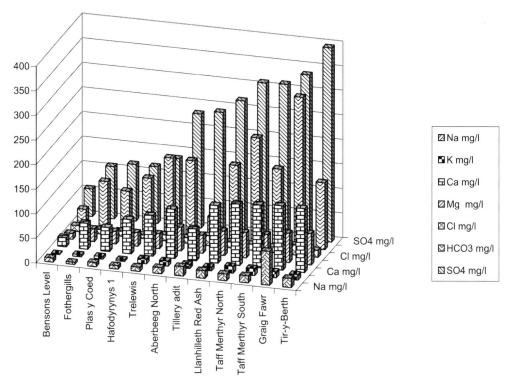

Figure 6. Bar chart showing water emissions considered to be associated with Upper Coal Measures strata.

sulphate and bicarbonate due to slightly different flow paths through the strata and mine workings. Increased conductivity of the water samples indicates a longer retention time within the water-bearing strata, and variations in iron content may be due to the amount of pyritic material in the flow path of each particular emission.

Figure 6 indicates all of the samples considered to be associated with the Upper Coal measures plotted on a 3-D column chart. The predominance of calcium in the cations and sulphate in the anions can be observed, as can the overall similarity in relative amounts of each of the major ions. The water emission sample from the Tillery adit has a low value for alkalinity (HCO_3), which may be due to the fact that this sample was tested for alkalinity in the laboratory rather than in the field, with a decrease in alkalinity occurring due to degassing of CO_2 while the sample was in transit/storage.

Figure 7 indicates the results from the Graig Fawr shaft, Six Bells, Vivian and Crumlin collieries.

Water samples from Six Bells and Vivian can both be classified as $NaSO_4$-type waters and Crumlin as $NaHCO_3$ type. The Graig Fawr emission lies between the $CaSO_4$ field and the $NaSO_4$ field.

To interpret these results, consideration needs to be made of the dynamic chemical processes that occur as water migrates through the strata. It has been shown that the typical mine water emission from the Upper Coal Measures has a $CaSO_4$ fingerprint, which is typical for sandstones, as found in the Pennant Series of the Upper Coal Measures. As water migrates downwards, exchange processes occur with sodium being released from cation exchange sites in preference to calcium and magnesium, which are held on these sites instead. This process occurs principally on the clay minerals present in argillaceous strata such as mudstones and shales, which have a high cation exchange capacity (Ineson 1967). The Middle and Lower Coal Measures have a higher proportion of argillaceous measures than the Upper Coal Measures, and thus it can be anticipated that water issuing from the Middle and Lower Coal Measures will have undergone a greater degree of cation exchange and have an elevated sodium content compared with calcium and magnesium. Increased cation exchange also

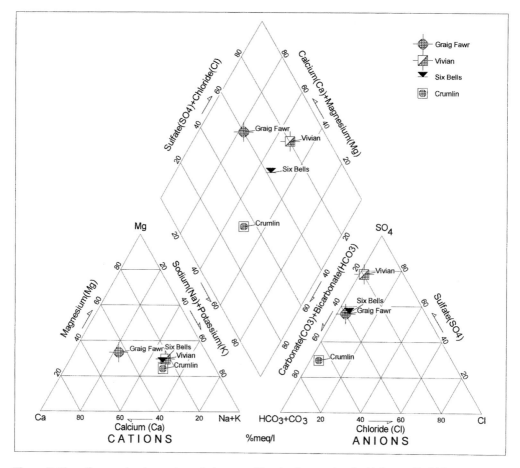

Figure 7. Piper diagram showing water emissions considered to be associated with Lower Coal Measures strata.

tends to give an increase in alkalinity in the form of bicarbonate. Increased alkalinity, particularly when associated with a decrease in sulphate, can also be an indication of activity by sulphate-reducing bacteria, which are likely to be found in relatively stagnant, anoxic groundwater.

Consideration of the above allows interpretation of the samples plotted on Fig. 7 as follows.

The Vivian emission has elevated sodium (see also Fig. 8, which shows these samples plotted on a 3-D column chart), high alkalinity and very high sulphate. When considered in conjunction with the temperature of the water at the point of emission, which was considerably elevated compared to ambient temperatures, and the low dissolved oxygen and redox potential, it is logical to propose that this emission is sourced from the Middle and Lower Coal Measures. The very high sulphate in the Vivian sample corresponds to a high iron content, which indicates that the water has been in contact with highly pyritic coal or rock strata that has been subject to oxidation.

The Six Bells emission has a very similar composition to Vivian, the main difference in major ions being in the amount of bicarbonate and sulphate, which is less extreme than at Vivian (see Fig. 8). The fact that the iron content is also considerably less at Six Bells suggests that this is due to less contact with highly oxidized, highly pyritic material. It is proposed that the major ion 'fingerprint' of the Six Bells emission also indicates that it has a Middle and Lower Coal Measures source.

The sample from the Graig Fawr Shaft at Celynen North (Fig. 7) plots between the sodium–sulphate field of the Middle and Lower Coal Measures and the calcium–sulphate field of the Upper Coal Measures, which suggests a degree of mixing of water from both Upper Coal Measures and Middle and Lower Coal Measures sources.

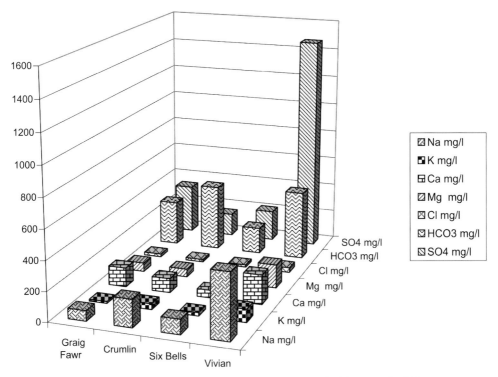

Figure 8. Bar chart showing water emissions considered to be associated with Lower Coal Measures strata.

The water emission at Crumlin Navigation Colliery has a separate hydrochemical facies to all the other samples, being a NaHCO$_3$ type. Considering the very low redox potential of the water at the point of emission, and the strong smell of hydrogen sulphide, it seems probable that this is due to the activity of sulphate-reducing bacteria in the stagnant water within the tailings-filled shaft. This would account for the relative decrease in sulphate and increase in bicarbonate compared to the other Middle and Lower Coal Measures waters.

Monthly Sampling Analysis

The results of sampling at Six Bells, Graig Fawr and Llanhilleth are shown in Fig. 9. The plots for all three sites fall in similar positions on the Piper diagram to the initial baseline sampling in the summer of 2000. This indicates that the three emissions have not significantly changed in chemistry during the five month sampling period and therefore the source of the water has probably not changed. This is an important observation when investigating the recovery of the mine water in the coalfield. The similar plots also indicate that the samples taken in the baseline sampling were representative of the mine water emission and give confidence in sampling precision. Slight variations in the water chemistry at Llanhilleth may be due to different flow paths through the mine workings. The results indicate that the Six Bells emission is responsive to rainfall. During the months of October and November, the chemistry was diluted compared to previous results and this corresponds to a period of high rainfall in the locality.

Conclusions and summary of sample analyses

Consideration of the water chemistry data from this mine water emission sampling exercise has allowed identification of two principal hydrochemical facies within the eastern sector of the South Wales Coalfield.

The first of these characterizes emissions that are sourced from the Upper Coal Measures and can be described as a CaSO$_4$ type water. The variation in iron content between the samples within this group is an indication of contact between the water and pyrite-bearing coal seams or rock strata, and the ionic strength of

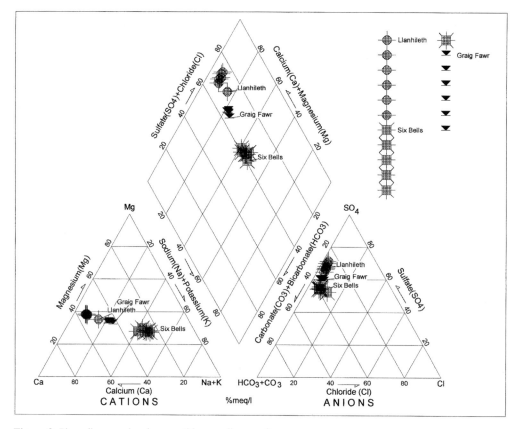

Figure 9. Piper diagram showing monthly sampling results.

the samples gives an indication of residence time of the water within the strata (longer residence time giving a higher ionic strength).

The second of these hydrochemical facies is characteristic of water sourced from the Middle and Lower Coal Measures and is typically sodium–sulphate type water, although anoxic, stagnant conditions may cause a change to $NaHCO_3$ type water due to the activity of sulphate-reducing bacteria.

These two separate hydrochemical facies can be considered as the 'end-points' of a sliding scale of water chemistry, with mixing between water sourced from the Upper, Middle and Lower Coal Measures giving an intermediate water type. Thus, Graig Fawr was interpreted as currently comprising a mixture of Upper, Middle and Lower Coal Measures water.

The interpreted observations made as a result of this study will constitute a useful 'yardstick' against which to assess other mine water emissions. Also, consideration of further mine water emission samples, both from future emissions and compared with historical data, will allow refinement of the definition of these hydrochemical facies and increase the confidence that can be placed in their interpretation.

References

INESON J. 1967. *Ground-water Conditions in the Coal Measures of the South Wales Coalfield.* Water Supply Papers of the Geological Survey of Great Britain Hydrogeological Report, **3**.

STRONG, W.J. 1984. Water – a necessity of life? Geological and hydrogeological and implications of pumping problems. *Mining Engineer*, **144**, 159–166.

Alkaline mine drainage from metal sulphide and coal mines: examples from Svalbard and Siberia

DAVID BANKS[1], VALERY P. PARNACHEV[2], BJØRN FRENGSTAD[3,4], WAYNE HOLDEN[5], ANATOLY A. VEDERNIKOV[6] & OLGA V. KARNACHUK[7]

[1]*Holymoor Consultancy, 86 Holymoor Road, Holymoorside, Chesterfield, Derbyshire S42 7DX, UK*

[2]*Department of Dynamic Geology, Tomsk State University, Prospekt Lenina 36, 634050-Tomsk, Russian Federation*

[3]*Norges Geologiske Undersøkelse, N-7491 Trondheim, Norway*

[4]*Institutt for Geologi og Bergteknikk, Norges Teknisk-Naturvitenskapelige Universitet, N-7491 Trondheim, Norway*

[5]*URS Dames & Moore, St. George's House (2nd Floor), 5 St. George's Road, Wimbledon, London SW19 4DR, UK*

[6]*State Committee for Environmental Protection of the Republic of Khakassia, Abakan, Republic of Khakassia, Russian Federation*

[7]*Department of Plant Physiology and Biotechnology, Tomsk State University, Prospekt Lenina 36, 634050-Tomsk, Russian Federation*

Abstract: Not all water from coal or metal mines is acidic. Circum-neutral or alkaline mine drainage may be due to: (i) a low content of sulphide minerals; (ii) the presence of monosulphides rather than pyrite or marcasite; (iii) a large pyrite grain-size limiting oxidation rate; (iv) neutralization of acid by carbonate or basic silicate minerals; (v) engineering factors (introduction of lime dust for explosion prevention; cement or rock flour during construction works); (vi) neutralization of acid by naturally highly alkaline groundwaters; (vii) circulating water not coming into effective contact with sulphide minerals; and (viii) oxygen not coming into direct contact with sulphide minerals or influent water being highly reducing.

The evolution of mine drainage water chemistry

The oxidation of pyrite is known to release dissolved acid, iron and sulphate:

$$2FeS_2 + 2H_2O + 7O_2 \rightarrow 2Fe^{2+} + 4SO_4^{2-} + 4H^+. \quad (1)$$

This is often followed by oxidation of iron (II) to iron (III) and precipitation of ferric oxyhydroxides (ochre) or, in low-pH, sulphate-rich environments, iron hydroxysulphates such as jarosite:

$$4Fe^{2+} + O_2 + 4H^+ \rightarrow 4Fe^{3+} + 2H_2O$$
(oxidation)

$$Fe^{3+} + 3H_2O \rightarrow FeOOH + H_2O + 3H^+$$
$$\rightarrow Fe(OH)_3 + 3H^+$$
(hydrolysis/precipitation). (2)

The net result of these reactions is the generation of protons:

$$4FeS_2 + 14H_2O + 15O_2 \rightarrow 8SO_4^{2-} + 16H^+$$
$$+ 4Fe(OH)_3. \quad (3)$$

The oxidation of many other monosulphides of the form MeS (where Me is a divalent metal) releases dissolved metal and sulphate, but not acid:

$$MeS + 2O_2 \rightarrow Me^{2+} + SO_4^{2-}. \quad (4)$$

Nevertheless, if pyrite is also present, the acid environment generated by pyrite oxidation will enhance the solubility of many metals released by sulphide oxidation (Cd, Zn, Pb, Cu) and others released by acid hydrolysis of carbonates, oxides or hydroxides (e.g. Al, Mn). Thus, one might expect water in contact with sulphide-rich ores or coals, in an oxidizing environment, to have a high loading of many potentially environmentally toxic metals, a low pH and a high concentration of dissolved sulphate. This phenomenon is often known as 'acid mine drainage'.

It is, however, not uncommon to observe water draining from pyrite-rich mines or mine wastes that is circum-neutral or even alkaline, with modest loadings of potentially toxic metals (e.g. carbonate-hosted lead mines of the English Pennines: Banks *et al.* 1996; Nuttall & Younger 1999; many coal mines in the UK and some colliery spoil tips receiving wastes from deep-mined strata: Banks *et al.* 1997*a*, *b*). In such cases one must assume that: (i) mine water is not coming into effective contact with sulphide minerals; (ii) an oxidizing environment is not present in the mine; or (iii) subsequent neutralization of acidic species by other minerals or alkaline dissolved species is taking place.

This paper presents previously unpublished data from coal mines on the Arctic archipelago of Svalbard, coal mines in the Abakan basin of Siberia, and iron ore mines in the Sayan and Kuznetsk–Alatau Mountains of Siberia (Fig. 1) to illustrate the chemistry of alkaline mine drainage waters and to discuss the possible mechanisms involved in their generation.

Sampling and analysis methods

The methods of analysis employed at all the sampled mines in this study are broadly similar and are described in Banks (1996) for Svalbard, and Banks *et al.* (1998, 2001) for the Khakassian, sites. In brief:

- pH, temperature and (where appropriate) reduction–oxidation (redox) potential were measured in the field using appropriate electrodes or thermometers;
- alkalinity was measured in the field by acid titration to a pH end-point of *c*. 4.3 using either AquaMerck 11109 or Hanna Instruments HI 3811 alkalinity test kits;
- samples for laboratory analysis were taken in clean polyethylene bottles, repeatedly rinsed with filtered sample water before sampling. Samples were filtered using syringe-mounted 0.45 μm Millipore filter units;
- analysis for anions (NO_3^-, SO_4^{2-}, Cl^-, Br^-, PO_4^{3-}, F^-) was carried out at the Geological Survey of Norway, by Dionex ion chromatography, using a small unacidified portion of the collected filtered sample;
- another quantum of each filtered sample was acidified in the original flask with Ultrapure nitric acid to remobilize sorbed or precipitated metals. This portion was analysed at the Geological Survey of Norway by inductively coupled plasma-atomic emission spectrometry (ICP-AES) for a range of *c*. 30 metals and cations. In some cases, inductively coupled plasma-mass spectrometry (ICP-MS) was also employed for analyses of selected trace elements.

Case study 1 – the coal mines of the Abakan–Chernogorsk region, Khakassia, Siberia

The coal mines of the Abakan–Chernogorsk region lie in the central part of the Republic of

Fig. 1. Map of Russia showing the location of Khakassia (case studies 1, 2 and 3) and Svalbard (Spitsbergen – case study 4).

Khakassia in southern central Siberia, a part of the Russian Federation. Geologically, they are found in the Minusinskii Basin (Parnachev et al. 1999), an extensional basin filled with sedimentary (ranging from mudstones to sandstones) and volcanic rocks of Devonian–Permian age, derived from erosion of the mountains produced by the Silurian–Devonian Western Sayan and Kuznetsk–Alatau orogenic episodes. The coal-bearing Carboniferous and Permian sequence forms a circular synclinal structure bisected by the Abakan River, a tributary of the Yenisei. According to Matveev (1960), coal was first extracted from 1904 onwards in the Izikskii and Chernogorskii deposits, although it was only in 1917 that extraction commenced on an industrial scale. In the Carboniferous–Permian sequence, 80 seams of coal are known, of which between 7 and 40 are regarded as being of workable thickness, ranging from 0.7 to 14 m. A total coal reserve of 37 billion tonnes (bt) has been estimated. Regarding hydrogeology, Matveev (1960) notes that water inflow to mine shafts in the area reaches up to $350\,m^3\,h^{-1}$, the dominant inflow features being permeable (sandy) strata, coal horizons, bedding-plane fractures and overlying Quaternary alluvial deposits. Table 1 illustrates the typical properties of the coal. Three samples of mine water were acquired from this region in June 1999:

- A sample of inflow water in the base of each of the two opencast pits (Izikhskii I and Izikhskii II) of the Byelii Yar complex, on the southern bank of the Abakan River. In the walls of the pits, the lower part of the Permian sequence comprises dark, coal-bearing strata. The upper part comprises lighter, coal-free strata, including carbonate-rich mudstones. The accumulated water in the Izikhskii I and II pits is reported to be pumped at rates of c. 220 and $18\,m^3\,h^{-1}$, respectively. In both pits, the water was free from iron coloration and ochre. The inflow water was dominantly derived from just above the base of the lighter-coloured coal-free strata. Testing with dilute hydrochloric acid in the field revealed the mudstones of the sequence to be carbonate rich.
- Mine water from the interconnected underground mine complex comprising the Khakasskaya and Yeniseiskaya shafts on the northern bank of the Abakan River. Water from the two mines, and reportedly also some other opencast mines in the west of the area, drains towards the easternmost Khakasskaya pit, from where the water is pumped to the surface at $50\,m^3\,h^{-1}$. The coal-bearing sequence in this complex is of Upper Carboniferous age, and the mined coal is from a single seam of thickness 2.2 m. The sampled water was collected underground in the Khakasskaya Shaft at 140 m depth. No iron coloration or ochre deposits were noted in the mine drainage, nor any H_2S smell.

Water analyses are presented in Table 2, and the following characteristics should be noted:

- the high pH;
- the low concentrations of many pH-sensitive metals (including Fe);
- elevated sulphate and alkalinity (especially in Izikhskii II the and Khakasskaya Shaft).

Three hypotheses may be put forward for the alkaline nature of the mine drainage:

(i) the low pyrite content of the coals (but, in this case, one might expect low SO_4^{2-});
(ii) neutralization of acid by carbonate-rich mudstones (here, one might expect elevated Ca concentrations;
(iii) neutralization of acid by ambient groundwater, which is known to be often highly alkaline and sodium–sulphate–chloride rich in this steppe environment (Parnachev et al. 1999).

The rather modest Ca concentrations suggest that (ii) may not offer a full explanation for the water quality. The high SO_4^{2-} and Na concentrations in the water suggest that inflow of alkaline sodium–sulphate–chloride ambient groundwater (explanation iii) may be responsible for neutralizing any products of pyrite oxidation. The elevated sulphate in the mine drainage is unlikely to be solely derived from influent saline groundwater, however. The significant sulphate:chloride excess noted in two of the samples

Table 1. *Properties of typical coals from the Minusinskii Basin (after Matveev 1960)*

Parameter	Value
Loss of moisture on ignition	up to 13%
Ash content	7–21%
Sulphur content	Up to 1%
Loss of volatiles	35–46%
Heat content	7600–8200 kcal kg^{-1}
Working yield	6840 kcal kg^{-1}
Loss of primary tars on coking	12–16%

Table 2. *Analyses of drinking water borehole (32 m deep) at Byelii Yar opencast mine in alluvial deposits of River Abakan (for comparison, column 1), water from the opencast pits Izikskii I and II (second and third columns) and water from the underground Khakasskaya shaft at 140 m depth (column 4). Samples taken 13 June 1999 and field-filtered at 0.45 μm. Analysis of anions by ion chromatography, Si, Mg, Ca, Na, K, Fe, Mn, Cu, Zn and Ba by ICP-AES and Pb, Cd and Al by ICP-MS. Eh readings are parenthesized due to probable unreliability caused by exposure to air and turbulence during sampling. SI = saturation indices calculated by the speciation program MINTEQA2 (US EPA 1991). The equivalent ratio of Cl^-/SO_4^{2-} is calculated by converting concentrations of these two anions to meq l^{-1}*

	Unit	Borehole	Izikskii 1	Izikskii 2	Khakasskaya Mine
pH		7.92	8.39	7.76	8.18
Temp	°C	10.1	8.6	11.2	10.1
Eh	mV	(+130)	(+140)	(+160)	(+170)
t-alkalinity	meq l^{-1}	4.1	6.6	24.3	22.1
F^-	mg l^{-1}	0.58	0.54	0.11	0.29
Cl^-	"	54	93	305	307
NO_3^-	"	15.2	7.2	7.1	2.8
SO_4^{2-}	"	89	115	567	1135
Cl^-/SO_4^{2-}	equivalent ratio	0.82	1.10	0.73	0.37
Si	mg l^{-1}	6.27	4.99	4.04	3.49
Mg	"	27.8	38.9	11.7	132
Ca	"	49.1	45.2	29.8	69.6
Na	"	67.5	129	812	858
K	"	1.68	2.76	3.77	10.8
Fe	μg l^{-1}	<10	<10	<10	116
Mn	"	1.2	13	18	7.6
Cu	"	<5	<5	<5	<5
Zn	"	4.6	<2	<2	<2
Pb_MS	"	<0.02	0.049	0.048	0.026
Ba	"	35	25	19	23
Cd_MS	"	<0.01	<0.01	<0.01	<0.01
Al_MS	"	<0.6	<0.6	6.07	1.29
Barite	SI	+0.17	+0.10	+0.40	+0.69
Calcite	"	+0.27	+0.83	+0.42	+1.05
Chalcedony	"	+0.05	−0.03	−0.15	−0.20
Dolomite	"	+0.37	+1.65	+0.52	+2.46
Gypsum	"	−1.55	−1.54	−1.32	−0.81
Magnesite	"	−0.38	+0.34	−0.38	+0.93

does indicate that sulphide oxidation is occurring.

Case study 2 – the Abaza Magnetite Mine, Khakassia, Siberia

The Abaza magnetite is one of several significant iron mineralizations in the Western Sayan and Kuznetsk Alatau Mountains that form the southern and western border areas, respectively, of the Republic of Khakassia (Polyakov 1971; Indukaev 1980). The town of Abaza lies in the Western Sayan Mountains on the main road from Abakan to the Tuva Republic. The ore is located in Cambrian metasediments (comprising sandstones, siltstones and carbonates) adjacent to a faulted boundary with Devonian volcanic and sedimentary rocks and a Devonian granite intrusion. It is estimated that 10–15% of the Cambrian sequence comprises calcium carbonate rocks. The ore mineralogy includes magnetites, pyrite, chlorite, actinolite and calcite (G.B. Knazev, Tomsk State University pers. comm.). The ore is mined both by a major opencast and a complex of deep, underground mines of depth of up to almost 800 m below ground.

The pumped mine water is chlorinated and recycled to the mine as process water. Analyses of the abstracted mine water quality (carried out on unfiltered samples by the staff of the mine's laboratory, rather than by the authors) are found in Table 3. Regrettably, the analyses are not complete. The following features will be noted:

- a rather high pH of 8.3;
- high nitrate (and ammonium) concentrations, believed to be derived from use of explosives;
- elevated concentrations of sulphate (relative to chloride), suggesting pyrite weathering;
- modest iron concentrations;
- rather high concentrations of calcium and magnesium.

These features suggest that acid generated by pyrite oxidation is being effectively neutralized by either:

- calcium carbonate minerals in the host rocks; or
- naturally alkaline ambient groundwater that has evolved in contact with the carbonate-rich rocks.

In 1995–1996 a new shaft was constructed and connected to the mine complex. At this time, staff at the mine claim to have noted a change in water chemistry of the pumped water. Table 3 indicates increased concentrations of calcium (and iron) in the 1999 analyses, offering some support to the anecdotal evidence of the mine staff. While changes in laboratory techniques cannot be ruled out as a possible source of this observation, mine staff believe that the changes are related to the large amounts of cement used in shaft construction and the quantities of rock flour (which would contain small, and hence reactive, particles of both sulphide and carbonate minerals). Any change in water chemistry could also simply be ascribed to the fact that the newly opening part of the mine may have been part of a somewhat different geological–hydrochemical regime to the existing mine.

Case study 3 – the Vershina Tyëi Magnetite Mine, Khakassia, Siberia

The Vershina Tyëi magnetite deposit lies in the Kuznetsk Alatau Mountains in the western border area of Khakassia at the head of the valley of the River Tyeya. The geology of the area is described by Polyakov (1971). The main mineralized deposits lie close to the junction of outcrops of a Cambrian carbonate sequence and an Early Palaeozoic complex of granitoids, volcanics and gabbros. Lithologies present include trachybasalt, trachyrhyolite, granosyenite, serpentinite, explosion breccias, limestones and dolomites. The magnetite mine itself is a large opencast complex, surrounded by extensive piles of relatively coarse mine wastes. The wastes exhibit little sign of extensive oxidation and contain both macrocrystals of pyrite and small disseminated pyrite crystals. The wastes are uncovered. No obvious damage to vegetation is observed at the toe of the wastes. Two samples of water were taken from the mine:

- sample Xa85, from a stream that rises above the mine complex and then flows through the mine waste area, and seeps through the waste itself to re-emerge as a spring at the toe of the waste. The water appears clear, supports benthic flora and

Table 3. *Chemical composition of pumped water from Abaza iron mine, compiled from data analysed by the mining company's on-site laboratory on monthly samples, and supplied by the Abaza epidemiological authority*

Year	1989 N	Average	1992 N	Average	1999 N	Average
Temp. (°C)	11	12	12	11	12	10
pH	12	8.3	12	8.4	12	8.2
Dissolved O_2 (mg l^{-1})	12	10.1	1	11.0	0	na
NH_4^+ (mg l^{-1})	12	1.4	12	0.3	12	1.0
NO_3^- (mg l^{-1})	12	58	12	50	12	36
Fe (tot) (mg l^{-1})	12	0.8	12	0.8	12	2.6
Ca (mg l^{-1})	12	86	12	74	12	112
Mg (mg l^{-1})	12	32	12	35	12	28
Cu (mg l^{-1})	4	0.13	0	na	0	na
SO_4^{2-} (mg l^{-1})	0	na	0	na	12	237
Cl^- (mg l^{-1})	0	na	0	na	12	120
Cl^-/SO_4^{2-} (equivalent ratio)		na		na		0.69

N = number of samples.
Samples were not filtered on collection. There is some confusion as to whether the nitrate and ammonium concentrations are cited as N or as NH_4^+/NO_3^-.
na, not analysed.

exhibits no sign of iron oxyhydroxide staining. The sample was taken at the point of re-emergence of the stream from the toe of the mine waste;
- sample Xa86 of the pumped mine water, which accumulates in the base of the opencast pit. Normally this water is discharged via a series of settlement ponds but, at the time of sampling, these were non-operational and the sampled water was untreated mine water. The water was clear and showed no signs of iron oxyhydroxide coloration or precipitation. The pumped flow was estimated as $21 \, s^{-1}$, and pumps operate for between 3 and $24 \, h \, day^{-1}$ depending on weather conditions.

From the analyses (Table 4), the following should be noted:

(i) slightly alkaline pH;
(ii) elevated sulphate, with low chloride (indicates sulphide oxidation);
(iii) elevated concentrations of Ca and, particularly, Mg;
(iv) low concentrations of potentially ecotoxic metals (including Fe);
(v) high concentrations of nitrate (again believed to be derived from explosive use).

Points (ii) and (iii) indicate that pyrite oxidation is occurring, but that acid is being effectively neutralized by carbonate phases or basic magnesium-rich silicate phases associated with mafic–ultramafic rocks. The low iron (and other metal) concentrations are likely to be due to low metal solubility in the alkaline, oxidizing environment of the mine water.

Case study 4 – the coal mines of Longyearbyen, Svalbard

The coals of Svalbard, an Arctic archipelago governed by Norway, are mined both by Russian and Norwegian concerns (Gram 1923; Hjelle 1993). The mines that are the subject of this study are Norwegian and located near Longyearbyen at latitude c. 78.2°N. The Longyearbyen mines have been active for around a century and worked in the Longyear Seam, the thickest (varying from 0.6 to 2 m) of several coal seams of the Lower Tertiary sequence of the area. Mining-related activity has occurred in the area since the early years of the twentieth century. Since the 1950s, production of coal on Svalbard has been between 250 000 and 500 000 t year^{-1}. The coal has a relatively low sulphur content of some 1% S. The Tertiary host rocks consist of series of sandstones, siltstones and shales. Mining conditions are further described by Amundsen (1994) and World Coal (1995).

As of 1996 (the time of sampling), only two mines actively produced coal in the Longyearbyen area: Mine 3 near the airport and Mine 7 at Foxdalen east of Longyearbyen. The Sverdrupbyen Mine site is derelict. Production of coal is planned to be increased in the future in the so-called Central Field, near Svea (Utsi & Myrvang 1992).

During July 1996, samples were taken of mine discharge water from the coal mine known as Mine No. 3 at Bjørndalen, near Longyearbyen, Svalbard. Samples were also taken of run-off water (leachate) from mine spoil tips at Mine No. 3 in Bjørndalen and at the derelict Sverdrupbyen Mine, Longyearbyen.

Table 4. *Analyses of: (i) stream passing through mine waste area, sampled immediately following its emergence from the toe of a major waste tip; and (ii) pumped mine water from the Vershina Tyei opencast. The first (horizontal) section of the table contains field analyses, the second anions by ion chromatography, the third chloride/sulphate equivalent ratio, the fouth elements (in mg l^{-1}) by ICP-AES, and the fifth Pb and Al (in μg l^{-1}) by ICP-MS*

Sample Location	Xa85 Mine stream, V. Tyëi	Xa86 Mine water, V. Tyëi
pH (field)	7.50	7.85
Temp. (field) (°C)	3.3	11.0
t-alkalinity (meq l^{-1})	2.34	2.34
Cl$^-$ (mg l^{-1})	12.7	19.1
NO$_3^-$	93.0	31.6
SO$_4^{2-}$	988	312
Cl$^-$/SO$_4^{2-}$ (equivalent ratio)	0.02	0.08
Si (mg l^{-1})	3.66	3.81
Al	<0.02	<0.02
Fe	<0.01	0.014
Mg	140	51.9
Ca	232	114
Na	11.0	14.5
K	7.03	10.8
Mn	0.005	0.031
P	<0.1	<0.1
Cu	<0.005	<0.005
Zn	0.002	<0.002
Mo	1.02	0.50
Cd	<0.005	<0.005
Pb (μg l^{-1})	<0.02	<0.02
Al	0.79	1.09

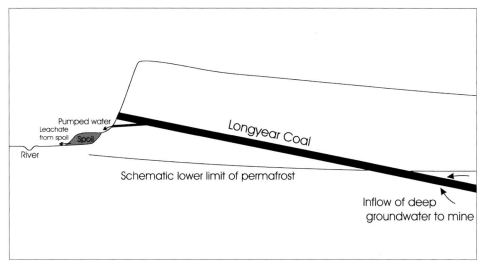

Fig. 2. Schematic cross-section through a mine at Longyearbyen, Svalbard.

Mine spoil leachate

Despite the coal's modest sulphur content, it has a strong potential to generate acid mine drainage, as indicated by analyses (Table 5) of leachate from two spoil tips. These are rather hydrochemically aggressive (note, in particular, the extremely low pH and high metals loadings of the Sverdrupbyen leachate) and not dissimilar to coal mine spoil waters recorded elsewhere in the world (Banks *et al.* 1997*a*, *b*).

On rock fragments at both the sampled spoil tips, yellowish-orange coatings were observed, which are most likely to represent jarosite or another ferric sulphate mineral (intermediate oxidation products of pyrite weathering). In the stream draining the Sverdrupbyen spoil tip, gelatinous precipitates of iron oxyhydroxide (ochre) were observed.

The low temperatures prevalent on Svalbard might be expected to reduce the rate of chemical and bacteriological sulphide oxidation. In fact, the exothermic nature of the oxidation reaction is probably sufficient to keep the internal temperature of the spoil high enough to allow the reaction to progress. It is unknown whether the spoil tips continue to produce leachate during the Arctic winter, although such an investigation might be of interest in the context of heat pumps utilizing heat generated by pyrite oxidation for domestic space-heating.

Pumped mine water – Mine 3

The Longyear Seam dips gently southwards, and mining takes place largely within the permafrost (which can reportedly exceed 300 m depth under the mountain areas) and partly in the unfrozen strata beneath it (see Fig. 2). The permafrost cap means that leakage of water from the surface into the deeper parts of the mine is low, although relatively small amounts of deep groundwater from saturated strata below the permafrost do enter the mine. The permafrost cap also permits the build-up of pockets of methane, necessitating the use of lime dust and water in the mine to bind coal dust and hinder explosions (Hjelle 1993; Amundsen 1994). The mine operators, Store Norske Spitsbergen Kulkompani, have observed that the quantities of water pumped, particularly from Mine 7, do show a seasonal variation (J. Utsi pers. comm.). It is thought that this may be indicative of fractures extending up through the permafrost layer to the zone of seasonal thaw.

At Mine No. 3, the pumped mine discharge occurs from a 200-mm diameter pipe just below a double access/ventilation tunnel. A small area of ground below the discharge pipe is stained orange, which probably indicates that the discharge of pumped water and the content of iron may, at certain times, be greater than that observed in the current study. The water pumped from the mine is believed largely to be derived from inflow of limited amounts of subpermafrost groundwater leaking into the mine's deepest levels. It is possible, however, that a component of the water may be derived from very shallow, supra-permafrost summer thaw water or from water introduced into the mine for dust damping. The water's chemistry may be influenced by

Table 5. *Composition of the three sampled waters from the coal mining area of Longyearbyen, Svalbard. FA implies a filtered (0.45-μm) sample, acidified in the laboratory with conc. HNO_3. FU implies a filtered, unacidified sample*

Sample site	Mine water, Mine 3	Spoil leachate, Mine 3	Spoil leachate, Sverdrupbyen
Elements by ICP-AES (FA)			
Ca (ppm)	15.5	15.6	48.2
K	2.8	<0.5	<0.5
Na	925	3.4	18.0
Mg	3.5	4.5	48.6
Fe	<0.01	1.6	179
Al	<0.02	1.8	27.5
Mn	0.0043	0.402	3.2
Si	2.9	0.974	6.8
B (ppb)	516	<10	46.7
Ba	3700	19.7	3.9
Cd	<5	<5	<5
Cr	<10	<10	<10
Cu	<5	14.0	168
Ni	<20	37.4	393
Sr	3100	260	1100
Zn	54.7	487	1300
Anions by ion chromatography (FU)			
Cl^- ppm	236	4.54	7.04
F^-	<0.05	<0.05	0.063
NO_3^-	1.43	0.569	2.43
SO_4^{2-}	7.43	76.6	1077
Cl^-/SO_4^{2-} equivalent ratio	43.03	0.08	0.01
Field measurements (all unfiltered)			
Flow ($l s^{-1}$)	c. 0.056	c. 0.1	c. 0.25
pH	8.12–8.20	3.73	2.70
t-alkalinity (meq l^{-1})	36.1	0	0
Eh (mV)	na	+420	+481
Temp. (°C)	4.7	0.0	1.9

na, not analysed.

chemicals introduced into the mine, for example lime dust.

The mine water from Mine 3 has a completely different chemistry to that of the spoil leachates, the most notable features being:

(i) an alkaline pH and extremely high alkalinity;
(ii) low concentrations of pH-sensitive metals such as Fe, Al, Mn, Cd and Cu;
(iii) low-concentrations of calcium and sulphate;
(iv) elevated concentrations of Ba;
(v) elevated concentrations of B, Sr, Cl and Na, with a significant excess of Na over Cl;

It might be argued that features (i) and (ii) are caused by a typical acidic mine drainage water having been neutralized either by carbonate-containing rocks within the sequence or by lime dust introduced to the mine. However, low calcium and sulphate (feature iii) indicate that this is not the case, and that sulphide oxidation has not occurred within the mine to a significant degree. Moreover, the elevated concentrations of B, Sr, Cl and Na, taken together with the low Mg and Ca concentrations and the high pH, are characteristic of the mature, deep, saline groundwaters present in many sedimentary aquifer sequences at depth (e.g. the British Coal Measures: Banks 1997). These are believed to have evolved their characteristic chemistry (excess Na over Cl) and low concentrations of Ca either by cation-exchange processes or a combination of silicate weathering coupled with calcite precipitation (Frengstad & Banks 2000).

The very low concentrations of sulphate (compared with the conservative, soluble anion, chloride) and the elevated concentrations of barium are characteristic indicators of sulphate reduction (Banks 1997). Sulphate reduction is also alkalinity generating:

$$2CH_2O + SO_4^{2-} \rightarrow H_2S + 2HCO_3^-. \quad (5)$$

The alkaline mine drainage from Mine No. 3 at Longyearbyen can thus be ascribed to the following factors:

- permafrost is present throughout the majority of the mined sequence, hindering active flow of water and oxygen to pyrite-bearing strata;
- the only significant groundwater flow is at great depth in the mine, below the permafrost. Such groundwater is hydro-chemically mature and has a naturally high alkalinity;
- the deep groundwater inflow appears to have a highly reducing character and is likely to be devoid of dissolved oxygen. Under such redox conditions, sulphide oxidation will not occur.

Conclusion

Not all water from coal or metals mines is acidic. Circum-neutral or alkaline mine drainage may be due to:

- low content of sulphide minerals (possibly *Izikhskii*);
- presence of monosulphides (MeS) rather than pyrite or marcasite;
- large pyrite grain-size limiting oxidation rate;
- neutralization of acid by carbonate or basic silicate minerals (*Abaza, Vershina Tyëi*);
- engineering factors: introduction of lime dust for explosion prevention; introduction of large quantities of cement or rock flour during construction works (*Abaza*);
- neutralization of acid by naturally highly alkaline groundwaters (*Izikhskii/Khakasskaya ??* and *Svalbard*);
- circulating water not coming into effective contact with sulphide minerals (*Svalbard*, due to permafrost);
- oxygen not coming into direct contact with sulphide minerals *or* influent water being highly reducing (*Svalbard*).

Finally, the data in this paper have demonstrated that nitrogen species (ammonium, nitrate), presumably derived from use of explosives, can occur at concentrations of environmental concern in mine waters.

The samples described in this paper were collected during the course of larger projects aimed at: (i) characterizing thermal springs on Svalbard's north coast; and (ii) regional hydrogeochemical mapping of Khakassia. Analytical costs, staff time and direct expenses in connection with these projects were supported by the British Council, Tomsk State University, the Norwegian University of Science and Technology, the Norwegian Geological Survey, URS Dames & Moore, Holymoor Consultancy, Chesterfield, UK, the State Committee for Environmental Protection, Khakassia and Norges Forskningsråd. While B. Swensen, M. Heim, A.Y. Berezovskii and Store Norske Spitsbergen Kulkompani were of assistance in sampling and provision of information.

References

AMUNDSEN, B. 1994. Dit sollyset aldri når – lengst inn i Verden i Gruve 3 [Where the sun doesn't shine: in the depths of the earth in Mine 3.]. *Svalbardboka*, 4th edn. Mitra forlag, Svalbard, 8–49 (in Norwegian).

BANKS, D. 1996. *The Hydrochemistry of Selected Coal Mine Drainage and Spoil-tip Run-off Waters. Longyearbyen, Svalbard*. Norges Geologiske Undersøkelse Rapport, **98.090**.

BANKS, D. 1997. Hydrochemistry of Millstone Grit and Coal Measures groundwaters, south Yorkshire and north Derbyshire, UK. *Quarterly Journal of Engineering Geology*, **30**, 237–256.

BANKS, D., YOUNGER, P.L. & DUMPLETON, S. 1996. The historical use of mine-drainage and pyrite-oxidation waters in central and eastern England, United Kingdom. *Hydrogeology Journal*, **4**, 55–68.

BANKS, D., YOUNGER, P.L., ARNESEN, R.-T., IVERSEN, E.R. & BANKS, S.B. 1997*a*. Mine water chemistry: the good, the bad and the ugly. *Environmental Geology*, **32**, 157–174.

BANKS, D., BURKE, S.P. & GRAY, C.G. 1997*b*. Hydrogeochemistry of coal mine drainage and other ferruginous waters in north Derbyshire and south Yorkshire, UK. *Quarterly Journal of Engineering Geology*, **30**, 257–280.

BANKS, D., PARNACHEV, V.P., BEREZOVSKY, A.Y. & GARBE-SCHÖNBERG, D. 1998. *The Hydrogeochemistry of the Shira Region, Republic of Khakassia, Southern Siberia, Russian Federation – Data Report*. Norges Geologiske Undersøkelse Rapport, **98.090**.

BANKS, D., PARNACHEV, V.P., FRENGSTAD, B., HOLDEN, W., KARNACHUK, O.V. & VEDERNIKOV, A.A. 2001. *The Hydrogeochemistry of the Altaiskii, Askizskii, Beiskii, Bogradskii, Shirinskii, Tashtipskii & Ust' Abakanskii Regions, Republic of Khakassia, Southern Siberia, Russian Federation – Data Report*. Norges Geologiske Undersøkelse Rapport, **2001.006**.

FRENGSTAD, B. & BANKS, D. 2000. Evolution of high-pH Na-HCO$_3$ groundwaters in anorthosites: silicate weathering or cation exchange? In: SILILO, O. et al. (eds) *Groundwater: Past Achievements and Future Challenges, Proceedings of the XXXIInd Congress* of the International Association of Hydrogeologists, Cape Town, South Africa. A.A. Balkema, Rotterdam, 493–498.

GRAM, J. 1923. *Den kemiske sammensætning av Spitsbergen–Bjørnøykul. [The chemical composition of coal on Spitsbergen–Bjørnøya.]. Norges Geologiske Undersøkelse*, **112** (in Norwegian).

HJELLE, A. 1993. *Geology of Svalbard*; Polarhåndbok nr. 7, Norsk Polarinstitutt, Oslo.

INDUKAEV, JU.V. 1980. *Rudnie formatsii kontaktovo-metasomaticheskikh mestorozhdenii Altae-Sayanskoi oblasti – Chast' 1 [Contact metasomatic ore formations of the Altai-Sayan region, part 1]*, Tomsk University Press, Tomsk (in Russian).

MATVEEV, A.K. *Geologiya ugol'nikh mestorozhdenii CCCP [Geology of the Coal Formations of the USSR.]* Gosudarstvennoe Nauchno-Tekhnicheskoe Izdatel'stvo Literaturi po Gornomu Delu, Moscow (in Russian).

NUTTALL, C.A. & YOUNGER, P.L. 1999. Reconnaissance hydrogeochemical evaluation of an abandoned Pb–Zn orefield, Nent Valley, Cumbria, UK. *Proceedings of the Yorkshire Geology Society*, **52**, 395–405.

PARNACHEV, V.P., BANKS, D., BEREZOVSKY, A.Y. & GARBE-SCHÖNBERG, D. 1999. Hydrochemical evolution of $Na-SO_4-Cl$ groundwaters in a cold, semi-arid region of southern Siberia. *Hydrogeology Journal*, **7**, 546–560.

POLYAKOV, G.V. 1971. *Paleozoiskii magmatizm i zhelezoorudenenie juga srednei Sibiri. [Palaeozoic Magmatism and Iron Ore Formation in southern-central Siberia.]*, Nauka Press, Moscow (in Russian).

US EPA 1991. *MINTEQA2/PRODEFA2, A Geochemical Assessment Model for Environmental Systems: Version 3.0 User's Manual*. US Environmental Protection Agency, **EPA/600/3-91/021**.

UTSI, J. & MYRVANG, A. 1992. Planning of a new mine in Svalbard. *In*: BANDOPADHYAY S. & NELSON M.G. (eds) *Mining in the Arctic*. A.A. Balkema, Rotterdam, 29–34.

World Coal. 1995. *Northern Exposure. Spitsbergen Coal*, World Coal.

Assessment, prediction and management of long-term post-closure water quality: a case study – Hlobane Colliery, South Africa

R. P. HATTINGH[1], W. PULLES[1], R. KRANTZ[2], C. PRETORIUS[1] & S. SWART[3]

[1]*Pulles Howard & de Lange Inc., PO Box 861, Auckland Park, 2006, South Africa (e-mail: phd@phd.co.za)*
[2]*Rison Consulting (Pty) Ltd, PO Box 1811, Rivonia, 2128, South Africa*
[3]*Kumba Resources Limited, PO Box 28, Hlobane, 3145, South Africa*

Abstract: Hlobane Colliery, located in the Vryheid Coalfields of South Africa, is to close down its operations after more than a century of mining activities. This paper presents a review of the assessments that were undertaken in order to determine the long-term water quality risks after mine closure and evaluate the effects of various water management actions thereon. An integrated assessment approach was adopted that incorporated hydrological, hydrogeological, mineralogical and geochemical assessment, and modelling techniques to predict the volumes and qualities of water discharging from various points on the mine for the base case situation where no water management options were implemented. Various water management options were identified that are primarily aimed at preventing the contamination of clean water, and the effects of these strategies were predicted and compared with the base case situation in order to provide a rational basis for the selection of the most appropriate strategies.

In South Africa, the most important environmental impact associated with the mining of sulphide-containing ores is poor quality seepage. Research results dealing with the multidisciplinary approach required to effectively manage water quality issues on mine closure are rare. The findings reported here form part of an integrated effort involving detailed situation analysis, hydrology, hydrogeology, mineralogy, predictive geochemical modelling and systems environmental management. The primary objectives of the investigation focus on the remediation of current and anticipated discharges, and the identification of suitable management strategies aimed at addressing residual impacts after mine closure. The case study involves the Hlobane Colliery, which has been mined for the past century.

The colliery is located in the Vryheid Coalfield in a mountainous region of the Kwazulu–Natal province of South Africa, some 20 km to the east of the town of Vryheid (Fig. 1). The Vryheid Coalfield covers an area of 2500 km^2, 15% of which is underlain by coal seams. It forms part of a highly dissected plateau with a typical Karoo topography.

Mining in the Vryheid Coalfield commenced in 1898, and developed into a major anthracite producer, with an annual output of 1500 kt of anthracite and 900 kt of bituminous coal from nine collieries at the peak of mining operations in the 1980s. Since then the production has decreased considerably due to the depletion of reserves. Large-scale mining at Hlobane Colliery continued until 1998, and small-scale mining is anticipated to continue until at least 2005. Three major coal seams were mined. In view of the impending closure, planning and implementation of final environmental management actions is currently underway. In its present form, the underground mining complex at Hlobane covers an area of 720 hectares (ha) projected to surface, with 32 opencast pits and five waste residue deposits (coal plant discard, washing plant slurry, sandstone pickings and shale partings).

Integrated water management plan for mine closure

On completion of mining activities, Hlobane Colliery is required under South African legislation to undertake a process that should culminate in the granting of a mine closure certificate. In this process the mine must satisfy various requirements relating to long-term residual and latent environmental risks and liabilities, with specific reference to the prediction of long-term volumes and qualities of various

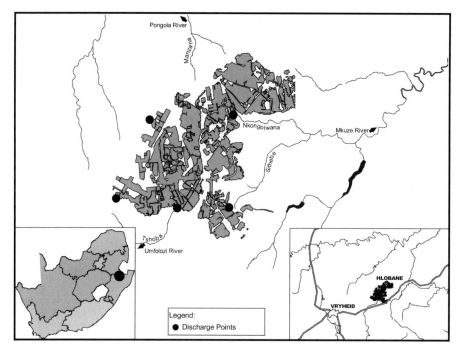

Fig. 1. Locality map of Hlobane.

discharges and their impacts on the receiving water courses. Where such impacts are unacceptable to the State and downstream water users, management strategies that can practically be applied to mitigate these impacts must be identified, and the effect on volumes and qualities of the various discharge points and the receiving water courses must be predicted. Management actions should preferably aim at preventing or minimizing water pollution, with water treatment and discharge to be considered as a last resort. Where long-term maintenance of management actions is required, the costs of the maintenance must be determined and the mine must provide such funds.

The resulting 'integrated water management plan' (IWMP) is essentially:

A practical and defensible strategy that optimizes the allocation of a finite financial resource to the application of management actions in a manner that will give the lowest risk of unacceptable impacts on the water resource now and in the future.

The process that was undertaken therefore included detailed assessments of system hydrology, hydrogeology, mineralogy, geochemistry and catchment management, and the integration into a coherent integrated water management plan.

Physical description

Summers are relatively hot and wet, and winters are cold and dry. The mean annual rainfall is 754 mm year^{-1}, and the mean annual evaporation is 1500 mm year^{-1}. The rainfall episodes are normally of high intensity and short duration (electric storms), with a 1:50 return period intensity of 134 mm in 24 h. Snow occurs rarely. The locations of the three main river systems affected by Hlobane are depicted in Fig. 1.

The Karoo landscape has dolerite sill-capped mountains approximately 500 m above the plains. The sills show typical columnar jointing in outcrop. These sills overlie and intrude a sandstone succession. According to Edgecombe (1998), the 96-m thick Zungwini Sill forms the crest of Hlobane Mountain, while the approximately 50-m thick Mashongololo Sill intruded the Karoo sediments some 15 m lower down. The sediments in which the coal seams occur form the Vryheid Formation, which is a sequence in the Middle Ecca beds of the Karoo Supergroup. The main components of these sediments are massive, coarse-grained arkosic sandstones. The coal seams at Hlobane Colliery occur some

200-250 m below the hill tops, and were deposited in fluvial and lacustrine environments. Several seams are present in the Vryheid Coalfield, with the Fritz, Alfred, Gus and Dundas seams (in order of increasing age) comprising the mineable reserve. In-seam partings lead to seam splits. The thicknesses of the massive sandstones between the seams are between 12 and 16 m.

The three seams mined at Hlobane I Colliery consisted of the Alfred, Gus and Dundas, and at Hlobane II Colliery of the Gus, Upper Dundas and Lower Dundas. The Gus Seam (0.5-2 m thick) is the most important in the area and, where it has been partially devolatilized, consists of high-grade anthracite (Bell & Spurr 1986). The Lower Dundas Seam may be up to 2.5 m thick and consists of interbanded bright and dull coal. The Upper Dundas Seam reaches a thickness of 1-1.2 m at Hlobane II Colliery. The Alfred Seam (generally less than 1 m thick) tends to be of lower grade than the other seams.

The Lower Dundas Seam often has a thin shale and sandstone parting in the top portion of the Seam, reaching a thickness of 1 m. The roof strata comprise fine-grained, shaley sandstones, for the most part with a thin mudstone band on the coal contact. The floor is generally a micaceous mudstone or laminated sandstone. The Upper Dundas Seam is generally situated 1.5-6.5 m above the Lower Dundas. The roof and floor generally consist of fine-grained micaceous sandstones. The sulphur content of the Upper and Lower Dundas Seam is, in general, less than 1%, with pyrite associated with the bright bands being the main constituent. The Gus Seam sulphur content is less than 1%. The roof strata are normally well-bedded, coarse-grained sandstones. The seam floor is generally fine-to medium-grained sandstone, often interbedded with thin shaley bands. The roof of the Alfred Seam generally contains mudstone and inferior quality coal. The floor lithology comprises predominantly medium-grained, well-bedded sandstone. The sulphur content may be as high as 2%. In all seams the most common mineral impurities include clay minerals (mainly illite), sulphide minerals (pyrite, marcasite and pyrrhotite) and quartz.

Dolerite sills in the area have been classified into at least five sill types. The earlier intrusions are generally concordant and persistent, while the younger intrusions are typically erratic in their occurrence, are commonly dyke-like and have a sinuous form. The sill phase was followed by a period of dyke intrusion (up to 10 m in thickness), which is associated with minor faulting. Karoo dolerite is predominantly ophitic to subophitic in texture. Minerals include plagioclase, clinopyroxene, and subordinate orthopyroxene and iron oxides, with varying proportions of alteration minerals, such as sericite and chlorite.

Conceptual system models

An initial conceptual model of the hydrological, hydrogeological and geochemical processes that defines the physical conditions was developed at the outset of the project in order to assist with planning. While this conceptual model has been refined and updated during the course of the project, the fundamental elements remain valid.

System hydrology and hydrogeology

The pre-mining condition

The low relief of the Karoo topography causes run-off to consist mainly of sheet flow. Stream beds on the mountains are shallow due to the presence of the dolerite sill. The stream beds on the slopes are more deeply incised. Other mechanisms affecting run-off in these systems are the relatively high intensity rainfall and the evaporative surface formed by exposed dolerite on the top of the mountains. The vegetation has been considerably modified by man, and mainly consists of savannah-type plains. On the slopes the soil profile is deeper (up to 15 m) than on top of the mountain and, therefore, infiltration is anticipated to be greatest on the slopes.

From a hydrogeological perspective, the net infiltration through the columnar dolerite jointing on the top of Hlobane Mountain tends to form small, seasonal springs at the dolerite-massive sandstone contact under pre-mining conditions. This is attributed to the relatively low permeability of the underlying sandstones. Although a significant number of dykes are present in the study area, very few were found to be water bearing. The main hydrogeological mechanism at or near the top of the mountain, therefore, consisted of vertical flow through the dolerite sill, followed by lateral flow on the dolerite-sandstone contact, with very little infiltration through the latter.

The post-mining condition

The conceptual flow model is depicted in Fig. 2. The controlled collapse of the overlying strata after total extraction of the coal seams led to two types of subsidence features. In deeper undermined areas, where total extraction of the coal seams was the favoured mining method, and where the dolerite sill comprises the surface of the

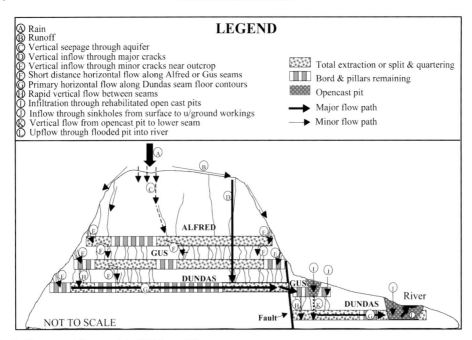

Fig. 2. Conceptual flow model of Hlobane Mine.

mountain top, fissures formed that extend to surface. In shallow undermined areas, the roof above the bord-and-pillar mining is unstable due to the depth of weathering of some 4–20 m below surface. This has resulted in areas of surface subsidence around the base of Hlobane Mountain. The dykes in the Hlobane Colliery have been mined through in most cases and do not act as hydraulic barriers.

The mining-induced fractures are particularly well developed along the perimeter of the Hlobane Mountain. These fractures vary in length from a few metres to several hundred metres, and are generally narrow. The orientation and location of the observed fractures on the mountain top correlate well with the extrapolation of the superposition of common boundaries of superimposed high extraction areas.

In contrast to the pre-mining condition, mining-induced subsidence features control the hydrology and hydrogeology of the system. Fractures intersect all major stream beds on the mountain top, and virtually all streams on slopes show evidence of shallow subsidence in their lower reaches. Ingress into the mine workings is therefore governed by the spatial location of subsidence features, where these are superimposed on the natural drainage features of the system. Besides the larger fractures and subsidence structures, the three major seams are vertically interconnected due to the additional floor and roof failures within the mine workings themselves. These failures are due to a number of circumstances. Pillars from the underlying seams may punch through the 15–17 m parting into the floor of overlying workings in instances where the pillars have not been precisely superimposed. High extraction on the lowermost Dundas Seam inevitably results in the collapse of the overlying seam floors. Additional hydraulic connection between the various seams has also been created by mining infrastructure, such as vent shafts and box holes. These features were located in the coal floor depressions of the overlying seams to assist with the drainage of water from the mining operations. The effect of the interconnection between the various seams is that any water within the upper mine workings can only flow laterally for short distances before reporting to the lowest seam. The coal floor contours and the layout of the mine plan on the Dundas Seam therefore generally control the rate and direction of the mine discharge out of the underground workings.

There are isolated areas where intact dykes form flooded compartments, particularly where the major interconnections to other workings only occur on the Gus Seam. The latter give rise to large ponds in the system, in addition to localized ponding caused by minor differences in elevation as a result of floor heave or the numerous dykes and faults in the mined area. The underground workings, therefore, have a substantial water storage capacity that is distributed throughout the mine, resulting in a balancing effect that attenuates flow and quality variations at the various discharge points.

The coal seams dip in an easterly direction at approximately 1:50, with the lowest point in the underground workings a few metres below the surface of the adjacent plains. Most of the water accumulating in the underground mined areas discharges from a central point in the coastal plains (Sithebe discharge). Apart from seeps and minor discharges, two significant discharge points (Tshoba and Manzana discharges) occur in the SW and NW in localized depressions.

Additional flow into the system is caused by infiltration through backfilled opencast pits. Most of these opencast pits are connected to the underground workings, and yield additional water to these areas. Three out of 32 opencast pits discharge directly to surface.

System geochemistry

There are two major types of water–rock interaction that dictate seepage quality emanating from Hlobane. First, the dissolution, precipitation and ion-exchange processes taking place in the overlying sedimentary rocks enrich the groundwater in elemental concentrations such as sodium. Secondly, the oxidation of sulphide minerals in the coal seams contributes towards the deterioration of water quality in terms of acidity, sulphate and metals. This process includes the dissolution of neutralizing minerals within the mine workings resulting in the removal of acidity and precipitation of secondary minerals along the flow paths.

Ion-exchange reactions predominate in the undersaturated zone overlying the coal horizon. The elevated sodium concentrations that are characteristic of the Hlobane Colliery discharges probably originate from the dissolution of albite in the arkosic hanging wall. A positive correlation exists between the thickness of the overlying arkosic sandstone and the sodium concentrations in the water found in the mine workings. Furthermore, honeycomb textures in the hanging wall indicate dissolution of albite and concurrent precipitation of quartz and formation of kaolinite.

In terms of sulphide mineral oxidation the dominant mineral is pyrite, occurring mainly in massive euhedral grains. However, the mode of occurrence often involves framboidal clusters of pyrite, exposing large reactive surface areas per volume of pyrite. Sensitivity analyses that were performed as part of the predictive geochemical modelling exercise have shown that the available surface area is a critical parameter in determining the future water quality that would flow from the underground environment. Furthermore, the fine-grained sulphide mineral grains that form part of the 'dust' accumulated on the footwall contribute in a significant manner towards the water quality. This mechanism is affected by the type of mining that has occurred, insofar as normal bord-and-pillar mining results in a relatively unobstructed flow path with minimum source term of sulphide minerals available for reaction. High extraction, however, results in a tortuous flow path with numerous small ponds and a larger source term of sulphide minerals (due to the longer flow path and collapse of sulphide-bearing hanging wall into the flow path). Coupled to this is the process that effectively neutralizes acidity and removes the bulk of the heavy metals through precipitation as secondary minerals. The type of mining affects this mechanism in a very similar manner to that described for the sulphide oxidation process. As Hlobane was classified a 'fiery mine' (prone to methane explosions) in 1953, the use of low silica 'stone dust' was mandatory since that date. This calcium–magnesium carbonate-rich material was added as a very fine-grained powder on the sidewalls of mine workings during the operational phase to inhibit the potential combustion of coal dust during methane explosions. This gave rise to a short-lived buffering effect on the reactions that participate in lowering the pH. Historic accounts depict a scenario of dramatic lowering of the pH prevailing shortly after the exposure of fresh sulphide material.

Assessment process for the quantification of residual impacts

The mine discharges contaminated water into three important river catchments and is expected to continue discharging water into these catchments in perpetuity. In order to devise the most appropriate management strategies, the residual or long-term impact of these discharges required quantification. Three phases were required to complete this component of the project: data collation and conceptualization; source term quantification; and catchment hydrology.

Data collation

Although data were historically collected at Hlobane Colliery, the quality and type of data were not directly suitable for use in the source term quantification process, except for data relating to the spatial layout and geology of the mine. A georeferenced geographical information system (GIS) database was created, containing the topography, surface hydrology, surface infrastructure (such as discard dumps), layout of the three seams mined (differentiating between the mining methods employed) and the geological features (dykes and faults). Additional data collected included the surveyed positions of the fractures extending to surface as well as shallow subsidence areas. This database was used to generate floor contours for the Dundas Seam at 1 m intervals. From the floor contours, the positions of subsidence features and a field verification exercise, the flow paths for water in the mine workings were derived. This information was used in turn to inform and devise a water quantity and quality monitoring programme. A number of samples were also collected underground and from the outcrops for geochemical assessment as described below.

Source term quantification

The main objective of this component is to predict the future quantity and quality of the water discharges emanating from the mine. As the mine closure policy under South African legislation is tending towards quantitative assessment techniques, it was decided to use numerical modelling techniques to produce quantitative predictions. The basic approach was, therefore, to conduct the following steps in sequence:

- calculation of water and salt balances across the mine;
- prediction and quantification of the flow into the mined areas;
- prediction and quantification of the flow in and out of the mined areas;
- prediction and quantification of the impact of the geochemical processes in the mined areas on water quality;
- prediction and quantification of the impact of the water discharges on the receiving streams.

These individual numerical modelling components are described below. The context of and philosophy behind the various models requires some consideration. Arguably, the most important obstacle to a quantitative (numerical) prediction of water quality and quantity is that this is a function of the prediction of long-term rainfall trends. The prediction period must be of the order of 50–100 years, and long-term rainfall predictions are available at best for a three month horizon. Issues such as the impact of global warming on rainfall patterns may play a significant role in the prediction of future impacts. It was, therefore, decided to approach this specific issue from two different perspectives, and to reconsider the basic objective of the investigation.

First, for the purpose of an impact assessment, the critical questions relate to the worst-case scenario. A cause and effect relationship between rainfall and water quality, therefore, needs to be established, i.e. what is the impact of deviations from average rainfall on water quality and quantity? Should lower annual rainfall lead to worse water quality than the average rainfall scenario, then this low rainfall scenario becomes the worst case, and vice versa.

The second important aspect relates to fluctuation within wet and dry cycles, and whether they are seasonal or longer-term cycles (dry and wet periods at different scales). For the models, averaged monthly rainfall data for the full dataset (1915–present) were used. In broad terms, the long-term trends in terms of annual rainfall show that: (a) annual rainfall has been increasing since 1915; (b) there are longer-term wet and dry periods, ranging from 7 to 11 years; and (c) the annual rainfall has been erratic since 1974.

With regard to the impact of seasonal and longer-term cycles on water quality, the critical questions are: (a) whether there is a short-term correlation between flux (flow rate) and water quality; and (b) whether a short-term correlation (if it exists) can be extrapolated and holds true for variations in rainfall patterns. Considering that changes in rainfall cause changes in flow volumes through the mine workings, these questions also refer to the relationship between reactive surface area and water quality, and the relationship between water quality and wetting and drying cycles within the reactive minerals.

The conclusions from this study are that the critical variables in terms of the geochemical processes are the flow rate and reactive surface area. This must, however, be read in the context of aspects such as oxygen availability and modal mineralogical composition assumed to be a constant within the variation in flow rate. There appears to be a direct correlation between these parameters in the short-term and the longer-term cycles and the impact on water quality. This then leads to the conclusion that a worst-case scenario would be represented by dry weather flow, whether it is within a year (seasonal fluctuation) or within longer-term cycles (typically 7–11 years in this case). The shape of the predictive curve therefore stays the same, but the location of the curve on the graph in terms of concentration and time will differ as a function of rainfall (and therefore flow rate and reactive surface area).

Calculation of water and salt balances

The basic principle is to calculate water and salt balances, complete a load and impact apportionment process on the various point and diffuse sources, and use the results to then optimize and guide the numerical modelling components of the source term quantification phase. A thorough understanding of the mechanisms involved is paramount to completing these balances. The risk in completing this aspect early on in the assessment process lies in these processes not being understood adequately, with the result that impact apportionment and, consequently, the focus on specific sources may be incorrect.

A number of water flow and quality monitoring programmes have been undertaken at Hlobane. These include detailed river profiles in the Tshoba, Manzana and Sithebe–Nkongolwana systems. Detailed daily flow monitoring programmes have been undertaken at specific sites (such as the top of Hlobane Mountain) to provide information for use in the calibration of models. A monthly water quality and flow monitoring programme has been underway at selected sites since around 1997 (although much of these flow data have low reliability). Reliable daily flow and quality data were collected over a six month period. These data were used to construct a water and salt balance, using recorded flows and total dissolved solids measurements. Average monthly salt loads and relative contributions from the various discharge points were calculated on this basis.

The averaged salt balance for the 14 measured discharge points at Hlobane Colliery is shown in Table 1, which shows that 96% of the salt load is discharged from Hlobane I Colliery. From the salt balance it is evident that management action efforts should primarily focus on the Sithebe and Tshoba rivers, and that the underground complex is the largest contributor towards adverse water quality. The water and salt balance provides a first-order indication of the current status of the system; however, given the kinetic aspects of the geochemical processes, the predictive component still needed to be addressed. To this end, a series of numerical modelling exercises were undertaken on a number of individual aspects relating to the Hlobane I underground complex.

Quantification of flow into the mined areas

Three hydrologically distinct components were recognized and addressed individually: the top of the mountain, the slopes and the plains. Surface hydrology was modelled using HSPF Version 11 (Hydrological Simulation Program Fortran (HSPF), US EPA). HSPF is a set of computer codes that simulates the hydrological and water quality processes on pervious and impervious land surfaces and in streams. The software is capable of dynamically processing data of various time steps by way of automatic aggregation and disaggregation of time-series data. GenScn Version 1.0 was used to enhance the data processing and graphical presentation of the HSPF modellfing package.

Two rainfall datasets were used. The daily rainfall data measured at Hlobane from 1915 to 2001 were used to calibrate the hydrological parameters for each individual catchment area. A monthly rainfall set representing average hydrological conditions was used to evaluate various potential mitigation strategies aimed at reducing the impact of contaminated mine water on the receiving water environment.

The objective of the hydrological model developed for the mountain top areas was to obtain a water balance for the mountain top system in terms of the apportionment of mean annual precipitation (MAP) to evapotranspiration, stream flow and loss to the underground mine

Table 1. *Salt balance data for Hlobane discharge points*

Stream	Minimum salt load		Maximum salt load		Average salt load	
	(kg day^{-1})	(%)	(kg day^{-1})	(%)	(kg day^{-1})	(%)
Tshoba discharges	6762	21.2	9309	14.4	7947	15.8
Manzana discharges	766	2.4	4543	7.0	2769	5.5
Sithebe discharges	24 427	76.4	50 820	78.6	39 503	78.7
Total underground	29 948	93.7	61 063	94.4	47 456	94.5
Total other*	2007	6.3	3609	5.6	2763	5.5
Total excluding Hlobane I	31 267	97.4	61 140	94.2	48 234	95.7
Total excluding Hlobane II	688	2.1	3532	5.4	1985	3.9
Total Hlobane	31 955	100	64 672	100	50 219	100

*Includes opencast pits and discard facilities.

workings. A lack of reliable flow data for the streams on top of the mountains, as well as the difficult access to the area, complicated this exercise. A gauging structure consisting of a system of pipes fitted with mechanical flow meters was constructed in a key subcatchment on the Hlobane I mountain top. The structure was limited in its capacity for recording the high flows present after storm events but gave good data for base flow conditions. These data allowed for satisfactory calibration of parameters associated with base flow conditions, such as the interflow recession rate and the active groundwater recession constant. Hlobane II Colliery is different in that no defined stream beds exist in the area and no mining-induced fractures and subsidence areas were observed. This model could, therefore, not be directly calibrated to observed surface flow. The approach taken in this case was to apply the same parameter set obtained for Hlobane I Mountain and compare the resultant ingress of water into the associated mine workings with that derived from flow measurements taken at the discharge from the workings. The resultant water balances obtained for the two systems are shown in Table 2. The results show that model-derived ingress rates correlate well with the difference in land surface characteristics observed between the two areas.

For the hydrological model of the slopes around the Hlobane I and II mountain areas, a distinction was made between slopes upstream of potential shallow subsidence areas and slopes not impacted by mining activities. The significance of impacted slopes is that run-off generated in these areas flows towards subsidence areas where it is intercepted and flows into the underground mine complex. The impact of interception on slopes is primarily on water quality, as the total ingress volumes comprise a minor percentage of the river catchments (0.3% in the Manzana catchment, 1.4% in the Tshoba catchment and 2.6% in the Sithebe–Nkongolwana catchment).

The model was calibrated for a slope catchment area on the southern slopes of Hlobane I Mountain, where a suitable daily flow monitoring

Table 2. *Water balances for the tops of the mountains, Hlobane I and II collieries*

Water sink component	Hlobane I (% of MAP)	Hlobane II (% of MAP)
Evapotranspiration	42	61
Surface run-off	13	23
Ingress into mining compartments	45	16

Table 3. *Water balances for mountain slopes, Hlobane I and II collieries*

Water sink component	Normal slope (% of MAP)	Impacted slope (% of MAP)
Evapotranspiration	78	78
Surface run-off	16	3
Ingress into mining compartments	6	19

site provided data that could be applied to assess the impact of ingress in subsidence areas. For areas not impacted by subsidence, the model was run with parameter sets similar to those used for the catchments on the plains, except for actual land surface slope values that were calculated using the GIS database. The resultant water balances calculated for the two slope types are shown in Table 3.

Quantification of hydrogeology in the mined areas

The ambient groundwater gradients within the Karoo fractured aquifers have been disturbed by the preferential paths associated with mining-induced fractures and subsidence. In certain instances, groundwater ingress into Hlobane Mountain has increased from approximately 3% of the MAP to 45% of the MAP. Ingress into the mountain along these preferential features tends to be relatively rapid, in contrast to the slower infiltration rates expected on the slopes of the mountain through the soil profile.

Field observations and water-balance calculations undertaken during the course of this investigation indicate that the ingress along the preferential flowpaths into the workings is rapid. Although there is some lateral flow, the disruption of the mine workings, particularly due to high extraction, results in rapid downward vertical migration and the accumulation of the groundwater on the lowermost Dundas Seam, where the water flows along the coal floor contours to ponds located within the depressions in the coal floor. Numerous flow paths converge in the vicinity of the main discharge into the Sithebe River. Changes in the Dundas coal floor morphology have also created localized depressions on the perimeter that have resulted in discrete discharge points. These latter discharges are relatively insignificant with respect to the total volume of water under consideration.

Although there is a noticeable difference between the various forms of water ingress on the mountain top, these differences are dampened over the length of the flow path. Some of the flow paths within the mine workings are of the order of 4 km long, and the seasonal variations in the flow volumes have been attenuated to such an extent that the long-term average of the flow volume is a reasonable approximation. This dampening effect is enhanced in the vicinity of high extraction or isolated flooded compartments that act as large reservoirs. A constant flow of discharge water has been observed during the drier winter months (March–September), which confirms this conceptual scenario.

Given the modified hydrogeological environment and the inaccessibility of the mine workings, it was decided to establish water ingress factors for areas where there were relatively good data as an initial estimate. These values were then used to determine more specific transient water ingress volumes for areas that were less clearly defined. To this end the following strategy was employed:

- A global water balance for the long-term periods was first established for the entire mine. The measured discharge volumes provided the net groundwater ingress into the hydrogeological regime.
- A GIS database was used to interpolate observed fractures and subsidence areas to other portions of the mine. These interpolated mining-related features were verified by means of field observations.
- The GIS system was also used to define catchment areas for water ingress into the various preferential flow paths. These ingress points were allocated to a flow path associated with the Dundas floor contours and were subsequently assigned to the various discharge points at the edge of the workings.
- The infiltration factors for the mountain top and the mountain slopes were calculated from observed discharge flow measurements and changes in water level within the mine workings, where available. This process enabled an independent verification of the model input parameters used during the hydrological modelling.

The groundwater ingress factors (Table 4) provide a reasonable global water balance for the Hlobane Mountain. The relatively high ingress for the mountain top is attributed to the mining-induced fractures that cut across the stream beds at intermittent points along the water course. Groundwater ingress on the mountain slopes is associated primarily with subsidence features

Table 4. *Water balance infiltration parameters*

Component	Effective infiltration (% of MAP)
Mountain top with fractures	45
Mountain slopes	23
Undisturbed strata	3–5

that are located at the base of the mountain. Run-off is directly intercepted by these features that encircle the perimeter of the Hlobane Mountain where shallow mining has occurred.

The mountain was subdivided into the various discharge points to be analysed in more detail. In certain instances, a detailed evaluation was problematic given the changing groundwater flow paths and infiltration due to the current mining activities. In such cases the evaluation was limited to a broad water balance for the area.

A more detailed water balance was constructed for the individual discharge points for the transient rainfall records accumulated during the course of this investigation. It was evident that the seasonal variation of water ingress into workings was not only severely dampened but was also subjected to varying lag times between the rainfall event and the discharge of water from the mine.

Simple MODFLOW models were set up for the individual discharges to simulate the response of the discharge volumes to rainfall patterns. An example of the discharge model calibration for the Boomlager IV area is shown in Fig. 3, where a lag time of approximately 7 weeks between the rainfall event and the measured discharge is evident. There appears to be a variable dampening effect within portions of the workings, as the simulated volumes have a greater variation than the observed discharge levels particularly during the early periods. This has also been attributed in part to the uncertainty in the initial conditions of the water level and rainfall for the model simulations, as the workings are not accessible.

The variation in the discharge volume ranges from 2000 to 7000 m^3 $week^{-1}$ for Boomlager IV (Fig. 3). This variation is attributed to the short flow paths and the small catchment associated with this portion of the mine.

Longer flow paths and larger catchment areas tend to produce more dampened discharge volumes, such as in the case of the main Sithebe discharge. This portion of the mine was calibrated to a downstream surface monitoring point. Although this monitoring point also includes the measurement of surface run-off from undisturbed

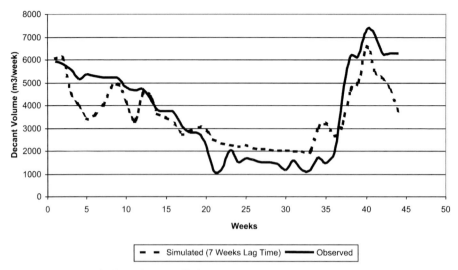

Fig. 3. Calibration curve for Boomlager IV discharge.

upstream catchments, it is still considered to be a useful indication of the discharge predictions, particularly during the low surface flow periods when the water quality is expected to be at its worst. The calibration of the Sithebe discharge volume is shown in Fig. 4. Clearly, the groundwater discharge does not approximate the surface run-off during the rainy summer season from September through to March. However, there is reasonable agreement between the flow measurements and the simulated discharge volumes during the drier periods. The discharge volumes range from 15 000 to 32 000 m^3 day^{-1}. The total simulated discharge volumes to the various river systems from Hlobane Colliery are presented in Table 5.

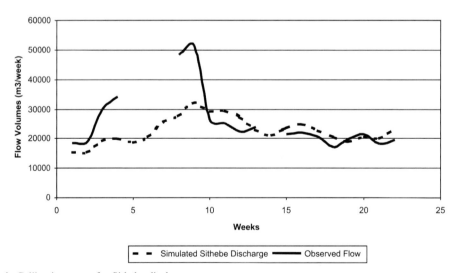

Fig. 4. Calibration curve for Sithebe discharge.

Table 5. *Total simulated discharge volumes from Hlobane I and II collieries into the various river systems*

Recipient system	Average discharge volume (m^3 week^{-1})
Tshoba River	37 100
Manzana River	23 920
Sithebe	159 320
Total discharge volume	220 340

Quantification of geochemical processes in the mined areas

Flushing of connate water containing high sodium and bicarbonate concentrations from freshly disturbed overlying sandstones will proceed at a different rate to the generation of acidity and sulphates in the coal measures (Glynn & Brown 1996). The generation of new sodium and bicarbonate through the ongoing albite dissolution process, although proceeding slowly relative to the other two mechanisms, will also have an effect on discharge water quality. The numerical modelling exercises for the two mechanisms were undertaken separately, and the results integrated in order to present a coherent prediction in terms of future water quality.

There are a number of parameters and principles common to the mechanisms. Water–rock interaction during the vertical migration of the infiltrating water is small, because of the high flow rates observed. Horizontal flow is controlled by the floor contours and the mining method (Fig. 5). In bord-and-pillar mined areas, flow paths essentially consist of open channels.

CONCEPTUAL GEOCHEMICAL MODELS OF DIFFERENT MINING SCENARIOS

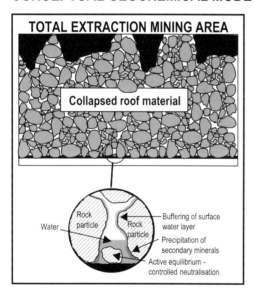
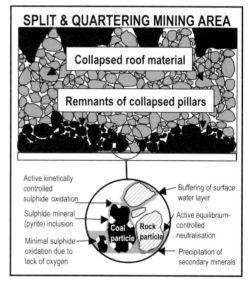
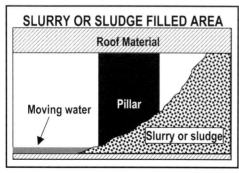
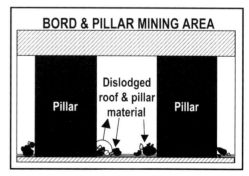

Fig. 5. Conceptual geochemical models of different mining scenarios.

In totally extracted areas, as well as in areas where pillars were split and quartered followed by subsequent roof collapse, the flow paths are more tortuous and may be braided. It is estimated (based on visual observations at the edges of collapsed areas) that the latter flow paths could be up to 50 times the width of the equivalent channels in bord-and-pillar areas. Localized ponds in the system serve to provide large wetted areas and have an attenuation effect as discussed in the previous section.

The irregularity of flux into the system is attenuated along individual flow paths as a function of tortuosity, floor contours, barriers, hydraulic head loss and ponding. This has the effect that, along an individual flow path, the variation in total wetted surface decreases from the main ingress points to the discharge point. Along the longer flow paths the lowest areas in the flow path should, therefore, experience very little seasonal variation in wetted area.

The sulphide oxidation and neutralization processes require reactive surface areas. The reactive surface does not differ between the different types of mining or mining methods, but flow rates and, consequently, contact periods may differ as a result of head losses in the system. The amount of debris and dust may, however, be higher in bord-and-pillar mined areas than in subsided areas, with concomitant impacts in terms of surface area.

The mineralogical composition of the components of the system potentially contributing to acid mine water generation was determined by means of extensive outcrop sampling, as well as sampling from areas underground where access was still possible. In terms of the modal distribution of minerals along the full spatial extent of the coal seams, extrapolations had to be made due to the inaccessibility of large portions of the mine. The mineralogical database consists of 92 samples, which have been analysed by means of X-ray diffraction (XRD) and X-ray fluorescence (XRF) techniques in order to produce a semi-quantitative indication of the mineral phases that are present (Fig. 6).

The strategy followed for conceptualizing kinetic reaction paths was to divide mineral reactions into three groups. The first group contains the reactions that proceed quickly over the time span of the calculation. It was assumed that these minerals remain in equilibrium with the fluid. A second group consists of minerals that react negligibly over the calculation and may be ignored or suppressed. The reactions for the remaining minerals fall in the third group, for which it is necessary to account for reaction kinetics.

By considering a known modal proportion of the minerals in the mine or hypothetical scenario and expressing that as a volume, it is possible to calculate a characteristic turnover time for each mineral by dividing the total amount of mineral by its respective dissolution rate (Strömberg & Banwart 1994, 1999a, b). Because geochemical processes controlling dissolution rates may change significantly over time, these characteristic times are not necessarily related to the lifetime of the minerals. The elements included as components in the model calculations, Na^+, K^+, Mg^{2+}, Ca^{2+}, Cu^{2+}, SO_4^{2-}, are considered conservative solutes whose turnover time is of the same order as the hydraulic residence time. On the other hand, elements dominated by internal cycling (participate in secondary mineral formation), H^+, Fe^{2+}, Fe^{3+}, Al^{3+}, $H_4SiO_4(aq)$, $O_2(aq)$, have turnover times of the order of days (Lasaga 1984).

Different reaction mechanisms can predominate in fluids of differing composition, since species in solution can serve to catalyse or inhibit the reaction mechanism. In studying dissolution and precipitation, it is commonly considered that a reaction proceeds in five generalized steps (Bethke 1996):

(1) diffusion of reactants from the bulk fluid to the mineral surface;
(2) adsorption of the reactants onto reactive sites;
(3) a chemical reaction involving the breaking and creation of bonds;
(4) desorption of the reaction products;
(5) diffusion of the products from the mineral surface to the bulk fluid.

The adsorption and desorption processes (steps 2 and 4) are almost certainly rapid, so two classes of rate-limiting steps are possible. If the reaction rate depends on how quickly reactants can reach the surface by aqueous diffusion and the products can move away from it (steps 1 and 5), the reaction is considered to be transport controlled. If, on the other hand, the speed of the surface reaction (step 3) controls the rate, the reaction is surface controlled. Reactions for common minerals fall in both categories, but many important cases tend, except under acidic conditions, to be surface controlled. For this reason and because of their relative simplicity, we have considered rate laws for surface-controlled reactions.

Sodium proved to be problematic in terms of elevated concentrations in the mine effluent. The features that were observed in the overlying sedimentary package, as well as small-scale alteration features (e.g. silicification of the

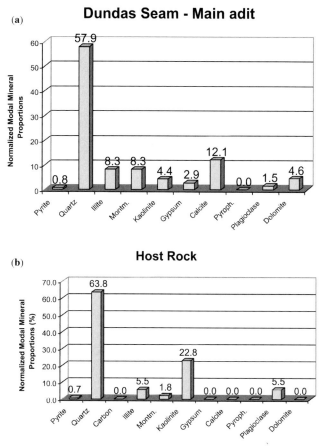

Fig. 6. Normalized modal mineralogical distribution of (**a**) the Dundas Seam and (**b**) host rock samples to the Gus and Dundas seams.

'honey-comb textures' observed in the arkosic sandstone), explained the high sodium values. It is assumed that the connate fluid is characterized by Na–Ca–Cl that is initially in equilibrium with kaolinite, quartz, muscovite or illite, and calcite, and in contact with a small amount of albite. However, feldspar cannot be in equilibrium with quartz and kaolinite, as the minerals will react to form a mica-like clay mineral. Silica activity decreases causing the albite to become undersaturated. It begins to dissolve, consuming kaolinite in the process while producing paragonite and quartz. The reaction continues until all of the kaolinite in the system is consumed, after which point the albite and quartz quickly reach equilibrium with the fluid. During the reaction progress the albite saturation level fluctuates, which results in sodium being liberated into solution. The behaviour of Na^+ is consistent with cation exchange, with preferential uptake of Ca^{2+} and displacement of Na^+ on exchange surfaces during dilution of the saline groundwater.

The predictive geochemical modelling component for the seepage quality emanating from the Hlobane Colliery made provision for the treatment of chemical reactions in a kinetic manner (Boer pers. Comm.). The mine was divided into various geochemical compartments or nodes, each with defined specific geochemical conditions. The seepage emanating from the various compartments were modelled using *The Geochemist's Workbench* (Bethke 1998). The profiles depicted in Figs 7 and 8 could be related in a direct manner with the modal mineralogy. First, the pH turns very acidic and recovers after a relatively short period. The fact that the effluent qualities become very acidic corresponds with the indication given by acid–base accounting (ABA) results. However, the pH recovers completely and remains slightly above

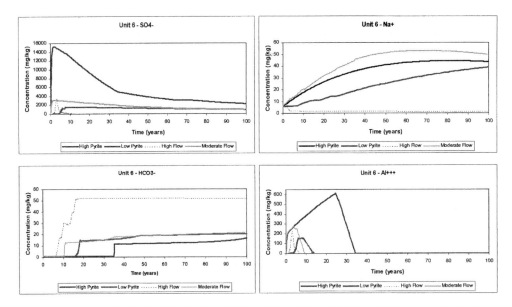

Fig. 7. Predictive geochemical profiles depicting high (1.0%) and low (0.25%) modal percentages of pyrite, as well as relative high and low flow rates.

neutral for the remaining period. This is a result of the relatively small modal proportion of minerals that are able to produce low pH conditions. In addition, the relatively large proportions of neutralizing minerals, including the stone dust, play a significant role in raising the pH after the initial acidic period.

It is clear from Fig. 7 that the ultimate effluent emanating from Unit 6, which was used as an example, is found to be of Ca–Na–HCO_3^- type with high pH. In summary, two plausible models were considered to explain the effluent composition: (1) a cation-exchange model; and (2) a silicate weathering model. The latter proved to be a practical explanation for the typical water composition encountered at Hlobane. The silicate-weathering model involves an open silicate and carbonate weathering system with a carbon

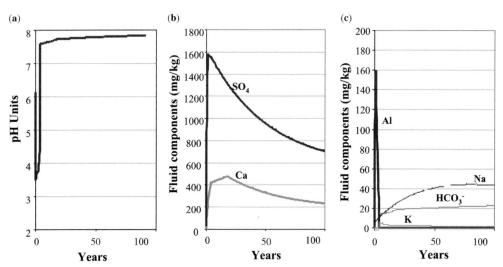

Fig. 8. Predictive geochemical profiles depicting (**a**) pH, (**b**) components in fluid and (**c**) components in fluids (enlarged view).

dioxide partial pressure (pCO$_2$) of 10^{-3} atm. Such a system is capable of generating Ca–Na–HCO$_3^-$ type water with a pH of up to 8.2. The reaction involving andesine feldspar, on which this model is primarily based, could be summarized as follows:

$$4NaCaAl_3Si_5O_{16} + 4CO_2 + 10H_2O + 4H^+$$
$$= 4CaCO_3 + 4Na^+ + 8SiO_2$$
$$+ 6Al_2Si_2O_5(OH)_4 \quad (1)$$

The mineralogy of the aquifer is compatible with such reactions. Furthermore, there is a relatively large proportion of kaolinite present in the samples analysed.

Quantification of the impact of water discharges on the receiving streams

The modelling of the catchments on the plains around Hlobane included hydrology, as well as water quality parameters using the HSPF model. The impacts of mining-induced water contamination were simulated by modelling total dissolved solids (TDS), sulphate (SO$_4$) and sodium (Na) concentrations as conservative substances in all three river catchments (Manzana, Tshoba and Sithebe). The approach that was followed was to first calibrate the hydrological component of the catchment by superimposing the effect of point source flows, originating at discharge points from the underground mine, onto natural catchment hydrology. The model was calibrated for low flow conditions. After a satisfactory calibration was obtained, the water quality component was calibrated against the observed water quality database by defining load rates for each substance at each of the discharges. The model was then used to assess the benefit of various proposed management strategies.

Given that the rainfall data used for the predictive models represent a synthetic dataset, and that rainfall is the driving force in terms of the geochemical processes, the predicted geochemistry of discharge water can never be a unique solution. This implies that the predictions cannot easily be verified against monitoring data. However, if it is accepted that a mathematical equation describing the behaviour of specific chemical parameters over time can be derived and that this will comprise a unique solution, its validity can be verified against observed data. The integration of the numerical models should be considered within this context.

The objective of the numerical modelling is to provide an indication of the relative impact of various management options. Given that the seasonality in rainfall leads to variation in stream flows, and that the attenuation in the underground complex results in less variation in flow from discharge points than in the streams, it follows that the worst-case scenario in terms of the impact of point source discharges on downstream users will be in the dry periods. From the perspective of the evaluation of management strategies it follows that impacts should be assessed with specific reference to the dry periods (winter months).

The hydrological model components dealing with flow into the mine were used to generate hydrographs for the average year, in weekly time steps. These hydrographs were used to generate discharge hydrographs for an average year for the various discharge points, in weekly time steps. The average annual flows were used to model the impact on water quality over a 100-year period. In all instances the predicted water qualities improved over time, with the result that the current monitoring results for the key parameters involved (sodium, TDS and sulphate) represent the worst case in terms of future scenarios.

Evaluation of water management strategies

The management strategies considered for the closure of Hlobane Colliery were identified in terms of the water management hierarchy preferred by South African regulators:

- pollution prevention or minimization;
- re-use of contaminated water;
- water treatment;
- licensing of discharges.

The assessment process was used to evaluate various generic management options. Based on application of the pollution prevention principle, the primary management options that were evaluated related to the reduction of inflow of clean run-off water into fractures, subsidence features and backfilled opencast pits. From the salt balance it is clear that the effort (and available funds) should be focused on the discharge from the underground complex, with the result that this aspect became the focus of modelling residual impacts after application of management strategies.

The physical constraints of the terrain and the fact that less than 20% of the mine workings can be accessed, constrain potential pollution prevention management strategies for the underground complex to the following:

- minimization of ingress from mine-induced fractures;
- minimization of ingress from shallow subsidence areas;

- minimization of ingress from opencast pits.

In addition to these pollution prevention measures, other potential measures that have been considered are:

- control of the point of discharge by sealing of adits and/or water tunnels;
- passive treatment of certain discharges that have significant problems with acidity and/or metals.

The effect of minimized flow into the system, irrespective of the specific engineering solution applied to achieve it, needed to be modelled. Minimization measures implemented on an experimental basis at Hlobane I in 2000 resulted in a 25% decrease in flow into the system in a specific area. Given the uncertainty regarding the potential success of minimization measures, it was decided to use potential reductions in flow of 25, 50 and 75% in the modelling. The physical catchment area for the underground mine complex was divided into subcatchments on the basis of flow paths and discharge points, and each individual subcatchment was considered in terms of the potential minimization strategies and, consequently, the impact of these strategies on water quality. A total of 16 scenarios were considered in this manner. An example is presented in the next section.

Reduction in flow at Boomlager IV discharge

The Boomlager IV discharge on the northwestern perimeter of Hlobane I Colliery contributes, on average, 2.4% of the total salt load. Given the configuration of the groundwater catchment, two potential management strategies relating to the minimization of flux into the system were identified:

- achieve a 50% reduction in ingress to the workings from the mountain top by means of the diversion of streams across crevasses on the mountain top (only address the mountain top);
- achieve a 50% reduction in ingress to the workings from the mountain top by means of the diversion of streams across crevasses on the mountain top, in addition to a 75% reduction in ingress to the workings from the slopes by means of the diversion of run-off on slopes (address the mountain top and slopes simultaneously).

The ingress hydrographs generated for the top and slopes were used to calculate the reductions in flows, in weekly time steps, and these data were used as inputs to the calibrated groundwater model. There is some uncertainty regarding initial conditions, with the result that the water balance was used to calculate initial steady-state

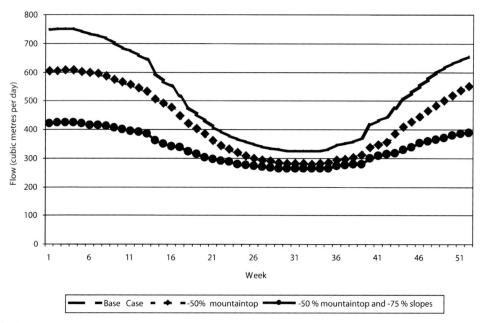

Fig. 9. Annual hydrographs for the various management strategies at Boomlager IV.

Table 6. *Impact of Boomlager IV management strategies on the Manzana River*

Strategy	Flow (l s^{-1})	TDS (mg l^{-1})	SO$_4$ (mg l^{-1})	Na (mg l^{-1})
	5th percentile	95th percentile	95th percentile	95th percentile
Base case	132.1	509.9	230.2	32.1
Base case -50% ingress from top	131.4	507.8	202.1	30.5
Base case -50% ingress from top and -75% ingress from slopes	131.0	507.5	201.9	30.0

groundwater conditions. The resultant annual hydrographs are presented in Fig. 9.

It is evident that the main impact on flow occurs in the wet season. The outputs from the discharge quality and quantity models were then used to run the river model for the worst-case scenario (current monitoring data and dry weather flow). The results for the critical water user approximately 20 km downstream of the discharge point are presented in Table 6. In this particular instance, the relative difference between management strategies are negligible, and it is arguable whether the expenditure involved in executing these strategies would be justified. The negligible beneficial effect is largely due to the fact that this discharge accounts for a very small portion of the water reporting to the critical downstream water user and even total removal of the discharge would only have a minor effect on discharge water quality.

Discussion on overall management strategies

Management measures similar to those identified for the Boomlager IV discharge were also developed for the other discharge points, and a similar modelling and assessment approach was applied to these strategies. In all cases, the worst-case scenario was established to be the dry weather flow when there are still substantial discharges of contaminated mine water into the receiving streams with minimal storm water or base flow recharge into these streams to dilute the mine water.

Whereas the primary focus from the perspective of potential negative impacts on the receiving streams has been the dry weather flow conditions, the primary benefit in terms of improved water quality occurs in the wet weather season. The assessed management strategies result in significant reductions in infiltration into the mine workings with a concomitant increase in clean water run-off into the receiving streams resulting in improved water qualities and decreased total salt loading.

Conclusions

The integrated water management plan (IWMP) for Hlobane Colliery presents the mine, downstream water users and the regulators with a quantified assessment of long-term, post-closure water pollution risks and the beneficial impacts of various proposed water management strategies. This has assisted all parties in making informed decisions when deciding how to allocate a finite financial resource to arrive at the best environmental option.

The assessment was undertaken at the end of mine life when most of the mine workings were inaccessible due to prior high extraction mining. This made it impossible to obtain data throughout the mine, and only bord-and-pillar sections close to the open adits could be accessed. The inability to access large parts of the mine also restricts the application of water management measures within the mine (e.g. sealing holings through dykes in critical locations in order to redirect water to a preferred discharge point). A key conclusion, therefore, is that this type of closure planning process should start well before the mine undertakes mining activities that restrict underground access.

The authors wish to acknowledge Kumba Resources Limited for commissioning the project and giving permission to publish this paper. The information presented in this paper is based on the work and contributions of many people, including the following: J. Lake, M. Chelin, U. Pfafferot, L. Coetser, W. Lemmer and P. Younger.

References

BELL, K. & SPURR, M.R. 1986. The Vryheid Coalfield of northern Natal. In: ANHAEUSSER, C.R. & MASKE, S. (eds) *Mineral Deposits of Southern Africa*. Geological Society of South Africa, Johannesburg, 2023–2032.

BETHKE, C. M. 1996. *Geochemical Reaction Modelling – Concepts and Applications*, Oxford University Press, New York.

BETHKE, C. M. 1998. *The Geochemist's Workbench. A User's Guide to Rxn, Act2, Tact, React and GTPlot.* Hydrogeology Program, University of Illinois, Illinois, USA.

EDGECOMBE, R. 1998. *The Constancy of Change: A History of Hlobane Colliery, 1898–1998,* The Vryheid (Natal) Railway, Coal and Iron Company Limited, Hlobane, South Africa.

GLYNN, P. & BROWN, J. 1996. Reactive transport modelling of acidic metal-contaminated ground water at a site with sparse spatial information. In: LICHTNER, P.C., STEEFEL, C.I. & OELKERS, E.H. (eds) *Reactive Transport in Porous Media.* Reviews in Mineralogy, **34**, 377–436.

LASAGA, A.C. 1984. Chemical kinetics of water–rock interactions. *Journal of Geophysical Research*, **89**, 4009–4025.

STRÖMBERG, B. & BANWART, S. 1994. Kinetic modelling of geochemical processes at the Aitik mining waste rocksite in northern Sweden. *Applied Geochemistry*, **9**, 583–595.

STRÖMBERG, B. & BANWART, S. 1999*a*. Development and fluctuations of sulphidic waste rock weathering at an intermediate physical scale: column studies. *Journal of Contaminant Hydrology*, **39**, 59–89.

STRÖMBERG, B. & BANWART, S. 1999*b*. Experimental study of acidity consuming processes in mining waste rock: some influences of mineralogy and particle size. *Applied Geochemistry*, **14**, 1–16.

Integrated hydraulic–hydrogeochemical assessment of flooded deep mine voids by test pumping at the Deerplay (Lancashire) and Frances (Fife) Collieries

C. A. NUTTALL, R. ADAMS & P. L. YOUNGER

Hydrogeochemical Engineering Research Outreach (HERO) Department of Civil Engineering, University of Newcastle, Newcastle upon Tyne, NE1 7RU, UK
(e-mail c.a.nuttall@ncl.ac.uk)

Abstract: To provide the basis for the design of two Coal Authority mine water management schemes, IMC Consulting Engineers (IMC) carried out step-drawdown pumping tests at the Deerplay (Lancashire) and Frances (Fife) abandoned collieries in the summer of 2000. Supplementary hydrochemical investigations were funded by NERC and undertaken by the University of Newcastle and Queen's University Belfast (QUB).

The results of the step-drawdown tests can only be interpreted by invoking a substantial component of turbulent flow in large open voids. Overall, the Deerplay system behaves in a manner analogous to natural aquifers, lending itself to modelling (using VSS-NET) to obtain effective hydraulic parameters that may be applicable in similar systems of flooded bord-and-pillar workings elsewhere.

The hydrochemical results for both sites showed some similarities, for example there was evidence of depth stratification of water quality in both cases, but also contrasts. For instance, although the total iron in the mine water pumped from the Deerplay Colliery rose gradually to a plateau at around 30 mg l^{-1}, the water remained net-alkaline throughout the test. By contrast, not only did the total iron in the Frances waters rise in abrupt steps to as much as 600 mg l^{-1}, but the water also switched from being net-alkaline at the beginning of the test to become strongly net-acidic by the end.

Mine abandonment often causes water pollution through the release of acidity and metals to receiving water courses. Pollutant loadings decline following the 'first flush' from the workings, although they can persist at ecologically damaging levels for many centuries (Younger 1997).

Frances Colliery (NT 309938) was abandoned in 1995, the closure of this colliery marked the end of dewatering in the East Fife Coalfield (see Fig. 1 for location map). Since 1995, groundwater has been recovering through a complex of multiple coal seams (Fig. 2b) formerly worked by Frances and adjoining collieries (Fig. 2a). Coal was exploited at Frances via older shallow bord-and-pillar workings and deeper more recent longwall workings. At Frances, test pumping took place from the mine shaft itself. The shaft is elliptical in section with a with a long-axis diameter of 6.86 m and a short-axis diameter of 3.05 m. The response to the test pumping was monitored by dataloggers placed in peripheral shafts and boreholes (Fig. 2a). The purpose of the test pumping by IMC Consulting Engineers (IMC) was to investigate how groundwater recovery could eventually be controlled by pumping and treating mine water in a designated site rather than letting the water recover naturally and treating mine water discharges as they emerge.

Coal mining in the North East Lancashire Coalfield was first recorded in the late thirteenth century. Nineteen seams within the Lower Carboniferous Coal Measures were exploited. The seams were thin, generally having a mined coal thickness of less than 1.5 m (Williamson 1999). Deerplay Colliery (SD 810267) (Fig. 1) was abandoned in the 1960s, since when a long-established polluting ferruginous discharge has impacted Black Clough and its receiving water course – the River Calder (Fig. 3a). The workings comprise a relatively simple system of bord-and-pillar workings in one major seam (the Lower Mountain Seam) (see Fig. 3a). Minor workings in other seams are largely above the water table. IMC drilled an abstraction borehole (Fig. 3b shows the log section of the Deerplay borehole) to intersect a roadway within the Deerplay workings. The location of the peripheral boreholes used to measure the response to the drawdown induced by test pumping is shown in Fig. 3a. The purpose of the subsequent test pumping was to abstract mine

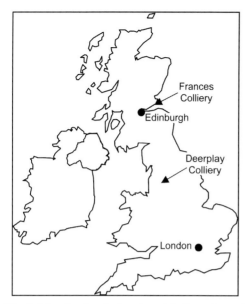

Fig. 1. Map showing the location of the Deerplay and Frances collieries.

water at a sufficient rate to intercept and, hence, stop the discharge entering the Black Clough.

Sampling and data collection

Automatic weather stations (AWS) were installed at both sites in order to record temperature, humidity, net solar radiation, rainfall, and wind speed and direction during the test period and the subsequent recovery phase. The weather data were then used for the hydrogeological modelling described later. When the test pumping began, samples were taken frequently over the first few days (i.e. every hour) but reduced after a few days to daily sampling for metals and weekly for a full analysis (i.e. metals and major cations and anions). Samples were also collected for isotope analysis by Queen's University Belfast (QUB). In the field, measurements of pH, reduction–oxidation poterhal (Eh), temperature and conductivity were taken using a Camlab Ultrameter, and alkalinity was measured using a Hach (HACH AL-DT) digital titrator with the range $0-4000\,mg\,l^{-1}$ calcium carbonate.

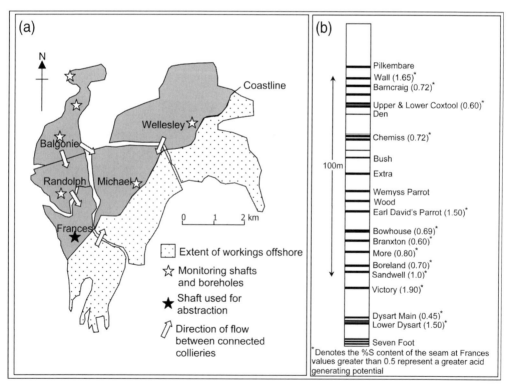

Fig. 2. (**a**) Schematic diagram showing how Frances Colliery is connected with adjacent collieries (after Sherwood 1997). (**b**) Generalized section showing the sequence of numerous coal seams throughout the East Fife Coalfield (after Knox 1954).

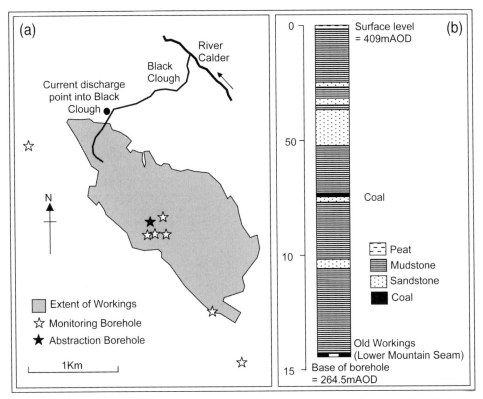

Fig. 3. (a) Schematic diagram showing the Deerplay Colliery workings and the site of the Black Clough discharge. (b) Log section of the abstraction borehole, which was driven to intercept a roadway within the Deerplay workings.

Dissolved oxygen was measured when a YSI 95 DO meter was available. In the laboratory, samples were analysed for major ions (calcium, sodium, magnesium and potassium) and major anions (sulphate and chloride) using a Dx-100 ion chromatograph. Dissolved silica was measured using a Hach colorimeter, and metals (iron, aluminium and manganese) were analysed by atomic absorption on a Unicam 929 AA spectrometer. Water level data were collected from peripheral boreholes and shafts by IMC Consulting Engineers, using a combination of manual dips and dataloggers.

Hydrogeological results

Stepped test pumping at Deerplay ran for 10 weeks, pumping at discharges rates (Q) of 7.25–30 l s^{-1} (see Table 1). The total drawdown (s in the following calculations) at Deerplay was around 20 m, and the gravity discharge was successfully intercepted and ceased flowing after one week of pumping. The stepped test at Frances took place over a six week period at rates of 38–76 l s^{-1} (see Table 2). During this time water levels were lowered by around 3 m. Figure 4a and b were achieved by plotting discharge (Q)

Table 1. *Summary of the test pumping data gained from Deerplay Colliery*

Step	Q (m^3 day^{-1})	Period	Water level (m aOD)	s/Q	% laminar flow
1	626.4	10/07 13.00–17/07 13.00	333.29–332.12	0.001485	53
2	1296	17/07 13.00–31/07 13.00	332.12–329.72	0.001937	35
3	1900.8	31/07 13.00–14/08 12.00	329.72–327.52	0.003346	27
4	2635.2	14/08 12.00–18/09 13.00	327.52–311.56	0.003639	21

Table 2. *Summary of the test pumping data gained from Frances Colliery*

Step	Q (m³ day⁻¹)	Period	Water level (m aOD)	s/Q	% laminar flow
1	3283.2	07/08 11.00–04/09 16.30	−54.235 to −55.449	0.00037	18
2	5616	04/09 16.30–19/09 11.15	−56.449 to −55.377	0.000381	11
3	6566.4	19/09 11.15–24/09 14.00	−55.377 to −56.733	0.00038	10

against the specific capacity (s/Q), and can be used to estimate the predominant flow regime (i.e. laminar or turbulent) within both systems. Hydrological evidence suggests that when the large, open voids within flooded collieries are pumped, turbulent flow becomes the dominant flow regime at increasing discharge rates and the amount of laminar flow decreases. The evidence for the existence of turbulent flow when pumping water from areas containing mined voids was highlighted as follows. When the equation of the line from the graphs in Fig. 4a and b are known, it is possible to calculate a very rough approximation of the percentage of laminar flow (L_p) using the slope (C) and the y-intercept (B) in equation (1) (Driscoll 1987):

$$L_p = \frac{BQ}{BQ + CQ^2} \times 100. \qquad (1)$$

The rough percentage of laminar flow at Deerplay decreased from 50 to 20% over the course of the test (i.e. turbulent flow increased from 50 to 80% over the same period) (Table 1). A more accurate assessment of the amount of turbulent flow occurring at the end of the Deerplay test is provided in the modelling section. A rough calculation of the percentage of laminar flow present in the system at Frances indicates that there is a large component of turbulent flow. This increases as the mined voids are pumped at higher discharge rates until, at the end of the test, over 90% of the flow is estimated as being turbulent (Table 2).

The second piece of evidence for the presence of a turbulent flow regime comes from the calculation of Reynolds numbers (Re) (Driscoll 1987). For Frances, a detailed section of the shaft was available and it was possible to calculate Re for the four submerged roadways using equation (2):

$$Re = \frac{\rho d v}{\eta} \qquad (2)$$

where ρ is the fluid density, d is the pipe diameter (or roadway diameter in this case), v is the flow velocity (calculated from Q and the cross-sectional area of the roadway) and η is the viscosity (in this case the viscosity of water was used, 1.14×10^{-5} m² s⁻¹).

In general, turbulent flow begins to occur at a Re value exceeding 10. Table 3 shows that the Reynolds numbers gained for the Frances Colliery roadways greatly exceeds the value of 10 and, therefore, turbulent flow must be significant within these roadways. The presence of a dominant turbulent flow regime may have

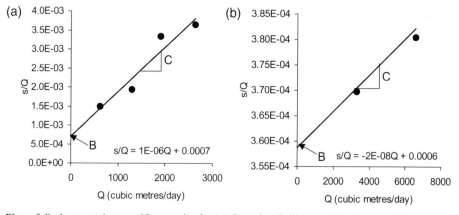

Fig. 4. Plots of discharge against specific capacity for (**a**) Deerplay Colliery and (**b**) Frances Colliery.

Table 3. *Calculation of Reynolds numbers (Re) for each of the submerged insets at Frances Colliery*

Inset name	Inset depth (m aOD)	Re
Lower Sandwell	66.96	350
Lower Dysart	140.11	190
Lethemwell	221.89	242
Pit Bottom	233.35	242

long-term implications for the efficiency with which the large open voids can be pumped.

Hydrogeological modelling

The complex hydrogeology of abandoned mines has been investigated under previous research that used physically based and lumped hydrological models to simulate groundwater flow in both the laminar and turbulent flow regimes (Younger & Adams 1999; Adams & Younger 2001). The VSS-NET model was developed as a new component of the physically based SHETRAN hydrological modelling system (Ewen *et al.* 2000), adding the capability of simulating turbulent subsurface flow in three dimensions (3D). This component links the VSS (*v*ariably *s*aturated *s*ubsurface) laminar flow component with the NET pipe network simulator. The mine workings are discretized as a system of conduits (representing the open roadways and shafts of abandoned mines) routed through heterogeneous, variably saturated porous media representing the surrounding rock (both intact rock and rock that has fractured in response to the mining below). The conduit network is shown in Fig. 5. The network was drawn using the US EPA's EPANET™ software. Deerplay Mine was deemed a suitable mine for a further application of VSS-NET because of the areal extent of the underground mine workings, the availability of detailed 1:2500 mine plans and the relative simplicity of the layout underground of the (mostly single seam) workings. The Frances system was deemed too complex for this type of modelling.

The principal aims of the hydrological simulations of flow in the Deerplay workings during the case study were:

- to assess the importance of turbulent flow in the hydrogeological response of the mine before, during and after test pumping;
- to reproduce by simulation the observed drawdown at the three deep boreholes into the Lower Mountain Mine (LMM) Seam

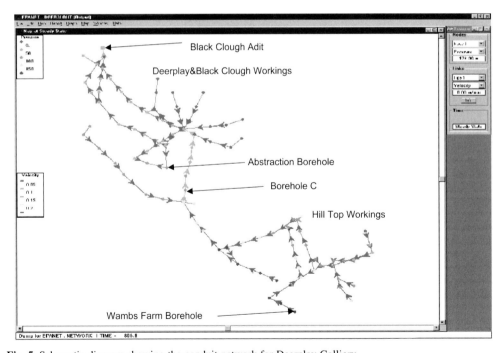

Fig. 5. Schematic diagram showing the conduit network for Deerplay Colliery.

(Wambs Farm, Abstraction and C) in the model (these boreholes are indicated in Fig. 5);
- to reproduce by simulation a discharge from the Black Clough adit of the correct order of magnitude. The model was also expected to concur with the field observations that pumping from Deerplay Abstraction borehole caused the discharge to cease flowing;
- to allow the relative contributions of turbulent and laminar to the pumped mine water to be quantified (for comparison with the analysis in the previous section).

The Deerplay hydrological simulations were run after a SHETRAN conceptual model was set up. The model domain comprised 202 125 × 125 m finite-difference elements, derived from the national grid (see Fig. 6 for illustration). The boundary of the model domain was determined from the 1:2500 mine plans of Deerplay and Hill Top collieries. The model domain was extended on the northern and northeastern side of the mines to the seam outcrop. Abandoned collieries to the west and south were considered to be unconnected to the modelled workings (IMC Consultant Engineers 2000). The base of the LMM Seam was considered to be the base of groundwater circulation, i.e. the impermeable lower boundary. The elevation of the LMM Seam was obtained from the mine plans. The strata between the LMM Seam and the ground surface was divided into two layers, a high-conductivity layer representing goaf and an upper layer extending to the ground surface representing the fractured, mixed Coal Measures strata above the goaf. The thickness of the goaf layer was estimated to be 4 times the LMM Seam thickness (1.2 m on average) (Younger & Adams 1999). Where the LMM Seam was not mined, the lower layer (effectively the unmined coal) was assigned a low hydraulic conductivity. The Upper Mountain Mine (UMM) Seam was also worked

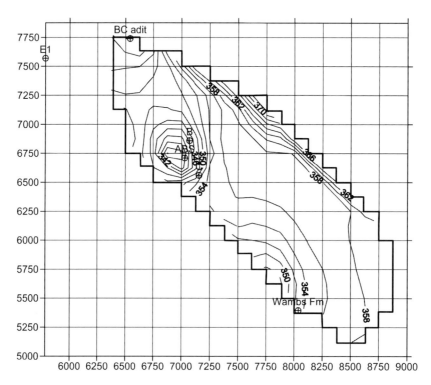

Fig. 6. (**a**) Contour plot of modelled groundwater levels without conduits. (**b**) Contour plot of modelled groundwater levels with conduits.

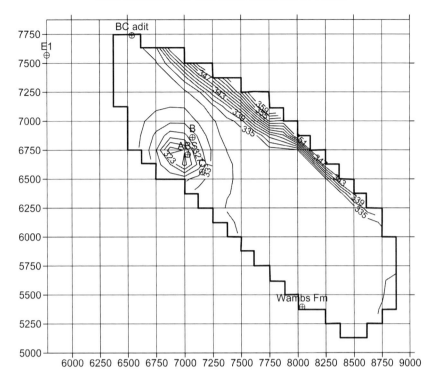

Units: m (AOD) for Water Levels
(b) m (for co-ordinates)

Fig. 6. (*continued*).

separately from the Black Clough adit, and a layer representing the goaf was included wherever workings were indicated on the mine plans. High-conductivity values were also assigned to grid squares where the mine plans indicated that either or both seams had been opencasted (surface mined). In the southern and eastern Deerplay workings the UMM and LMM seams combine to form the Union Seam, here the workings become single seam.

Initial conditions were obtained by running the model with steady-state recharge until groundwater levels were constant. The steady-state recharge value was calculated from an estimated average discharge from Black Clough adit of $16.5 \, \text{l s}^{-1}$ (based on the field measurements of discharge made before test pumping). Meteorological data collected by the weather stations installed on site were used to calculate recharge into the mine using equation (3), which accounted for both rapid infiltration through the opencasted LMM and UMM Seam outcrops and slower, delayed infiltration through the strata overlying the seams:

$$R = f_1 A_{SO} P_{ND} + f_2 A_{SM} (P_{NM}/D_i) \quad (3)$$

where R is the daily recharge into the model (mm), f_1 is a constant representing the percentage of the daily net rainfall, P_{ND}, infiltrating into the mine through the seam outcrop due to opencast mining (1.0), A_{SO} is the Area of the LMM and UMM Seam outcrop in the model (0.266 km^2), f_2 is a constant representing the percentage of the monthly net rainfall, P_{NM}, infiltrating into the workings (the remainder is assumed to either recharge the drift aquifer or form surface runoff) (0.2), A_{SM} is the Area of the mine workings (2.5 km^2) and D_i is the number of days in month i.

Pumping fluxes were abstracted from the model using a well element located at the grid element corresponding to the abstraction borehole. Daily rainfall totals and pumping rates are shown in Fig. 7. The total area of the model domain was equal to $A_{SO} + A_{SM}$ + unworked areas, and was, in fact, 3.156 km^2.

The daily and monthly net rainfall values were calculated from the daily time series of rainfall from the automatic weather station (AWS) and

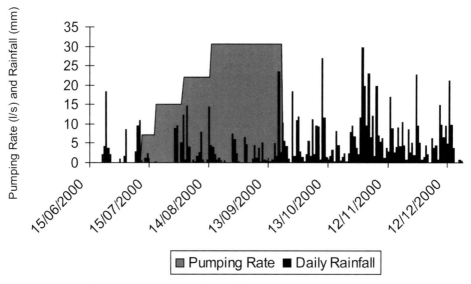

Fig. 7. Graph showing the pumping rate and daily rainfall.

monthly average potential evapotranspiration values were obtained from a national UK Meteorological Office database of potential evaporation supplied on a 40 km² grid (MORECS) (Thompson et al. 1981). The annual value for the grid square in the locality of the mines averaged over the period 1961–1990 was 558 mm year^{-1}. The AWS recorded daily rainfall for the period 15 June 2000–19 December 2000, which was also the SHETRAN simulation period. A time step of 4 h was used in the simulations, which proved to be stable. Minimal calibration was required in order to achieve the results described below.

Simulations were run using SHETRAN with and without the conduit network to determine the importance of turbulent flow during the test period. The conduit network simulated turbulent flow in the major roadways and the adits of the Deerplay and Hill Top workings (Fig. 5). The simulation without conduits only modelled laminar flow in the LMM Seam and the fractured strata above. This simulation failed to reproduce the observed drawdown at all three boreholes, and the spring representing the discharge did not dry up during pumping. Figure 6a shows a contour plot of water levels predicted by this simulation at the time of maximum drawdown in September 2000. It is clear that the cone of depression around the pumping borehole (A) does not extend to the workings draining towards the Black Clough (abbreviated to BC in Fig. 6) adit or Wambs Farm borehole (a small discharge was predicted to flow throughout the test at both locations). The predicted water levels during the pumping test were much higher than the observed water levels (for example, 354 m above ordnance datum (m aOD) at Wambs Farm compared to the observed water level of 328 m aOD on 18 September 2000).

The simulation with a conduit network reproduced the observed drawdown at the abstraction borehole within 1 m of the maximum value during the test (Fig. 8). The timing of the maximum drawdown was reproduced exactly. The observed drawdown at Borehole C was underpredicted by less than 3 m; but the model did not reproduce the observed drawdown at Wambs Farm (Fig. 8). Figure 6b shows a contour plot of water levels with the conduit network included. The groundwater flow direction was predicted to be towards the abstraction borehole (ABS). Furthermore, the absence of contours for a large region of the model area shows that the hydraulic gradient across the workings is very low due to the inclusion of the pipes. The steeper hydraulic gradient towards the NNE edge of the model is due to recharge flowing through the seam outcrop into the workings.

The simulation with a conduit network predicted that the discharge ceased flowing after around 54 days of simulation (7 August) and recommenced flowing after 96 days of simulation (approximately 16 h after pumping ceased). This matches the observed behaviour of the adit quite closely, although the modelled discharge recommences slightly early. The maximum discharge from the adit was predicted to be around 50 l s^{-1} after recovery in December 2000.

Analysis of the origin of the groundwater flow to the model element containing the abstraction

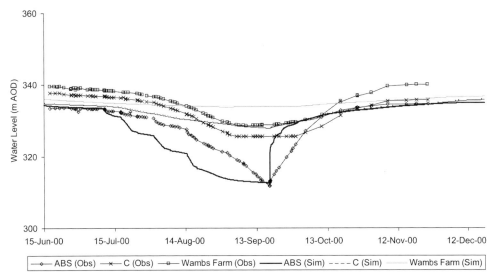

Fig. 8. Graph showing observed and modelled groundwater levels during the drawdown and recovery phase at Deerplay Colliery.

Hydrogeochemistry

The Deerplay pumped mine waters remained net-alkaline (i.e. their alkalinity exceeded their acidity, as described by Younger 1995) throughout the course of the test, pH remained around 7 and alkalinity around $300\,\mathrm{mg\,l^{-1}}$. The iron concentrations rose from $10\,\mathrm{mg\,l^{-1}}$ and levelled at $30\,\mathrm{mg\,l^{-1}}$ (Table 4).

The chemistry of the Frances pumped mine waters changed during the test (see Table 5). Initially, the waters were net-alkaline with an iron concentration of $6\,\mathrm{mg\,l^{-1}}$ and an alkalinity of around $400\,\mathrm{mg\,l^{-1}}$, but after 24 h of pumping the well (during pumping at the maximum rate in September) indicated that turbulent flow from the conduit network accounted for $17\,\mathrm{l\,s^{-1}}$ (56%) of the total flow.

Table 4. *Summary of the pumped mine water quality at Deerplay Colliery (compared to the Black Clough discharge)*

Determinand	Initial	End	Black Clough
pH	6.96	7.0	6.9
Alkalinity (mg l^{-1} CaCO$_3$)	302	326	260
Conductivity (μS cm^{-1})	1189	1225	1200
Fe (mg l^{-1})	17	27	40
SO$_4$ (mg l^{-1})	213	355	370

Table 5. *Summary of the pumped mine water quality during the test pumping at Frances Colliery*

Determinand	Initial	1 day	End of test
pH	6.34	5.22	4.8
Alkalinity (CaCO$_3$)	437	75	0
Conductivity (μS cm^{-1})	5557	25 550	26 500
Fe (mg l^{-1})	6.5	406.7	596.6
Al (mg l^{-1})	B.D.	14.65	51.6
SO$_4$ (mg l^{-1})	4975	4223	6254

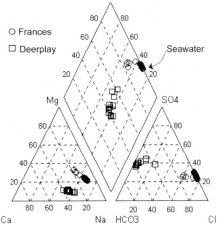

Fig. 9. Piper diagram plotting the pumped mine water chemistry at Frances and Deerplay Collieries.

quality deteriorated to 100 mg l^{-1} of iron and an alkalinity of around 40 mg l^{-1}. After a week of pumping, iron concentrations had risen to between 500 and 600 mg l^{-1}, and there was no alkalinity. There were also appreciable concentrations of aluminium (up to 50 mg l^{-1}) and manganese (up to 25 mg l^{-1}).

The chemistry of pumped mine waters from Frances and Deerplay Collieries are plotted on Piper diagrams (Fig. 9). The waters from Frances plotted in the sodium chloride field (Fig. 9) suggesting that there is a sea-water component (rather than a connate brine source) to the mine water (Younger *et al.* 1995; Younger 2001). The Deerplay pumped water plotted in the calcium bicarbonate field reflecting the carboniferous host geology (Fig. 9). The Black Clough discharge plotted nearer the calcium sulphate field, the sulphate probably sourced from pyrite decomposition.

Stratification within mine workings

The increase in iron concentrations over the duration of both tests can be attributed to stratification within the workings. Recovered (Deerplay) or slowly recovering (Frances) systems tend to stratify with better quality (mainly recharge) water remaining at the top of the system with poorer quality water beneath. As soon as these systems are disturbed by pumping (or by gravity discharge on completion of rebound) the turbu-

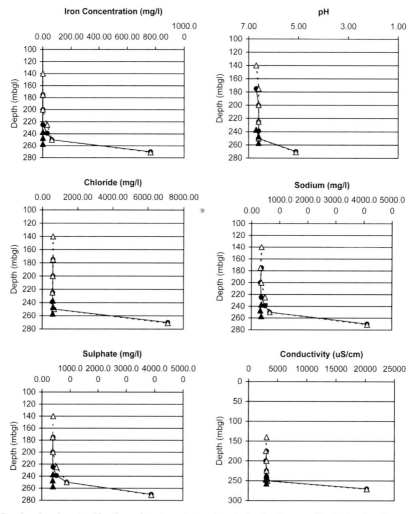

Fig. 10. Graphs showing stratification of various determinands in the Frances Shaft shortly after cessation of pumping (▲ = 12/12/95) and twice during the period of mine water recovery (● = 02/07/97; △ = 25/08/98).

lence induced causes mixing of the stratified water (Younger & La Pierre 2000).

Depth sampling carried out by IMC prior to the test pumping at the Frances Colliery showed that stratification was in existence in the shaft (Fig. 10). At Frances the top 200 m of the water column was of better quality (i.e. iron concentrations of less than $10\,\text{mg}\,l^{-1}$, pH of 7 and a conductivity of $3000\,\mu\text{S}\,\text{cm}^{-1}$). The deepest 70 m of the water column sampled contained the poorest quality water (i.e. iron concentrations greater than $500\,\text{mg}\,l^{-1}$, pH of around 5 and conductivity of $20\,300\,\mu\text{S}\,\text{cm}^{-1}$). Often the position down a stratified shaft where water quality may change is marked by an inflow from a previously worked roadway or tunnel, and the change in quality at Frances represented a connection with the adjacent Randolph workings (via a high-sulphur seam mentioned below). An appreciation of stratification is important because whole treatment systems have often been designed following analysis of the better quality water, which are unable to cope with the poorer quality water that follows.

The following theory explains how stratification may build up within a system and how it may subsequently be lost (refer also to Fig. 11):

- The shaft fills up to the point just below an intersecting roadway. Stratification of the water column takes place. Poorer quality water is found at the bottom of the shaft and better quality water remains above.
- The water level rises steadily and the water reaches a connection. As the water flows along this route and fills the adjacent workings stratification is lost due to turbulent flow along the roadway causing mixing within the main water column
- As the water fills the adjacent workings, via the connection, a more stable rise in water level continues. Stratification may then redevelop.
- A pump is installed within the shaft, when pumping begins the initial water quality reflects that found at the top of the shaft and this quality remains until approximately one shaft volume has been cleared. By this time the turbulence induced by the act of pumping has caused mixing of all the water in the shaft and poor quality water is accessed, as at Frances Colliery.

The source of the poor quality water at Frances

Frances Colliery is directly connected with the adjacent Randolph and Michael collieries. Water transfer between the collieries (Fig. 2a) is known from datalogger information from surrounding shafts and boreholes. The poor quality water at Frances is due to the presence of high-sulphur coals associated with marine bands. The Dysart Main Seam provides the connection between Frances and the adjacent Randolph pit. Mining records show that Randolph discharged very low pH, iron-rich water whilst it was operational and it would seem very likely that this colliery provides the source of the poor quality water.

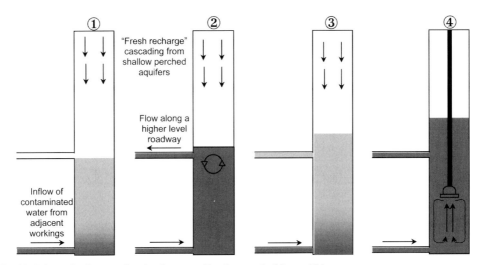

Fig. 11. Schematic diagram showing how stratification can build up within a system.

Conclusions

The behaviour of the Deerplay system was been successfully modelled with VSS-NET using a conduit-based model (which invokes a turbulent flow component). The model sucessfully reproduced the drawdown observed during the test and also managed to simulate the behaviour of the Black Clough discharge. These findings are supported by the data from the Frances and Deerplay step-drawdown test pumping, which was analysed by roughly calculating the percentage of laminar flow and also by calculating the Reynolds numbers as an indicator of the predominance of turbulent flow. These calculations show that there was a large component of turbulent flow occurring whilst both systems were pumped, due to the presence of large open voids within the workings.

Stratification is responsible for the changes in chemistry that occurred when both of these systems were pumped. Pumping a stratified system causes mixing of better and poorer quality water from the top and bottom of the shaft, respectively. The changes in chemistry that occurred at the Frances Colliery were rapid and unexpected, although previous depth sampling had shown the presence of contaminated water at depth. This exercise has outlined the importance of determining the extent of stratification within a system before designing any long-term treatment scheme.

The authors would like to express their gratitude to NERC who funded this project via an urgency procedure grant. We are also indebted to the Coal Authority and IMC Consulting Engineers for site and data access, namely, K. Parker, I. Burns, K. Whitworth, A. England and W. Kerr. We must also thank our colleagues at Queens University Belfast namely R. Kalin and T. Elliott, and also colleagues at Newcastle especially G.H. Karami for his assistance in the interpretation of the drawdown data. The views expressed represent those of the authors.

References

ADAMS, R. & YOUNGER, P.L. 2001. A strategy for modelling ground water rebound in abandoned deep mine systems. *Ground Water*, **39**, 249–261.

DRISCOLL, F.G. *Groundwater and Wells*, Johnson Division, Minnesota, St. Pauls.

EWEN, J., PARKIN, G. & O'CONNELL, P.E. 2000. SHETRAN: a coupled surface/subsurface modelling system for 3D water flow and sediment and contaminant transport in river basins. *American Society of Civil Engineer, Journal of Hydrologic Engineering*, **5**, 250–258.

IMC CONSULTING ENGINEERS, 2000. *Report on the Test Pumping for Minewater Control at Deerplay, Lancashire*, Coal Authority Report, Mansfield, Notts **D3621/U**.

KNOX, J. 1954. *The Economic Geology of the Fife Coalfields, Area III, Markinch, Dysart and Leven*. Memoirs of the Geological Society, Scotland, Edinburgh.

SHERWOOD, J.M. *Modelling Minewater Flow and Quality Changes After Coalfield Closure*. PhD Thesis, University of Newcastle upon Tyne.

THOMPSON, N., BARRIE, I.A. & AYLES, M. 1981. *The Meteorological Office Rainfall and Evaporation Calculation System MORECS*, Hydrological Memorandum, Meteorological Office, Bracknell, **45**.

WILLIAMSON, I.A. 1999. The Burnley Coalfield: Some geological influences upon the former mining exploitation and present-day development. In: *Memoirs of the Northern Mine Research Society*, NMRS Publications, Keighley, 5–27.

YOUNGER, P.L. 1995. Hydrogeochemistry of mine waters flowing from abandoned coal workings in County Durham, UK. *Quarterly Journal of Engineering Geology*, **28**, S101–S113.

YOUNGER, P.L. 1997. The longevity of mine water pollution: a basis for decision-making. *Science of the Total Environment*, **195/195**, 457–466.

YOUNGER, P.L. 2001. Mine water pollution in Scotland: nature, extent and preventative strategies. *Science of the Total Environment*, **265**, 309–326.

YOUNGER, P. L. & ADAMS, R. 1999. *Predicting Mine Water Rebound*. Environment Agency, R&D Technical Report, **W179**.

YOUNGER, P.L. & LA PIERRE, A.B. 2000. 'Uisge Mèinne': mine water hydrogeology in the Celtic lands, from *Kernow* (Cornwall, UK) to *Ceap Breattain* (Cape Breton, Canada). In: ROBINS, N.S. & MISSTEAR, B.D.R. (eds) *Groundwater in the Celtic Regions: Studies in Hard Rock and Quaternary Hydrogeology*. Special Publications, Geological Society, London, **182**, 35–52.

YOUNGER, P.L., BARBOUR, M.H. & SHERWOOD, J.M. 1995. Predicting the consequences of ceasing pumping from the Frances and Michael collieries, Fife. In: BLACK, A.R. & JOHNSON, R.C. (eds) *Proceedings of the Fifth National Hydrology Symposium*. Edinburgh, 4–7 September 1995. British Hydrological Society, London 2.25–2.33.

Hydrogeological and geochemical interactions of adjoining mercury and coal mine spoil heaps in the Morgao catchment (Mieres, NW Spain)

J. LOREDO, A. ORDÓÑEZ & F. PENDÁS

Departamento de Explotación y Prospección de Minas, Universidad de Oviedo, c/ Independencia, 13, 33004 Oviedo, Asturias, España (e-mail: jloredo@correo.uniovi.es)

Abstract: Asturias, NW Spain, has a long coal and metal mining tradition including mining for mercury, which was intermittently exploited from Roman times up to 1974. In the valley of the Morgao Stream, there is a spoil heap from the Los Rueldos mercury mine immediately uphill from another one from the Morgao coal washery, both situated less than 3 km away from the town of Mieres (30,000 inhabitants). At the Los Rueldos site, arsenic is abundant in the form of As-rich pyrite. Total arsenic concentrations in representative samples from the spoil heap range from 4746 to 62 196 mg kg^{-1}. Readily leachable Pb and Zn are respectively present at average concentrations of 3680 and 45 mg kg^{-1}. Leachate from this spoil heap drains on to the El Batán coal spoil heap, together with acid drainage from old underground workings of the Los Rueldos Mine, which has very low pH (2.5), with up to 2900 mg l^{-1} SO$_4$, and between 5.3 and 8.3 mg l^{-1} As. The total flow of polluted water from the Los Rueldos site that flows into the Morgao Stream is estimated to average 3200 m^3 year^{-1} (corresponding to only 1% of the total flow from the Morgao drainage basin). Analysis of the Morgao Stream downstream of both spoil heaps clearly shows the results of the dilution effects, with the As content being lowered by more than 95%.

Although mercury extraction in Asturias (NW Spain) is known to date back to the Roman period (centuries I and II BC), it was in the nineteenth and twentieth centuries that mercury mining began to be an important and prosperous industry (Dory 1894; Aramburu & Zuloaga 1899; Gutiérrez Claverol & Luque Cabal 1993). The Mieres district was the most productive area, with important mines and metallurgic operations in the valleys of the Morgao and San Tirso streams (both tributaries of the Caudal River, one of the most important rivers in the region). The Los Rueldos Mine lies in the valley of the Morgao Stream, about 2–3 km NE of the town of Mieres (30 000 inhabitants) and 20 km SE of Oviedo, the capital city of Asturias (Fig. 1). These old mining works are located on the slope of a mountain, which rises nearly to 500 m above sea level. Mining wastes were accumulated at the site of the mine, forming spoil heaps with an average inclination of 35°. At the toe of the mercury mine spoil heaps lies a rehabilitated coal spoil heap, formed from waste materials from a coal washery that were deposited at the site between the early 1960s and the early 1970s (Fig. 1). Between both spoil heaps, a small drainage channel has been excavated to drain any mercury mine water and spoil heap leachates that do not infiltrate towards the Morgao Stream.

The chemical instability of some toxic metal-rich residues from the abandoned mercury mine, which lies in a humid environment and is affected by surface waters, has led to the production of metal-rich leachates causing an obvious impact on the environment (Loredo 2002). Among the elements abundantly present in the mine water and spoil leachates draining from the Los Rueldos Mine, arsenic is probably the most important, as its toxic effects both to humans and other animals have long been known to effect almost every major bodily function and metabolic pathway. There is a long history of toxicological study of arsenic with regard to human health issues. Cases of arsenic contamination in groundwater leading to human arsenic poisoning have been widely reported in the literature (e.g. Borgono & Grebier 1971; Astolfi *et al.* 1981; Cebrian *et al.* 1983; Lu 1990; Das *et al.* 1996).

The toxicity of arsenic varies depending upon its chemical state: in general, inorganic forms are more toxic than organic complexes, and soluble forms more than insoluble. In aquatic systems, arsenic may be present in two main forms:

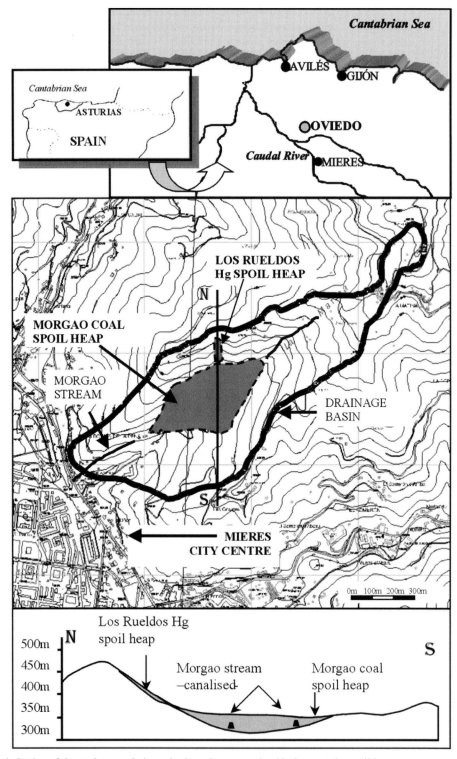

Fig. 1. Setting of the study area, drainage basin and cross-section N–S across the spoil heaps.

- as free or complexed ionic arsenites (As^{3+}) or arsenates (As^{5+}), with the arsenite form being approximately 60 times more toxic than arsenate (Ferguson & Anderson 1974);
- as covalent organic methylarsonic or di-methylarsinic acids, with the formation of methyl-arsenic compounds being essentially a detoxification process, as these compounds are less toxic than the ionic arsenic forms (Boyle *et al.* 1998). However, there is evidence to indicate that arsenic is an essential element in animals (Mertz 1981; Azcue & Nriagu 1994) and, thus, possibly for human metabolism, albeit in very low concentrations. Arsenic has also enjoyed a successful history as an effective medicinal element (as arsenicals) for some diseases (Boyle *et al.* 1998).

The purpose of this paper is to assess the impact of both mercury and coal mining spoil heaps on local hydrogeology and water quality, being particularly interesting as this is one of the rare examples of acidic mercury mine water pollution in Europe.

Characteristics of the studied area

Climate and precipitation

In contrast to most areas of Spain, Asturias has a humid climate characterized by abundant precipitation during much of the year. The annual average maximum and minimum temperatures in the study area are 17 and 8°C, respectively, with an annual average of 13°C over the past 20 years. Average yearly rainfall for the study area is 9710 mm. The annual relative average humidity is about 78%. Maximum daily precipitations are 82.1, 95.6, 99.8, and 103.8 mm for return periods of 10, 25, 50 and 100 years, respectively. Maximum hourly precipitation for a return period of 100 years is 40 mm h^{-1}.

The Morgao Stream drains a small drainage basin of $616\,300 \text{ m}^2$ (Fig. 1). This is the only perennial water course in the catchment: in dry periods the flow of the Morgao Stream varies between 5 and 10 l s^{-1}. The surface run-off in the drainage basin is intercepted by artificial trenches leading water to the stream.

Geology and mineralization

Geologically, the study area is located in the northwestern margin of the Asturian Central Coal Basin, in a zone of intense tectonic deformation (Fig. 2). It is included within the stratigraphical unit known as 'La Justa–Aramil', which consists of a thick series of shales with intercalation of sandstones, conglomerates and limestones of Carboniferous (Westphalian) age, with coal beds, which usually have thicknesses of less than 0.05 m. The Carboniferous sediments are overlain discordantly by Upper Stephanian–Lower Permian materials of the San Tirso Formation, comprising (from bottom to top) calcareous conglomerates, clay lutites, and tuffaceous and cineritic materials.

As with other epigenetic mercury mineralizations in the area (Loredo *et al.* 1988), the Los Rueldos deposit occurs only within the La Justa–Aramil unit. The Hg-bearing zones comprise conglomeratic-brecciated bodies, with siliceous clasts, disposed according to a complex anticlinal structure with a N–S trend, which is faulted at the hinge (Luque *et al.* 1991). The mineralization is closely associated with the intensely deformed zone of the La Carrera Fault (Luque 1992). Typically, the mineralization is buried under colluvial and alluvial materials derived from the weathering and lateral mass movement of shales in the sequence.

At the Los Rueldos site, mercury is most commonly present in the form of cinnabar, although metacinnabar and native mercury are also occasionally found. Cinnabar is irregularly distributed both in veinlets inside the conglomeratic-brecciated bodies and scattered inside the matrix of the conglomerate, and it also appears in contact zones between Carboniferous (sandstones and sandy lutites) and Permian materials (tuffaceous materials) (Luque 1985). Other metallic minerals that are present in the paragenesis of the ore deposit are pyrite, melnikovite, sphalerite, marcasite, chalcopyrite, galena and stibnite. The arsenic is present in the ore deposit as arsenopyrite, realgar and As-rich pyrite. Smithsonite, hemimorphite, cerusite, goethite, malachite, jarosite, melanterite and gypsum are present as secondary minerals. The gangue constituents are quartz, carbonates (ankerite, calcite and dolomite) and argillaceous minerals (kaolinite and dickite). The weathering of arsenic-bearing minerals in the mine and in the spoil heap results in the transport of dissolved As into underlying aquifers and superficial waters, which eventually discharge to streams and rivers.

Mining and wastes impoundments

Los Rueldos Mine

Asturias was, from the 1950s until the 1970s, an important mercury producer, with an average

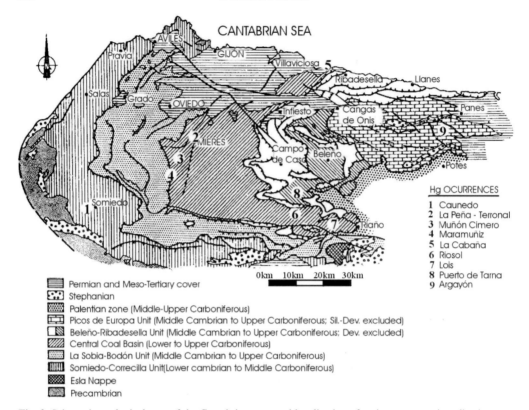

Fig. 2. Schematic geological map of the Cantabrian zone and localization of main mercury mineralizations.

annual production of 15 000 flasks (SADEI 1968–1991). A great part of this production corresponded to 'La Peña' and 'El Terronal' mines, which are known to have been exploited since ancient times, and which are less than 3 km away from the Los Rueldos Mine. This mineralization at Los Rueldos was exploited by drift mining up to 1972, and two drift portals are known at different elevations on the mountain side. The lowest of the two existing drifts is partly flooded and discharges mine water.

Mercury mining wastes

One of the more conspicuous and environmentally important legacies of the historical mining activities remains in the form of significant quantities of mine wastes, totalling some 3000 m^3 (Baldo 2000). The wastes are lying on the mountain side, downhill from the mine entrances, without any preventative measures to impede leaching. Consequently, land adjacent to this site is contaminated by mining activities, either directly as a result of chemical and mechanical erosion of the spoil heap, or indirectly through airborne dispersion. This is in addition to the naturally high mercury and arsenic concentration in the area (local background levels), related to mercury mineralization. The steep slope of the spoil heap (38% average) promotes vigorous mechanical and chemical dispersion of toxic elements.

The mining wastes from the Los Rueldos Mine are located approximately 2 km away from Mieres town centre, where geochemical arsenic anomalies in urban soils have been detected (Loredo *et al.* 1999). The microscopical study of samples from the spoil heap shows that iron sulphides (pyrite, marcasite and pyrrhotite) are very abundant, and are typically in an advanced state of oxidation, with haematite and amorphous iron hydroxide coatings (Baldo 2000). Jamesonite frequently replaces the sulphides. Some cinnabar, arsenopyrite and sphalerite have also been found.

Systematic sampling of waste materials and their chemical characterization was performed by analysis of the soluble fraction resulting from an acid digestion. In the laboratory, soil samples were slowly dried at temperatures lower than 40°C (to prevent evaporation of mercury; Alloway 1995). After drying, samples were crushed and homogenized. They were sieved through a 147 μm mesh and quartering was carried out using an aluminium rifler, each 0.5 g sample was then digested with 3 ml 2-2-2 HCl–HNO$_3$–H$_2$O at 95° C for 1 h and then diluted to 10 ml with water. Concentrations of major and trace elements (Ag, Al, As, Au, B, Ba, Bi, Ca, Cd, Co, Cr, Cu, Fe, K, La, Mg, Mn, Mo, Na, Ni, P, Pb, Sb, Sr, Th, Ti, Tl, Zn, U, V and W) were determined by inductively coupled plasma (ICP) and Hg by flameless atomic absorption (FAA), at ACME Analytical Laboratories in Vancouver, Canada. Quality control was maintained by analysis of duplicates and the use of certified reference standards, thus ensuring sufficient accuracy and precision in the results.

Table 1 shows concentrations of selected elements from chemical analyses of representative mining wastes samples from the Los Rueldos spoil heap. Arsenic concentrations encountered in nine samples regularly distributed along the whole the spoil heap range from 4746 to 62 196 mg kg^{-1}, with an average concentration of 36 540 mg kg^{-1} (Loredo et al. 1999). The readily leachable metals Pb and Zn are present at average concentrations of 3680 and 45 mg kg^{-1}. The Sb average concentration is 2391 mg kg^{-1}. The pH values found in the samples (measured using a soil pH-meter) are particularly low, ranging from 3.3 to 4.6.

Coal mining wastes

The Morgao coal spoil heap has a surface area of 175 000 m^2, and an estimated volume of 3 000 000 m^3 of wastes from the El Batán coal washery, with up to 50 m in thickness of mining spoils. They are disposed such that they fill and flatten the eastern part of the Morgao Valley. The base of the coal spoil heap is underlain by shales, and between fresh shales and spoil materials there are weathered shales (variable in thickness under different parts of the spoil heap) and clay beds also of variable thickness.

The beginning of the coal waste disposal at this ate occurred in 1961, and the topographic difference between the top and the bottom of the spoil heaps reached 120 m. After some years of uninterrupted works a landslide, caused by intense rainfall and surface run-off from springs, flooded some houses located downstream of the spoil heap and cut the national road, which is about 1 km away. As a consequence, a new drainage network consisting of two underground galleries, excavated under the coal washery spoil heap and covered with concrete, were constructed. Figure 3a shows a schematic view of both spoil heaps and surface and the underground drainage systems. These galleries are 1.5 m high and 1 m wide, and meet at the bottom of the spoil heap. Figure 3b shows a cross-section of the coal spoil heap. The galleries are connected to another secondary drainage that guides the surface waters (including the Morgao Stream) beneath the spoil heap. Surface run-off from the mercury mine and its spoil impoundment is collected in drainage channels along the southern margin of the impoundment and conducted to the Morgao Stream. The drainage system and the deposit of spoil materials into the heap was finished in 1972.

After that, the conditioning of the topography of the spoil heap was changed to prevent new slides, and revegetation of the spoil heap, including the planting of apple trees, was made in the raised area of the coal spoil heap in 1977.

Table 1. *Concentrations of selected elements in representative samples from the Los Rueldos mercury mine spoil heap*

	Samples								
	X1	X2	X3	X4	X5	X6	X7	X8	X9
pH	3.6	4.1	3.5	3.3	3.4	3.5	3.7	3.4	4.6
Ag (mg kg^{-1})	0.9	1.4	<0.3	0.7	1.2	1.1	1.1	14.2	<0.3
As (mg kg^{-1})	26 574	43 240	4746	32 594	34 518	41 535	62 196	36 368	22 807
Cd (mg kg^{-1})	<0.2	0.3	<0.2	1.2	0.5	0.3	<0.2	1.6	<0.2
Cu (mg kg^{-1})	17	23	13	23	20	29	39	67	14
Total Fe (%)	2.98	4.05	4.44	5.06	3.14	4.15	5.34	2.81	2.10
Hg (mg kg^{-1})	183	383	14	155	256	2224	160	1302	53
Mn (mg kg^{-1})	30	3	171	70	9	<2	15	2	6
Pb (mg kg^{-1})	1922	4319	92	1545	947	2254	1457	22 667	177
Sb (mg kg^{-1})	660	1518	23	369	375	656	467	17 396	51
Zn (mg kg^{-1})	25	34	13	21	29	27	25	251	5

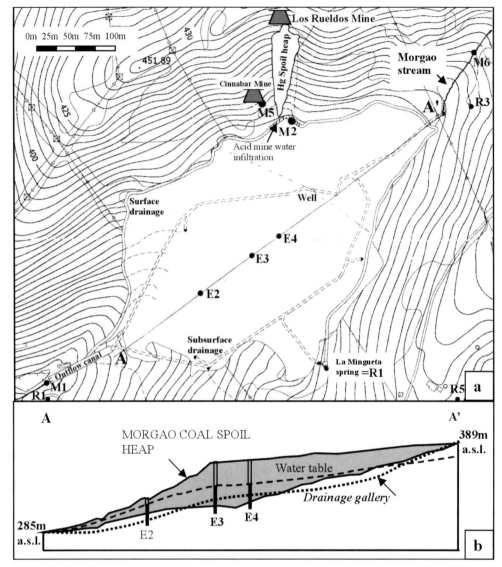

Fig. 3. (a) Schematic view of the spoil heaps and surface and subsurface drainage systems. (b) Cross-section of the coal spoil heap.

The coal spoil heap wastes are tailings from a coal washery, so the material is relatively homogeneous, with an average density of 2.15 g cm^{-3}, average moisture content 5.9% and permeability 1.21×10^{-5} cm s^{-1} (ENADIMSA 1972). Mineralogical and chemical analyses of representative samples of materials of the Morgao coal spoil heap show that their main constituents are SiO_2 (44%), Al_2O_3 (24%), Fe_2O_3 (5%) and other oxides. Other physico-chemical characteristics of these materials are: 10.9% fixed carbon, 81.1% ash, 8.0% volatile, 0.33% sulphur, 19.6% calcination fines and 1056 kcal kg^{-1} calorific value (Medina 1999).

Hydrogeology and water quality

Water movement

The groundwater flow system in the mining wastes spoil heap and underground galleries may be considered to behave as a pseudokarstic aquifer, in contrast to conventional porous media flow (characterized by slower-moving diffuse flow through the intergranular pore spaces and/or fractures in the rock). Groundwater flow in pseudokarstic aquifers is essentially characterized by multiple flow paths, extreme ranges of hydraulic conductivity and a high degree of unpredictability. The spoil heap regulates the flow of Morgao Stream.

From the geological data we can obtain an idea about the overall groundwater circulation in the Morgao Basin. Given that the shales constituting the substrate have low permeability, the water that does not evaporate nor flows superficially is infiltrated on the more permeable colluvial materials and weathered shales, giving rise to small springs, such as the La Mingueta Spring (R1 in Fig. 3). Where the thickness of the weathered materials and weathered shales is large, water infiltration is favoured, as occurs in the northwestern slope (where boreholes reveal thickness of 11 m of weathered rocks, whereas the thickness of weathered material in the southern slope revealed by drilling is only 1.5 m). Variations of piezometric level are closely correlated with infiltration from rainfall, following its seasonal changes. In Fig. 4, the relationship between rainfall and water table in piezometer E4 (for location see Fig. 3) is shown for the period 1983–1984. According to these data, a significant delay between strong storm events and the rising of piezometric levels can be observed. This fact, related to the low permeability of spoil heap materials, constitutes an important limitation for underground drainage in the case of intense storm events. Then the superficial drainage, including the channelling of the Morgao Stream and other superficial run-off through the underground galleries, is fundamental to the stability of the spoil heap.

Surface water flowing through the slope where the Los Rueldos Mine and spoil heap are located is intensively polluted by contact with leachable materials. Considering annual rainfall and evapotranspiration, the flow of polluted water from the Los Rueldos area can be roughly estimated at around $3200 \, m^3 \, year^{-1}$ in the affected area ($6000 \, m^2$), which is less than 1% of the area of the Morgao Basin, estimated to be $616\,300 \, m^2$. This water is mainly infiltrated through the still permeable side trenches around the Morgao spoil heap, and it only flows superficially after storm events. Considering an evapotranspiration coefficient in the region of 0.55 (Medina 1999) and an average yearly rainfall of 971 mm, the theoretical yearly water flow from the coal spoil heap is $329\,135 \, m^{-3} \, year^{-1}$. This flow corresponds also to both canalized and infiltrated water.

Water quality

Data of ICP chemical analysis of surface and groundwater collected in different points of the studied area are given in the Table 2, where (Fig. 3):

- M1 – effluent from the Morgao spoil heap;
- M2 – effluent from the Los Rueldos Mine–spoil heap/influent to Morgao spoil heap;
- M5 – acid mine water at the pit heads of Los Rueldos Mine;
- M6 – Morgao Stream upstream of both spoil heaps
- R1–R5–groundwater from the same local formation that arises in different springs around the spoil heaps.

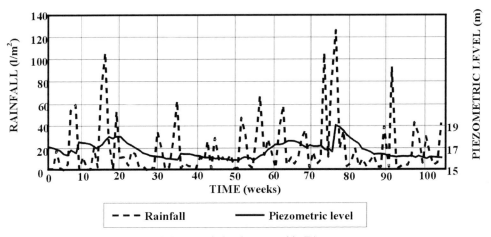

Fig. 4. Relation between rainfall and piezometric level measured in E4.

Table 2. *Chemical analysis of superficial and ground waters sampled in the area*

		Superficial water				Groundwater R1–R5
		M1	M2	M5	M6	
No. of samples		10	3	10	3	4
pH	Mean	7.71	2.44	2.26	6.94	7.2
	Range	6.65–8.29	2.15–2.66	1.66–2.60	6.78–7.15	7.1–7.3
Electrical Conductivity ($\mu S\,cm^{-1}$)	Mean	1033	5600	5510	N/A	N/A
	Range	950–1230	5500–5800	500–5700	N/A	N/A
Sulphate ($mg\,l^{-1}$)	Mean	340	3700	5000	39	N/A
	Range	220–480	2900–4600	4300–6000	36–42	N/A
As ($\mu g\,l^{-1}$)	Mean	125.5	4702	6845	0.7	9.6
	Range	5–400	1407–9200	2400–17742	<1–1	<1–23
Hg ($\mu g\,l^{-1}$)	Mean	0.8	6.0	0.9	<1	0.8
	Range	<1–6.1	<1–14	<1–3.7	<1	0.6–1
Total Fe ($mg\,l^{-1}$)	Mean	6.23	460.5	700.0	N/A	804
	Range	1.31–11.15	345–1048	120–1085	N/A	12–2581
Pb ($\mu g\,l^{-1}$)	Mean	20	250	77	N/A	4.5
	Range	18–22	30–480	12–330	N/A	<2–15
Mn ($\mu g\,l^{-1}$)	Mean	660	1140	2434	N/A	108.4
	Range	392–940	286–2650	610–3742	N/A	8.8–183.2
Sb ($\mu g\,l^{-1}$)	Mean	N/A	N/A	29.3	N/A	0.06
	Range	N/A	N/A	19.0–57.0	N/A	<0.05–0.12
Zn ($\mu g\,l^{-1}$)	Mean	<10	618	6208	N/A	14.4
	Range	<10	520–730	2700–7572	N/A	1.4–26.6

N/A, not available.

Samples of surface waters collected upstream of the Los Rueldos and Morgao spoil heaps (M6) show pH values between 6.78 and 7.15, sulphate concentrations of 36–42 ppm and less than $1\,\mu g\,l^{-1}$ of total arsenic or mercury.

Effluents from the Los Rueldos Mine (M5) show low pH values (1.66–2.60), an average sulphate concentration of $5000\,mg\,l^{-1}$, lead concentrations between 0.01 and $0.33\,mg\,l^{-1}$, arsenic concentrations ranging from 2.4 to $17.7\,mg\,l^{-1}$, and a maximum mercury content of $3.7\,\mu g\,l^{-1}$. Analyses of leachates from the 'Los Rueldos' mercury mine spoil heap (M2) show an average pH of 2.44, an average sulphate concentration of $3700\,mg\,l^{-1}$, arsenic concentrations between 1.4 and $9.2\,mg\,l^{-1}$, lead concentrations between 0.03 and $0.48\,mg\,l^{-1}$, and mercury concentrations between <1 and $14\,\mu g\,l^{-1}$. Part of the mining effluents and leachates from the Los Rueldos mercury mine flow through a peripheral channel to reach the Morgao Stream downstream of the coal spoil heap, while the rest infiltrates to the Morgao coal spoil heap. In both cases these As-rich effluents are strongly diluted. The above mentioned infiltration process might be the reason why some of the sampled groundwater (local springs) show higher As concentrations ($<1-23\,\mu g\,l^{-1}$) than superficial water upstream the spoil heaps (Table 2).

Effluents from the Morgao coal spoil heap (M1) show a neutral pH, an average sulphate content of $340\,mg\,l^{-1}$, an arsenic content between 0.005 and $0.400\,mg\,l^{-1}$, lead concentrations between 0.018 and $0.022\,mg\,l^{-1}$, and maximum mercury concentrations of $6.1\,\mu g\,l^{-1}$.

Drinking water limits in Canada and the United States for total As are set at 25 and $50\,\mu g\,l^{-1}$, respectively (Federal-Provincial Subcommittee on Drinking Water 1996; US EPA 1998), but in June 2000 the US EPA proposed a rule that would lower the existing standard for arsenic in drinking water from 50 to $5\,\mu g\,l^{-1}$, and sought comments on standards of 3, 10 and $20\,\mu g\,l^{-1}$ (Suzukida 2000), although according to the American Water Works Association (AWWA) a final standard below $20\,\mu g\,l^{-1}$ cannot be scientifically justified. These low regulatory concentrations reflect the toxicity and carcinogeneity of As. The Spanish legislation for mining/industrial effluents limits the total arsenic content to $0.50\,mg\,l^{-1}$, and $0.05\,mg\,l^{-1}$ for superficial waters destined for the production of drinking water (BOE 1985). Both limits are exceeded by Los Rueldos mine water, but, after dilution in the Morgao Stream, the concentrations are seriously reduced. Moreover, water sampled in the Morgao Stream 500 m downstream at sampling point M1 showed very low arsenic concentrations.

Alternatively, as a link between flow processes and chemistry, dilution effects produced by rainfall events can also be considered. In Fig. 5 a strong inverse correlation between rainfall and As concentration in Los Rueldos mine water (M5) can be observed. This opposite relationship was found to be a maximum considering the accumulated rainfall for periods of 3 weeks before the sampling day, which correspond to the estimated delay time taken by the rain water to infiltrate and show up out of the mine.

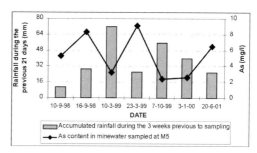

Fig. 5. Relation between As concentration in the Los Rueldos mine water and the rainfall during the previous 3 weeks.

Conclusions

Geochemical and mineralogical data for the Los Rueldos site suggest that weathering of sulphide minerals contributes to the high arsenic contents found in mine effluents and spoil heap leachates. The potential environmental impacts of elevated concentrations of arsenic and heavy metals in soils and waters could well be important.

Hydrogeological and geochemical studies at the Los Rueldos Mine site have determined the extent and distribution of areas affected by acid water generated from mine wastes. It was determined that groundwater flow and metal transport on the site is controlled by the filling of a part of the valley with tailings derived from coal mining. Acid drainage produced from mercury-rich mine effluents, and from the waste rock pile on the slope of the mountain, flows through bedrock and towards local stream tributaries of the Caudal River.

As-rich effluents and leachates from the Los Rueldos mercury mine enter the Morgao Stream downstream of the Morgao coal spoil heap where they are diluted such that an As content of $1.4-9.2\,mg\,l^{-1}$ upstream of the Morgao spoil heap is lowered to $0.005-0.4\,mg\,l^{-1}$ downstream. However, the contribution of polluted water from Los Rueldos Mine and spoil heap to

Morgao Basin constitutes approximately 1% of the total annual flow in the Morgao Stream. Local springs have a neutral pH water with an average As content of 0.009 mg l^{-1}.

A strong inverse correlation between the As concentration in the Los Rueldos mine water (M5) and the accumulated rainfall during the three weeks previous to sampling can be observed. This corresponds to the estimated time from when the rain falls until the same water goes out of the mine.

References

ALLOWAY, B.J. 1995. *Heavy Metals in Soils*, 2nd edn. Chapman & Hall, London.

ARAMBURU ZULOAGA, F. DE 1899. Monografía de Asturias, Oviedo.

ASTOLFI, E., MACCAGNO, A., FERNÁNDEZ, J.C.G., VACCARA, R. & STIMOLA, R. 1981. Relation between arsenic in drinking water and skin cancer. *Biological Trace Element Research*, **3**, 133–143.

AZCUE, J.M. & NRIAGU, J.O. 1994. Arsenic: historical perspectives. In: NRIAGU, J.O. (ed.) *Arsenic in the Environment. Part 1: Cycling and Characterization*. John Wiley & Sons, London, Vol. 26, 1–15.

BALDO, C. 2000. Impacto ambiental en areas afectadas por minería antigua de mercurio en el Concejo de Mieres (Asturias). Thesis Doctoral Universidad de Oviedo.

BOE 1985. *Ley de aguas 29/1985*; BOE, Madrid, **189**.

BORGONO, J.M. & GREBIER, R. 1971. Epidemiological study of arsenicism in the city of Antofagasta. *Trace Substances in Environmental Health*, **5**, 13–24.

BOYLE, D.R., TURNER, R.J.W. & HALL, G.E.M. 1998. Anomalous arsenic concentrations in groundwaters of an island community, Bowen Island, British Columbia. *Environmental Geochemistry and Health*, **20**, 199–212.

CEBRIAN, M.E., ALBORES, A., AGUILAR, M. & BLAKELY, E. 1983. Chronic arsenic poisoning in the north of Mexico. *Human Toxicology*, **2**, 121–133.

DAS, D., SAMANTA, G. KUMAR MANDAL, B. et al. 1996. Arsenic in groundwater in six districts of west Bengal, India. *Environmental Geochemistry and Health*, **18**, 5–15.

DORY, A. 1894. Le mercure dans las Asturies. *Revue Universelle des Mines et Metallurgie*, **32**, 145–210.

ENADIMSA. 1972. Estudio de la estabilidad en la escombrera de Morgao. ENADIMSA.

FEDERAL-PROVINCIAL SUBCOMMITTEE on DRINKING WATER 1996. *Guidelines for Canadian Drinking Water Quality*, 6th edn, Canadian Government Publishing.

FERGUSON, J.F. & ANDERSON, M.A. 1974. Chemical forms of arsenic in water supplies and their removal. In: RUBIN, A.J. (ed.) *Chemistry of Water Supply, Treatment, and Distribution*. Ann Arbor Science, Michigan, USA, 137–158.

GUTIÉRREZ CLAVEROL, M. & LUQUE CABAL, C. 1993. *Recursos del subsuelo de Asturias*. Servicio de Publicaciones, Universidad de Oviedo.

LOREDO, J. 2002. Historic unreclaimed mercury mines in Asturias (northwestern Spain): environmental approaches. In: *Assessing and Managing Mercury From Historic and Current Mining Activities*. EPA, San Francisco, CA, 175–180.

LOREDO, J., LUQUE, C. & GARCÍA IGLESIAS, J. 1988. Conditions of formation of mercury deposits from the Cantabrian Zone (Spain). *Bull. Minéral*, **111**, 393–400.

LOREDO, J., ORDÓÑEZ, A., GALLEGO, J., BALDO, C. & GARCÍA IGLESIAS, J. 1999. Geochemical characterisation of mercury mining spoil heaps in the area of Mieres (Asturias, northern Spain). *Journal of Geochemical Exploration*, **67**, 377–390.

LU, F.J. 1990. Blackfoot disease: arsenic or humic acid? *Lancet*, **336**, 115–116.

LUQUE, C. 1985. Las mineralizaciones de mercurio de la Cordillera Cantábrica Thesis Doctorat Universidad de Oviedo.

LUQUE, C. 1992. *El mercurio en la Cordillera Cantábrica. En: Recursos minerales de España. García Guinea y Martínez Frías (Coords.)*, C.S.I.C. Textos Universitarios, Madrid, **15**, 803–826.

LUQUE, C., MARTÍNEZ GARCÍA, E., GARCÍA IGLESIAS, J. & GUTIÉRREZ CLAVEROL, M. 1991. Mineralizaciones de Hg–As–Sb en el borde occidental de la Cuenca Carbonífera Central de Asturias y su relación con la tectónica: El yacimiento de El Terronal–La Peña. *Bol. Soc. Esp. Mineral.*, **14**, 161–170.

MEDINA, A. 1999. *Estudio hidrogeológico en la Escombrera de Morgao. Proyecto Fin de Carrera*. E.U.I.T.M.M., Universidad de Oviedo.

MERTZ, W. 1981. The essential trace elements. *Science*, **213**, 1332–1338.

SADEI, 1965–1991. *Datos y cifras de la economía asturiana*. Sociedad Asturiana de Estudios Económicos e Industriales, Oviedo.

SUZUKIDA, I. 2000. Arsenic in drinking water regulation. In: *EPA Mining Waste Scientist-to-Scientist Meeting*, Environmental Protection Agency, Cincinnati, USA.

US EPA. 1998. *Arsenic in Drinking Water – Regulatory History*, US EPA, Cincinnati.

Structural influence on plume migration from a tailings dam in the West Rand, Republic of South Africa

ILKA NEUMANN[1] & KARIM SAMI

Council for Geoscience, Private Bag X112, Pretoria 0001, Republic of South Africa
[1]*Present address: British Geological Survey, Maclean Building, Wallingford, 0X10 8BB UK.*

Abstract: Gold mining activity in South Africa is a potential cause of groundwater contamination. In particular, leachate infiltrating from tailings dams can affect surface water, as well as groundwater quality. This study was aimed at assisting an individual mine to investigate the groundwater flow system downstream of a tailings dam in order to predict contaminant plume migration emanating from the decommissioned dam. During the investigation, evidence was found that the contaminant plume emanating from the dam was affecting the groundwater quality in the area, predominantly by impacting on chloride and sulphate concentrations. It was suspected that structural features influenced plume migration rather than lithology, hence a strong emphasis was placed on the structural geology. The orientation of tensional and compressional couples were derived through tectonic analysis and verified through field observations. Incorporation of tensional structural features, acting as preferential groundwater flow paths, into a numerical model enabled the simulatation of the groundwater flow regime in the area to predict plume migration.

Gold mining activity is a potential cause of groundwater contamination. In particular, leachate infiltrating from tailings dams can affect surface water, as well as groundwater quality. Under recent laws, the South African Government and the mining industry have reached an agreement on the preparation of an Environmental Management Programme to regulate the impact of mining on the environment. The mining industry has accepted that environmental rehabilitation must form an integral part of mining operations, and use of predictive models is becoming more commonplace in assisting the mining industry with the implementation of these environmental programmes. This study was carried out for a mining house to help understand the groundwater flow system downstream of a tailings dam and to predict contaminant plume migration emanating from the decommissioned dam. Strong emphasis was laid on a firm understanding of the hydrogeological system underlying the tailings dam, as it was suspected that structural features might influence plume migration more than lithology due to the complex nature of the fractured rock aquifer. The study is presented on the merit of the adopted methodology, and names and locations are withheld for reasons of confidentiality.

Site locality

The study area is situated in the West Rand of South Africa and includes a gold mine tailings dam areas downstream and potentially impacted by it. The study covers an area of approximately 18 km^2.

Topography ranges from 1760 m above mean sea level (amsl) in the north, which is the position of the local watershed, to 1560 m amsl in the south (Fig. 1). Generally, the area is drained by N–S running streams, which form the eastern and western boundaries of the area of interest (Fig. 2). The area is dominated by the tailings dam, which has an area of about 1.2 km^2.

Methodology

The overall aim of the study was to investigate the groundwater flow system downstream of the tailings dam in order to predict contaminant plume migration emanating from the dam. The study required an investigation and re-interpretation of the hydrogeology, structural geology and water chemistry data for the area. The results were incorporated into a groundwater flow and contaminant transport model. In detail, the investigation procedure consisted of the following aspects:

- review of all existing data and, where necessary, field studies regarding geology, hydrogeology, structural geology, water chemistry and climate in the area;
- sediment investigation of tailings dam material. To adequately characterize the chemistry and quantity of leachate emanating

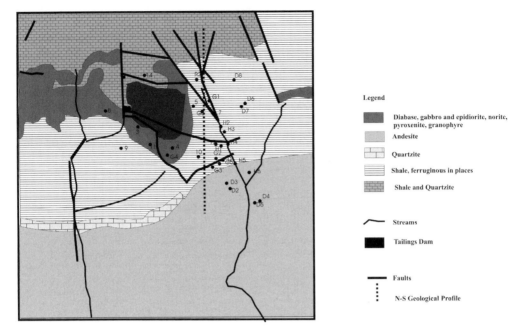

Fig. 1. Geological and structural features in the study area.

from the tailings dam, and its variations in time, it was necessary to estimate the percolation rate through the tailings into the underlying aquifer. Augered borings were established across the tailings dam to characterize the chloride, moisture content and saturation profile in the vertical profile;
- constant discharge tests were carried out to characterize hydraulic properties of the penetrated lithologies to assist with developing the conceptual model. The transmissivity (T) was calculated using the Cooper–Jacob method (Cooper & Jacob 1946). To verify the validity of permeability estimates, transmissivity estimates were compared with air discharge yields established during drilling;

- estimation of regional groundwater recharge by the chloride mass-balance approach;
- simulation of groundwater flow and contaminant transport using the Groundwater Modelling System GMS™ model.

Geology and structure

The study area is located in the Witwatersrand Basin, South Africa. The southern portion of the area is underlain by andesites, whereas silicified shales, quartzites and diabases belonging to the Pretoria Group crop out in the northern portion of the study area. Figure 1 shows a 1:10 000 scale geological map of the study area with a N–S geological section presented in Fig. 2.

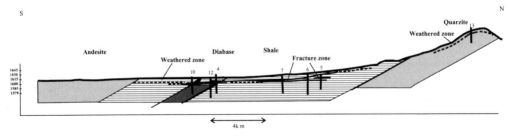

Fig. 2. N–S geological section through the study area.

Generally, the rocks are highly fractured and dip gently to the south. Fracturing and faulting are believed to be related to uplift of the Vredefort Dome to the south of the study area. The Vredefort Dome is the remnant of the central-uplift of a 250–300 km diameter impact structure dated at 2.02×10^9 years BP. The dome comprises a 40 km wide core of predominantly granitoid gneisses of Archean age, which are surrounded by a wide collar of sharply upturned Witwatersrand strata, enhancing a pronounced semi-circular structure (Reimold & Gibson 1996). The structural uplift associated with the impact event for the centre of the dome is, according to geophysical modelling, of the order of 8–12 km (Henkel & Reimold 1998). However, uplift resulted in beds moving northwards in a wave, or as nappes, resulting in northward thrusting component stresses and extensional stresses radiating outward from the dome as a ring of concentric strike-slip and normal faults. In the study area this would have resulted in extensional stresses being oriented approximately E–W, resulting in N–S dip-slip faulting.

Structural investigations in the study area revealed vertical slickenslides on a N–S-trending fault north of the tailings dam, supporting the hypothesis that these features were extensional faults or shears. They connect with E–W-running features, regarded as thrust faults of a compressional nature.

Field investigations, aided by aerial photograph interpretation, identified several E–W and N–S structural features in the study area. The streams to the west and east of the tailings dam follow the two most prominent N–S faults. Several E–W-running thrust faults occur in the centre of the area, and these are believed to be of compressional nature and unlikely to influence plume migration.

Hydrogeological regime operating in the study area

Hydrostratigraphic units

Geological logs from 26 boreholes indicate a fairly consistent picture of the subsurface. Irrespective of the lithology, a 14 m thick zone of low-permeability weathered overburden exists, and this acts as a confining layer for the underlying fractured aquifer. Bedrock is encountered approximately 20 m below ground surface.

Water strikes in the exploration boreholes are all associated with fracture zones. However, seepage has also been encountered in some boreholes within the weathered overburden. There is no evidence of any water strikes associated with lithological contacts, emphasizing the importance of the structure to the groundwater regime. Borehole yield distribution indicates that N–S-trending extensional faults appear to have a significant influence on the groundwater system, resulting in localized high borehole yields and the occurrence of springs along the fault lines. A constant discharge test indicates a transmissivity of $75 \, m^2 \, day^{-1}$ is typical for these fault zones.

Besides the weathered overburden, other hydrostratigraphic units can be distinguished on structural grounds in the confined fractured aquifer. The central, fractured part of the aquifer has a median transmissivity of $9 \, m^2 \, day^{-1}$ and is separated from poorly fractured shales and quartzites in the north and south of the area, which have a transmissivity of between only 0.6 and $1.9 \, m^2 \, day^{-1}$.

The boundaries of the different hydrostratigraphic units seem to be oriented along vertical E–W thrust faults, which are of a compressional nature.

The different hydrostratigraphic units are listed in Table 1. Hydraulic parameters for the units derived from pumping test data and were compared with airlift estimates made as drilling progressed. A linear relationship was established between the two sets of permeability estimates.

Flow system

The hydrogeological system in the study area is isolated and can be defined by hydraulic boundaries that include groundwater divides and flow paths. A topographic divide coincides with the northern boundary of the flow system, with hydraulic gradients directed to the south. Two streams form the eastern and western boundaries of the system. The southern boundary of the system is a topographic high, which acts as a water divide, while the base of the aquifer system is formed by impervious solid bedrock.

Water level measurements were available from 22 boreholes in the area and indicate that the general water movement is directed from north to SSW and from north to SSE into the eastern and western streams under influent conditions. This hydraulic situation is probably typical for three-quarters of the year. However, field observations and the groundwater chemistry reveal that at least the eastern stream is contributing some water into the groundwater system in the late winter period when groundwater levels are low.

Water sources in the area are rainfall recharge, the seepage from the tailings dam and the river under effluent conditions. A recharge estimate has been made using the chloride mass-balance technique. The average chloride concentration

Table 1. *Hydrostratigraphic units and their transmissivities*

	Initial parameters T ($m^2\,day^{-1}$)	After calibration T ($m^2\,day^{-1}$)
Low conductivity zone in the north	0.6	0.6
Low conductivity zone in the south	0.7	0.8
Highly fractured aquifer in the centre of the study area	8.9	8.9
Transiton zone between N–S faults and adjacent rock formations	23.1	23.1
N–S strike-slip faults	75	75
Weathered zone	4.5	5.2

for background water quality is $10\,mg\,l^{-1}$, derived from boreholes upstream of the tailings dam, and the chloride concentration in rainfall is $0.45\,mg\,l^{-1}$ in the wider area. The mean annual precipitation is 710 mm, so the regional recharge was estimated as $32\,mm\,year^{-1}$, or 4.5% of rainfall. Seepage from the dam was calculated according to water balance figures supplied by the mining house for the period that the dam was in use. During operation of the dam, water levels and the seepage flux are assumed to have been constant $4500\,m^3\,day^{-1}$ of water were pumped onto the dam and, subsequently, lost due to evaporation and/or seepage through the dam. Considering the water deficit between rainfall and open water evaporation, water loss due to seepage is about $708\,mm\,year^{-1}$.

After decommissioning, no water or tailings were added to the dam. The tailings dam is now contributing a significant volume of leachate to the aquifer system due to rainfall recharge and drainage of pore water storage over time as the water stored in the dam drains away. Hence, the hydraulic head imposed by the dam will have decreased over time and, consequently, seepage from the dam into the aquifer will have decreased accordingly. The decreasing hydraulic head was calculated on the basis of the tailings volume of $24\,000\,000\,m^3$, the initial head in the tailings dam (20 m above ground surface), a specific yield of 0.18, obtained from moisture and saturation vertical profiles of the dam to various depths, and the seepage rate of $708\,mm\,year^{-1}$. The seepage flux over time was then calculated on the basis of declining heads using a Darcian approach. The calculations show that the expected rate of seepage through the tailings decreases exponentially and reaches the regional recharge rate of $32\,mm\,year^{-1}$ some 19 years after the decommissioning of the dam. These figures will need to be verified in the future when time-series water level measurements in the dam become available.

Water sinks in the area consist of base flow discharge to rivers under effluent conditions and evaporation from zones with a shallow water table, which occur primarily in the vicinity of the tailings dam.

A conceptual model of the hydraulic system in the study area is shown in Fig. 3. The different hydrostratigraphic units represent the structural and geological set-up of the deeper-seated aquifer. The model domain boundaries are defined by watersheds, groundwater flow paths and streams.

Sources of contamination

Investigations into the water chemistry in the area revealed that the groundwater quality in boreholes in the vicinity of the tailings dam suffers adversely high chloride and sulphate concentrations. Whereas to the west of the dam the concentration gradients are rather steep, a more gradual gradient is observed towards the east. The observation of the elongated plume confirms the presence of a highly permeable N–S fault to the east of the tailings dam. Owing to the presence of carbonates in the tailings, which buffer the pH of the solution to between 6 and 7, no indication of acid mine leachate was found and mobility of other contaminants, like heavy metals, is not of concern at present.

Chloride originates from the concentration of dissolved salts in the dam by evaporation of water from the dam and the addition of mineralized mine water. Chloride can be considered as non-reactive and, hence, exhibits conservative behaviour. It can be assumed that no retardation occurs; hence, the chloride front represents the advective and dispersive front as it would appear if no retardation takes place.

During operation of the tailings dam, chloride concentrations in the leachate were around $500\,mg\,l^{-1}$, as determined from the median concentration of dam run-off water. For the period after decommissioning, the expected release of chloride into the aquifer was calculated on the basis of the calculated seepage flux and the initial chloride concentration in the tailings. It would

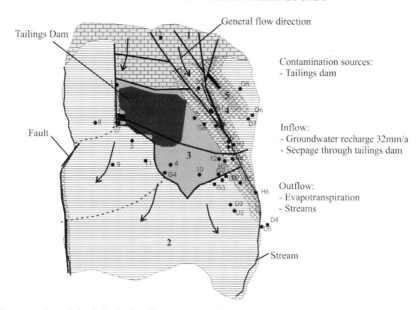

Fig. 3. Conceptual model of the hydraulic system in the study area together with an outline of the hydrostratigraphic units of the confined deeper-seated, fractured aquifer. 1 and 2, poorly fractured, low-conductivity zone; 3, well-fractured central zone; 4, transition zone between N–S faults and adjacent rock formations; 5, N–S strike-slip faults. The aquifer is overlain by a weathered zone 14 m thick. Model domain boundaries are defined by watersheds, streamlines and streams.

take more than 200 years for the chloride concentration in the leachate to fall to the background water concentrations of $10\,mg\,l^{-1}$ chloride.

Sulphate concentrations originate as a by-product of the oxidation of pyrite and other sulphide minerals in the tailings both aerobically and anaerobically.

Although sulphate is an anion basically not subject to sorption under neutral pH conditions, it is subject to precipitation as gypsum in the presence of calcium and magnesium once its concentration exceeds saturation. As gypsum is soluble, this process would lead to an apparent retardation of the sulphate plume, which is observed in the study area. In addition, the reduction of sulphate may also occur by organic matter and this could also lead to a reduction of sulphate in solution.

The extent of sulphate generation in the tailings dam is dependent on the thickness of the unsaturated zone in which oxygen is present, the addition of dissolved oxygen and water by recharge, and the ingress of gaseous oxygen due to barometric pressure variations. In order to quantify temporal variations in the sulphate concentrations in the leachate, a pyrite oxidation spreadsheet model developed by Sami (1999) was utilized and run for each year from the decommissioning of the dam until sulphate concentrations reached saturation. The computer program calculates water chemistry emanating from pits, tailings dumps and rehabilitated or exposed soils, taking into account parameters such as affected rock mass, recharge, saturation of tailings, porosity of tailings, acid potential, oxidation potential and moisture content of the tailings. Model results for steady-state sulphate saturation conditions indicate that the oxygen diffusion is the limiting factor on sulphate generation. The model was calibrated so that sulphate concentrations matched observed concentrations at the run-off water monitoring point at the base of the dam, which are assumed to be representative of the leachate (Fig. 4). It was found that 9 years after decommissioning the dam the sulphate concentration would reach saturation.

Numerical model

Discretization

A finite-difference numerical model was constructed on the conceptual analysis. To simulate the groundwater flow, the US Department of Defence GMS™ Groundwater Modelling System was utilized. GMS™ includes the US Geological Survey MODLFOW three-dimensional (3D)

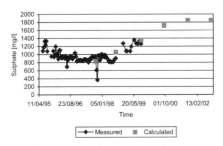
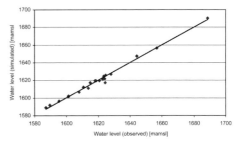

Fig. 4. Simulated and observed sulphate concentrations for leachate from the tailings dam over time.

Fig. 5. Simulated vs observed groundwater levels in observation boreholes.

finite-difference model (Mcdonald & Harbaugh 1988) and MODPATH 3D particle tracking model (Pollock 1989), and these were used to generate the groundwater flow model of the area. A two-layered model was established to represent the upper 14 m thick weathered zone and the underlying confined bedrock aquifer, which was considered to be 7 m thick. The tailings dam was situated in the centre of the model domain (Fig. 3). The boundaries of the system comprised water divides, groundwater flow paths and streams, and were represented in the numerical model using no-flow and constant head boundaries.

A steady-state model was established and parameters calibrated to simulate pre-decommissioning conditions. This model was subsequently used to establish initial conditions for the time-dependent transient model used to simulate the changes in hydraulic head resulting from the decline in seepage from the tailings dam. Time was therefore discretized into stress periods dependent on tailings dam seepage conditions. On the basis of the calibrated flow model, a solute transport model using MT3DMS (Zheng 1990) was set up. To verify the flow model and to predict future contaminant plume migration, the chloride plume emanating from the tailings dam was simulated, applying the advection and dispersion modules of MT3DMS to describe the transport behaviour of chloride.

Calibration

The calibration of the steady-state model was based on 22 water level measurements (Fig. 5). Owing to the absence of long-term water level measurements in observation boreholes, the flow simulation represents the hydraulic situation in the study area that is probably typical for the majority of the year, when groundwater discharges into the streams (Fig. 6). The transport model was calibrated on the basis of the borehole chemistry data that were available for the area.

However, there is a lack of sufficient transient water chemistry data and the model had to be calibrated on the basis of the plume development between the time before the tailings dam influenced the water quality (background water quality) and 1998, when most of the observation boreholes were sampled. The 1998 simulation shows the plume location as it is to date, and this coincides very well with the observed field data (Fig. 8).

Sensitivity analyses of various parameters revealed that the extent of the N–S-striking highly permeable fault zones and the fracture zone in the confined deeper aquifer strongly influence the simulation results. Errors in the geometry of the faults will affect model results. In addition, it should be noted that the volume of leachate that is produced in the area is a function of the amount of water percolating through the tailings dam. Hence, the estimation of the seepage emanating from the dam and of its chemistry is critical to the numerical model.

Results

Flow model

Figure 6 represent the steady-state hydraulic head in the model domain for the year 1998. The groundwater flow direction is generally from north to SSW and SSE. Figure 7 shows E–W cross-sections of the hydraulic head, identifying where the piezometric surface intersects the ground surface, so that base flow to streams may take place. The area underlain by the tailings dam is characterized by a saturated mound where piezometric heads are above ground level. This results in seepage into the lower aquifer and modification of the natural hydraulic gradient, so that strong easterly and westerly gradients are established towards the streams. In addition, near saturation is achieved in the immediate vicinity of the tailings dam, and surficial saturation results

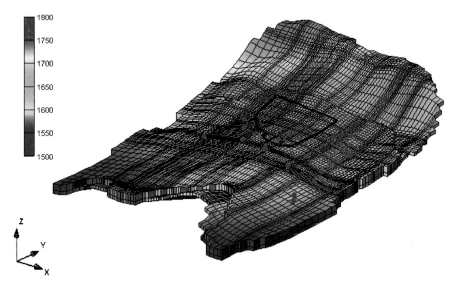

Fig. 6. Steady-state hydraulic heads in the model domain for 1998. Red point, observation borehole; black line, outline of tailings dam.

in the re-emergence of leachate near the dam and evaporation from the water table.

Transport model

Figure 8 represents the chloride plume migration up to 1998. The simulated chloride plume migration over time is shown in Fig. 9. Included are the plume positions in 2013 (15 years after this study), 2028 (30 years later) and 2058 (60 years later). The transport pattern of the plume was found to be affected by heterogeneities in the permeability of the subsurface, with a somewhat complex migration pattern.

The impact of structure on plume migration is severe, resulting in non-coincidence of the hydraulic gradient and the major axis of the contamination plume. The chloride plume is elongated in a southeasterly direction due to a high permeability zone related to the N–S extensional fault running along the eastern valley and the fracture zone in the deeper aquifer. These create a preferential pathway along narrow zones of higher hydraulic conductivity. The plume

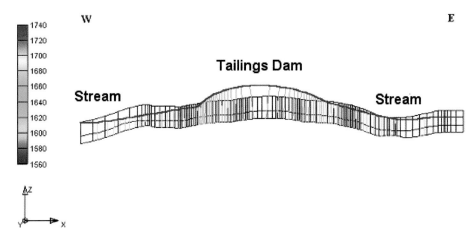

Fig. 7. E–W cross-section through the model layers featuring the water mound underneath the tailings dam. Blue line, water table; vertical lines, equipotential lines.

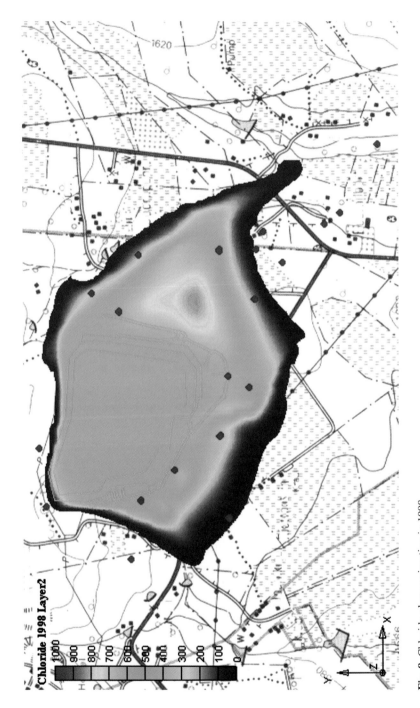

Fig. 8. Chloride plume migration in 1998.

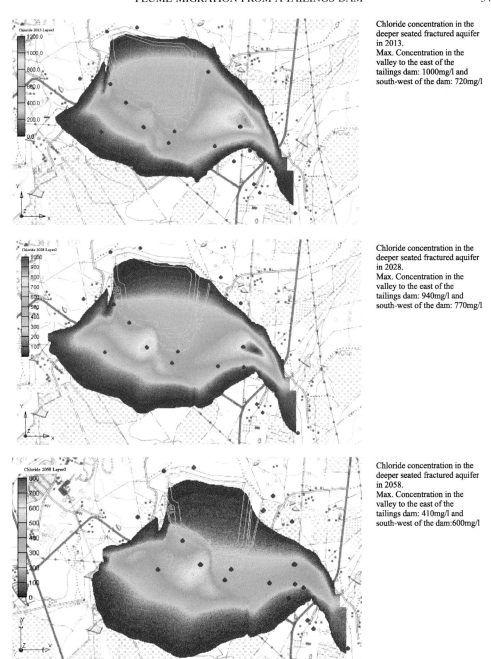

Fig. 9. Chloride plume migration in (**a**) 2013, (**b**) 2028 and (**c**) 2058.

follows the extensional fractures, not necessarily the hydraulic gradient.

With time, the plume spreads into the more homogeneous, but less permeable, southern part of the study area, where the primary control on contaminant distribution is the hydraulic gradient. As the gradient is greater in the westerly direction, the contaminant plume migrates to the SW. In this zone contamination migration follows the potentiometric surface.

The chloride concentration will decrease as the plume spreads over a larger area and the input load also reduces as the rate of leachate generation declines. Consequently, dilution of the contaminant away from the source area is observed. The model results also indicate a substantial reduction in the chloride concentration with time, resulting in improved groundwater quality with time and with increasing distance downgradient from the tailings dam. Natural attenuation alone would require 58 years before chloride levels of between 200 and 400 mg l^{-1}, required by statute, would be reached.

The model results strengthen the field observations by showing that evapotranspiration has a strong effect on the contamination of the shallow aquifer in the immediate vicinity of the tailings dam. However, the shallow water table around the dam encourages evaporation and locally increases the salinity.

This study shows that the existing monitoring boreholes had been located on the assumption that groundwater movement conforms to the surface topography. On the basis of the structural work and the groundwater flow model developed during this study, it was possible to locate new monitoring boreholes that would detect the actual migration of the contaminant plume and deliver time-dependent information on the water quality and the potentiometric surface elevation in the future. The new boreholes are concentrated in the vicinity where maximum plume migration is expected, which is along the two major N–S-trending faults and the southern boundary of the confined fractured aquifer. In addition, a monitoring network of the tailings dam was proposed, which should deliver data on the water level decline in the dam, the acid and base potential of the tailings and the redox conditions in the tailings at various depth to determine oxygen influx and oxidation state.

The authors would like to thank Mr E. Erasmus and Mrs J. Cameron, who have contributed to this work.

References

COOPER, H.H. & JACOB, C.E. 1946. A generalized graphical method for evaluating formation constants and summarizing well field history. *American Geophysical Union Transactions*, **27**, 526–534.

HENKEL, H. & REIMOLD, W.U. 1998. Integrated geophysical modelling of a giant complex impact structure: anatomy of the Vredefort Structure, South Africa. *Tectonophysics*, **287**, 1–20.

MCDONALD, M.G. & HARBAUGH A.W. 1988. *A Modular Three-dimensional Finite-difference Ground-water Flow Model*. US Geological Survey, Techniques Water-Resources Investigations, **06-A1**.

POLLOCK, D. W. 1989. *Documentation of Computer Programs to Compute and Display Pathlines Using Results From the US Geological Survey Modular Three-dimensional Finite-difference Groundwater Model*. US Geological Survey, Open File Report, **89-381**.

REIMOLD, W. & GIBSON, R. 1996. Geology and evolution of the Vredefort impact structure, South Africa. *Journal of African Earth Science*, **23**, 125–162.

SAMI, K. 1999. *Screening Programme to Calculate Decant Water Chemistry Emanating from Mining Open Pits, Discard Dumps and Rehabilitated or Exposed Spoils*. Council for Geoscience, Pretoria, SA. Unpublished Report.

ZHENG, C. 1990. MT3D. *A Modular Three-dimensional Transport Model for Simulation of Advection, Dispersion and Chemical Reaction of Contaminants in Groundwater Systems*. Report to the US Environmental Protection Agency, Robert S. Kerr Environmental Research Laboratory, Ada, OK.

Hydrogeological and geochemical consequences of the abandonment of Frazer's Grove carbonate hosted Pb/Zn fluorspar mine, north Pennines, UK

K. L. JOHNSON & P. L. YOUNGER

Hydrogeochemical Engineering Research & Outreach (HERO), Department of Civil Engineering, University of Newcastle, NE1 7RU UK (e-mail: k.l.johnson@ncl.ac.uk)

Abstract: The problems associated with predicting where mine water will emerge and what the quality will be in a post-closure situation are recognised world-wide. The closure of Frazer's Grove, a fluorspar mine in the North Pennines in the UK has given the opportunity to study in detail the relationship between rising groundwater and the strata/mineworkings through which it is rising. Detailed sampling and surveys both above and underground were carried out before, during and after rebound. During the rebound phase the mine water was stratified. Since the mine water emerged in August 1999, stratification has broken up and re-established itself twice to date. The possible causes of the break-up of stratification are examined with the aid of hydrogeochemical data and geophysical techniques. The main contaminants present in the mine water are zinc, manganese, iron and sulphate. A general exponential decrease in dissolved metal concentration in the mine water is seen with time. The hydrogeochemical data also establishes the origin of the contamination in the mine water discharge with zinc and manganese originating from Frazer's Grove mine itself and iron from several sources. The Frazer's Grove mine investigation provides insight into water quality and its likely development with time in abandoned mines.

Mine water discharges from abandoned workings cause a great deal of problems world-wide and their environmental impacts are well documented (for example, Hamilton *et al.* 1999). The problems associated with the prediction of both discharge quality and mine water emergence are also widely appreciated. Frazer's Grove has provided an excellent opportunity to study an abandoned mine both before and after abandonment. Frazer's Grove mine is of international interest as there are many similar ore bodies which are currently being mined such as the Olkusz region in Poland (Adamczyk *et al.* 2000). The implications of neutralisation of acidic mine drainage by limestone country rock are of particular relevance to other geologically similar workings. Observations on the temporal and spatial geochemical nature of mine water within the mine workings have allowed a greater understanding of the process of stratification.

Frazer's Grove mine is situated in the North Pennines in County Durham, UK (see Fig. 1). It consists of four interconnected mines: Frazer's Hush, Rake Level/Firestone Incline workings, Grove Rake and Greencleugh, as shown in Fig. 2. The mine has been worked since the 9th century for lead, zinc, barium and fluoride (in the form of fluorite, locally termed "fluorspar"). More recently only fluorspar has been worked in the deeper sections of Frazer's Hush and the shallower workings of the Rake Level/Firestone Incline mine. However, when fluorspar prices plummetted in the nineties, mining operations were severely reduced. Mining of the deep workings was stopped at the end of 1998. The dewatering pumps for the deeper sections (which are situated in Grove Rake No.2 Shaft) were switched off and removed on March 5th 1999, at which point the groundwater started to rise. Mining of the shallower workings in the Rake Level/Firestone Incline workings continued until July 1999 when all mining operations ceased and this, the last mine in the North Pennines was closed.

During mining, the average pumped discharge from both the deeper and shallower workings of Frazer's Grove mine was 1895 m^3/d (Younger 2000*a*). The majority of this water came from the Great Limestone (an extensive minor aquifer) and was strongly net-alkaline, with circum-neutral pH. This pumped discharge was monitored from January to April 1998 and the only contaminant present at problematic concentrations was zinc (average concentration ~12.3 mg/l) (Younger

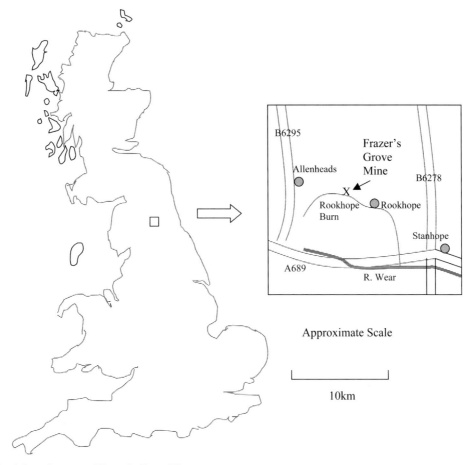

Fig. 1. Location map of Frazer's Grove Mine.

2000a). Since discharges from recently abandoned mines are usually of worse quality than pumped discharges from working mines, a sampling routine was devised in conjunction with the Environment Agency (EA) to monitor any environmental implications of the closure of Frazer's Grove Mine (Johnson & Younger 2000). The monitoring programme ran from April 1999 to February 2001 when it was interrupted by the outbreak of foot and mouth disease in the local area.

Geological setting

The Carboniferous rocks of the North Pennines are cut by numerous mineral veins. The veins are associated with the Caledonian Weardale granite. The granite batholith is not thought to have been the principal source of heat for mineralisation but it may have directed hot mineralising fluids through the Carboniferous sediments (Dunham et al. 1965). Many veins are little over 1 m wide but in places widths of over 10 m have been recorded; veins of up to 3 m wide can be seen in the Firestone Incline at Frazer's Grove mine today. There is a marked zonation of the constituent minerals within the orefield, especially between the non-metalliferous or gangue minerals. Deposits in the central zone carry abundant fluorite (CaF_2), with quartz, chalcopyrite ($CuFeS_2$), galena (PbS) and sphalerite (ZnS) also present. Surrounding the fluorite zone is a wider zone of deposits in which barium minerals including baryte ($BaSO_4$) and witherite ($BaCO_3$) are the characteristic gangue minerals. The presence of galena and other sulphide minerals is less common in this zone. The mineral zonation reflects progressively lower temperatures of crystallisation from mineralising fluids as they flowed outwards from central 'emanative centres' above high spots (cupolas) on the underlying Weardale Granite (Dunham 1990).

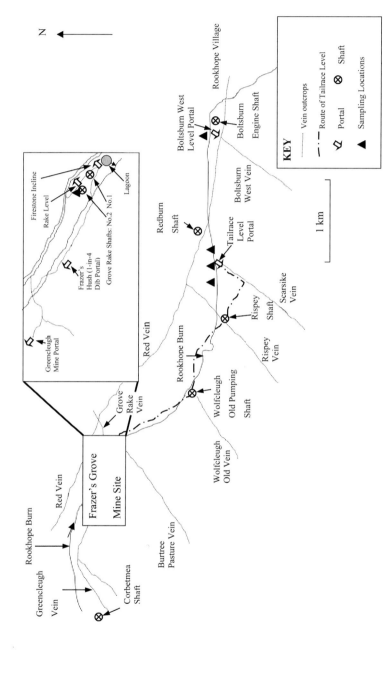

Fig. 2. Sketch map of Frazer's Grove Mine showing the sampling locations.

Frazer's Grove Mine is located above this emanative centre within the fluorite zone. A simplified cross-section of the mine and representative geological log for the mine are presented as Fig. 3 and Table 1 respectively.

Monitoring locations

Water samples have been taken on a monthly basis since April 1999 to February 2001 from both above ground and underground. Both the Boltsburn (NY 9365 4283) and Tailrace Levels (NY 9162 4271) were sampled at the same time as the depth sampling in the No.2 Shaft (NY 8949 4410). The No.2 Shaft was sampled using a SEBA 1L depth sampler. It was assumed that the shaft water at depth is at equilibrium with the minerals in the corresponding strata. Samples were taken every 10 m between the water surface (currently stable at ~ 373 m aOD in the No.2 Shaft) and a depth of 315 m aOD: it was not possible to sample below 315 m aOD due to a platform across the shaft.

The Boltsburn Level is situated on outcropping Great Limestone and its discharge is typical of Great Limestone groundwaters ($Ca \sim 100$ mg/l, $HCO_3 \sim 180$ mg/l, conductivity ~ 500 μS/cm) (Dunham 1990). However, it does drain some old mineworkings and although it contains <0.5 mg/l of both manganese and zinc it has variable iron concentrations ($1-10$ mg/l). The Level is not directly connected to Frazer's Grove mine and was monitored in order to assess any changes in flow or chemistry in the Great Limestone aquifer. The Tailrace Level drains old mineworkings but also carries a significant component of surface run-off which enters the level via crownholes. The water quality in the Tailrace Level was also fairly constant ($Ca \sim 60$ mg/l, $HCO_3 \sim 130$ mg/l, conductivity ~ 400 μS/cm) until the shaft water overflowed into it on the 25th August 1999.

The Rookhope Burn was also sampled approximately 200 m both up and downstream of the Tailrace discharge. All sampling locations are shown on Fig. 2.

Methodology

pH, temperature, Eh and conductivity were measured on site using a calibrated Camlab MY/6P meter and alkalinity was measured using a Hach digital titrator. Two 200 ml sample bottles (one being acidified with 2 drops of concentrated nitric acid) were filled and returned to the Environmental Engineering Laboratory at Newcastle University for anion and cation analysis. Dilutions were frequently necessary as sulphate was present at high concentrations. However, the dilution resulted in less accurate measurement of anions which are present at low concentrations in the mine water (such as chloride) as any percentage errors in analysis are increased dramatically on multiplication by the dilution factor. For this reason, the chloride results presented in Table 2 should be treated with caution.

The hydrochemical results from Frazer's Grove mine were modelled using WATEQ4F (Ball & Nordstrom 1991). This modelling was carried out in order to calculate the saturation indices of relevant minerals rather than determine the hydrogeochemical nature of the mine.

Robertson's geophysical sondes were used for geophysical logging in the No.2 Shaft in June 2000. Two sondes were used, one of which measured temperature and conductivity and the other measured the flow velocity using an impeller. The latter also measured natural gamma but as the shaft was wide the natural gamma signal was too weak for interpretation. A Portalog2 unit was used above ground to interpret the data received.

Rising groundwater was monitored both manually and with data-loggers. Manual measurements were taken in the No.2 Shaft every week or fortnight. Data-loggers were installed in the No.2 Shaft and in Frazer's Hush from March and February 1999 respectively. As the Hush is a 1-in-4 dib (inclined level) it was not possible to measure the rise in water level using a conventional pressure transducer. However, access to the Hush was also limited as the entrance to the workings was sealed in July 1999 for safety reasons. The Environment Agency, however, had designed a mechanism which allowed the rise of water in the Hush to be monitored from the surface. This consisted of a series of switches which were placed every 10 m along the length of the Hush. The switches were triggered on contact with the rising water and this event was recorded on a conventional data logger situated above ground. The location of the switches was translated into metres above Ordnance Datum. This allowed a comparison between groundwater rebound in the No.2 Shaft and the Hush to be made, thereby confirming that the water was rising evenly throughout the mine workings.

Lithological samples were taken in December 1998 which was the last opportunity to gain access to the deeper workings. Samples of acid generating minerals were taken for X-ray Diffraction (XRD) analysis. A small amount of the dried and ground sample (0.5 g) was equilibrated with 20 ml of de-ionised water and agitated for 30 minutes and repeated three

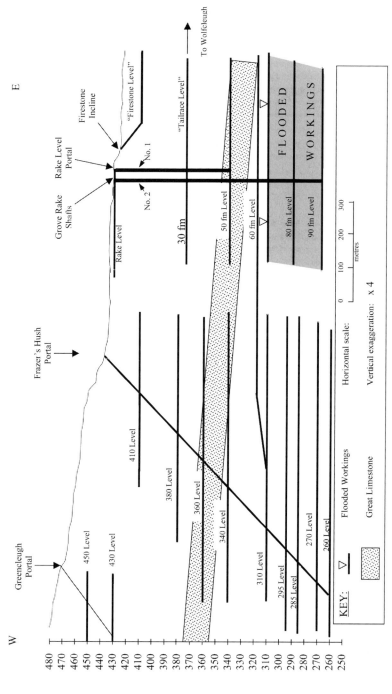

Fig. 3. Sketch diagram of cross-section of Frazer's Grove Mine.

Table 1. *Simplified Geological Succession for the Frazer's Grove Mine Site*

Unit Name (key units in bold typeface)	Approx. elevn. of (m aOD) at Grove Rake Shafts*	Lithology†	Thickness (m)	Hydrogeological Classification
Low Slate Sill	434.6	Sst(silty)	11.3	Aquitard
Knuckton Shell Beds	423.3	Slt/Mdst	19.8	Aquitard
Crag Limestone	403.5	Lst	0.15	Aquitard
Plate beds	403.4	Mdst/Sst	1.4	Aquitard
Firestone Sill	402	Sst	8.2	Minor aquifer
Plate beds	397.3	Mdst	4.7	Aquitard
White Sill	391.3	Sst	6	Minor aquifer
Plate beds	381.9	Mdst	9.4	Aquitard
Girdle Beds	377.5	Sst/Mdst	4.4	Aquitard
Pattinson's Sill	373.2	Sst	4.3	Minor aquifer
Plate beds	367.9	Mdst	5.3	Aquitard
Little Limestone	364.5	Lst	3.4	Minor aquifer
Sandstone beds	359.9	Sst	4.6	Minor aquifer
Plate beds	357.2	Mdst	2.7	Aquitard
High Coal Sill	355.6	Sst	1.6	Minor aquifer
Coal	355.45	Coal	0.15	Aquitard
Plate beds	350.05	Mdst	5.4	Aquitard
Coal	350	Coal	0.06	Aquitard
Low Coal Sill	349	Sst	1	Aquitard
Plate beds	346.2	Mdst	3.8	Aquitard
Great Limestone	325.8	Lst	20.4	Major aquifer
Tuft	323.7	Sst/Slt	2.1	Minor aquifer
Plate beds	322	Mdst (py)	3.5	Aquitard
Iron Post Limestone	320.3	Lst/Mdst	1.7	Aquitard
Quarry Hazle	308.9	Sst	11.4	Minor aquifer
Plate beds	297.3	Mdst	11.6	Aquitard
Four Fathoms Limestone	291.7	Lst	5.6	Minor aquifer
Plate parting	291.1	Mdst	0.6	Aquitard
Nattrass Gill Hazle	278.8	Sst	12.3	Minor aquifer
Grey beds	265.6	Slt/Sst	13.2	Aquitard
Three Yard Limestone	263	Lst	2.6	Minor aquifer
Six Fathom Hazle	254	Sst	9.0	Minor aquifer
Plate beds	245.9	Mdst	8.1	Aquitard
Five Yard Limestone	243.3	Lst	2.6	Minor aquifer
Grey beds	223.8	Slt/Sst	19.5	Aquitard
Scar Limestone	213	Lst/Mdst	10.5	Minor aquifer

* Note: OD elevations in Frazer's Hush workings are higher, because of the regional dip of ~ 1.5° ENE (ie about 0.025 m drop per m travelled ENE) eg the Great Limestone Base at Frazer's Hush (600 m W of Grove Rake shafts) should be $600 \times 0.025 - 15$ m higher than at Grove Rake shafts. This is indeed the case with the Great Limestone base being seen at the 340 Level in Frazer's Hush.

† Sst = sandstone; Slt = siltstone; Mdst = mudstone; Lst = limestone; (py) = pyritic

times so that zinc, manganese, iron and sulphate concentrations could be measured.

V-notch weirs were installed at the Tailrace Level and the Boltsburn Level as part of the monitoring programme and data-loggers were installed in order to record flow on a daily basis whenever possible. Unfortunately, flow measurement was not possible at the Tailrace Level after the mine water emerged on 25 August 1999 due to construction works for a mine water treatment system.

Frazer's Grove mine

Previous work has included an extensive sampling routine both above ground in the Rookhope Burn and underground within the mine (Younger 2000a). Underground inspections of the mine geology were carried out during 1998. The most significant features noted were the 20 m thickness of Great Limestone and the presence of yellow/grey highly weathered shales which contained acid generating minerals just

Table 2. *Typical key water quality data from Spring 1999 to Spring 2001 from Frazer's Grove Mine and surrounding area*

Name	Height (m aOD)	Date	Na (mg/l)	K (mg/l)	Ca (mg/l)	Mg (mg/l)	Fe (mg/l)	Mn (mg/l)	Zn (mg/l)	Cl (mg/l)	SO_4 (mg/l)	HCO_3 (mg/l)	pH	Cond (μS/cm)	Eh	Temp (°C)
No2 Shaft	337	22/4/99	35.8	33.7	386.8	100.5	12.0	31.7	35.1	114.7	1328.5	146	6.22	2376	117	9.1
No2 Shaft	330	22/4/99	35.7	25.2	404.5	101.6	11.7	32.7	34.1	90.2	1403.0	60	6.03	2380	112	8.9
No2 Shaft	320	22/4/99	38.0	20.8	446.7	122.4	33.0	65.4	61.8	37.5	1795.5	0	4.35	2861	250	10
No2 Shaft	315	22/4/99	34.2	18.6	464.8	119.3	30.0	66.1	63.1	173.4	1785.9	0	3.82	2874	344	8.9
Tailrace Level		22/4/99	11.8	5.7	52.3	9.8	4.1	0.1	0.1	17.5	34.1	180	6.00	335	146	8.2
Boltsburn Level		22/4/99	14.2	6.2	72.0	12.9	0.1	0.2	0.1	18.5	62.6	178	6.79	492	50	10.2
No2 Shaft	360	1/9/99	30.5	23.6	673.0	194.0	24.6	114.4	88.0	212.0	1997.0	58	5.09	3033	89	11.7
No2 Shaft	350	1/9/99	27.7	20.4	662.0	198.0	24.8	120.8	93.0	169.0	1662.0		5.30	2850	73	11.4
No2 Shaft	340	1/9/99	25.8	19.3	623.0	185.0	27.9	106.8	87.0	0.0	1844.0	56	5.58	1981	45	10.8
No2 Shaft	330	1/9/99	22.1	17.6	567.0	182.0	25.6	109.3	95.0	99.0	1808.0	32	5.36	3160	74	11.5
No2 Shaft	315	1/9/99	20.7	14.2	623.0	215.0	67.9	138.1	111.0	138.0	2178.0	26	5.05	3188	80	11.4
Tailrace Level		1/9/99	16.4	11.0	303.0	82.0	7.1	44.0	35.6	258.0	907.0	92	5.90	1805	50	10.3
Burn d/s TRL		1/9/99	11.4	6.8	176.0	22.0	3.4	18.2	20.0	305.0	518.0	60	6.32	1106	n/a	14.4
Boltsburn Level		1/9/99	11.5	6.3	123.0	15.6	1.2	0.3	0.0	29.0	82.0	243	6.62	604	n/a	11.2
No2 Shaft	370	26/2/00	10.9	6.6	66.8	15.2	0.0	3.0	5.6	24.0	191.0	30	6.36	488	n/a	8.4
No2 Shaft	360	26/2/00	10.7	9.2	88.1	23.2	0.0	6.3	4.1	24.0	342.0	80	6.67	659	n/a	7.6
No2 Shaft	350	26/2/00	24.0	25.0	451.0	94.1	14.3	61.1	22.2	28.0	1011.8	180	6.17	2364	n/a	8.3
No2 Shaft	340	26/2/00	28.0	30.0	565.0	115.0	18.3	75.6	26.6	24.0	1514.5	186	6.12	2486	n/a	9.1
No2 Shaft	320	26/2/00	26.4	27.3	540.5	113.7	19.3	75.9	27.7	24.0	2501.0	180	6.06	2610	n/a	9.0
Tailrace Level		26/2/00	14.7	12.1	283.9	67.9	3.9	36.5	16.2	24.0	842.4	106	6.52	1630	n/a	8.6
Burn u/s TRL		26/2/00	6.3	1.1	8.1	3.0	0.5	0.2	0.2	13.0	9.9	18	6.70	91	n/a	4.5
Burn d/s TRL		26/2/00	7.5	3.0	48.0	11.6	1.2	6.0	2.3	34.6	132.0	30	6.85	406	n/a	5.4
Boltsburn Level		26/2/00	11.2	6.2	77.0	15.3	0.4	0.8	0.2	27.2	76.4	180	7.57	548	n/a	7.5
No2 Shaft	365	24/1/01	9.1	6.3	68.0	10.4	0.1	1.5	2.9	97.7	185.1	n/a	6.37	483	152	6.4
No2 Shaft	350	24/1/01	9.1	6.3	70.2	10.3	0.1	1.5	2.9	130.0	186.9	n/a	6.51	513	155	7.8
No2 Shaft	340	24/1/01	9.1	6.4	64.6	10.3	0.1	1.5	2.8	98.8	210.9	n/a	6.62	516	164	7.8
No2 Shaft	330	24/1/01	12.0	16.9	238.0	41.2	21.0	27.3	7.7	218.6	818.0	n/a	6.35	1317	36	8.0

Table 2 – continued

Name	Height (maOD)	Date	Na (mg/l)	K (mg/l)	Ca (mg/l)	Mg (mg/l)	Fe (mg/l)	Mn (mg/l)	Zn (mg/l)	Cl (mg/l)	SO_4 (mg/l)	HCO_3 (mg/l)	pH	Cond (μS/cm)	Eh	Temp (°C)
Tailrace Level		24/1/01	10.1	9.2	131.8	17.9	4.7	11.7	3.6	101.0	328.4	118	6.69	917	16	7.4
Burn u/s TRL		24/1/01	5.0	1.6	6.1	1.8	0.6	0.5	0.2	23.7	17.9	16	7.90	89	13	3.7
Burn d/s TRL		24/1/01	4.8	1.7	7.3	2.1	0.6	0.7	0.2	23.6	22.6	14	7.18	103	18	2.6
Boltsburn Level		24/1/01	11.3	4.9	63.0	9.4	0.8	0.8	0.2	88.1	114.0	162	6.79	518	83	7.5

below the limestone (see Fig. 3 & Table 1). The Great Limestone is a prolific aquifer throughout the North Pennines and plays an important part in the hydrogeology of the mine.

The underground surveys suggested that around two thirds of the Frazer's Grove watermake comes from the Great Limestone aquifer with the remaining one third from the Firestone Sill and surface inflows (Younger 2000a). It was predicted that after abandonment the water table would rise above the aquifer and this source of head-dependent water would stop. A final post-rebound water-make of only ~680 m^3/d was expected. It was thought that most of the rising groundwater would dissipate into the Great Limestone and that mixing between mine and limestone water would result in satisfactory discharge water quality (according to EA regulations). However, it was also pointed out (in an unpublished and confidential report to the EA in October 1998) that if the limestone could not take up this water there would be a discharge at the Tailrace Level which is the lowest surface discharge point at 364 m aOD for the water. This discharge would end up in the Rookhope Burn. The Burn contains few fish (with the exception of some hardy brown trout) probably due to the mining legacy. The pumped discharge from the working mine contained significant dissolved zinc (up to 40 mg/l) which would have had a major deleterious effect on life in the Burn.

The highest zinc concentrations encountered in the surveys was 40 mg/l in water draining through a disused hopper deep within the Frazer's Hush workings at 285 m aOD (Younger 2000a). However, the highest zinc concentrations seen in the pumped discharge occurred in the spring of 1996 when the No.2 Shaft pumps temporarily failed. The water rose to 328 m aOD flooding the pyritic shales just below the Great Limestone. The water quality deteriorated rapidly as acid generating minerals in the shales dissolved resulting in a pH of 3.6 and most notably zinc concentrations of 27 mg/l (Younger 2000a).

Results and discussion

Lithological samples

Ground samples of the acid generating minerals found below the Great Limestone equilibrated with de-ionised water gave a resulting solution with a pH of 2.7 and concentrations of manganese ~1.9 mg/l, iron ~1.7 mg/l, zinc ~1.0 mg/l and sulphate ~100 mg/l.

Three X-Ray Diffraction (XRD) analyses of the acid generating minerals gave consistent results but identification of all the minerals present proved difficult. The XRD results are not presented here but it is clear from the presence of peaks in the 3.4 Å and 10 Å area that both quartz and illite were present. However the dissolution of quartz or illite would not give the metal concentrations or the pH which were observed in laboratory experiments. The unassigned peaks in the XRD analyses do suggest that another mineral is present but there are insufficient peaks present for an identification to be made. One possibility considered was the mineral franklinite which has the chemical formula $Zn_{0.6}Mn_{0.8}Fe_{1.6}O_4$. However, although franklinite is associated with the weathering of zinc-rich orebodies it is usually associated with limestone and not with shales/clays. It is very common for other iron and manganese oxides to be present in association with clay minerals (Reddy & Perkins 1976). Manganese and iron oxides are recognised as scavengers of cations and any dissolved zinc arising from sphalerite dissolution within the mine (Younger 2000a) is likely to have been adsorbed onto their surface (Jenne 1967). This could explain the metal concentrations seen in the laboratory experiments but would not explain the high sulphate concentrations and low pH which was observed. Sulphate is present in the mine from the oxidation of sphalerite (Younger 2000a) and could be sorbed onto the surface of iron-manganese oxides (Geelhoed et al. 1997) which would explain the observed acidity. There is at present, insufficient data to identify this mineral and with very little in the literature about the nature of these acid generating minerals, there is a real need for more research into this area.

Hydrogeology

Groundwater rose at varying rates depending on the available void space within the mine; the average rate of rise between March 1999 and August 1999 was between 25 and 30 cm/day (Fig. 4). The groundwater stopped rising on the 25th August 1999 at ~373 m aOD which is 1 m above where the Tailrace Level enters the No.2 Shaft (see Fig. 3). An increased discharge was noted at Tailrace Level on the same day and water quality in the Level deteriorated.

It was anticipated that the Great Limestone would accommodate the majority of the rebounding mine water. However this did not happen and in fact the majority of water discharged to the surface via the Tailrace Level. With hindsight a possible explanation has been formulated which is illustrated by Figs 5 and 6. The Great Limestone is a confined aquifer and so the peizometric head is above the top of the aquifer. It is the weight of the overburden above the aquifer which drives the water through the fractures of the Great Limestone. Before mining severely altered the nature of the Great Limestone aquifer in the Frazer's Grove area, the aquifer would easily have accomodated an extra ~680 m³/d. However the Great Limestone has been severely modified by the workings at

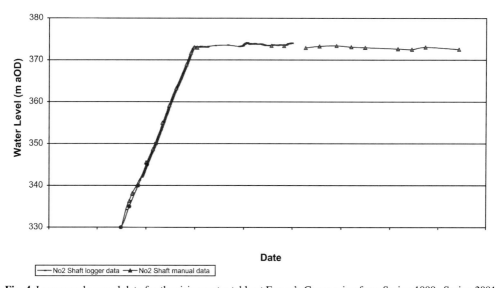

Fig. 4. Logger and manual data for the rising water table at Frazer's Grove mine from Spring 1999–Spring 2001.

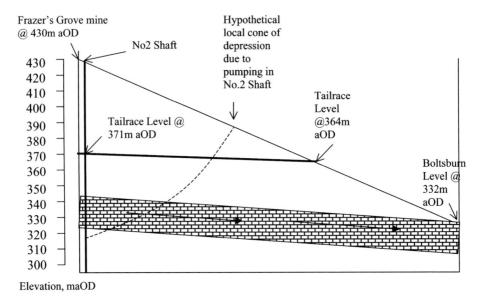

Fig. 5. Sketch diagram of hydrogeology of Great Limestone during mining.

Frazer's Grove mine. It is no longer a confined aquifer in the vicinity of Frazer's Grove mine as the overburden which creates the necessary environment for confined conditions has been breached. During mining the water table is artificially lowered and so there is no chance of a surface discharge (see Fig. 5). When the groundwater rebounded, the mine water rose above the aquifer and flowed in the channelised system of mine workings towards the Tailrace Level portal as there was insufficient pressure to drive the water through the aquifer (see Fig. 6).

Geophysics

The temperature and conductivity logs from June 2000 correlate with the stratigraphy in the No.2 Shaft as shown in Fig. 7. The data confirm that

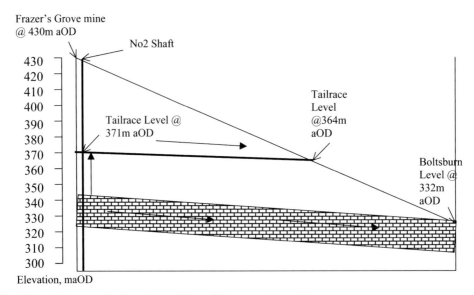

Fig. 6. Sketch diagram of hydrogeology of Great Limestone after mining.

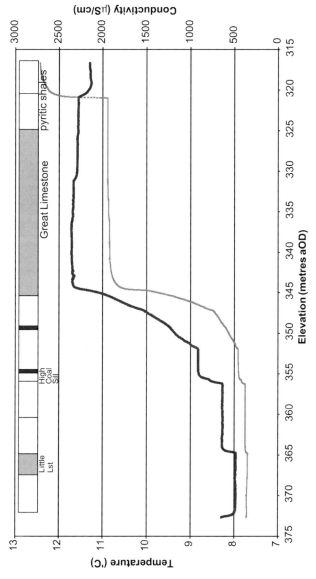

Fig. 7. Conductivity and temperature log of No.2 Shaft at Frazer's Grove Mine.

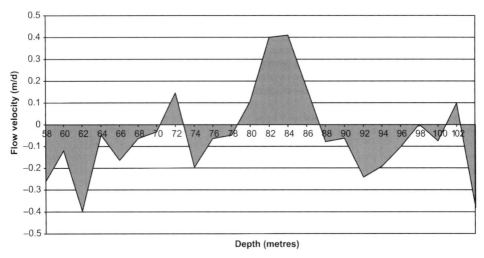

Fig. 8. Velocity flow log of water column in the No.2 Shaft at Frazer's Grove Mine.

the worst quality water (with the highest conductivity) is encountered where the pyritic shales are present at 350 m aOD. This poor Shaft water quality then persists throughout the Great Limestone formation.

The velocity flow log (Fig. 8) shows that water was mainly moving down the shaft as would be expected but that there was a significant component moving up towards 350 m aOD. This suggests that a proportion of the mine water is flowing at 350 m aOD which is the top of the Great Limestone where most fractures occur.

Hydrochemistry

A standard geochemical modelling program, WATEQ4F (Ball & Nordstrom 1991) has been used to model the water chemistry in the shaft. Results showed that goethite was supersaturated and that gypsum was close to saturation. The goethite saturation explains the orange ochreous precipitates in the samples taken and why, of the three metals measured, iron concentrations are by far the lowest.

Typical key hydrogeochemical data from both before and after mine water emergence are given in Table 2. The data are also presented graphically in Figs 9–12. Metal and sulphate concentrations increased in the No.2 Shaft as the water rose through the highly weathered acidic shales and then improved following contact with the Great Limestone. The water column within the shaft was stratified as it rose until 25 August when it overflowed into the Tailrace adit. The energy of the process caused complete mixing resulting in very homogenous shaft water geochemistry. Peak toxic metal concentrations seen within the shaft occurred on 25 August 1999 with iron, manganese and zinc concentrations as high as 50 mg/l, 130 mg/l and 120 mg/l respectively. It remained mixed until the end of 1999 when it once again became stratified. The current situation where the Tailrace water is more contaminated than the water at the top of the shaft can be explained if the Tailrace adit is blocked at the No.2 Shaft. Historically, water was pumped from lower workings into the adit and allowed to drain through it; however, the Tailrace adit was never a successful gravity drainage feature as it tended to silt up. If this adit entrance is blocked at the No.2 Shaft the water has found an alternative route.

The next route is along the 40 fathom level (340 m aOD) and up the No.1 shaft to the Tailrace Level (see Fig. 3). The water above the 340 m aOD level is then free to stratify on top of the turbulent and mixed water column beneath. A similar pattern of stratified mine water on top of a mixed column has also been observed at Wheal Jane in Cornwall (Adams & Younger, this volume). Hydrochemical evidence supports this theory as the correlation between metal concentration in the shaft and the Tailrace Level is best at the depth 340 m aOD as is shown in the matrix plot below (Fig. 13). The correlation is equally strong for zinc but not for iron suggesting that the relationship for iron is more complicated. As goethite is supersaturated in the shaft water there may, therefore, be more geochemical sinks for iron than for either manganese or zinc. There may also be iron oxides and oxyhydroxide deposits present within the Tailrace Level which could affect the geochemistry of the emerging discharge.

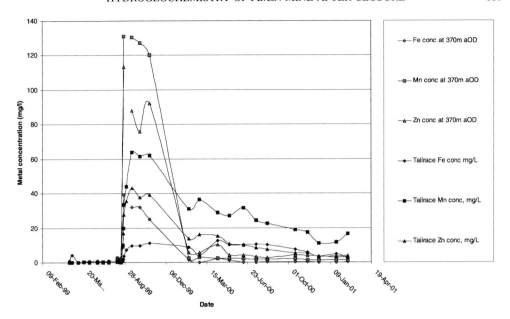

Fig. 9. Metal concentration at 370 m aOD in the No.2 Shaft at Frazer's Grove Mine compared with metal concentration in the Tailrace Level.

Three interesting points arise from the hydrogeochemical data presented in Figs 9–12. They are represented on Fig. 10 by arrows marking out points A, B and C:

(a) Firstly, the data show that between 22nd June and the 7th July 1999 there was an unexpected increase in metal concentration at depth 350 m aOD as highlighted in Fig. 10 by 'A'. The increase in metal concentration can also be seen at later dates throughout the water column at the depths 340 m and 330 m aOD but it occurs first and is most distinct at 350 m aOD suggesting that the source of the increase

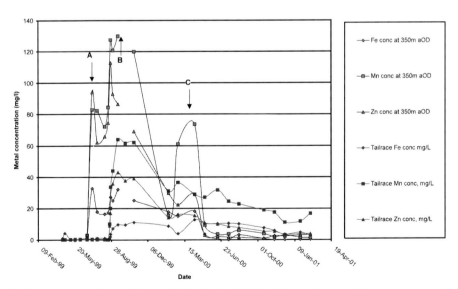

Fig. 10. Metal concentration at 350 m aOD in the No.2 Shaft at Frazer's Grove Mine compared with metal concentration in the Tailrace Level.

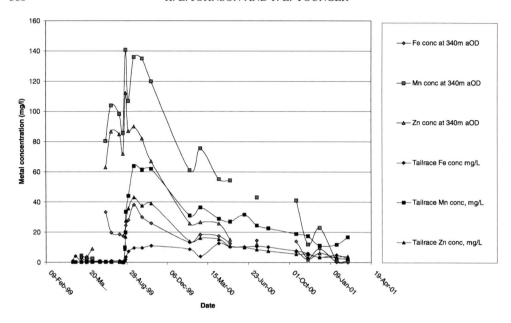

Fig. 11. Metal concentration at 340 m aOD in the No.2 Shaft at Frazer's Grove Mine compared with metal concentration in the Tailrace Level.

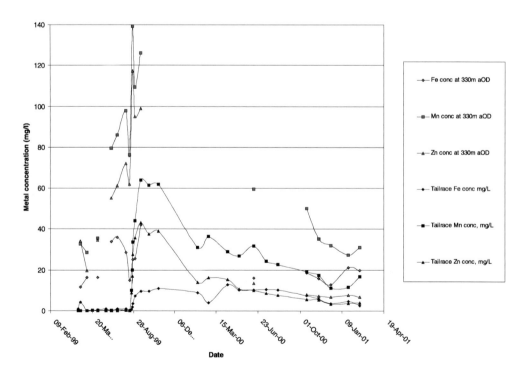

Fig. 12. Metal concentration at 330 m aOD in the No.2 Shaft at Frazer's Grove Mine compared with metal concentration in the Tailrace Level.

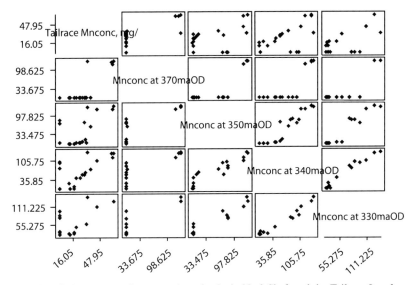

Fig. 13. Matrix plot of Mn concentrations at various depths in No.2 Shaft and the Tailrace Level.

came from this level. The water level during this period was between 355 m aOD and 359 m aOD and Table 1 shows that the strata present around this depth are 'Plate beds' and coal and sandstone beds beneath the Little Limestone. It was not possible to take samples from these beds during underground surveys. A literature survey has also revealed little information on their petrological nature. These beds are likely to contain acid generating minerals as they are probably of a similar nature to the marine cyclothems which are present beneath the Great Limestone. If this is the case then pyritic material present in these beds could offer an explanation for the dramatic decrease in shaft water quality between 22 June and the 7 July 1999.

(b) Secondly, water quality in the shaft deteriorates on 25 August when the water overflows into the Tailrace Level. Water quality in the Tailrace Level shows a peak value in metal concentrations ($\sim$40 mg/l Zn, 62 mg/l Mn, 10 mg/l Fe, conductivity $\sim$1700 μS/cm) on 15 of September and water quality has slowly improved since with concentrations in February 2001 at $\sim$4 mg/l Zn, 15 mg/l Mn, 3 mg/l Fe and conductivity $\sim$1060 μS/cm.

(c) Thirdly, at point C (between 31 January 2000 and 11 April 2000) metal concentration increases dramatically within the shaft between 330–350 m aOD with the clearest increase at the 350 m aOD level suggesting that a further change in the shaft stratification has occurred.

The water quality of the mine water discharge was expected to improve as flushing of flooded workings occurs. Younger (2000b) found that it was possible to relate the total length of this initial period of flushing (initial worst water quality) empirically to the rebound time:

$$t_f = 4t_r$$

where:

t_f is the total length of the first flush period
t_r is the 'rebound period' (ie the time taken for the mine to fill to overflowing)

In this case, the 'rebound period' was approximately five months and therefore the total length of first flush period was estimated to be 20 months. During this time, zinc and manganese concentrations decreased exponentially until they reached a relatively stable value of approximately 4 mg/l zinc in Spring 2001 (Johnson & Younger 2000). Since initial zinc concentrations in the Tailrace effluent were about 40 mg/l, this figure if $\sim$1/10th of the first-flush value (Younger 2000b).

The high zinc concentrations in the rebounding mine water may seem surprising given that the highest zinc concentration recorded during the underground surveys was only $\sim$40 mg/l (Younger 2000a). However, these values were recorded before the groundwater rebounded and dissolved the acid generating minerals present within the mine. Water levels rebounded to 328 m aOD when the pumps failed in 1996 and in this case zinc concentrations reached only 27 mg/l. However, on 25 August 2000 when complete groundwater rebound had occurred, the water level reached 373 m aOD and, therefore, (hydrochemical evidence suggests) passed through not just one but two separate lots of acid generating minerals. With the break-up of stratification at the end of August 1999, zinc concentrations peaked at 120 mg/l. Such high zinc concentrations have never been seen in a North Pennines' mine before. It is presumed that the high zinc concentrations have been caused by dissolution of acid generating minerals in the shales and mudstones both above and below the Great Limestone.

Conclusions

Frazer's Grove has provided an excellent opportunity to study an abandoned limestone-hosted metal mine both before and after abandonment. It has generally been assumed that the environmental impact from limestone-hosted metal mines would be minimal due to the neutralising effect of the limestone. However, although the post-abandonment discharge from Frazer's Grove mine is circum-neutral it had a very high dissolved metal content. The source of the dissolved metals was the acid generating minerals in the shales and mudstones present both above and below the Great Limestone within the mine. The weathered shales may contain a new mineral which has yet to be named and more research must be carried out into the nature of these acid generating minerals. As pyritic shales are often found in association with limestone since both are of marine origin, Frazer's Grove post abandonment hydrochemistry may be representative of other limestone-hosted mines.

Water quality has improved exponentially with time giving more credence to the empirical formula presented by Younger (2000b). This is a powerful tool for the prediction of water quality in a post-abandonment situation for a given mine. It was presumed at Frazer's Grove that mine water quality would be superior by further contact with the limestone strata if the rebounding mine water had travelled through the aquifer to its surface discharge point. However, an important observation arising from Frazer's Grove mine is that mining severely alters the hydrogeology of area. When mine workings have intercepted a confined aquifer, rebounding water is likely to take the easiest route to the surface along the mine workings rather than via the aquifer. This has serious implications for the prediction of both water quality and on where rebounding mine water will emerge.

Observations on the temporal and spatial geochemical nature of mine water within the mine workings have allowed a greater understanding of the process of stratification and of the development of water quality with time in abandoned mines. It is often assumed that water quality at the highest point within the mine for example at the top of a shaft will be representative of the emerging discharge water quality. However, Frazer's Grove mine has confirmed that this is not always the case as water in the mine can be stratified. Surface discharge water quality should be assumed to be closer in nature to the worst quality water in the mine rather than the best. This allows for the development of the most appropriate treatment technology for the resultant discharge and for any changes in its chemical nature. It is interesting to note that the pattern of break up of stratification and restratification at Frazer's during rebound could not have been predicted and that the hydrogeology of the mine is a dynamic process which evolves with time. However, it has highlighted the importance of careful monitoring and sampling which allowed us to ascertain the probable cause of restratification. Such an event may reoccur if the current exit point for the mine water also becomes blocked leading to a further possible change in discharge quality.

We would like to acknowledge the Engineering and Physical Sciences Research Council for funding this work at the University of Newcastle upon Tyne (Award Reference No.98316317). We would also like to thank Paul Allison who owns Frazer's Grove Mine for his continued patience and assistance. Thanks goes also to Dr Adam Jarvis who allowed us to use some of his diagrams (Figs 2 and 3).

References

ADAMCZYK, Z., MOTYKA, J. & WITOWSKI, A.J. 2000. Impact of Zn-Pb ore mining on groundwater quality in the Olkusz region. *Mine Water and the Environment, 7th International Mine Water Association Congress, Ustron, Poland, September 2000.*

ADAMS, R. & YOUNGER, P.L. 2002. A physically based model of rebound in South Crofty tin mine, Cornwall. *This volume.*

BALL, J.W. & NORDSTROM, D.K. 1991. User's Manual for WATEQ4F, with revised thermodynamic database and test cases for calculating speciation of major, trace and redox elements in natural waters. *US Geological Survey Open-file Report*, 91–183.

DUNHAM, K.C., DUNHAM, A.C., HODGE, B.L. & JOHNSON, G.A.L. 1965. Granite beneath the Viséan sediments with mineralisation at Rookhope, Northern Pennines. *Quarterly Journal of the Geological Society, London*, **121**, 383–417.

DUNHAM, K.C. 1990. *Geology of the North Pennine Orefield, Volume One – Tyne to Stainmore*, HMSO, London, 118–150.

GEELHOED, J.S., VAN RIEMSDIJK, W.H. & FINDENEGG, G.R. 1997. Effects of sulphate and pH on the plant-availability of phosphate adsorbed on goethite. *Plant and Soil*, **197**, 241–249.

HAMILTON, Q.U.I., LAMB, H.M., HALLETT, C. & PROCTOR, J.A. 1999. Passive treatment systems for the remediation of acid mine drainage at Wheal Jane, Cornwall. *J.C.I.W.E.M*, **13**, 93–103.

JENNE, E. A. 1967. Controls on Mn, Fe, Co, Ni, Cu and Zn concetrations in soils and water: the significant role of hydrous Mn and Fe oxides in *Trace Inorganics in Water*, Advanced Chemistry Series 73, 1968.

JOHNSON, K.L. & YOUNGER, P.L. 2000. Abandonment of Frazer's Grove Fluorspar Mine, North Pennines, UK: Prediction and observation of water level and chemistry changes after closure. In: *Mine Water and the Environment: 7th International Mine Water Association Congress, Ustron, Poland, 11–15 September 2000.*

REDDY, M.R. & PERKINS, H.F. 1976. Fixation of manganese by clay minerals. *Soil Science*, **121**, 21–24.

YOUNGER, P.L. 2000a. Nature and practical implications of heterogeneities in the geochemistry of zinc-rich, alkaline mine waters in an underground F-Pb mine in the UK. *Applied Geochemistry*, **15**, 1383–1397.

YOUNGER, P.L. 2000b. Predicting temporal changes in total iron concentrations in groundwaters flowing from abandoned deep mines: a first approximation. *Journal of Contaminant Hydrology*, **44**, 47–69.

Paradise lost? Assessment of liabilities at a Uranium mine in the Slovak Republic: Novoveska Huta

DAVID HOLTON, MARTIN KELLY & ADRIAN BAKER

Serco Assurance, 424 Harwell, Didcot Oxfordshire OX11, 0QJ, UK

Abstract: In central and eastern Europe in the 1970s and 1980s, prevailing economic and political conditions resulted in a rapid closure of many uranium mining and processing activities. These closures have left a long-lived human health impact. In countries and regions of relatively sparse economic resources, it is essential to understand the true significance of arising impacts and the financial consequences of their mitigation. A pilot study performed at a former site of uranium mining in the Slovak Republic illustrates a methodology to evaluate human health and environmental impact. The main findings are:

- The former mining site shows complexity typical of an area in which there is diffuse contamination arising from leaching from waste heaps, and uncontrolled discharges from adits into water courses. In particular, significant hazards occur at the sites due to the presence of uranium ore and its progeny at the surface, which may result in radiological exposure via direct irradiation, ingestion and inhalation of dust or radon.
- Site characterization considered both traditional areas of sampling and analysis (rock, soil, dust, radon and water) and identification of those activities and groups or individuals directly or potentially affected by exposure to contamination at the site. These ranged from workers occupying offices and workshops on one of the waste rock heaps, to house builders using waste rock (and potentially ore) for construction purposes, to a range of people exposed due to their recreational activities at the sites (hikers, mineral collectors, rock climbers and gatherers of wild produce).
- Historical recultivation measures performed at the site in the 1980s were generally ineffective at curtailing the whole range of radiological hazards. Measures were taken at most sites to bury exposed ore to minimize external irradiation. However, in those cases in which recultivation of heaps was successful, it did little to reduce the impact of radon emanation. Instead, recultivation appeared to have the surprising consequence of reducing the potential dose to the public, via an unrelated route, by making waste rock/ore more difficult to remove for the purpose of construction.
- When the importance of all the hazards were ranked, the most significant risk factor arose from inhalation of radon emanating from foundations built from waste rock material.
- Of all of the liabilities, a partially water-filled waste rock pit resulted in the highest dose rates. When time spent at this relatively remote site was taken into account, the potential doses received remained to be comparatively high. The most at risk groups were those working, in buildings, on the waste rock heaps and those people who have removed waste rock/ore for building purposes.
- Mitigation measures to reduce doses experienced by the exposed groups can be summarized as:
    - prevention of the use of waste rock material for the purpose of house building;
    - reducing the overall accessibility to the sites, using barriers;
    - restricting the recreational value of the sites, by placement of warning signs/fencing (short term);
    - relocation of offices and laboratories on the heaps or improvement of the overall ventilation of the working areas.

Uranium mining has taken place in many different ways and has typically resulted in the deposition of radiologically contaminated materials at, or near to, the ground surface. At most redundant uranium mining sites some remediation of the contamination has been attempted, with varying degrees of success. The work described here investigates the environmental consequences, in particular the evaluation of risk to human health following remediation performed in the 1980s at Novoveska Huta, located in central–eastern Slovakia. The Novoveska Huta region is of national importance because it is located in the so-called Slovakian 'Paradise' National Park (Slowakisches Paradies), a region of significant natural beauty. In addition to the recreational uses of the area, waters from the site drain into the River Hornad, the primary source of drinking water for the region. This paper describes the work performed as part of a pilot project on behalf of the European Commission between 1998 and 2000. The work develops a data collection methodology and a generalized environment impact assessment (EIA) to quantify the risk arising from a range of liabilities at the site.

Geological and historical context

In excess of 40 uranium occurrences within seven districts have been identified in Slovakia, all of them within the West Carpathians. All known uranium occurrences were subeconomic, except for the Novoveska Huta area located in the Spiš-Germer Ore Mountains located in central–eastern Slovakia, part of the northern segment of the Slovakian Ore Mountains (Sloneské Rudehorie) (Fig. 1). Three zones of uranium mineralization have been distinguished by Arapov (1984), from north to south: namely the North Germer, Central Germer and South Germer zones. Significant uranium deposits are restricted to the North Germer zone.

The geology at the Novoveska Huta site consists primarily of Early and Late Palaeozoic meta-sedimentary and meta-volcanic rocks. Superficial deposits (consisting of sand and gravels) are local in extent, primarily in river valleys, with a mean thickness of approximately 3 m. The uranium deposits of Novoveska Huta are located within meta-sedimentary rocks of Permian age. These are subdivided into three litho-stratigraphic formations:

- Knola – comprises terrigenous material, manly conglomerate, sandstone and aleurite. The formation is subdivided into Muran Conglomerate and Markusovce Sandstone. The veins of copper sulphides and rare uranium minerals are developed in Markusovce Sandstone;
- Pterova – comprises meta-volcanic rocks, tuff and agglomerate;
- Novoveska Huta – comprises terrigenous clastic sedimentary rocks and sandstone with aleurite intercalations. Thick layers of variegated shale containing variable quantities of gypsum and anhydrite are typical of the upper part of the formation.

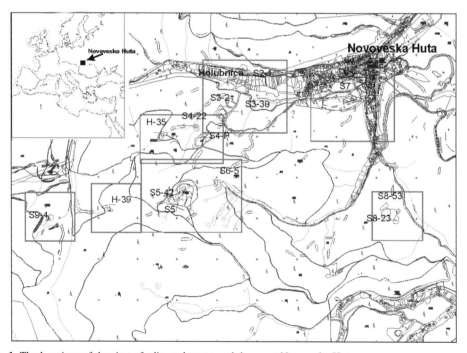

Fig. 1. The locations of the sites of adits and waste rock heaps at Novoveska Huta.

At Novoveska Huta, bannerite, U–Ti oxides and tuyamunite are the primary uranium minerals.

Mining in the Novoveska Huta region, located in the North Germer Zone, goes back to the 1630s, although archaeological finds indicate that even these dates may not be the earliest evidence of mining activities. Copper exploitation in the area finally ended in 1896 with a total production of copper ore in excess of 500 000 t. Systematic exploration for uranium started in Slovakia, then a territory of the CSSR, in 1952. It was intensified after 1964 after discovery of significant uranium mineralization in former copper deposits at Novoveska Huta. Ore recovery was largely by underground workings and a few openpit mines. Recent mining activities in the area have been restricted to gypsum recovery.

Uranium production was restricted to deposits in the Novoveska Huta–Muran–Hnilcik area. The village of Novoveska Huta is located some 5 km SW of the town of Spišská Nová Ves (SNV). Four small deposits occur in the vicinity of SNV on the northern slopes of the Maly Muran Mountain. They have yielded several hundred tonnes of uranium from low-grade ore. Mining grades were less than 0.1% uranium. Deposits in the vicinity of Novoveska Huta are located in a group about 10 km long and 3.5 km wide. There are various principle types of mineralizations at the site, which have been exploited either by underground exploration or shallow pits. In particular, stratiform (uranium–molybdenum) U–Mo mineralization appears in discontinuous lenses in two horizons. The lower horizon extends for 4 km in length, is 60–200 m wide and is as much as 80 m thick. Mineralized lenses are from several metres to tens of metres thick, cover from tens to tens of thousands of square metres, and average 0.06% U. In the upper horizon, ore lenses are up to several metres thick, cover between tens and hundreds of square metres, and contain an average of about 0.08% U. In addition to these types of mineralization, structure-hosted U-mineralization occurs in the form of 0.1–2 m wide lodes with stockwork structure in faults and in cleavage-controlled ore shoots within stratiform ore. Grades attain higher values in excess of 0.1% U.

Hydrogeological setting

The Novoveska Huta mining region is located in the regional basin of the Hornad River. The Novoveska Huta site drains through a series of streams and water discharging from old copper and uranium mine adits into the Holubnica River. The total Holubica catchment is of the order of 32 km^2, with an annual mean discharge of 340 l s^{-1} into the Hornad. Groundwater flow broadly follows the topography of the river valley. However, the groundwater flow pattern is significantly disrupted by the numerous mine adits in the region (each acting as a local drain of the rock mass).

There are three significant water abstractions. First, a series of springs located on the western edge of the uranium mining region of Novoveska Huta and piped to Skišska Nová Ves where it is used for drinking water, the average abstraction rate being 8.2 l s^{-1}. The spring flow is correlated to rainfall and varies between 3.6 and 12.7 l s^{-1}; however, it is located on the line of the most significant fault system in the region and it is possible some of the base flow is provided from depth. Secondly, water is abstracted from the Holubnica Stream for public drinking water (average abstraction of 5 l s^{-1}). Thirdly, adit #35 is used by Uranpres both for industrial processes and for drinking water. The discharging water is captured and piped to heap S7 where the workshops and storage buildings of Uranpres are located (adit flow rate of 2.5 l s^{-1}).

Historically, the overall water chemistry sampling in the Novoveska Huta mining region has focused on the concentrations of copper, uranium and ^{226}Ra. Before 1982, low concentrations (maximum 0.08 mg l^{-1} for Cu and U, and 0.3 Bq l^{-1} for ^{226}Ra) were measured (not at the above public supply). After 1984, concentrations increased (Cu, 0.3–0.6 mg l^{-1}; U, 0.02–0.04 mg l^{-1}; ^{226}Ra, 0.3–16 Bq l^{-1}). These concentrations exceeded limits specified by regulations for drinking and surface water in Slovakia.

Liabilities and remediation measures

Uranium ore recovery at Novoveska Huta was largely by underground workings and a few openpit mines seeking to exploit the zones of significant ore deposits described above. With the fall in the price of uranium ore below subeconomic levels in the 1970s, came the end of many uranium mining activities worldwide, especially in central and eastern Europe. This, combined with the relatively low-grade ore, resulted in the closure of mining for uranium at Novoveska Huta in the late 1970s. Shortly after cessation of mining activities at the site, a series of revitalization measures were implemented largely related to the treatment of waste ore and rock heaps. The types of action undertaken at the larger waste rock sites can be summarized into broad safety-related categories:

- disassembly and demolition of structures associated with former mining activities;

- cover of shaft entrances with concrete slabs;
- explosives destruction of adit collars and cover of entrances by bulk material;
- mechanical reshaping of heaps;
- liquidation of ventilation shafts (capping);
- removal of exposed low-grade ore to burial in local heaps followed by covering with waste rock cap;
- soil cover to most heaps and revegetation in some cases;
- dewatering of adits by pipelines;
- filling of open pits;
- stabilization of pit walls;
- provision of fencing and warning notices to control access to dangerous areas, e.g. pits with overhanging walls.

The remediation activities performed in the early 1980s, covering an overall area of approximately 7 by 3 km, were focused largely on short-term site restoration and safety considerations. There was some evidence that the measures were taken within a radiological hazard assessment, if not a broader 'environmental impact assessment' (EIA) context. However, in addition to the safety-related activities described above, extensive and very detailed radiological surveys were performed, which afforded the ability to compare with the current distribution of radiological contamination and how this may have changed over a period of almost 20 years.

In addition to the basic systematic approach adopted for the remediation of the larger waste rock sites, some variation exists at the large open pit at Muran (site S5) and the largest heap, S7. The difference between the standard remediation methods and those at these two sites are:

- the open pit Muran has not been refilled and differs from other sites in that it is large, steep sided and permanently contains a large pond at its base, which is fed by several adit discharges;
- at heap S7, extra safety measures were performed: a local deposit of high-grade uranium ore was located, wrapped in plastic foil, and covered by a cap of waste rock soil and vegetation.

There is also anecdotal information suggesting that uranium ore, rather than waste rock, may have on occasion been removed from the Muran pit for road construction.

The types of action undertaken at the smaller waste rock sites can be summarized into three categories:

- heaps left without attention;
- heaps shaped and levelled mechanically;
- heaps shaped levelled mechanically, then covered by a cap consisting of waste rock, soil and vegetation.

In total, the closure resulted in the creation of a variety of liabilities, including two mine shafts, 27 adits, 17 gin pits and one openpit. Related to these liabilities are numerous waste rock heaps. Some of these contain uranium ore, whereas others do not, but nevertheless can be contaminated by significant amounts of uranium and its radioactive progeny. The major sites identified at Novoveska Huta are groups of liabilities that contain areas of waste rock heaps, associated adits, pits and shafts. The most significant areas are indicated in Fig. 1.

A summary of the key sites at Novoveska Huta containing a significant collection of liabilities is given in Table 1. This map illustrates the close proximity of some of the liabilities to the village of Novoveska Huta and the Holubnica River, which forms part of the river Hornad catchment.

Methodology: tiered data collection

In 1998, a European Commission project entitled 'Efficiency of Former Revitalization after Uranium Mining – Slovakia', was initiated. The overall aim of the project was to establish the environmental impact of the former uranium ore exploitation at the Novoveska Huta site, and consequently to establish a view of the effectiveness of historical remedial actions.

The data collection and evaluation methodology was designed principally to establish the likely pathways by which contamination may interact with key groups and potentially cause a risk to health. Typically, these included information covering surface water, groundwater, dust migration, radon migration, direct dose, ingestion of soil and ingestion of wild crops that have been in direct contact with the naturally occurring radioactive material (NORM).

A large number of historical reports exist (approximately 1200) containing data relating to mining operations at Novoveska Huta that have helped to establish the overall picture of the site and the associated liabilities. The project required not just general information related to former mining operations, but heap- and adit-specific information, including shape and stability of the waste rock heaps, geological and hydrogeological information in the heap's environs, radiological survey information (dose rate measurements), soil and rock analyses, stream flow rates, and surface and groundwater chemistry in the immediate vicinity of the heap, data related to radon exhalation, ecological information, land

Table 1. *Illustration of the major sites of significant groupings of liabilities, indicating waste rock dimensions (areal extent multiplied by range of thickness) and the presence of associated adits, pits and shafts*

Site identifier	Waste rock heap identifier and dimensions (length × width × thickness (m))	Adits	Pits (P#) and shafts
S2	S2-1 (160 × 150 × 10–50)	#5, #52	P#61, Shaft-1
S3	S3-2 (100 × 100 × 5)	#2, #30	P#32, P#35
	S3-30 (160 × 10 × 10–40)		P#27, P#31
S4	S4-22 (50 × 30 × 1–10)	#22	P#25, P#26
S5	S5-H1 (30 × 17 × 1–3)	#3, #3a, #3b, #3c, #3d, #42	Muran open pit
	S5-H2 (50 × 17 × 2–4)		
	S5-H3 (70 × 40 × 5–10)		
	S5-H4 (160 × 60 × 10–30)		
	S5-H42 (70 × 30 × 3–10)		
	S5-H5 (200 × 100 × 10–30)		
	S5-H6 (30 × 30 × 2–10)		
S6	S6-5 (50 × 10 × 1–3)	#5	P#34
S7	S7 (600 × 250 × 10–50)		Shaft-3
S8	S8-23 (90 × 50 × 5–30)	#23, #53	
	S8-53 (50 × 30 × 5–20)		
S9	S9-4 (60 × 70 × 3–20)	#4	

use, establishing human behaviour in the vicinity of the site and identification of major water abstractions in the region.

As there were potentially so many liabilities at the site, a preliminary data collection strategy was adopted by performing a screening study covering all the sites with basic information derived from a walkover survey, with 17 features recorded including details of soil and vegetation cover. Soil and rock samples were collected from all sites in an initial screening exercise with 12 water samples from adit discharges and streams. The rock and soil samples were analysed for long-lived radionuclides (U-natural, ^{226}Ra, ^{232}Th, ^{40}K) and a suite of toxic metals (arsenic, boron, cadmium, chromium, copper, lead, mercury, nickel and zinc). Water samples taken from adit discharges and streams were analysed for radionuclides in the uranium decay chain (U-natural, ^{226}Ra, ^{222}Rn) for a suite of toxic metals and for additional components (sulphate, sulphide, total alkalinity, pH, electrical conductivity, carbon dioxide, ammonium ion, calcium, magnesium, chlorine anion, iron and manganese).

The overall methodology for data collection was to collect baseline information for all uranium mine liabilities and then to select three sites for more detailed examination. The original methodology assumed that 'typical' sites would be chosen for more detailed investigation. In practice, there were no completely typical sites in the region, each one having its own idiosyncrasies. Instead the sites were chosen based on the likelihood that a contaminant presented a health risk. The approach used is basic to an environmental risk assessment: the presence of a *source* of contamination, a *pathway* through which contaminant may migrate and a *receptor* that may be affected by the contamination.

Data used for the source evaluation were dependent on the comprehensive radiation maps produced in the mid-1980s post-remediation, the radiation measurements during the walkover survey, recorded and anecdotal history of the site, and results of soil and water analyses. The initial pathway analysis took data from the walkover survey, hydrogeological and hydrological information, and climatic information. The receptor's information was drawn from the walkover survey reports, which enabled an evaluation of the time people (or critical groups) spent on a site, people spending time in the vicinity of the site and people within the water supply catchment. In addition, land-use maps and water supply information were considered. Once these various risk factors were taken into account, it was decided that there were three sites most likely to potentially pose a health risk to the public or workers. These were the Muran pit (site S5), site S7, and the closely grouped sites of S2 and S3. In particular, radiation maps and subsequent follow-up radiation measurements identified some areas of the pit in which very high doses occur, which were not indicated in the original radiation maps.

Site descriptions

The focus of the remainder of this paper will be used to describe these key sites in greater detail, and to explain the data and results of

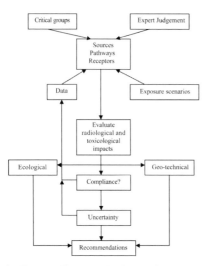

Fig. 2. The overall approach adopted for the environmental impact assessment.

the structured environmental impact assessment (EIA). The EIA approach is illustrated in Fig. 2. Its overall philosophical aims are to focus the data collection on the identification and quantification of sources of contamination, key pathways, to evaluate the consequences of human behaviour at the site and to evaluate the overall risk to key critical groups. In addition to the evaluation of radiological and toxicological impacts, the approach used includes consideration of geotechnical hazards, such as slope stability of waste rock heaps and ecological factors. A summary of the averaged measurements at the key sites is given in Table 2. This comprises the aggregate and average of many measurements recorded at each site for the most important contaminants. Compliance, in the radiological context, refers to the exposure limit for the public of 1 mSv year^{-1}.

Table 2. *Tabulation of averaged values of uranium concentration at each of the sampling sites or in the case of small number of samples taken the largest value was assumed. The bracketed number indicates the number of measurements collected*

Liability	Uranium concentration in rock (Bq/kg^{-1})	Uranium concentration in water (Bq/l^{-1})
S2-1	686 (9)	0.499 (2)
S3-2	524 (3)	4.87 (1) adit #2
S3-30	986 (2)	0.0998 (1) adit #30
S5	2650 (9)	0.998 (5)
S7	112 (4)	0.0749 (8)

Site S2–S3

Spoil heap S2-1 had been recultivated by grass and trees (silver spruce and larch). After 20 years of growth the spruce are typically 1 m high and the larch 2 m. The spoil heap had been formed from waste rock from shaft #1 and adit #52. A wood storage area is located on the flat top, and there are scree sides located to the western, northern and southern edges, with slopes approximately 40–50°. Part of the heap is located in the Holubnica Valley in direct contact with the river. It is located approximately 1 km from Novoveska Huta with a solid dusty road for trucks and tractors used by forest workers and locals. There is a residence directly downslope of the heap. Adit #52 had formerly been discharging but no water has now been recorded for some years. An area of marshy ground is located downslope of the heap.

Spoil heap S3-2 located upgradient of S2-1 has similar characteristics to S2-1, except that an active outflow from adit #2 infiltrates through the heap, which has a marshy area at its base. Heap S3-30 was created from adit #30, from which water is now discharging. Significant scree is present, and large boulders have been observed sliding down into the forest. An illustration of the overall instability of these slopes is illustrated in Fig. 3, which shows evidence of rockslides from the S3-30 waste rock heap. All sites show evidence of being visited by forest workers, campers, mineral and mushroom collectors, with tracks crossing all heaps arising from both human activity and by wild pigs. There is evidence that some waste rock material has been used for the construction of roads in the near vicinity.

Fig. 3. Rock slides in the vicinity of site S3-30, showing evidence of erosion of former recultivated waste rock heap, illustrating the possibility of exposing material of higher activity.

The site was selected because the heaps (S2-1, S3-3 and S3-30) are similar and close together, with a significant amount of interaction between pathways and receptors. The survey indicated that there were intermediate levels of radiation at the site. The water flowing from adits was contaminated and adit water flowing through some of the heaps may have been leaching uranium salts. The main pathway to the environment would be direct dose, dust transport, surface and groundwater transport. The receptors arise because of the number of tourist trails crossing the site used by locals or campers. In addition, site S2-1 is the place of work for a few forestry workers, and there is the possibility of adit water being consumed by hikers.

Site S5

Area S5 comprises a series of waste rock heaps, a series of adits and a flooded waste pit. Most of the waste rock heaps at this site consist of loose rock material with very little soil. As a consequence of the impoverished cover material, the spoil heaps are very poorly vegetated, the trees being generally yellow in colour and clearly underdeveloped. The most important liability in this area is a flooded waste pit (water depth of 4–5 m), which now forms a 'natural' valley fed by several adit discharges.

High levels of radiation were recorded at the base of the pit due to boulders and fragments of uranium ore, registering levels greater than 3000 counts s^{-1}.

A detailed radiation map of each area at Novoveska Huta was established. An example of such mapping is shown in Fig. 4 for the environs of the Muran pit. It should be noted that the pit area itself is located at the unmapped centre of the large area and was not systematically measured due to the steep sides and general inaccessibility of the pit area.

There are tracks in the vicinity of the Muran pit used by forestry workers, and mushroom and raspberry collectors. A scout hide is located on the cliff overlooking the pit. In addition, there is a series of ropes, used by climbers, to the pit floor. Some of the pipes dewatering the adits are either broken or in a very poor state of repair. Some relics of pit fencing are evident. However, it is either in a very poor state of repair or almost completely inadequate. The northern margin of the cliff is very dangerous, with forest extending to the edge of the cliff margin.

This site is of concern because radiation mapping shows high radiation levels and has an exposed ore body as well as buried ore within

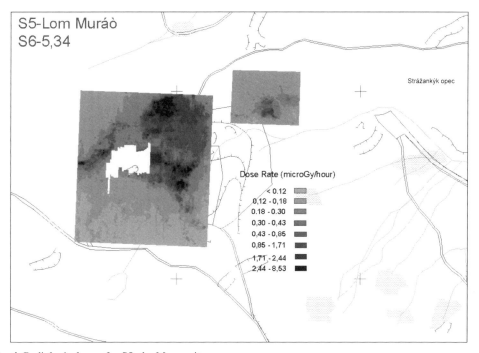

Fig. 4. Radiological map for S5, the Muran pit area.

heaps round the site. Similar exposure pathways to those identified in S3 would be relevant here. The site is a natural focus for tourists, visited by climbers, scouts, walkers and mushroom collectors. Access to the site, via dirt roads, is good. Water discharging from the site is likely to be consumed by walkers. In addition, there was anecdotal evidence that locals had wanted to use the water from the pit for the purpose of producing snow for ski runs as part of tourist development for the area!

Site S7

Site S7 is the largest of the recultivated waste heaps with a range of buildings located at the top. The slopes of the heap are very steep (approximately 55°), with evidence of minor rockslides. The site is accessible by road, however the area is guarded and fenced by Uranpres security. Uranpres is the Slovak organization formerly with the responsibility for the mining of uranium. The fencing on the western and northern side is partially damaged with evidence of part of the western margin of the heap having recently been excavated (possibly used for house and road construction). The buildings on the site range from the gatehouse, which is permanently occupied by a security guard, to Uranpres laboratory buildings, used to store uranium-ore samples, to office buildings for the sawmill, which is located on S7.

The outdoor storage and sawmill is located directly on a concrete slab that covers an area in which uranium ore was disposed of after closure. At the western edge of the site, located at the top of the heap, there is an apple orchard covering approximately 100 × 100 m that is also the home of a herd of sheep. Uranpres staff consumes both the apples and the lamb. In all, there are approximately 50 people spending up to 8 h day^{-1} at the site. There is no direct discharge of water from the heap; however, there are two adits dewatering the interconnecting mine system below the site. There is an array of water wells located in Novoveska Huta directly below site S7. Much of this is used as household water (washing, garden watering and drinking water for domestic animals). Water used on the heap for drinking is derived from adit #35 (connected to the copper mine) and is regularly monitored.

The radiation maps at site S7 show intermediate doses, one area of uranium ore is known to have been buried, and leaching of uranium salts could give rise to groundwater contamination. The overall pathways are the same as the other sites; however, this site is the place of work for a number of people and is very close to the village of Novoveska Huta, home to 800 people. The access to the site is good and water wells are located within a few hundred metres of the site.

Environmental impact assessment

The mathematical basis of the EIA will not be described in detail. However, the equations for evaluating radiological impacts can conveniently be characterized under the following headings:

- external irradiation;
- inhalation of ^{222}Rn and its progeny;
- inhalation of dust;
- ingestion of soils and dust;
- ingestion of wild and cultivated plants;
- ingestion of animal products;
- ingestion of drinking water.

The external irradiation is calculated on the basis of the measured dose rate at the sites multiplied by the time spent at the site.

Exposures to ^{222}Rn and its short-lived daughters are likely to be of greatest significance in enclosed spaces such as adits, buildings and domestic dwellings. Radon is very significant in the context of uranium mining, and various special quantities and units have been adopted for characterizing concentrations of short-lived daughters of radon in air and the resulting inhalation exposures. This information is summarized in ICRP (1993).

Contaminated materials containing uranium-series radionuclides may be present as an aerosol of dust particles and subject to inhalation. This assessment of the annual effective dose due to inhalation of radioactive aerosols requires consideration of the mass loading of radioactively contaminated material in air; the radionuclide content of the resuspended contaminated material; the breathing rate of the exposed individual, the period during the year for which the individual, is exposed; and the committed effective dose per unit intake value for the radionuclide.

Accidental ingestion of contaminated dust by members of the public may occur. The relevant literature on rates of soil consumption is given in Simon (1998). The annual effective dose from ingestion of contaminated dust is derived from the rate of consumption of contaminated dust, averaged over the year of interest; the average concentration of the radionuclide in the material ingested, estimated from the concentration in bulk material inferred from the results of the gamma dose rate survey; and the committed effective dose per unit intake value for the radionuclide.

The ingestion of wild and cultivated plants is calculated in much the same way as calculated above, i.e. the annual effective dose from ingestion of a specific contaminated plant product is derived from the rate of consumption of that contaminated foodstuff, averaged over the year of interest; the average concentration of the radionuclide in the foodstuff, as ingested; and the committed effective dose per unit intake value for the radionuclide.

The basic equations governing the ingestion of animal products are basically the same as those for wild and cultivated crops.

In the current study, drinking waters may be abstracted from wells, the nearby river or streams, or from surrounding adits. In all three cases, the basic formulation to calculate the annual effective dose from ingestion of a specific contaminated water is derived from the rate of consumption of water, averaged over the year of interest; the average concentration of the radionuclide in the ingested waters; and the committed effective dose per unit intake value for the radionuclide.

Concentration data for radionuclides

Concentration data are typical values based on the gamma survey measurements and sample assays. An example tabulation of results is given in Table 2 and some simple illustrations of the consequences are described below.

The elevated levels of uranium measured at site S5, shown in Table 2, result in proportionately higher levels of external irradiation and, hence, individual dose at the site. For example, annual individual dose from external irradiation at site S7 will be approximately 20 times less than S5. Also, this is true of the users of waste rock material removed for the purpose of construction. In an analogous way the high levels of uranium in water at site S3-2 translate to higher levels of dose for groups, such as hikers and campers, drinking water from adit #2.

Outdoor ^{222}Rn concentrations are low at all the liabilities except for the Muran pit, S5. At S5, concentrations at the base of the pit can be high, a concentration of 6.8×10^4 Bq m^{-3} was measured. This was the highest concentration measured.

Indoor ^{222}Rn concentrations are only relevant to S7, where office and storage buildings are located on the heap. An indoor concentration of 1.05×10^3 Bq m^{-3} was adopted. This was the highest concentration measured indoors.

Occupancy times

Annual occupancy times for the various exposure or critical groups can be estimated, based on accepted generic times or and understanding of working or leisure patterns. The critical groups considered are: 'house builders', namely those who have used waste rock material for the construction of their property; 'inside workers', who work in buildings on the large S7 waste rock heap; 'outside workers', who may typically spend significant time on any of the heaps such as the forest workers; 'campers', who may come in continuous contact with heaps for relatively short times; 'hikers', who walk the hills in the region; 'mineral collectors', who may collect various ore samples; 'swimmers', who may bath at the Muran pit; 'rock climbers', who may spend some time climbing in the pit and vicinity; and other groups who may consume mushrooms (pickers), apples and lamb.

Note that these occupancy times apply only to the computation of external exposure, dust inhalation and exposure to ^{222}Rn and its progeny. It should be noted that some critical groups are defined only for a single liability. For example, inside workers are found only in buildings located on top of heap S7.

The value for house builders actually relates to house occupancy and assumes that a house could be constructed in the vicinity of, or using materials from, any of the liabilities. House occupancy is taken as 100%. This is slightly cautious for most occupants, for whom a value of 70–95% would be more appropriate. However, it is noted that some people do spend almost 100% of their time indoors at home, for example for health or religious reasons.

A standard working year of 2000 h is taken for outside workers at all liabilities, as there is nothing to preclude full-time work in the vicinity of a liability, for example for those working in forestry.

Campers are taken to spend 1 week year^{-1} in the vicinity of a liability, and hikers are taken to make five 1 h visits year^{-1}, assuming that the liability lies on the route of one of their preferred walks. Mineral collectors are assumed to visit the area of the liability several times each week in good weather, collecting specimens for themselves and others. The high occupancy is justified by consideration of the accessibility of the fragmented waste material and the potential for interesting minerals to be identified.

Swimmers are associated only with the lake at the base of the pit at S5. The occupancy is based on daily 1 h visits throughout the summer months. Similarly, rock climbers are associated only with the pit walls at S5. The various waste heaps do not provide an interesting challenge, whereas the pit walls constitute a useful training and practice location.

Mushroom pickers also cover other forms of regular recreation to a single area, for example dog walkers. No discrimination is made between liabilities.

An apple orchard is present only on S7. In its present state, it would not require much effort for picking. The occupancy adopted includes a recreational component.

It is emphasized that the occupancies and other aspects of behaviour adopted are not intended to include all possibilities. Rather, they are selected to illustrate the range of radiological and toxicological impacts that may arise and to assess the degree to which different pathways are emphasized by alternative assumptions concerning human behaviour.

For the purpose of calculation, reasonable water and foodstuff consumption rates are assumed consistent with data given in ICRP (1975). Daily consumption rates of dust for house occupiers were set at the cautious value of 200 mg. For the buildings a ventilation rate was taken as $5.8 \times 10^{-4} \, s^{-1}$, corresponding to a mean air residence time of just under 30 min. In order to compute radon inhalation rates, a standard breathing rate of $1.2 \, m^3 \, h^{-1}$ was adopted, corresponding to light activity. Intake to effective dose factors was taken from ICRP (1996), adopting the largest of the available factors in each case.

Results

The highest individual doses are received by those people who remove heap material and use it for the construction of buildings that will be occupied on a regular basis. The principal exposure pathways are inhalation of radon emanating from the foundations, and external irradiation. These doses range from $2 \, mSv \, year^{-1}$ for heap S7, through to $50 \, mSv \, year^{-1}$ for heap S5.

Very high doses may result to those critical groups exposed to the radon trapped at the bottom of the open pit at S5. Hikers, who are taken to spend just $5 \, h \, year^{-1}$ in the pit, are assessed to receive a dose that exceeds the public dose limit of $1 \, mSv \, year^{-1}$ (ICRP 1990) by a factor of 2.

For the heaps at S5, outside workers, house builders and mushroom pickers are assessed to receive individual doses in excess of $1 \, mSv \, year^{-1}$. For the outside workers, this is $2.75 \, mSv \, year^{-1}$, for house builders it is $34.7 \, mSv \, year^{-1}$, and for mushroom pickers it is $3.16 \, mSv \, year^{-1}$. The individual dose for outside workers is based on occupancy of 2000 h per year. As site S5 is relatively remote, this occupancy can be considered to be very pessimistic, and as such individual doses to outside workers may be considerably less than $2.75 \, mSv \, year^{-1}$. For the house builders, the individual dose is considerably higher than the dose limit of $1 \, mSv \, year^{-1}$. In this case, no substantial pessimistic assumptions were made. Therefore, action is required to prevent any use of waste rock heap material from S5 for housebuilding purposes.

The individual doses received by mushroom pickers exceed the public dose limit by a factor of about 3. The components of this dose are split approximately in the ratio 2:1 between dose received by ingestion of the mushrooms and the dose received in picking the mushrooms.

The occupants of the buildings on top of heap S7 are assessed to receive substantial doses due to the inhalation of radon that is seeping into these buildings through the foundations. The dose received could be over $10 \, mSv \, year^{-1}$. In arriving at this value, measured radon concentrations were used, and occupancy of $2000 \, h \, year^{-1}$ was assumed. This is not considered to be unduly pessimistic.

On heap S3-2, outside workers are assessed to receive individual doses that are in excess of $1 \, mSv \, year^{-1}$, as are mushroom pickers. Mushroom pickers are also assessed to receive doses in excess of $1 \, mSv \, year^{-1}$ on heap S3-30. As for heap S5, it is likely that the mushrooms will be consumed by more than one person, thus reducing individual doses by a proportionate factor.

The EIA for campers shows that doses are well below the public dose limit of $1 \, mSv \, year^{-1}$, provided that the campers are not located in the open pit at S5, which because of accessibility is unlikely. The EIA for hikers shows that doses are well below the public dose limit of $1 \, mSv \, year^{-1}$, provided that the hikers do not spend in excess of about $2 \, h \, year^{-1}$ in the open pit at S5.

The EIA for those members of the public who pick apples from the orchard located on S7 shows that doses to those people are well below the public dose limit of $1 \, mSv \, year^{-1}$. The same conclusion is obtained for people who consume lamb meat reared on S7. It should be noted that very high consumption rates were chosen for these two groups; in practice, individual doses received will probably be much lower than those obtained in the EIA ($0.3 \, mSv \, year^{-1}$ for apple pickers and eaters, and $0.3 \, mSv \, year^{-1}$ for lamb eaters).

For the rock climbers and swimmers located in the pit at S5, the preliminary EIA shows that doses could be as high as $80 \, mSv \, year^{-1}$ if these groups are exposed to the radon at the base of the open pit. Individual doses arising from exposure pathways other than the radon pathway are well below the public dose limit (assessed doses are about $0.5 \, mSv \, year^{-1}$ for both rock climbers and swimmers).

The EIA for mineral prospectors shows that individual doses could exceed the public dose limit on heaps S3-2 and S5, although in calculating the dose received at S5 it is assumed that the prospectors receive a dose from the radon trapped at the bottom of the pit and that the prospectors spend $200\,h\,year^{-1}$ at the heaps. The assessed doses for mineral prospectors at the other heaps are well below the public dose limit.

Recommendations to minimize health risks

On the basis of the calculations undertaken and the conclusions listed above, the following recommendations are made in respect of the three areas considered. In arriving at the conclusions, only the most significant hazards (all radiological) have been considered at each site.

Overall, the most significant hazards arise from radon and its inhalation. Consumption of contaminated water, dust and various foodstuffs were all of secondary importance with the exception of adit water located at S3-2.

For heap S2-1, locals should be prevented from using heap materials for building or constructing houses or other buildings to be occupied for significant periods. This may be achieved by ensuring that the amount of loose material is minimized. This, in turn, may be achieved by ensuring that there is additional recultivation, especially in those areas in which the historic revitalization has been unsuccessful, largely at the margins of the waste heaps. Therefore, the focus of any reshaping and recultivation should be in these areas. Reducing the accessibility of the heap by road is an additional measure that would reduce the possibility of removing waste rock material from the heaps. This could involve the placement of barriers to prevent casual access.

Additional measures for preventing access to the heap are the placement of warning signs and erection of fencing, but these can be considered only to be short-term measures – the futility of this has already been demonstrated with the fencing and warning signs at the open pit.

For heap S3-2, locals should, again, be prevented from using heap materials for building or constructing houses or other buildings to be occupied for significant periods. As such, the comments regarding reshaping and recultivation described above also apply to heap S3-2. The area on and around heap S3-2 shows considerable evidence of forestry work, so gaining access to heap S3-2 via paths and roadways constructed for forestry purposes to remove heap materials in bulk will not be difficult. Additional short-term measures include fencing and appropriate warning signs.

An additional hazard at S3-2 is the high level of uranium in waters emanating from adit #2 (around $0.4\,mg\,l^{-1}$). As these waters flow in sufficient quantities to act as a source of drinking water for workers, campers, hikers, etc., remedial action will be required to prevent consumption of this water. This could take the form of covering the source, so that it is no longer visible to those requiring a source of water.

For heap S3-30, locals should, once more, be prevented from using heap materials for building or constructing houses or other buildings to be occupied for significant periods. As for heaps S2-1 and S3-2, reshaping and recultivation of the heap is desirable, but it can be seen from Fig. 3 that the steep slopes on the sides of S3-30 may make this course of action very expensive.

The principal radiological hazards at S5 arise from the very high radon concentrations in air at the bottom of the pit. This arises from exposed uranium ore at the floor and walls of the pit, along with the fact that the air is stationary at the bottom of the pit. In this context it is noted that increased ventilation at the bottom of the pit is not a practical solution to this problem. An additional source of radon in the pit is loose waste and ore material that could be covered, but this is unlikely to reduce radon concentrations to the required degree.

In carrying out the assessment of radiological doses to this radon, an extreme case of $2000\,h$ occupation per year was assumed. This led to doses of nearly $1\,Sv\,year^{-1}$, (1000 times over recommended limits), although it is not expected that this level of occupancy would ever be achieved by a single individual. In order to reach satisfactory levels of exposure, this value needs to be reduced by about three orders of magnitude. This requires a reduction of occupancy times to $2\,h$ or less per year, a value that may be impossible to enforce.

Of course, the ideal solution would be to fill the pit in completely. However, the size of the pit precludes this option on financial grounds, and so other options must be sought. As for the other liabilities, signage and fencing will provide a degree of protection in the short term. If it is assumed that occupancy of the pit will be principally for recreational purposes, further remedial measures may need to focus on making the pit unattractive as a location for recreation. Thus, existing facilities (e.g. climbing ropes) shown should be routinely removed, and, if replaced by enthusiasts, should continue to be removed. However, this would in turn lead to radiological exposure of the implementer of these measures, and this would require monitoring.

For the heaps at site S5, the conclusions from the preliminary EIA are broadly similar to those for heaps S2-1, S3-2 and S3-30. The recommended remedial actions are therefore similar. Reshaping and recultivation are the ideal schemes, especially as the lack of vegetation means that high surface uranium concentrations are obtained (about 210 ppm of uranium, or about 2500 Bq kg^{-1}), thus leading to high radiological doses in the EIA. Fencing and appropriate warning signage would act as useful short-term measures.

For heap S7, locals should be prevented from using heap materials for building or constructing houses or other buildings to be occupied for significant periods. As for the other heaps, reshaping and recultivation are appropriate remediation measures, although the areal extent of heap S7 ($c.$ 150 000 m^2) would make this an expensive option if it were applied to the entire heap. Appropriate warning signage and fencing, as ever, would be useful short-term measures.

Occupants of the offices and other buildings located on top of heap S7 also experience high individual and collective doses. These result from the build-up and accumulation of radon and its progeny in the buildings. In order to reduce these doses, either the concentration of radon and its progeny needs to be reduced, or the occupation times of the buildings by any particular individual needs to be reduced. Increasing the ventilation rate can reduce the radon concentration, as these two quantities are inversely proportional to one another. Reducing the occupation times may be a more easily (and cheaply) implemented option, as an order of magnitude increase in the ventilation rate is required to reduce doses to satisfactory levels. Alternatively, a combination of ventilation rate increases and lowering of occupation times can be considered. The most satisfactory option is simply to find alternative premises away from areas of contamination by uranium.

The historical radiological surveys performed post-recultivation showed a heterogeneous, but diffuse, distribution of radiation (as illustrated in Fig. 4) that compared well with the subsequent transects of each of the sites using gamma spectrometry. It did not suggest that there had been a significant increase in radiation due to a reduction in the effectiveness of screening provided by the erosion of soil cover.

Conclusions

A pilot study performed at a former site of uranium mining in the Slovak Republic illustrates a methodology to evaluate human health and environmental impact. The main findings and lessons are:

- The former mining site shows complexity typical of an area in which there is diffuse contamination arising from leaching from waste heaps and uncontrolled discharges from adits into water courses. In particular, significant hazards occur at the sites due to the presence of uranium ore and its progeny at the surface, which may result in radiological exposure via direct irradiation, ingestion and inhalation of dust or radon.
- Site characterization considered both traditional areas of sampling and analysis (rock, soil, dust, radon and water), and identification of those activities and groups or individuals directly or potentially affected by exposure to contamination at the site. These ranged from workers occupying offices and workshops on one of the waste rock heaps, to house builders using waste rock (and potentially ore) for construction purposes, to a range of people exposed due to their recreational activities at the sites (hikers, mineral collectors, rock climbers and gatherers of wild produce).
- Historical recultivation measures performed at the site in the 1980s were generally ineffective at curtailing the whole range of radiological hazards. Measures were taken at most sites to bury exposed ore to minimize external irradiation. However, in those cases in which recultivation of heaps was successful, it did little to reduce the impact of radon emanation. Instead, recultivation appeared to have the surprising consequence of reducing the potential dose to the public, via an unrelated route, by making waste rock/ore more difficult to remove for the purpose of construction.
- When the importance of all the hazards were ranked, the most significant risk factor arose from inhalation of radon emanating from foundations built from waste rock material.
- Of all of the liabilities, a partially water-filled waste rock pit resulted in the highest dose rates. When time spent at this relatively remote site was taken into account, the potential doses received remained to be comparatively high. The most at risk groups were those working, in buildings, on the waste rock heaps and

those people who have removed waste rock/ore for building purposes.
- Mitigation measures to reduce doses experienced by the exposed groups can be summarized as:
  - prevention of the use of waste rock material for the purpose of house building;
  - reducing the overall accessibility to the sites, using barriers;
  - restricting the recreational value of the sites, by placement of warning signs/fencing (short term);
  - relocation of offices and laboratories on the heaps or improvement of the overall ventilation of the working areas.

We would like to thank Uranpres and Koral of Slovakia for access to historical data and assistance with collation of data, and in particular Mr Daniel of Uranpres and T. Grand for their generous hospitality. We would also like to thank our former colleague, Dr M. C. Thorne, for his advice throughout the project.

References

ARAPOV, J. 1984. *Czechoslovak Uranium Deposits.* SNTL, Praha, 365 (in Czech).

ICRP. 1975. *Report of the Task Group on Reference Man*, ICRP Publication 23. International Commission on Radiological Protection, Pergamon Press, Oxford.

ICRP. 1990. *Recommendations of the International Commission on Radiological Protection*, ICRP Publication 60. International Commission on Radiological Protection, Pergamon Press, Oxford.

ICRP. 1993. *Protection Against Rn-222 at Home and at Work*, ICRP Publication 65. International Commission on Radiological Protection, Pergamon Press, Oxford.

ICRP. 1996. *Age-dependent Doses to Members of the Public from Intake of Radionuclides: Part 5 – Compilation of Ingestion and Inhalation Dose Coefficients*, ICRP Publication 72. International Commission on Radiological Protection, Pergamon Press, Oxford.

SIMON, S.L. 1998. Soil ingestion by humans: a review of history, data and etiology with application to risk assessment of radioactively contaminated soil. *Health Physics*, **74**, 647–672.

Influence of mine hydrogeology on mine water discharge chemistry

S. B. REES[1], R. J. BOWELL[1] & I. WISEMAN[2]

[1]SRK Consulting, Windsor Court, 1–3 Windsor Place, Cardiff CF10 3BX, UK
(e-mail: brees@srk.co.uk)
[2]Environment Agency Wales, Penyfai House, Furnace, Llanelli, UK

Abstract: Using data for 81 coal mine discharges in the UK, the influence of discharge hydrogeology on discharge chemistry is assessed and typical chemical parameters derived for five discharge types. A combination of modified and new classification schemes is used to differentiate between the various discharge sources. Drainage from spoil tips generally has a pH below 5 and net-alkalinity values as low as $-2500\,\text{mg}\,l^{-1}$ $CaCO_3$. Drainage from flooded workings and pumped discharges are net-alkaline, while drainage from flooded and free draining workings are either moderately net-alkaline or net-acidic. Iron is the major contaminant of concern, although many mine waters contain less than $30\,\text{mg}\,l^{-1}$ and Fe/SO_4 ratios are less than unity. The classification schemes developed can be used to assess mine water treatment requirements and processes operating in passive treatment systems.

The process of mining can lead to the exposure of potentially reactive mineralogy to the atmosphere. During underground coal mining below the water table, the creation of void space in the coal seam and the intervening strata allows the ingress of atmospheric oxygen to otherwise saturated areas. As a result, aerobic sulphide oxidation can occur. In such circumstances, sulphide oxidation results in the formation of secondary salts on the surface of exposed void areas such as the floor, wall and roof rocks. These secondary salts can contain ferric Fe and SO_4, as shown by the formation of potassium jarosite:

$$3FeS_2 + \frac{9}{2}O_2 + \frac{15}{2}H_2O + K^4$$
$$= KFe_3^{3+}(SO_4)_2 \cdot (OH)_6 + 4SO_4^{2-} + 9H^+. \quad (1)$$

Other potential secondary salts include copiapite, melanterite, jarosite and shwertmannite (Bowell et al. 2000).

Following the cessation of pumping, the secondary salts are 'flushed' into solution as the water table recovers due to their high solubility. As a result, the concentration of total $Fe_{Unfiltered}$ and SO_4 can significantly increase. This process has recently been cited as an explanation for the trend in water chemistry observed at the former Lindsay Colliery, South Wales (Younger & Banwart 2002). Prior to February 2000, the discharge from the abandoned colliery contained less than $150\,\text{mg}\,l^{-1}$ total $Fe_{Unfiltered}$. However, by April of the same year, the Fe content peaked at $236\,\text{mg}\,l^{-1}$. Subsequently, the Fe content decreased and by the end of 2000 the discharge contained less than $100\,\text{mg}\,l^{-1}$.

Following water table equilibration, the processes of sulphide oxidation and secondary salt formation can only potentially continue within and above the zone of seasonal water table fluctuation, as aerobic sulphide oxidation within the now flooded workings is not possible due to limited oxygen diffusion (Perry 2001). Beyond the initial flush event, the long-term chemistry of the discharge is governed by the recharge water chemistry, the relative abundance of acid-generating and buffering mineral assemblages in the mine, and their relative reactivity (Younger & Banwart 2002).

As underground workings above the water table are free draining, they never become completely flooded and oxygen ingress can still occur following mining. Therefore, sulphide oxidation can potentially continue generating acidity. Depending on the availability and morphological nature of the sulphides, this can lead to a long-term source of secondary salts. Also, as there is no groundwater 'flush', any secondary salts produced can accumulate, potentially providing a long-term source of secondary constituents to any interflow passing through the workings.

In addition to the effect on the groundwater regime during and post-mining, material classified as 'waste' is generated. In coal mining areas, this is often present at surface as loose tipped spoil tips or waste rock piles. Like the flooded and free draining settings, spoil tips also have their own unique hydrogeological properties that can potentially affect the chemistry of any associated drainage. Depending upon the environment, mine spoil can exhibit characteristics of both porous medium and double-porosity aquifers (PDEP 1998). For example, relatively continuous water tables and multiple water tables have been observed in spoil tips and, as a result, the water contact time can vary considerably. Coupled with the exposure of greater mineral surface area within a spoil tip, the processes of sulphide oxidation and secondary salt formation can therefore be more prevalent in spoil tips than in underground workings.

A summary of the different properties of each mine setting described are conceptually illustrated in Fig. 1, along with the main processes occurring that can affect the chemistry of associated discharges. Although discharge sources can, in some cases, be readily identified from mine plans and the history of the discharge, an understanding of the processes that affect discharge chemistry can prove a useful tool where such information is lacking or inconclusive. From a management perspective, the identification of the discharge source is vital for successful treatment, as this can influence long-term treatment requirements (Younger & Banwart 2002). As the discharge chemistry must be known for treatment purposes, it is, therefore, cost-effective to also use this to help characterize the discharge source.

Using chemical data for 81 coal mine discharges from the UK, typical chemical properties for each discharge type are derived and the reason for the differences explained. The discharges assessed are located in the South Wales, Scottish and County Durham coalfields, as summarized in Table 1. The time since first emergence of each discharge used in the assessment ranges from less than 5 years to over 100 years. Based on the author's knowledge, published information and contact with other workers, five different types of discharge source have been identified. These are flooded workings, flooded and free draining workings, pumped discharges, free draining workings, spoil heap drainage and discharges from an unknown source. Many of the discharges have not received detailed historical monitoring, so the amount of chemical data available is often limited. Therefore, a limited chemical dataset has been purposely selected to demonstrate how such data can still be used to characterize discharges. As summarized in Table 2, the chemical parameters used are Ca, Mg, Na, K, Cl, SO_4, HCO_3, acidity, pH, redox potential and Fe.

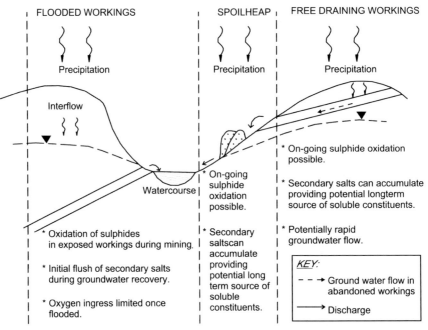

Fig. 1. Summary of processes occurring in flooded workings, spoil tips and free draining workings that can affect discharge chemistry.

Table 1. *Mine water discharge name, source and time since first emergence (age)*

Discharge point	Drainage source	Age (years)	Discharge point	Drainage source	Age (years)
SOUTH WALES			Douglas	Flooded	c. 40
Craig Yr Aber	Free draining	40	Pool Farm	Flooded and free draining	c. 40
Cedfw	Flooded	?			
Cwmgors	Unknown	?	Blairenbathie	Unknown	c. 35
Mountain Gate	Unknown	?	W. Colquhally	Unknown	c. 35
Lash	Unknown	?	Cardenden	Unknown	34
Tan-y-garn	Flooded	10	Parson's Mill	Unknown	34
Llechart	Flooded	20	New Carden	Unknown	33
Garwed	Unknown	?	Kinglassie	Unknown	34
Ynysarwed	Flooded	7	Balgonie Bing	Spoil heap	c. 50
Nant y Fedw	Flooded	?	Randolph Bing	Spoil heap	c. 30
Goytre	Unknown	?	Frances Colliery	Pumped	
Corrwg	Flooded and free draining	c. 40	Michael Colliery	Pumped	
			Dalquharran Mine	Flooded and free draining	22
Gwynfi	Flooded and free draining	?	Roughcastle	Flooded	5
Gwenffrwdd	Flooded and free draining	c. 30	**COUNTY DURHAM**		
			Broken Banks	Flooded	45
Whitworth No.1	Flooded and free draining	c. 30	Quaking Houses	Spoil heap	c. 25
			Stoney Heap	Flooded	20
Whitworth A	Flooded and free draining	c. 30	Tindale Colliery	Flooded	33
			Kibblesworth	Pumped	
Whitworth B	Flooded and free draining	c. 30	Lumley 6th Pit	Pumped	
			Chester Moor	Pumped	
Garth Tonmawr	Flooded and free draining	c. 30	Kimblesworth	Pumped	
			Nicholson's Pit	Pumped	
Dunvant Square	Flooded	?	Sherburn hill	Pumped	
Dunvant Clyne Trib	Flooded	65	Ushaw Moor	Pumped	
Lindsay	Flooded	2	Page Bank	Pumped	
Rhymney	Flooded and free draining	10	Vinovium	Pumped	
			Helmington Row	Flooded	21
Maerdy	Unknown	?	Burnopfield	Flooded	>10
Ynyswen	Unknown	?	Milkwell Burn	Flooded	>10
Six Bells	Flooded and free draining	c. 50	Pont Waterlevel	Flooded	>10
			West Kyo	Flooded	>10
Taff Merthyr	Flooded	6			
Abersychan	Unknown	26			
Penrhiwfer	Spoil heap	?			
Ynysybwl	Unknown	?			
Blackwood	Partially flooded workings and seeps	8			
N. Celynen	Flooded	10			
Tram Road	Flooded	10			
Trosnant Brook	Unknown	?			
SCOTLAND					
Lathallan Mill	Flooded	100			
Star Road	Flooded	106			
Elginhaugh	Flooded	35			
Blackwood	Flooded	117			
Cairnhill	Flooded	20			
Pennyvenie No 3	Flooded	18			
Brora No 1	Flooded	c. 40			
Macrihanish	Flooded	c. 50			
Baads Bing	Spoil heap	c. 40			
Minto No 2	Flooded	34			
Kames No 1	Flooded	28			
Fordell Day Level	Unknown	c. 40			
Cuthill No 1	Free draining	c. 40			

Data for the Welsh discharges come from the Environment Agency Wales mine waters database. The County Durham and Scotland data are taken from Younger (1995a, 1998, 2001) and Wood *et al.* (1999).

Mine water characterization

Piper plot

Comparison of the major ion chemistry of all the discharges in Table 2 using a Piper diagram indicates that cation proportions (Na, K, Ca and Mg) are broadly similar between the different discharge types, with the exception of most pumped discharges and a small number of other types of discharge (Fig. 2). As a proportion of total cations (expressed as meq l^{-1}), most discharges contain less than 20% Na, and between 40 and 60% Ca and Mg. In comparison, anion

Table 2. *Mine water discharge data for Wales, Scotland and Country Durham**

Units	pH	Redox Potential (mV)	Acidity (mg l⁻¹) CaCO₃	Alkalinity (mg l⁻¹) CaCO₃	Cl (mg l⁻¹)	SO₄ (mg l⁻¹)	Na (mg l⁻¹)	K (mg l⁻¹)	Mg (mg l⁻¹)	Ca (mg l⁻¹)	Total Fe Filtered (mg l⁻¹)	Total Fe Unfiltered (mg l⁻¹)
Wales												
Craig Yr Aber	6.87	177	37.40	122.18	21.53	573.37	57.47	11.72	64.09	124.95	20.33	22.13
Cedfw (3)	6.74	171	24.00	101.75	nm	15.10	nm	nm	7.66	31.43	19.92	17.46
Cedfw (2)	7.03	215	15.20	115.25	nm	3.63	nm	nm	10.48	27.40	8.85	8.39
Cedfw	6.33	196	42.25	55.75	nm	73.78	nm	nm	13.75	21.60	17.10	17.29
Cwmgors	7.48	186	15.35	418.00	nm	266.67	nm	nm	29.60	41.53	6.36	6.67
Montain Gate	6.94	269	39.15	376.33	14.70	131.33	22.57	29.70	56.10	75.43	3.29	3.75
Lash (Brooklands)	7.35	270	21.49	361.67	21.83	245.67	17.53	28.53	74.33	96.80	1.61	2.70
Lash (Derwen)	6.90	292	38.25	212.67	16.60	192.67	12.83	8.96	49.87	73.90	8.86	22.57
Tan Y Garn (Mountain Colliery)	6.32	256	18.16	34.45	nm	679.17	nm	nm	71.47	151.83	20.08	20.60
Tany Garn (Cathan)	6.13	145	103.22	35.15	12.70	346.73	11.76	6.07	36.93	53.65	61.66	68.03
Llechart (2)	5.99	388	18.31	15.26	9.75	66.85	7.63	1.58	11.52	11.96	1.04	1.55
Llechart	6.72	nm	7.69	31.12	9.98	142.48	nm	nm	21.53	32.19	3.15	3.70
Garwed (Lower)	3.52	473	55.21	nm	nm	193.61	nm	nm	19.37	25.06	5.57	5.62
Garwed (Upper)	3.05	490	205.00	nm	12.70	683.43	12.43	8.44	63.01	80.50	48.83	49.99
Ynysarwed	5.55	74	396.83	45.81	15.86	2364.97	143.21	26.39	184.64	299.33	237.33	256.38
Nant Y Fedw (Gelli Farm)	7.44	360	4.64	80.33	8.70	28.57	5.92	2.18	11.96	22.43	1.56	1.75
Nant Y Fedw (lower)	7.23	335	12.86	74.33	9.00	68.63	3.36	3.36	17.10	27.70	2.77	3.36
Nant y Fedw (Middle)	7.26	377	23.00	99.50	nm	207.50	nm	nm	34.90	59.50	25.20	25.40
Goytre	7.30	353	15.59	92.00	21.25	94.80	19.67	4.21	18.50	35.00	5.13	8.92
Corrwg (Lower)	6.82	174	26.23	80.90	nm	277.80	nm	nm	40.84	66.16	18.06	18.75
Corrwg (Upper)	6.48	262	35.65	52.00	nm	165.71	nm	nm	23.94	37.89	8.43	9.63
Corrwg Fechan	6.96	239	28.44	121.78	nm	149.56	nm	nm	32.20	52.39	12.87	13.56
Gwynfi	6.61	286	13.21	38.62	9.78	102.95	6.51	4.33	16.43	27.41	5.06	5.57
Gwenffrwd	5.22	319	17.10	10.45	12.00	82.78	3.21	0.00	9.63	17.33	3.66	4.09
Whitworth No. 1	6.33	155	38.65	41.02	10.93	343.47	12.24	13.05	39.59	71.67	22.84	24.62
Whitworth B	5.90	222	134.76	9.73	17.96	97.14	5.62	2.94	13.04	19.86	5.51	5.48
Whitworth A	5.94	164	344.75	23.66	12.75	334.86	11.90	9.44	34.38	56.95	57.87	59.53
Garth Tonmawr	5.59	231	286.46	14.09	12.11	271.91	7.64	8.56	31.33	47.19	27.10	28.49
Nantyffyllon	6.94	337	22.50	103.00	14.16	100.00	nm	nm	21.30	38.00	6.50	6.50
Dunvant Square	7.72	212	3.21	128.20	nm	75.58	nm	nm	17.02	52.10	2.03	2.62
Dunvant Clyne Trib	7.53	nm	5.68	149.33	nm	51.27	nm	nm	18.23	52.73	1.51	2.14
Rhymney	7.04	225	16.21	225.28	16.31	449.08	21.95	12.28	69.66	141.81	5.27	7.08

Site												
Maerdy	6.84	287	18.33	64.00	6.13	22.13	4.45	1.96	10.44	16.23	4.15	8.00
Ynyswen	7.64	167	1.93	100.33	9.53	15.77	9.09	5.75	12.53	18.30	1.08	1.27
Six Bells	7.60	63	35.35	877.25	33.48	1391.25	464.00	90.93	171.75	193.75	28.58	35.30
Taff Merthyr	7.93	104	11.61	247.33	12.57	471.00	15.63	15.97	67.63	165.00	14.26	20.33
Abersychan	7.55	180	12.53	230.00	12.33	250.67	15.67	24.06	50.20	100.07	2.69	3.24
Penrhiwifor	6.21	250	3.27	8.30	13.43	11.77	6.58	0.91	2.33	5.46	1.37	1.72
Ynysybwl	6.79	359	2.04	23.00	12.97	24.37	5.96	1.14	6.49	8.58	0.21	0.38
Blackwoodsunningdale	6.71	302	24.55	149.33	22.23	594.67	17.90	8.76	90.93	149.33	3.85	4.08
North Celynen	7.07	124	21.91	276.33	17.74	427.67	26.47	9.29	71.27	161.00	10.53	11.77
Tram Road	6.98	236	14.58	222.00	22.51	494.00	26.93	11.70	72.40	153.00	5.77	6.56
Trosnant Brook	7.76	282	5.25	132.76	12.43	80.43	9.17	4.55	21.63	50.17	0.64	1.10
Scotland												
Lathallan Mill	6.10	17	20.80	182	35	214	16	6	41	87	nm	11
Star Road	6.50	−57	10.00	173	34	53	13	12	21	61	nm	4
Elginhaugh	5.70	6	192.00	207	22	1100	16	23	188	256	nm	93
Blackwood	7.20	−40	3.00	265	30	37	17	12	27	76	nm	1
Cairnhill	7.60	−76	27.00	80	23	1346	28	72	88	385	nm	7
Pennyvenie No 3	6.90	−7	2.00	854	21	142	302	24	42	58	nm	7
Brora No 1	3.60	nm	90.71	0	123	340	72	20	18	125	nm	0
Macrihanish	6.10	nm	153.51	285	57	51	26	8	20	46	nm	8
Randolph Bing	3.70	nm	1565.75	0	103	2475	26	6	115	229	nm	60
Michael colliery	6.90	nm	67.34	410	2662	1713	1304	78	318	312	nm	43
Frances Colliery	6.90	nm	29.55	162	5880	1240	4290	60	320	160	nm	34
Baads Bing East	2.80	nm	2420.00	0	16	3077	19	7	32	407	nm	12
Baads Bing West	3.50	nm	190.00	0	10	304	10	3	36	84	nm	550
Minto No 2	6.80	nm	32.00	670	nm	1550	nm	nm	nm	nm	nm	6
Kames No 1	5.80	nm	28.83	232	9	247	9	8	60	130	nm	13
Fordell Day Level	6.40	nm	29.00	270	nm	1200	nm	nm	nm	nm	nm	14
Cuthill No 1	5.50	nm	68.35	247	40	932	75	15	97	345	nm	16
Douglas	6.00	nm	75.00	190	15	595	11	12	73	180	nm	37
Pool Farm	5.60	nm	18.00	98	26	240	7	4	27	66	nm	40
County Durham												
Helmington Row	4.80	264	214.47	0	65	810	22	7	93	185	nm	79.80
Burnopfield	6.50	−40	14.44	146	73	301	39	9	55	97	nm	5.20
Milkwell Burn	6.70	−45	14.54	216	29	35	15	2	14	97	nm	5.40
Pont Waterlevel	6.30	−15	20.76	319	49	724	37	16	123	174	nm	1.80
West Kyo	6.60	−45	25.08	187	27	104	40	3	26	46	nm	9.30
Kibblesworth	7.02	nm	2.36	755	909	395	782	27	52	162	nm	0.88
Lumley 6th Pit	6.88	nm	11.28	895	462	420	577	27	61	149	nm	4.21
Chester Moor	7.58	nm	2.28	1050	132	672	683	20	35	79	nm	0.85

Table 2 – *(continued).*

Units	pH	Redox Potential (mV)	Acidity (mg l⁻¹) CaCO₃	Alkalinity (mg l⁻¹) CaCO₃	Cl (mg l⁻¹)	SO₄ (mg l⁻¹)	Na (mg l⁻¹)	K (mg l⁻¹)	Mg (mg l⁻¹)	Ca (mg l⁻¹)	Total Fe Filtered (mg l⁻¹)	Total Fe Unfiltered (mg l⁻¹)
Kimblesworth	7.22	nm	4.50	835	138	358	446	16	34	90	nm	1.68
Nicholson's Pit	6.96	nm	16.15	735	112	1052	530	26	71	173	nm	6.03
Sherburn hill	6.71	nm	20.63	550	96	967	238	23	107	221	nm	7.70
Ushaw Moor	6.69	nm	1.49	700	51	232	178	23	67	108	nm	0.56
Page Bank	6.71	nm	2.29	654	113	601	350	30	73	142	nm	0.85
Vinovium	6.81	nm	1.35	625	79	510	216	23	85	149	nm	0.51
Broken Banks	6.50	39	4.89	364	60	137	27	11	61	101	nm	1.80
Crook	4.80	264	214.48	0	65	810	22	7	93	185	nm	79.80
Quaking Houses	4.10	327	49.90	0	1012	1358	464	57	103	255	nm	18.00
Stoney Heap	6.30	36	70.51	188	102	325	28	7	50	84	nm	26.30
Tindale Colliery (Brusselton)	6.40	−50	4.89	357	75	890	80	13	107	262	nm	1.80

* Data for the Welsh discharges come from the Environment Agency Wales mine waters database. The County Durham and Scotland data is taken from Younger (1995a, 1998, 2001) and Wood *et al.* (1999). The data for the Welsh discharges is based on average values since monitoring was initiated, while the County Durham and Scottish discharge data are based on data for a specific date and are not averaged. nm, not measured; nd, not detected.

proportions show greater variation with HCO_3 and SO_4, the dominant end-members, and only a small number of discharges contain greater than 20% Cl.

The pumped discharges contain a greater Na, K and Cl component, as they have been shown to either contain a sea-water component or to have interacted with deep basin brines and undergone extensive ion exchange (Younger 1995a, 1998). However, a similar explanation is not applicable for the other discharge types that plot in a similar position to the pumped discharges. The discharge classified as arising from flooded and free drainage workings is the discharge at Six Bells, South Wales, and it may be that this water has interacted with evolved Devonian formation water in the mine. The higher proportion of Na, K and Cl in two spoil tip discharges may reflect rapid water flow in these settings and also a lack of carbonate buffering.

Owing to the evolved nature of most of the pumped mine waters, they plot in a distinct position of the Piper-plot and are generally characterized by either a $Na–HCO_3$ or $Na–SO_4$ hydrochemical signature. Unfortunately, the other discharge types are not so readily distinguished as there is overlap between their hydrochemical signatures. For instance, discharges from flooded workings have a $Ca–Mg–SO_4$ and a $Ca–Mg–HCO_3$ hydrochemical signature. Similarly, discharges from flooded and free draining workings also generally have a $Ca–Mg–SO_4$ signature. As a result, the Piper diagram cannot be used alone to distinguish between different discharge types, although the characterization of major ions is useful. The reason the Piper plot is insufficient in classifying mine waters can be seen by comparing the plotting positions of the majority of the spoil tip discharges and the discharges from flooded and free draining workings.

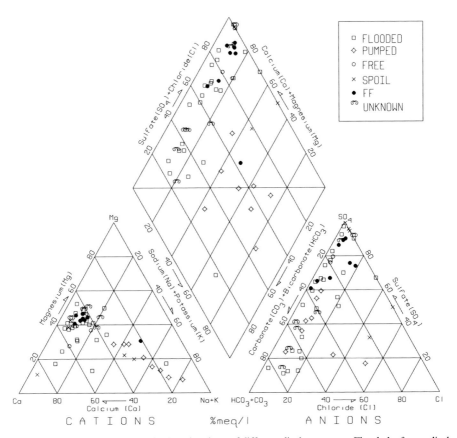

Fig. 2. Piper diagram summarizing major ion chemistry of different discharge types. Flooded refers to discharges from flooded workings; FF refers to drainage from flooded and free draining workings; Spoil refers to discharges from spoil tips; Pumped refers to discharges that are pumped; Free draining refers to discharges from free draining workings; Unknown refers to discharges for which is a source is not known.

Many of the spoil tip discharges do not contain alkalinity. However, their hydrochemical signature, according to the Piper plot, is the same as the flooded and free draining discharges, and so the distinct chemical difference is lost.

Net-alkalinity versus *Cl/Cl + SO₄*

Such limitations of the Piper plot have been noted previously by Younger (1995a), who subsequently proposed a different method of classifying mine water discharges (Fig. 3). In this plot, an earlier classification system proposed by Hedin *et al.* (1994) is modified slightly and plotted against the ratio of Cl to the sum of Cl and SO_4 expressed in $meq\, l^{-1}$.

With such a diagram, net-acidic, high-SO_4 waters plot to the lower left, waters affected by bacterial sulphate reduction (BSR) to the upper right and waters affected by carbonate dissolution to the upper left. Younger (1995a) concluded that such a diagram would enable the hydrological source of a discharge to be identified as these basic geochemical processes controlling mine water chemistry are presented.

As expected, the spoil tip discharges are located in the bottom left-hand corner. However, a number of other discharges also plot in a similar position. Therefore, as with the Piper plot, this figure does not enable the differentiation of the discharge sources.

Proposal for a modified classification scheme

In the scheme proposed by Hedin *et al.* (1994), mine waters were classified as being either net-alkaline or net-acidic based on the difference between alkalinity and combined proton and mineral acidity content. Younger (1995a) modified the classification scheme and plotted the alkalinity as a percentage of total alkalinity and acidity. As a result, mine waters with zero alkalinity but different acidity (proton and mineral) content are not differentiated and all plot with a zero *y*-axis value on Fig. 3. Therefore, a modified classification scheme is proposed that uses the absolute net-alkalinity values (Fig. 4).

In this diagram, differences between the mine water discharge sources are more readily identifiable as the absolute alkalinity and acidity values are illustrated. Based on this diagram, typical ranges in net-alkalinity for each discharge type can be identified. These are summarized in Table 3, along with the Piper classifications from Fig. 2 and also a summary of the range in pH values for each discharge type from Table 2.

Spoil tips are the most net-acidic waters encountered, with values almost as low as $-2500\, mg\, l^{-1}$ $CaCO_3$. This is because sulphide oxidation and secondary salt formation and mobilization are the dominant processes occurring due to the exposure of greatest mineral surface area and potentially rapid water throughflow.

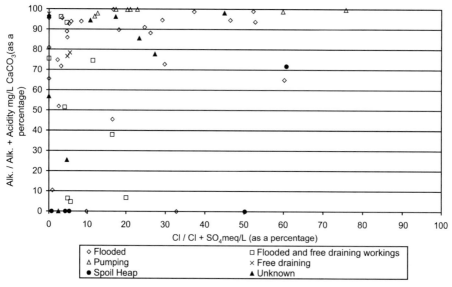

Fig. 3. Classification scheme used to originally assess Scottish mine waters according to Younger (1995a).

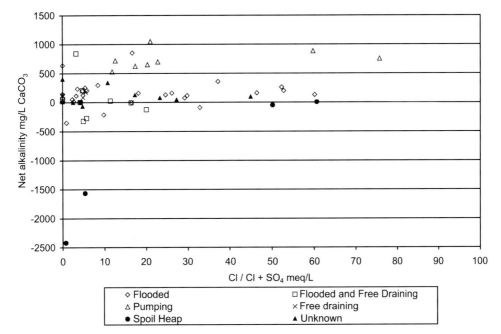

Fig. 4. Modified mine water classification scheme.

Drainage from flooded workings and flooded and free draining workings are more net-alkaline than spoil tip drainage, with flooded workings the most net-alkaline. In both settings, sulphide oxidation will be less prevalent compared to a spoil tip exposed to the atmosphere, and buffering reactions will play a more significant role in controlling water chemistry. Flooded workings generally give rise to more net-alkaline discharges compared to flooded and free draining workings as aerobic sulphide oxidation is not possible within the flooded workings, whereas the dissolution of buffering mineral assemblages, such as carbonates, can continue (Younger & Banwart 2002).

Iron chemistry

Owing to the oxidation of iron sulphides, such as pyrite, the major contaminant of concern in UK coal mine drainage is Fe (Table 2). The proposed classification scheme can be used to assist in identifying between discharges on the basis of net-alkalinity, but it does not include an assessment of potential differences in Fe content between the various types of discharge. Interestingly, regardless of the discharge source, many of the discharges do not contain more than approximately 30 mg l^{-1} total Fe$_{Unfiltered}$, as illustrated in Fig. 5. This has been noted to be particularly the case for discharges from flooded shafts that have been flowing for approximately 40 years (Wood et al. 1999). On this basis, a classification scheme including Fe may not be beneficial in the long term. However, there are discharges that do contain in excess of 30 mg l^{-1}, even though they have been flowing for more than 40 years (Tables 1 and 2). Consequently, a brief attempt has been made to develop a scheme to better define these waters.

Table 3. *Summary of typical chemical properties for each type of discharge*

Discharge source	pH	Net-alkalinity mg l^{-1} CaCO$_3$	Piper classification
Flooded workings	<5–8	0 to +500	Ca–Mg–SO$_4$/HCO$_3$
Spoil tip	<5	−2500 to 0	Ca–Mg–SO$_4$
Free draining workings	5–7	+80 to +180	Ca–Mg–SO$_4$
Flooded and free draining workings	>5 <8	−350 to +200	Ca–Mg–SO$_4$
Pumped	6.5–7.5	+500 to +1000	Na–HCO$_3$/SO$_4$

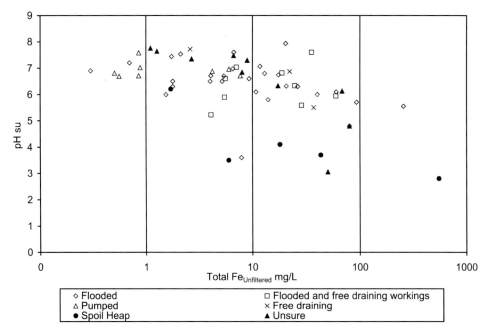

Fig. 5. Variation of total Fe$_{Unfiltered}$ with pH.

In many circumstances, historical information about Fe speciation is not available and only total Fe$_{Unfiltered}$ will have been measured. Although it is useful to have total Fe$_{Filtered}$ concentrations, Fig. 6 indicates that the majority of UK mine waters have total Fe$_{Filtered}$ values that account for over 80% of total Fe$_{Unfiltered}$. Therefore, total Fe$_{Unfiltered}$ values have been used.

The classification scheme of Younger (1995a) indicated the benefit of using elemental

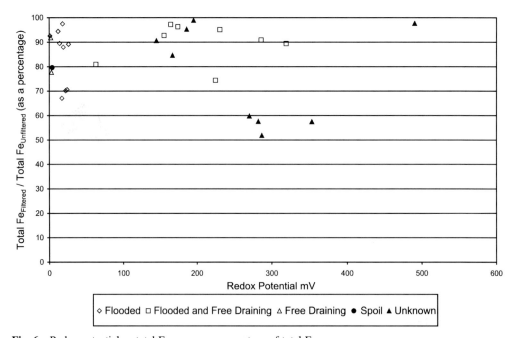

Fig. 6. Redox potential vs total Fe$_{Filtered}$ as a percentage of total Fe$_{Unfiltered}$.

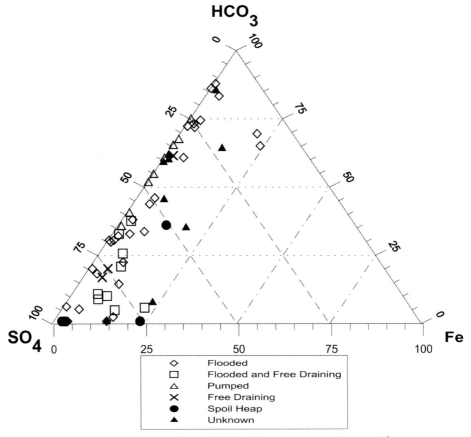

Fig. 7. Ternary diagram comprising Fe, SO_4 and HCO_3. All values expressed as meq l^{-1}.

ratios to differentiate between basic geochemical processes. Therefore, if Fe is to be incorporated into a classification scheme it was considered best that it was represented as part of the overall chemistry and that absolute values were not used. In addition to the use of ratios, this can be achieved by using a ternary diagram (Fig. 7). On this diagram, three important mine water variables are presented as meq l^{-1}. As noted from the Piper plot (Fig. 2), and reflected in the net-alkalinity values in Fig. 4 and Table 3, there is a degree of overlap between the different discharge types in terms of the proportions of HCO_3 and SO_4. Also, the inclusion of Fe does not assist in source characterization using this method. This is because the majority of discharges have total $Fe_{Unfiltered}$ less than 30 mg l^{-1}, and the ratio of Fe/SO_4 (in terms of meq l^{-1}) is less than unity, as Fe is susceptible to oxidation and precipitation to a greater extent that SO_4. Also, during complete aerobic oxidation of pyrite (FeS_2), for every 1 mol of Fe released, 2 mol of sulphate are released.

Although Fig. 7 is limited in assisting discharge source characterization, it can certainly be used to assess mine water treatment requirements and processes operating in passive treatment systems, such as wetlands. For instance, spoil tip drainage is typified as strongly net-acid with elevated SO_4 and Fe and low pH (Table 2). Therefore, successful treatment requires the lowering of acidity (increase of alkalinity) and the removal of Fe and SO_4. Such improvements in water quality can be achieved using a passive system such as the Successive Alkalinity Producing System (SAPS), which comprises aerobic and anaerobic Fe and SO_4 removal and carbonate dissolution (Younger 1995b; Rees et al. 2001). As these reactions influence the three components illustrated in Fig. 7, changes in water quality occurring during the different treatment stages could potentially be readily traced.

Summary

Hydrogeological conditions of a mine discharge play a significant role in controlling discharge chemistry. Within the coalfields studied, five types of mine settings have been identified and differentiated on the basis of a number of lines of chemical evidence. The observations made should assist future workers aiming to characterize the source of a coal mine discharge. In order to assist with source characterization a number of modified and new classification schemes have been proposed. These schemes can also be used for assessing treatment requirements and processes operating within passive treatment systems.

References

BOWELL, R.J., REES, S.B. & PARSHLEY, J.V. 2000. Geochemical predictions of metal leaching and acid generation: geologic controls and baseline assessment. In: CLUER, J.K., PRICE, J.G., STRUHSACKER, E.M., HARDYMAN, R.F. & MORRIS, C.L. (eds) *Geology and Ore Deposits and Ore Deposits 2000: The Great Basin and Beyond. Geological Society of Nevada Symposium Proceedings, Reno/Sparks, May 2000.* 799–823.

HEDIN, R.S., NAIRN, R.W. & KLEINMANN, R.L.P. 1994. *Passive Treatment of Coal Mine Drainage.* US Bureau of Mines Information Circular, **9389**.

PDEP. 1998. *Coal Mine Drainage Prediction and Pollution Prevention in Pennsylvania.* World wide web address:http://www.dep.state.pa.us/dep/deputate/minres/districts/CMDP/main.htm.

PERRY, E.F. 2001. Modelling rock–water interactions in flooded underground coal mines, Northern Appalachian Basin. In: *Geochemistry: Exploration, Environment, analysis*, Vol. 1, 61–70.

REES, S.B., BOWELL, R., DEY, M. & WILLIAMS, K. 2001. Performance of a Successive Alkalinity Producing System (SAPS). *Proceedings of the International Conference on Mining and the Environment: Securing the Future, Skellefteå, Sweden,* The Swedish Mining Association, Vol. 2; 703–708.

WOOD, S.C., YOUNGER, P.L. & ROBINS, N.S. 1999. Long-term changes in the quality of polluted mine water discharges from abandoned underground coal workings in Scotland. *Quarterly Journal of Engineering Geology,* **32**, 69–79.

YOUNGER, P.L. 1995a. Hydrogeochemistry of mine water flowing from abandoned coal workings in County Durham. *Quarterly Journal of Engineering Geology,* **28**, S101–S113.

YOUNGER, P.L. 1995b. Design, construction and initial operation of full-scale compost-based passive systems for treatment of coal mine drainage and spoil leachate in the UK. In: *Proceedings of the International Mine water Association Symposium, Johannesburg, South Africa.* Vol. II; 413–424.

YOUNGER, P.L. 1998. Coalfield abandonment: geochemical processes and hydrochemical products. In: NICHOLSON, K. (ed.) *Energy and the Environment: Geochemistry of Fossil, Nuclear and Renewable Resources.* MacGregor Science, 1–29.

YOUNGER, P.L. 2001. Mine water pollution in Scotland: nature, extent and preventative strategies. *Science of the Total Environment,* **265**, 39–326.

YOUNGER, P.L. & BANWART, S.A. 2002. Time-scale issues in the remediation of pervasively contaminated groundwaters at abandoned mine sites. In: OSWALD, S. & THORNTON, S. (eds) *Groundwater Quality 2001.* International Association of Hydrological Sciences, in press.

Index

Abakan–Chernogorsk region, Khakassia, Siberia, alkaline coal mine drainage 287–90
    analysis of drinking water at Byelii Yar opencast mine 290
    properties of typical coals from Minusinskii 289
    sampling and analysis methods 288
Abaza Magnetic Mine, Khakassia, Siberia, alkaline coal mine drainage 287–8, 290–1
    chemical composition of pumped water 291
    sampling and analysis methods 288
acid mine drainage
    CarnoulŠs, France, arsenic 267–74
    South Nottinghamshire Coalfield closure 99–104
acid–base accounting (ABA) 145
acidic spoil drainage
    acidity values 252
    Cleveland Ironstone Field 260–2
    vestigial/juvenile acidity 139, 205
Adrio Valley, SW Spain, Aznalicollar mine spill 187–204
Aitik mine, Sweden, mineral weathering rate prediction 151–5
    albite weathering 153–5
    reactive surface area at field scale 143
alkaline coal mine drainage
    Siberia
        Abakan–Chernogorsk region 287–90
        Abaza Magnetic Mine 287–8, 290–1
        Vershina Ty‰i Magnetic Mine 287–8, 291–2
    Svalbard, Longyearbyen ?87–8, 292–5
Altiplano aquifer complex, Bolivia 215–39
anorthite weathering 153–5
    reactive surface area at field scale 143
aquifers overlying coal mines 17–45
    hydraulic head drops 20–3
    hydraulic tests 28–30, 34–5
    transmissivities 22–3
    water level recovery 23–4
    see also longwall coal mining (UK and USA)
Arctic region *see* Longyearbyen, Svalbard
arsenic
    acid mine drainage system, removal by oxidizing bacteria 267–74
    mine pit lakes 163
    potable water limits 335
    Spain, NW, Mieres, Asturias, pollution from mercury and coal mine spoil in Morgao catchment 327–36
Aznalicōllar *see* Spain, SW

bacteria, arsenic removal 267–74
base flow index (BFI) 130
berthierine 254
biotite weathering 151–5
    reactive surface area at field scale 143
Black Clough discharge, Deerplay Colliery *see* Lancashire, UK
Bolivia (San Jos, Mine, Oruro), contaminant sources 215–39
    cross-section and site 218
    geological and hydrogeological setting 219–23
        bedrock 219

    chloride concentrations 220
    contour maps of aquifer complex 220, 222
    sedimentary complex 219–23
    groundwater chemistry in wells 237
    mine water hydraulics 223–8
        ingress of surficial waters 224
        MIFIM model 223–6
        mine flooding modelling 224–7
        MODFLOW model, ground water leakage, flooded mine 223–8
        possible water outflows 224
        void distribution 223
        water influxes/inflows 223–4
    mine water quality 228–33
        composition of pumped mine water at Santa Rita entrance 230–1
        hydrochemical characteristics 228–9
        results of analyses of water 232
        speciation modelling 229–32
    pumping rates from mine 219
    risk assessment 234–6
        leachate analyses 235
        map of copper concentrations 236
        mine wastes 233–4
        risk source characterization 234
        secondary mineral efflorescences 234

Cantabrian zone, map 330
CarnoulŠs, Gard *see* France
chalcopyrite weathering 147–55
    reactive surface area at field scale 143
chamosite 254
chemical stratification, mine pit lakes 167
Chile (Escondida Copper Mine), depressurization of north wall 107–19
    conceptual model of groundwater flow 113–14
        altered (argillic) porphyry 113–14
        rhyolites 113
        silicified Escondida porphyry 113
    conceptual model of recharge mine 108
    current pore pressures 110–13
    depressurization system design 116–18
    field programme and database 108–10
    location plan 108
    numerical modelling 114–16
chloride plume migration, gold mine tailings dam 337–46
Cleveland Ironstone Field, UK, pyritic rook strata in aquatic pollutant release 251–66
    current mine water discharges 255–7
        hydrochemistry 258, 260
        known discharges of polluted mine drainage 256
        metals 259
        WATEQ4F modelling 261
    data collection methods 252–3
    Eston Mine, major discharge 257–60
    mining history 253–5
    geological framework 254
    pollutant generation and attenuation reactions 254
    sketch map of area 253
New Marske Mine, acidic spoil drainage 260–2

Saltburn, new discharge 262–3
Skinningrove, two overflowing mines 263–4
clubmoss (*Lycopodium clavatum*) spores, underground minewater tracing 52–3
Coal Measures
permeability values 66
Upper, Middle and Lower *see* South Wales Coalfield
coal mining, *see also* longwall coal mining
CODE-BRIGHT model, flow and heat 192–4
column, sulphide oxidation in unsaturated soil 189–90
ConSim model 234
contaminant sources
mine water pollution 138
vestigial/juvenile acidity 139, 205
contaminant transport, modelling 208–9
copper *see* Bolivia (San Jos, Mine); Chile (Escondida Copper Mine); Sweden (Aiitik Mine)
Cumbria, UK (Nenthead), abandoned mines as sinks for pollutant metals 241–50
geological setting 243–4
map of Nent Valley, main inputs of contaminants 242
mine water chemistry, absence of sinks for zinc 249
other mineral sinks for zinc 245
Rampgill Mine 244–5
sampling methods 244
zinc concentration in River Nent 243
zinc deficits in Nent Valley mine waters 245–8
results of calculations 248
x-y plots showing lack of correlation between zinc and sulphate 248–9

deep mine voids, test pumping for assessment 315–26
Deerplay Colliery *see* Lancashire, UK
depressurization systems, design 116–18
dolomite 255
Donana National Park, SW Spain, Aznalicollar mine spill 187–204
Durham County, UK, iron release from spoil heap 205–14
conceptual model 209–13
contaminant sinks 210
contaminant sources 209–10
input parameters 210
laboratory and modelled iron and sulphate 211
long term sulphate concentrations 212
results and discussion 210–13
historical and geological overview 206
laboratory methods/results 206–8
location map 206
modelling methods 208–9
contaminant transport 208
oxygen diffusion 208
weathering reactions 208–9
Durham County, UK, mine water recovery records 64–7

East Fife *see* Fife, UK
environmental impact, South Nottinghamshire Coalfield closure 99–104
environmental impact assessment (EIA)
gold mining in Ghana 121–34
uranium mine, Slovak Republic 370
Escondida Copper Mine *see* Chile

Eston Mine, Cleveland Ironstone Field, UK 257–60
European Commission, uranium mine liabilities, Slovakia project 368
evapotranspiration, actual (AE) 127

Ferrobacillus ferrooxidans 209
Fife, UK, East Fife coalfield, monitoring mine water recovery 62–4
Fife, UK, Frances Colliery, test pumping deep mine voids 315–26
hydrogeochemistry 323–4
Piper diagram, plotting pumped mine water chemistry 323
pumped mine water quality 323
hydrogeological results 317
calculation of Reynolds numbers (RE) for submerged insets 319
plots of discharge against specific capacity 318
test pumping data 318
map of location 316
mine water levels 64
sampling and data collection 316–17
schematic diagram showing connection with adjacent collieries 316
source of poor water quality 325
stratification within mine workings 324–5
diagram showing build-up 325
various determinants 324
Fife, UK, Lochhead Colliery, mine water levels 64
fingerprinting minewater emissions, South wales Coalfield 275–86
flooded workings, mine water discharge chemistry 379–90
flow measurement 252
fluorspar mine (Frazer's Grove, North Pennines, UK)
consequences of abandonment 245, 347–63
geological setting 348–50
geophysics 356
hydrochemistry 356–62
metal concentration data 359–61
hydrogeology 355–6
map of Great Limestone during/after mining 357
lithological samples 355
map of location 348
methodology 350–2
mine description 352–5
geological succession 352
sketch diagram 351
monitoring locations 349, 350
water quality data 353–4
fluorspar mine (Strassberg-Harz), underground minewater tracing 49–57
France (CarnoulŠs, Gard), arsenic removal by oxidizing bacteria 267–74
acid mine drainage system 268–72
mean values and SD, acidic waters at 40m and 1500m 270
seasonal variations in soluble and particulate As concentrations 271
bio-oxidation 272–3
sampling and analytical methods 268
site description and map 268–9
Frazer's Grove, North Pennines, UK *see* fluorspar mine

Germany, Strassberg-Harz underground minewater traching   49–57
Ghana, gold mining, environmental impact assessment (EIA)   121–34
   geological map   122
   hydrogeological data collection
      appropriate phases of mineral exploration   133
      geology   124–5
      hydrogeochemistry   125
      hydrology   125
      pedology   124
      physiography   124
      use   132
   Tarkwa   125–32
      map   123
      mean chemical characteristics summary   131
      monthly water balance   128
      soil test summary   126
      Tarkwaian System   126–7
gold mining *see* Ghana; South Africa, plume migration from gold mine tailings dam
Guadiamar River, SW Spain, Aznalicollar mine spill   187–204

Hlobane Colliery *see* South Africa, post-closure water quality
hydrological simulation program Fortran (HSPF)   303

Illinois *see* longwall coal mining
impact structure, Vredefort Dome   339
iron
   mine water discharge chemistry   387–9
   release from spoil heaps, Country Durham   205–14
   vestigial/juvenile acidity   139, 205
   see also pyrite
iron ore bodies
   pyritic roof strata   251–66
   see also Cleveland Ironstone Field; Durham

juvenile acidity   139, 205

Khakassia *see* Siberia

Lancashire, UK (Deerplay Colliery), test pumping deep mine voids   315–26
   hydrogeochemistry   323–4
      Piper diagram   323
      pumped mine water quality   323
   hydrogeological modelling   319–23
      contour plots, modelled groundwater levels, with/without conduits   320–1
      diagram of conduit network   319
      pumping rate and daily rainfall   322
   hydrogeological results   317–19
      plots, discharge vs specific capacity   318
      test pumping data   317
   map of location   316
   sampling and data collection   316–17
   schematic diagram of colliery and site of Black Chough discharge   317
   stratification within mine workings   324–5
lead–zinc mines(former)
   Cumbria, UK (Nenthead), abandoned mines as sinks for pollutant metals   241–50
   France, arsenic contamination   267–74

Frazer's Grove, North Pennines   245, 347–63
limestone-hosted metal mine *see* fluorspar mine (Frazer's Grove)
longwall coal mining, aquifer effects (UK)
   Sherwood Sandstone, Selby Coalfield   75–88
      background to study   76
      data analysis   82–3
      geology   77–9
      groundwater abstraction at Unitriton-BOCM   81–2
      hydrogeology   79–80
      piezometer installation   81
      results   83–6
      site description   76–7
      subsidence   80–1
longwall coal mining, aquifer effects (USA)   17–45
   head drops   21–23
   Jefferson County site, Illinois   26–33
      geochemical changes   30
      hydraulic tests   28–30
      potentiometric changes   30
   mechanisms of hydrogeological effects   17–24
      deformation zones   18–19
      drainage   17–18
      permeability changes, previous field studies   20–1
      subsidence   17, 18
   Saline Country site, Illinois   33–43
      geochemical results   38–40
      hydraulic tests   34, 35–7
      potentiometric responses   34–5, 37–8
      subsidence and strata deformation   33–4, 35
   water level recovery after mining   23–4
Longyearbyen, Svalbard, alkaline coal mine drainage   287–8, 292–5
   composition of three sampled waters   294
   mine spoil leachate   293
   pumped mine water   293–5
   sampling and analysis methods   288
   schematic cross-section of mine   293
Los Rueldos Mine, NW Spain, pollution from mercury and coal mine spoil in Morgao catchment   329–31
Lycopodium clavatum spores, undeground minewater tracing   52–3
LydiA technique (*Lycopodium clavatum*/microspheres), mine water tracing   52–6

Magnesian Limestones, Upper and Lower   79
mercury pollution, Morgao catchment, Spain, mining wastes   330–1
metal cations, mine pit lakes   163, 164
metal sulphide mines *see* alkaline coal mine drainage, Siberia and Svalbard
Mieres *see* Spain, NW
MIFIM model, mine water hydraulics   225–8
mine pit lakes, hydrogeochemical dynamics   159–85
   chemical stratification   167
   chemistry   161–73
      arsenic variation   163
      divalent metal cations variation   163
      representative analysis   162
      Younger diagram   163
   concentration processes   173–5
   conceptual model   160–1
   deep mine voids test pumping for assessment   315–26

geochemical controls 167–73
    attenuation processes 171–3
    grouping of minerals according to neutralization potential 172
    release processes 169–71
    stratified pit lakes 168
geochemical processes 160
geochemical trends over time 177–8
geological controls 175–83
    Nevada case study 179–83
hydromorphic properties, vs natural lakes 160
limnological processes 165–7
    vs natural lakes 160, 165
thermal stratification 166, 167
mine wastes
    mercury, Los Rueldos Mine, Spain 330–1
    NW Spain, pollution from mercury and coal mine spoil in Morgao catchment 331–2
    as risk source 233–6
    Slovak Republic, Novoveska Huta, uranium mine 366
mine water discharge chemistry 379–90
    iron chemistry 387–9
        ternary diagram comprising Fe, SO_4 and HCO 389
    mine pit lakes 161–73
    mine water characterization 381–6
        discharge data for Wales, Scotland and County Durham 382–4
        net-alkalinity vs $Cl/c1 + SO_4$ 386
        Piper diagrams 381–6
    names of mines/sources and times 381
    proposal for modified classification scheme 386–7
    summary of processes affecting discharge chemistry 380
mine water discharges, list 381
mine water inflow, general conceptual models 69–72
mine water pollution
    assessment by risk-based methods 139–44
        RBCA guidelines 139–44
    contaminant sources 138
    mine wastes as risk source 233–6
    sinks for metals (Cumbria) 241–50
    South Nottinghamshire Coalfield closure 102–4
mine water recovery in UK coalfields 61–73
    modelling 66–72
        area-related flow model 71
        average permeability model 71–2
        coal measures inflows 69
        inflow data 68–72
        logarithmic flow model 72
        recovery curves 66–8
        results 72
        shaft water 68
        shallow workings water 69
    monitoring 61–6
        dams 64–5
        East Fife coalfield 62–4
        mining connections, permeability 65–6
    predicting mineral weathering rates 137–57
mine water risk assessment *see* mine water pollution
mine water tracing 47–60
    artifical/natural tracers 49
    LydiA tecnique (*Lycopodium clavatum/microspheres*) 52–6
    sodium chloride 52
    Strassberg-Harz underground mine 49–57

mine description 49–52
tracer amount 52
tracer sampling and analyses 52–3
tracer test aims 48
mineral weathering rate prediction 137–57
    Aitik mine, Sweden 151–5
        laboratory vs field conditions and data 152
        predicted vs measured weathering rates at mesoscale and field scale 154–5
        scaling procedure, column and field rates from laboratory rates 152–4
        site assessment 151–2
    aqueous chemical methods 144–50
        batch reactors 144–7
        column reactors 147–50
        mine water risk assessment methodologies 145–6
        solute mass flows from field sites 150
    extrapolation from laboratory to field scale 144, 150–5
    mine water pollution
        assessment by risk-based methods 139–44
        environmental risk 138–9
        resolving scale dependence 150–5
        scaling parameters 142–3
        scaling weathering rates between similar field sites 143–4
minerals, neutralization potential 172
MINTEQA2 (US EPA) 233
modelling
    geochemical
        Aitik mine, Sweden 151–5
        aqueous chemical methods 144–50
        Durham, iron release from spoil heap 208–9
        Frances Colliery, Fife 315–26
        Lancashire (Deerplay Colliery) 319–23
        mine pit lakes, conceptual model 160–11
    groundwater flow
        alkaline coal mine drainage 287–91
        Bolivia (San Jos, Mine, Oruro) 229–32
        Chile (Escondida Copper Mine) 229–32
        Cleveland Ironstone Field 261
        conceptual models 113–14, 299–300, 307, 341
        contaminant transport 208–9
        mine water recovery in UK coalfields 66–72
        plume migration from gold mine tailings dam 341–2
        pyritic roof strata, aquatic pollutant release 261
        SW Spain, Aznalicollar mine spill 189–202
        water quality, speciation 229–32
models
    CODE-BRIGHT, flow and heat 192–4
    ConSim 234
    groundwater inflow 69–72
    hydrological simulation program Fortran (HSPF) 303
    MIFIM 223–6
    MINTEQA2 (US EPA) 233
    MODFLOW 114, 223, 226–8, 305, 342
    MODPATH, particle tracking 234–6, 342
    MT3DMS solute transport 342
    SHETRAN/VSS-NET 93–5
    WATEQ4F 356
    see also Piper diagram/plot
MODFLOW model, simulation of ground water leakage from flooded mine 114, 223, 226–8, 305, 342
MODPATH particle tracking model 234–6, 342

Morgao catchment, Spain, pollution from mercury and coal mine spoil   327–36
MT3DMS solute transport model   342

Nenthead *see* Cumbria, UK (Nenthead)
Nevada, case study, mine pit lakes   179–83
New Marske Mine, Cleveland Ironstone Field, UK   260–2
North Pennines, UK
   geology   243–4
   map   242
   see also fluorspar mine
Nottinghamshire, South Nottinghamshire Coalfield closure   99–104
Novoveska Huta *see* Slovak Republic

oxygen diffusion, coefficient   208

Packer hydraulic tests   28–30, 34–5, 110
piezometers, installation   81
Piper diagram/plot   381–6
   calcium sulphate   281
   mine water discharges   385
   pumped mine water chemistry   323
   water emissions, Upper/Lower Coal Measures strata   281, 282, 284
plume migration
   gold mine tailings dam   337–46
   see also South Africa
porphyry copper system, Escondida Copper Mine, Chile   113–14
prediction of weathering rates *see* mineral weathering rates
pyrite oxidation   287–8
   Durham County   208–9
   sulphide-containing ores, post-closure mine water, S Africa   297–314
   SW Spain, Aznalicollar mine spill   187–204
pyrite weathering   151–5
   reactive surface area at field scale   143
   see also iron
pyritic roof strata, Cleveland Ironstone Field, UK   251–66

Quaternary Altiplano aquifer complex, Bolivia   215–39
   maps   220, 222

radon, uranium mine, Slovak Republic   365–77
Rampgil Mine, North Pennines, UK   244–5
rebound *see* tin mine (South Crofty), Cornwall; fluorspar mine (Frazer's Grove, North Pennines)
RETRASO, reactive transport model, sulphide oxidation in unsaturated soil   193–5
rholites, Chile   113
risk assessment *see* mine water pollution
r" merite   255
Russian Federation *see* Siberia

Saltburn, Cleveland Ironstone Field, UK   262–3
Sherwood Sandstone, Selby Coalfield   79–80
   see also longwall coal mining, aquifer effects (UK)
SHETRAN/VSS-NET model, groundwater rebound   93–5
Siberia, alkaline coal mine drainage
   Abakan–Chernogorsk region   287–90
   Abaza Magnetic Mine   287–8, 290–1

   Vershina Ty‰i Magnetic Mine   287–8, 291–2
siderite   254
silver–tin mines *see* Bolivia
sinks for pollutant metals (Cumbria)   241–50
Skinningrove, Cleveland Ironstone Field, UK   263–4
Slovak Republic, Novoveska Huta, uranium mine liabilities   365–77
   concentration data for radionuclides   373
   environmental impact assessment (EIA)   372–3
   geological and historical context   366–7
   hydrogeological setting   367
   liabilities and remediation measures   367–9
   map of locations of adits and waste rock heaps   366
   methodology, tiered data collection   368–9
   occupancy times   373–4
   recommendations to minimise health risks   375–6
   results   374–5
   site descriptions   369–73
      overall approach for environmental impact assessment (EIA)   370
      radiological map of Muran pit area   371
      rock slides   370
      uranium concentration values at each site   370
smithsonite   250
soil, sulphide oxidation   187–204
South Africa, plume migration from gold mine tailings dam   337–46
   flow model   342–3
      E-W cross-section   343
      steady-state hydraulic heads in 1998 model   343
   flow system   339–40
   geology and structure   338–9
      map and section   338
   hydrogeological regime   339–40
   hydrostratigraphic units   339
   methodology   337–8
   numerical model   341–2
      calibration   342
      discretization   341–2
   site locality   337
   sources of contamination   340–1
      conceptual model of hydraulic system   341
   transport model   343–6
      chloride plume migration (1998–2058)   344–5
South Africa, post-closure coalmine water quality   297–314
   assessment process   301–11
      calculation of water and salt balances   303
      data collection   301–2
      flow into mined areas   303–4
         water balances for tops and slopes of mountains   304
      geochemical processes in mined areas   307–11
         conceptual models   307
         normalized modal distributions   309
         predictive geochemical profiles   310
      hydrogeology in mined area   304–7
         calibration curves for discharges   306
         water balance infiltration   305
      impact of water discharges on receiving streams   311
      source term   302
   conceptual system models   299
   physical description   298–9
      map of locality   298
   system geochemistry   301
   system hydrology and hydrogeology   299–301

conceptual flow model   300
    pre-mining and post-mining conditions   299–301
  water management strategies   311–13
    annual hydrographs   312
    impact on river water   313
    plan for mine closure   297–8
    reduction in river flow   312–13
South Nottinghamshire Coalfield closure   99–104
  borehole and shaft penetrations   101
  environmental consequences   104
  mine water risk evaluation   102–4
  water balance   103
South Wales Coalfield, Eastern sector, fingerprinting mine water   275–86
  methodology   277–8
    map of river catchments in Eastern sector   278
  Middle–Lower Coal Measures   277, 282
  mining position   277
  monthly sampling analysis   285–6
  outline geology   276
  presentation of data   278–85
    bar charts   283, 285
    Piper diagrams   281–4
  Upper Coal Measures   276–7, 279–82
Spain, NW, Mieres, Asturias, pollution from mercury and coal mine spoil in Morgao catchment   327–36
  characteristics of studied area   329
  climate and precipitation   329
  coal mining wastes   331–2
    impoundments   329–32
    schematic view of spoil heaps and drainage systems   332
  geology and mineralization   329
    map of Cantabrian zone   330
  hydrogeology   332–3
  Los Rueldos Mine   329–31
    elements from mercury spoil heap   331
    mercury mining wastes   330–1
  map of study area and drainage basin   328.
  water movement   332–3
  water quality   333–5
Spain, SW, Aznalicollar mine spill, sulphide oxidation in unsaturated soil   187–204
  column experiment
    leachates   188–90
    mass fraction of each mineral in sludge   189
  geochemical model   195–7
  location of mine and area affected by sludge   188
  modelling   189–202
    results   197–202
  reactive transport model   193–5
    clayey-soil mixture   201
    sandy-soil mixture   197
  transient flow and heat transport model   192–3
    clay-sludge mixture   194
    sand-sludge mixture   193
sphalerite   245
spills see Spain, SW, Aznalicollar mine spill
spoil drainage
  Cleveland Ironstone Field   260–2
  iron release   205–14
  leachate, alkaline mine drainage   293
  mine water discharge chemistry   379–90
  pollution from mercury and coal mine, Spain   327–36
Strassberg-Harz underground minewater tracing   49–57
  possible tracers   48–9

subsidence see longwall coal mining, aquifer effects (UK and USA)
sulphate mass flow   143
sulphide oxidation, SW Spain, Aznalicollar mine spill   187–204
sulphide–water reactions, mine pit lakes   170
sulphide-containing ores, post-closure mine water quality, S Africa   297–314
Summer Camp, Nevada, mine pit lakes   177–83
Svalbard, Longyearbyen, alkaline mine drainage   287–8, 292–5
Sweden, Aitik mine, mineral weathering rate prediction   151–5

tailings dam, plume migration   337–46
Tarkwaian System, Tarkwa, hydrogeological data collection   126–7
test pumping, deep mine voids   315–26
thermal stratification, mine pit lakes   166
Thiobacillus ferrooxidans
  arsenic removal   267–74
  iron oxidation   209
tiered risk assessment, mine water pollution   139–42
tin mines (Cornwall, UK)
  Wheal Jane, water quality   90
  see also Bolivia
tin mines (Cornwall, UK) groundwater rebound model   89–97
  calculation of infiltration   92
  meteorological data   92
  mine layout   91
  relationship of pumping data with rainfall   92–3
  SHETRAN/VSS-NET model   93–5
  simulations   94–6
transient flow and heat transport models, sulphide oxidation in unsaturated soil   192–3
transmissivities   22–3, 338

uranium mine see Slovak Republic, Novoveska Huta

Vershina Ty‰ei Magnetic Mine, Khakassia, Siberia, alkaline mine drainage   287–8, 291–2
  analyses of stream and pumped mine water   292
  sampling and analysis methods   288
Vredefort Dome, impact structure   339
Vryheid Coalfield see South Africa, post-closure coalmine water quality

WATEQ4F model   356
water see mine pit lakes; mine water discharge chemistry; mine water tracing; mine water recovery
West Rand see South Africa, plume migration from gold mine tailings dam wetlands, constructed   206
Whitemoor Common Fault   78–9

Xray diffraction   355

Younger diagram, chemistry of mine pit lakes   163

zinc
  Cumbria, UK (Nenthead), abandoned mines as sinks for pollutant metals   241–50
  deficits   245–8
  potable water   241
  see also lead–zinc mines (former)